MOLECULAR ANTHROPOLOGY

Genes and Proteins in the
Evolutionary Ascent of the Primates

ADVANCES IN PRIMATOLOGY

THE PRIMATE BRAIN
Edited by Charles R. Noback and William Montagna

MOLECULAR ANTHROPOLOGY: Genes and Proteins
  in the Evolutionary Ascent of the Primates
Edited by Morris Goodman and Richard E. Tashian

MOLECULAR ANTHROPOLOGY

Genes and Proteins in the
Evolutionary Ascent of the Primates

Edited by

Morris Goodman
Wayne State University School of Medicine
Detroit, Michigan

and

Richard E. Tashian
University of Michigan Medical School
Ann Arbor, Michigan

Associate editor

Jeanne H. Tashian
University of Michigan Hospital
Ann Arbor, Michigan

PLENUM PRESS • NEW YORK AND LONDON

Library of Congress Cataloging in Publication Data

Main Entry under title:

Molecular anthropology.

(Advances in primatology)
"Based in part on the papers presented at the Symposium on Progress in Molecular
Anthropology held at Burg Wartenstein, Austria, July 25-August 1, 1975."
Includes bibliographies and index.
1. Primates—Evolution—Congresses. 2. Chemical evolution—Congresses. I. Good-
man, Morris, 1925-   II. Tashian, Richard E. III. Series.
QL737.P9M59   599'.8   76-45445
ISBN 0-306-30948-3

Based in part on the papers presented at the
Symposium on Progress in Molecular Anthropology
held at Burg Wartenstein, Austria,
July 25-August 1, 1975

©1976 Plenum Press, New York
A Division of Plenum Publishing Corporation
227 West 17th Street, New York, N.Y. 10011

Printed in the United States of America

Contributors

P. Altevogt
Max-Planck-Institut für experimentelle
Medizin
Göttingen, West Germany

N. A. Barnicot (deceased)
Department of Anthropology,
University College London
London, England

H. U. Barnikol
Max-Planck-Institut für experimentelle
Medizin
Göttingen, West Germany

S. Barnikol-Watanabe
Max-Planck-Institut für experimentelle
Medizin
Göttingen, West Germany

Jan M. Beard
Department of Anthropology
University College London
London, England

J. Bertram
Max-Planck-Institut für experimentelle
Medizin
Göttingen, West Germany

Celia Bonaventura
Department of Biochemistry
Duke University Medical Center
and Duke University Marine Laboratory
Beaufort, North Carolina

Joseph Bonaventura
Department of Biochemistry
Duke University Medical Center
and Duke University Marine Laboratory
Beaufort, North Carolina

Christopher N. Cook
Department of Anthropology
University College London
London, England

John E. Cronin
Departments of Anthropology and
Biochemistry
University of California
Berkeley, California

Howard T. Dene
Department of Biology
Wayne State University
Detroit, Michigan

L. Dreker
Max-Planck-Institut für experimentelle
Medizin
Göttingen, West Germany

M. Engelhard
Max-Planck-Institut für experimentelle
Medizin
Göttingen, West Germany

Robert E. Ferrell
Department of Human Genetics
University of Michigan Medical School
Ann Arbor, Michigan

Walter M. Fitch
Department of Physiological Chemistry
University of Wisconsin Medical School
Madison, Wisconsin

A. E. Friday
University Museum of Zoology
Cambridge, England

Lila L. Gatlin
  Genetics Department
  University of California
  Davis, California

Morris Goodman
  Department of Anatomy
  Wayne State University School of Medicine
  Detroit, Michigan

David Hewett-Emmett
  Department of Anthropology
  University College London
  London, England

N. Hilschmann
  Max-Planck-Institut für experimentelle
    Medizin
  Göttingen, West Germany

Richard Holmquist
  Space Sciences Laboratory
  University of California
  Berkeley, California

J. Horn
  Max-Planck-Institut für experimentelle
    Medizin
  Göttingen, West Germany

K. W. Jones
  Epigenetics Research Group
  Department of Animal Genetics
  Edinburgh University
  Edinburgh, Scotland

K. A. Joysey
  University Museum of Zoology
  Cambridge, England

M. Kopun
  Institut für Anthropologie und
    Humangenetik
  Universität Heidelberg
  Heidelberg, West Germany

A. Kortt
  Max-Planck-Institut für experimentelle
    Medizin
  Göttingen, West Germany

H. Kratzin
  Max-Planck-Institut für experimentelle
    Medizin
  Göttingen, West Germany

Charles H. Langley
  National Institute of Environmental Health
    Sciences
  Research Triangle Park, North Carolina

Gabriel W. Lasker
  Department of Anatomy
  Wayne State University School of Medicine
  Detroit, Michigan

H. Lehmann
  University Department of Clinical
    Biochemistry
  Addenbrooke's Hospital
  Cambridge, England

Genji Matsuda
  Department of Biochemistry
  Nagasaki University School of Medicine
  Nagasaki, Japan

G. William Moore
  Department of Anatomy
  Wayne State University School of Medicine
  Detroit, Michigan

Peter E. Nute
  Regional Primate Research Center and
    Department of Anthropology
  University of Washington
  Seattle, Washington

W. Palm
  Max-Planck-Institut für experimentelle
    Medizin
  Göttingen, West Germany

William Prychodko
  Department of Biology
  Wayne State University
  Detroit, Michigan

R. Rathenberg
  Institut für Anthropologie und
    Humangenetik
  Universität Heidelberg
  Heidelberg, West Germany

A. E. Romero-Herrera
University Department of Clinical
  Biochemistry
Addenbrooke's Hospital
Cambridge, England

E. Ruban
Max-Planck-Institut für experimentelle
  Medizin
Göttingen, West Germany

Vincent M. Sarich
Departments of Anthropology and
  Biochemistry
University of California
Berkeley, California

M. Schneider
Max-Planck-Institut für experimentelle
  Medizin
Göttingen, West Germany

R. Scholz
Max-Planck-Institut für experimentelle
  Medizin
Göttingen, West Germany

Elwyn L. Simons
Department of Geology and Geophysics
Peabody Museum, Yale University
New Haven, Connecticut

C. Staroscik
Max-Planck-Institut für experimentelle
  Medizin
Göttingen, West Germany

Bolling Sullivan
Department of Biochemistry
Duke University Medical Center
and Duke University Marine Laboratory
Beaufort, North Carolina

Robert J. Tanis
Department of Human Genetics
University of Michigan Medical School
Ann Arbor, Michigan

Richard E. Tashian
Department of Human Genetics
University of Michigan Medical School
Ann Arbor, Michigan

F. Vogel
Institut für Anthropologie und
  Humangenetik
Universität Heidelberg
Heidelberg, West Germany

Alan Walker
Department of Anatomy
Harvard Medical School
Boston, Massachusetts

Emile Zuckerkandl
Marine Biological Laboratory
Woods Hole, Massachusetts
Directeur de Recherche
CNRS, Paris
Nonresident Fellow
Linus Pauling Institute for Science and
  Medicine
Menlo Park, California

Preface

In 1962 at the Burg Wartenstein Symposium on "Classification and Human Evolution," Emile Zuckerkandl used the term "molecular anthropology" to characterize the study of primate phylogeny and human evolution through the genetic information contained in proteins and polynucleotides. Since that time, our knowledge of molecular evolution in primates and other organisms has grown considerably. The present volume examines this knowledge especially as it relates to the phyletic position of *Homo sapiens* in the order Primates and to the trends which shaped the direction of human evolution. Participants from the disciplines of protein and nucleotide chemistry, genetics, statistics, paleontology, and physical anthropology held cross-disciplinary discussions and argued some of the major issues of molecular anthropology and the data upon which these arguments rest. Chief among these were the molecular clock controversy in hominoid evolution; the molecular evidence on phylogenetic relationships among primates; the evolution of gene expression regulation in primates; the relationship of fossil and molecular data in the Anthropoidea and other primates; the interpretation of the adaptive significance of evolutionary changes; and, finally, the impact on mankind of studies in molecular anthropology.

Most of the papers in this volume were presented in a preliminary form at Symposium No. 65 on "Progress in Molecular Anthropology" held at Burg Wartenstein, Austria, from July 25 to August 1, 1975. These papers were subsequently revised and some additional papers related to the theme of the symposium were also contributed to this volume.

On behalf of all participants at the symposium we would like to thank the Wenner-Gren Foundation for Anthropological Research for sponsoring our meeting. In particular, we are indebted to Lita Osmundsen, Director of Research of the Foundation, and to her staff at Burg Wartenstein for providing a truly stimulating and unique situation in which scientific discourse can take place.

M.G.
R.T.

Contents

III. Primate Phylogeny and the Molecular Clock Controversy

IV.   Primate Evolution Inferred from Amino Acid Sequence Data

Background to Some
Key Issues

I

What Is Molecular Anthropology?

1

GABRIEL W. LASKER

Q. What is molecular anthropology?
A. The words mean "the molecular study of man."
Q. Is that the same thing as "the study of human molecules"?
A. Not quite, we would call that anthropochemistry.
Q. Would you say that anthropochemistry is the study of the interaction of human molecules by analogy with biochemistry, the study of the interaction of living molecules?
A. Yes, biochemistry provides an analogy, but molecular biology certainly has come to mean something different.
Q. What?
A. Well, molecular biology is a special area within biochemistry, one might say; it is the evolutionary story of the molecular constituents of living things studied in genetic perspective.
Q. Then doesn't molecular anthropology simply mean the study of the evolution of human molecular constituents in genetic perspective?

Evolutionists have turned to molecular studies because nucleic acids are genetic substances that carry evolving organisms from one cell generation to the next, and protein molecules are the mirrored reflection of some of the structure of these nucleic acids.

In human evolution, human DNA is the carrier of the species; the peptides of human proteins mirror the reflection of the DNA off human RNA. The study of these peptides is like studying DNA through opera glasses: one can hope to see part of the magnified structure accurately reflected off two highly polished, precise prismatic mirrors.

The analogy can be carried a little further. Sitting in the family circle with

GABRIEL W. LASKER • Department of Anatomy, Wayne State University School of Medicine, Detroit, Michigan 48201.

3

opera glasses focused on the prima donna on the stage, one gets to the heart of the opera, the whole purpose of the staging; but the barrels of the opera glasses cut off the wider context. The audience is invisible—even the well-dressed ladies in the golden horseshoe who made the performance possible. And one is likely to miss the prompter in the prompting box who keeps the singers on cue and the conductor in the orchestra pit who has the overall responsibility, especially for timing.

The molecular view of human variability has an analogously narrow field of vision. The substance of evolution is in focus. The substance is the molecular composition and structure of the organism. But it may be relatively meaningless without the context of the environment within which it functions—both the external environment (the audience) and the internal environment of control and timing of differentiation and of physiological functions (the prompter and conductor).

The development of molecular anthropology has already given us a good view of the singers in the center of the stage. The next task will be to pick up the conductor in the dim light of the orchestra pit. To do so, the molecular anthropologist may have to put down his glasses for a while so that he can see the conductor and the singers at the same time. Perhaps he will later be able to pick up his opera glasses again to scan the whole inside of the opera house and see the functional aspects also in molecular terms. A major question is, is he ready to try?

The term "molecular anthropology" was apparently first used publicly in 1962 at a Burg Wartenstein symposium (Zuckerkandl, 1963). The idea that the biochemical molecular variations are inherited in a simple way unmodified by environmental influences and are therefore the ideal material for the study of human origins and the interrelations of human populations is, of course, much older.

Dr. (later Sir) Archibald Garrod had already observed the familial occurrence of biochemical variants when he learned of Mendel's experiments with peas, and published case descriptions of recessive inheritance of biochemical anomalies in man in 1902. By the time of his famous Croonian lectures on inborn errors of metabolism (1908), the list had been extended to four. More or less simultaneously in 1900, Landsteiner (1945) discovered the first blood groups in man. Already by 1919 Hirszfeld and Hirszfeld had examined the ABO blood groups of 500–1000 individuals of each of 16 nationalities and reported that the biochemical race index—the ratio of blood groups $(A + B)/(B + AB)$—divided the nations into three groups: a European group (English, French, Italians, Germans, Austrians, Serbians, Greeks, and Bulgarians) with an index of 2.5 or more, an Asian-African group (Madagascans, Negroes, Indo-Chinese, and Hindus) with an index of 1 or less, and an intermediate group (Arabs, Turks, Russians, and Jews) with indices between 1 and 2.5. However, it was not until 1924 that the mode of inheritance of blood groups was properly understood (Bernstein, 1924).

Like the blood groups, the inborn errors of metabolism proved to have different frequencies in different racial groups. It was noted that pentosuria occurs almost solely in Jews, for instance, and my first exposure to anthropology (Lasker *et al.*, 1936) was to confirm that against 35 cases among Jews there was none definitely ascribed to other ethnic groups; we also confirmed a Mendelian recessive hereditary mechanism through analysis of pedigrees. The condition is now known also to occur in non-Jews in the Levant (Khachadurian, 1962; Politzer and Fleischmann, 1962) but not in other groups. The problems with the use of inborn errors of metabolism in studies of the relation between groups are the following: (1) these are known (or at first were known) only in homozygous form; (2) selective pressure against the homozygous form of many of them is high (but probably not very high for albinism and pentosuria); (3) associated with the preceding factor, the frequency of cases is low and huge numbers of individuals have to be screened for frequency data.

Almost simultaneously with the work of Landsteiner and Garrod, Nuttall (1904) tested the serum of various species of animals with antisera to the whole serum of different animals including primates and reported the blood relationships of these species to each other. Although his method superimposed the results in respect to different antigens present in serum, the relationships he demonstrated in his precipitin tests have been confirmed by Goodman (1962) and others. The application of the methods of what we now call molecular anthropology to the study of man's place among the primates is therefore essentially as old as its application to variation between peoples.

Extension of immunological studies of nonhuman primates was slow in coming. A few additional studies by others extended the precipitin tests of sera. Then in 1940 Landsteiner and Wiener produced an antiserum to macaque red blood cells and demonstrated that a rhesus blood group substance (Rh) occurs in some, but not all, human beings. By 1945, Wiener demonstrated the Rh-Hr blood groups in the chimpanzee.

In man, the study of the geographic distribution of blood groups became a well-established branch of anthropology (Wiener, 1943), and by 1950 Boyd proposed that only traits of well-established genetic mode could properly serve for the study of human races. The original reason for advocating the use of blood groups in anthropological studies had been the idea that they were selectively neutral. Many early studies suggested a preponderance of one or another blood group type with one or another disease, but Wiener and others were unconvinced. The studies were not always confirmed by further research and the possibility of false correlation was often present. For instance, it should be no surprise to find that individuals with pentosuria have a different frequency of blood groups than the general population; after all, most cases are in Jews and Jewish people have higher frequencies of type B than the other populations in Europe and America among whom they live. Boyd (1950) saw that one could not discard a character because it has selective value. It is reasonable to suppose that, under certain conditions at least, all biological characteristics influence natural selection. Furthermore, in a certain sense, every trait that has been accepted in any organism has been naturally selected for, whether or not we know or are interested in the circumstances of the selection. Of course, if we are interested in evolution over millennia we shall want to emphasize traits that are stable (or at least react in the same way) for millennia, but there is nothing unnatural about the discrete causal events that are treated as "random."

While the controversy over whether to use only selectively neutral traits was going on in anthropology, the one gene–one enzyme theory was proposed by Beadle and the one-to-one-to-one mechanism of genetic determination of proteins by coded DNA and RNAs was developed by Watson and Crick (1953) and others. Meanwhile, the study of evolution at the chemical level in other organisms also progressed (Florkin, 1944), so the time was ripe for phylogenetic cladograms based on the detailed structure of a protein. Fitch and Margoliash (1967) published one based on the amino acid sequence of cytochrome *c* in plants and animals.

The plotting of "distance" between human populations in terms of molecular components was another matter. The Hirszfelds' index was obviously unsatisfactory even for three antigens (for which a triangular plot serves to depict the distances). For larger numbers of molecules, there are several possible methods of weighting data to yield a matrix of interpopulation distances (Sanghvi *et al.*, 1971; Ward, 1973; Harpending and Jenkins, 1973). The more serious problem, however, is how to interpret a dendrogram constructed from a matrix of interpopulation differences. The all-important distinctions between similarities caused by similar ethnic origins, similarities caused by life under similar selective pressures, and similarities due to interbreeding are difficult to perceive. Manipulation of gene frequency data cannot make such distinctions (Livingstone, 1973).

The biochemical methods needed for the study of molecular variation among the primates in general and for the study of polymorphisms and rare variants in man are very similar, but few of the students of the one set of questions have concerned themselves with the other. In fact, some of the borrowing back and forth has led to distortions. A cladogram with definite branchings is legitimate between species—each fork representing the point at which interbreeding ceased. But whereas, once established, species do not interbreed (at least not effectively), the same is not true of subpopulations within a species. A dendrogram is at best an inexact way of representing the relative antiquities of common origins within a species. Looked at in distant perspective, all organisms are interrelated since they share (presumably by reason of common origin) the same mechanism of reproduction and of peptide encoding through nucleic acids (DNA and RNA, or RNA alone). By the same token, all organic protein molecules have a common origin if we assume that the original nucleic acid coded for only one, and duplicated and modified itself into the genetic message for all the forms life now assumes. In this distant perspective, all sexually reproducing organisms are completely inbred, too. They receive their genes from both parents only because both parents received them ultimately from the same ancestor. Sewall Wright (1931) demonstrated that in finite populations breeding at random, under conditions of selective neutrality, "random genetic drift" would fix one allele at each locus and eliminate all polymorphisms. Wright's random genetic drift is simply the random component of inbreeding; thus over a sufficiently large number of generations the inbreeding coefficient always becomes 1.0 (Jacquard, 1974). In this perspective, the concept of "inbreeding" loses its utility since it is invariant. But we know that there are biological inbreeding effects if one limits one's concern to close and intimate inbreeding. The change in perspective to recent and high levels calls for a change in the model applied. (The needed model should involve time—inbreeding rate per five generations or per 100 years, for instance.)

The same change of perspective and need for consideration of rates are true of the difference between the study of mammalian variability and the study of human variability. All human beings evolved ultimately from some progressive group of primates in a restricted place reproductively isolated from the habitat of our nearest nonhuman primate relatives. The population must have been much smaller than now and the circumstances of selection (random or not) over the many generations since have further narrowed the effective original populations so that almost all of our ancestry was from a very few individuals. But since the human ecological niche is for wide adaptability, the situation of "Adam and Eve" and their family must have been a very brief one as the "tribes" emerged and spread over the earth.

That brief early diaspora may have almost nothing to do with the present differences between the peoples of different regions. The structure of this species is not that of a branching tree or even that of a nerve: there are fascicles of related people, but unlike bundles of nerve fibers, which run from neuron to end organ without cross-connections, the lines of ancestors cross-connect every generation and as one traces ancestry back they multiply rapidly: two parents, four grandparents, and doubling each generation to 1024 in ten generations, over a million in 20, over a trillion in 40 generations a mere thousand years or so ago, and 10^{23} about the time of Christ. The threads of this network surely spread over the whole inhabited world. Although the cross-fibers may have been few or even absent for quite a few generations in geographically isolated spots, it is almost certain that all the larger fascicles (those with names like "Caucasoid," "Mongoloid," and "Negro," or like "Japanese" and "Filipino") have loose threads streaming out and in, and the concept of definable populations is as fuzzy at the edges as is the concept of "races."

The problem facing the anthropologist is not so much to discover the common origins of particular populations (which are elusive) as it is to explore the general processes of the causes and timing of diversification. For instance, several human hemoglobin polymorphisms seem clearly related to the distribution of malaria (Allison, 1954; Livingstone, 1967). The presence of the same molecular form of these hemoglobins in various parts of Africa, Europe, and the New World is almost certainly evidence of some of those loose threads I mentioned, whereas the presence of different ones in Asia implies that the duration of appropriate conditions (endemic falciparum malaria) must be relatively recent and the threads too few to have influenced Asiatic populations yet. Since hemoglobin S in Americans of African origin is much less common than in Africans, it is possible to weigh the relative influence of admixture from other populations (White and Amerindian) vs. change caused by selection (Workman *et al.*, 1963). True, these studies of mechanisms are still in their infancy, but we know enough to see that there are reasons polymorphisms become established. The notion of random mutation is still useful, but to believe in the randomness of acceptance of mutations is to tolerate continued ignorance to an extent probably unnecessary at the present state of the science. For the future, the unfolding of reasons for molecular change must be extended from the within-species perspective to the interspecific arena. The search for functional differences between the molecular forms in different species is likely to tell us more about the timing and circumstance of speciation that can mere multiplication of protein sequences fed into the hopper of the constant-clock hypothesis. That is not to say that the averages and the constant clock do not provide a first approximation, but it is an approximation subject to more refinement by building selection coefficients into the model than by merely adding more molecules.

Those who argue for a constant-clock theory believe that, within a single protein lineage, acceptance of mutations has been at a more or less constant rate. Attention to the evolution of proteins since the first protein was encoded by nucleic acid demonstrates that the clock cannot have been constant over the whole span but may well seem to be constant over the short haul.

Why does the whole span of protein evolution point to a variable clock?

It is because the system of encoding for proteins through DNA and RNA is so complex that it almost certainly stems from a unitary evolutionary event. The earliest organisms must either have had only one, perhaps variable, kind of protein or at most a few varieties. Most kinds of proteins in complex organisms today must have had common origins with many other such proteins. Different proteins with distinctly different evolutionary rates of change (faster in transferrin than in albumin, for instance) are no doubt related and have had their rates change differently from the same original source.

It may be argued that the constant clock applies only to a single kind of protein, and that a constant rate of acceptance of new mutations in a lineage is dependent in part on a constant rate of point mutations and does not apply to more drastic events such as inversions and translocations. If the restriction to homologous molecules is required, the processes by which one molecular species derives from another and by which all protein molecules can be said to have a common ancestry do not bear on the argument, but it is well to remember that the actual protein lineages must have crossed the "homology" boundaries, and that, even among homologous molecules, processes other than single base replacements in the DNA base sequence may sometimes be involved. From an evolutionary view, the founding events of a protein lineage are not completely distinguishable from its later differentiation. At the very least, therefore, since all students of the problem recognize different rates of evolution in different proteins, acceptance of the unity of life and of evolution rules out the constant clock as a general principal. In fact, those who use the clock hypothesis to project timing of evolutionary events from accepted dates to unknown ones claim no more than that the clock is constant enough to give good estimates for dates of phylogenetic splitting between species. They recognize that a split between molecular species may antedate the split between animal or plant species that possess the molecule but rate such proteins as "nonhomologous" and believe that few such would go undetected. In any case, this means that the application of the constant clock gives maximum dates for species divergences. The real problem is that these maximum dates when derived from such occurrences as the separation of chimpanzee from *Homo* are so recent that they seem implausible on paleontological and comparative anatomical grounds. Thus the clock may have been slowing down.

If one accepts the history of nonhomologous proteins to have been of the same kinds as of homologous ones, would examples of changes in evolutionary rates be inconsistent with the constant molecular clock?

Some molecules, such as pancreatic ribonuclease, have come to differ rapidly through evolution. For example, the amino acid sequences of the bovine and pig enzymes differ by 22% (Chirpich, 1975). Since bovine and pig phyletic lines diverged about 40–50 million years ago, and the best assumption is that there were changes of 11% each in the bovine and pig lines; this is a rate of about 24.4% substitutions per 100 million years. In the case of a slowly evolving protein such as histone IV, however, between plants and animals there are only two amino acid differences in 102 bases, i.e., 1.96%. The difference between plants and animals occurred some 1 billion years ago or more; thus the rate in a billion years of plant evolution plus a billion years of animal evolution must have been of the order of only 0.098% amino acid substitutions per 100 million years.

Presumably the differentiation in the rates of change in the two proteins occurred on earth. The idea that life began at the interface between the earth's primitive atmosphere and sea water has been enhanced by comparative studies of the chemical makeup of bacteria, fungi, woody plants, and land animals and of sea water (Banin and Navrot, 1975). The earth is estimated to be about 4.5 billion years old. On the other hand, the differentiation of such proteins could not have been much more recent than 3 billion years ago—at least not if we accept the existence of microfossils of bacteria and algae 3.1 billion years before present.

Suppose that the deceleration of a now slowly evolving protein began at that time and continued for 1.5 billion years (until 1.5 billion years ago, the maximum likely antiquity for the diversion of plant from animal eukaryotes). How dramatic would the deceleration have been?

Not at all dramatic: The deceleration is at a rate of $(24.5\% - 0.065\%)/15 = 1.43\%$ per 100 million years per 100 million years. Within the last 100 million years, i.e., during the evolutionary radiation of the placental mammals, there would be no hope of perceiving a constant rate of deceleration of so small a magnitude. The decelerations that are attributed to comparisons of paleontological data with those of a clock based on homologous proteins are at much higher rates. They may therefore be only recent or secular or episodic. A unitary origin of proteins on earth and constant deceleration from the fastest known to the slowest known rates would yield short-term changes almost as consistent with accelerations and constant rates as with deceleration, in that the rate of change would be so slow.

A recent far departure from the molecular method is very instructive. We probably know as much about variation in the histology of the skin of primates as about any other characteristics of the order (thanks to the work of William Montagna and his colleagues). But when the primates are classified on the basis of the characteristics of their skin, the classification is out of line with that based on other anatomical or on molecular traits. Furthermore, depending on which set of characters of the skin one considers—dermal and epidermal or those of hair follicles, for instance—the relationships vary greatly, as seen in the dendrograms of Grant and Hoff (1975). Why? The reasons must be sought in the way skin functions in different environments as well as in the phyletic origin of the animals. Just averaging (known) molecular variations without searching for functional meanings may leave anomalous branches in dendrograms as bothersome (even if not as extreme) as those of any morphological traits, such as characteristics of skin. After all, skin traits are inherited and based on (still unknown) molecular variations. I have already mentioned that selection against the sickle cell gene in nonmalarial America makes the American Black population more European in respect to sickle cell than in respect to blood group genes. Likewise, the chimpanzee is more like the orangutan than like the gorilla in the size of bones of the upper limb but more like the gorilla than like the orangutan in the shape of these bones. Although very little is known about differences in functional properties of any protein associated with different ecological adaptations of the species in which they are found, we should anticipate interspecies anomalies in rates of change in molecular configurations under natural selection. An indication of such variable rates due to natural selection is provided in the case of globin evolution (Goodman *et al.,* 1975; Hewett-Emmett *et al.,* this volume). Rates of evolutionary change in molecules will make sense when we know how the molecules work in the different species.

Q. Then where does molecular anthropology fit into the study of human evolution?
A. Three approaches have been used and the molecular approach is the last.
Q. Which is the first?
A. The classical method has been that of natural selection. It seeks morphological evidence for changing adaptive behavior: in the case of man, bipedal locomotion, manufacture and use of tools, cultivation and cooking of food, and spoken language—"the evolution of man's capacity for culture" Spuhler called it in 1959.
Q. This seems unobjectionable, hasn't the emphasis been on behavior from Darwin to Washburn? What possible alternative could there be?
A. Well there is something called numerical taxonomy. The idea is that if you have numerous different characteristics they will sample the whole array; but, as noted

in the discussion of primate skin, the taxonomy depends on the choice of traits studied.

Q. Then why make any choice, couldn't one use all the characteristics instead?

A. Some scholars have tried to do that with measurements. On fossil bones they take virtually all the dimensions from every clearly definable landmark and the same measurements on the bones of extant species. Then they apply multiple variable methods to extract a smaller number of independent abstract factors. The difficulty is that the criteria for the acceptance of measurements are not directly related to testing either specific genetic or specific functional questions. Saturation bombing may hit the target, but it hits so much else as well. When one is done, it is often a matter of pure opinion about what, for instance, a canonical variate represents.

Q. Am I right in supposing that molecular anthropology is the third approach? It doesn't seem to be in the same class.

A. Well, perhaps not. The original idea of molecular anthropology was to free the analysis from all phenotypic adaptations and to compare strictly cladistic relationships; but, of course, since evolution proceeds in large part through natural selection, attributes of DNA, RNA, and protein molecules are adaptive. The beauty of their use will come, however, when the adaptive value of molecules is better known, because the molecular approach distinguishes genetic adaptation through natural selection from adaptation during the lifetime of individuals.

Since you asked me "What is molecular anthropology?" I should make it clear that I don't see it as totally distinct from molecular biology. Quite the contrary, I would use the broadest perspective: in the many thousands of generations of their evolution, human beings—in fact, all organisms—have become completely inbred. They share their humanity—they share their biology—by reason of identity through descent. All the genetic variability has therefore been established through acceptance of mutations of various kinds and at various rates through mechanisms which, for the most part, still remain to be explored and explained.

References

Allison, A. C., 1954, The distribution of sickle-cell trait in East Africa and elsewhere, and its apparent relationship to the incidence of subtertian malaria, *Trans. R. Soc. Trop. Med. Hyg.* **48:**313–318.

Banin, A., and Navrot, J., 1975, Origin of life: clues from relations between chemical compositions and natural environments, *Science* **189:**550–551.

Bernstein, F., 1924, Ergebnisse einer biostatistichen zusammenfassenden Betrachtung über die erblichen Blutstrukturen des Menschen, *Klin. Wochenschr.* **3:**1495–1497.

Boyd, W. C., 1950, *Genetics and the Races of Man: An Introduction to Modern Physical Anthropology*, Little, Brown, Boston.

Chirpich, T. P., 1975, Rates of protein evolution: A function of amino acid composition, *Science* **188:**1022–1023.

Fitch, W. M., and Margoliash, E., 1967, Construction of phylogenetic trees, *Science* **155:**279–284.

Fitch, W. M., and Neel, J. V., 1969, The phylogenetic relationships of some Indian tribes of Central and South America, *Am J. Hum. Genet.* **21:**384–397.

Florkin, M., 1944, *L'Evolution Biochimique*, Masson et Cie, Paris.

Garrod, A. E., 1902, The incidence of alkaptonuria: A study of chemical individuality, *Lancet* **2:**1616–1620.

Garrod, A. E., 1908, Lecture IV: Inborn errors of metabolism, *Lancet* **2:**214–220. Cited from: Garrod, A. E., 1923, *Inborn Errors of Metabolism*, 2nd ed., Henry Frowde and Hodder and Stoughton, London.

Goodman, M., 1962, Evolution of immunologic species specificity of human serum protein, *Hum. Biol.* **34:**104–150.

Goodman, M., Moore, G. W., and Matsuda, G., 1975, Darwinian evolution in the genealogy of haemoglobin, *Nature* **253:**603–608.

Grant, P. G., and Hoff, C. J., 1975, The skin of primates. XLIV. Numerical taxonomy of primate skin, *Am. J. Phys. Anthropol.* **42:**151–166.

Harpending, H., and Jenkins, T., 1973, Genetic distance among southern African populations, in: *Methods and Theories of Anthropological Genetics* (M. H. Crawford and P. L. Workman, eds.), pp. 177–199, University of New Mexico Press, Albuquerque.

Hirszfeld, L., and Hirszfeld, H., 1919, Serological differences between the blood of different races: The results of researches on the Macedonean front, *Lancet* **2:**675–679.

Jacquard, A., 1974, *The Genetic Structure of Populations,* Springer-Verlag, New York.

Khachadurian, A. K., 1962, Essential pentosuria, *Am. J. Hum. Genet.* **14:**249–255.

Landsteiner, K., 1945, *The Specificity of Serological Reactions,* Harvard University Press, Cambridge, Mass.

Landsteiner, K., and Wiener, A. S., 1940, An agglutinable factor in human blood recognized by immune sera for rhesus blood, *Proc. Soc. Exp. Biol.* **43:**223.

Lasker, M., Enklewitz, M., and Lasker, G. W., 1936, The inheritance of l-xyloketosuria (essential pentosuria), *Hum. Biol.* **8:**243–255.

Livingstone, F. B., 1967, *Abnormal Hemoglobins in Human Populations,* Aldine, Chicago.

Livingstone, F. B., 1973, Gene frequency differences in human populations: Some problems of analysis and interpretation, in: *Anthropological Genetics* (M. H. Crawford and P. L. Workman, eds.), pp. 39–66, University of New Mexico Press, Albuquerque.

Nuttall, G. H. F., 1904, *Blood Immunity and Blood Relationships,* Cambridge University Press, Cambridge.

Politzer, W. M., and Fleischmann, H., 1962, L-Xylulosuria in a Lebanese family, *Am. J. Hum. Genet.* **14:**256–260.

Sanghvi, L. D., Kirk, R. L., and Balakrishnan, V., 1971, A study of genetic distances among some populations of Australian Aborigines, *Hum. Biol.* **3:**445–458.

Spuhler, J. N., 1959, Somatic paths to culture, in: *The Evolution of Man's Capacity for Culture* (J. N. Spuhler, ed.), pp. 1–13, Wayne State University Press, Detroit. Reprinted from: *Hum. Biol.* **31:**1–13, 1959.

Ward, R. H., 1973, Some aspects of genetic structure in the Yanomama and Makiritare: two tribes of southern Venezuela, in: *Methods and Theories of Anthropological Genetics* (M. H. Crawford and P. L. Workman, eds.), pp. 367–388, University of New Mexico Press, Albuquerque.

Watson, J. D., and Crick, F. H. C., 1953, Molecular structure of nucleic acid, *Nature (London)* **171:**737–738.

Wiener, A. S., 1943, Anthropological investigations on the blood groups, in *Blood Groups and Transfusion,* 3rd ed., Chapter 18, Charles C Thomas, Springfield, Ill.

Wiener, A. S., and Wade, M., 1945, The Rh and Hr factors in the chimpanzees, *Science* **102:**177.

Workman, P. L., Blumberg, B. S., and Cooper, A. J., 1963, Selection, gene migration and polymorphic stability in a U.S. White and Negro population, *Am. J. Hum. Genet.* **15:**429–437.

Wright, S., 1931, Evolution in Mendelian populations, *Genetics* **16:**97–159.

Zuckerkandl, E., 1963, Perspectives in molecular anthropology, in: *Classification and Human Evolution* (S. L. Washburn, ed.), pp. 243–272, Viking Fund Publications in Anthropology No. 37.

Mutation and Molecular Evolution

2

F. VOGEL, M. KOPUN, and R. RATHENBERG

1. Introduction

The mutation process provides all the genetic variants from which evolution selects the most useful in order to produce the present-day diversity of living organisms. Until recently, evolution theoreticians have taken mutation more or less for granted, concentrating mainly on the theory of natural selection. The founders of the concept of molecular evolution on the one hand and Kimura (1968) with his theory of neutrality on the other have directed the interest of theoreticians toward mutation again. In view of the considerable amount of experimental work with different species, one would expect that theories in this field should be backed by a very detailed knowledge of the conditions under which mutations occur in nature. Surprisingly, however, this is not the case: most experimental work has been done on mutations which were induced either by ionizing radiation or by chemical compounds. From this type of work, only very limited information can be gained concerning spontaneous mutation. In addition, such research is usually carried out more or less as a sideline of other work. Molecular biology has taught us that the observed mutations, even if they can be traced down to the level of the polypeptide chain, are the end result of a complex interaction of primary events which are under genetic control. We should also consider the primary changes in the DNA, the probability of which depends on metabolic differences, the efficiency of polymerases, etc. How many of these changes will finally become visible as mutations

F. VOGEL, M. KOPUN, and R. RATHENBERG ● Institut für Anthropologie und Humangenetik der Universität Heidelberg, Heidelberg, West Germany.

depends on the efficiency of repair processes. However, polymerases, as well as repair processes, are under genetic control (cf. Drake, 1973).

These complexities make it very unlikely even on *a priori* grounds that something like an evolutionary clock exists. Even if many of the codon substitutions in genes coding for known proteins occurred at random, as the Kimura (1968) theory implies, the substitution rate would follow the mutation rate, and the mutation rate cannot be simply time dependent.* In view of this situation, one would tend to take a very pessimistic view about the possibility of including more realistic assumptions on mutation into our model of evolution. However, the prospects may not be so poor, provided that we do not expect to develop a final and altogether perfect concept. Our model should be refined step by step. After the first few steps, the picture admittedly may still be oversimplified, but at least the most obvious flaws of the old one are corrected. On the basis of present-day knowledge, such a correction seems to be possible in three ways:

1. Time-scale of mutation. There is enough evidence from different sources, lower organisms like phages as well as humans, that many of the spontaneous gene mutations occur in close connection with DNA replication. Even in the first papers of Watson and Crick (1953), mispairing of bases during replication was suggested as an obvious way to change the base sequence. Experiments with chemical mutagens as well as with base analogues have shown that this is actually possible. Evidence on single base replacements from protein and RNA sequencing is now abundant, and the best way to change a base pair is always to make a copy error. Furthermore, according to Drake (1973), many of the frameshift mutations analyzed in phages are also replication dependent. Statistical analyses on spontaneous mutations which have been done in humans and partially in mice point toward a close association between mutation and germ cell division. Of course, these data have to be regarded with restraint; the mutations have been analyzed on the phenotypic not the molecular level; heterogeneity between different mutations seems to exist; and the statistical data cannot be explained completely by simple replication dependence of the mutation process. (For a discussion, see Vogel and Rathenberg, 1975.) In spite of these difficulties, consideration of the mutation process as replication dependent, rather than time dependent, would help to create a much more realistic picture.

2. Direction of mutations. Current evolutionary theory regards the mutation process as random in direction. However, taking into account the different events needed on the molecular level to produce different

---

*It is, of course, true that the concept of the evolutionary clock does not necessarily imply random replacement of amino acids (see Sarich and Cronin, this volume). However, random replacement and purely time-dependent mutation rates are the only obvious causes for such a clock. Without it, the clock would be an unexplained phenomenon which has no connection whatsoever with other aspects of evolution theory.

base substitutions, one can assume on an *a priori* basis that this is an oversimplification. Therefore, different types of base replacements should be considered separately.

3. Inclusion of data from chromosome evolution. Everybody knows that genes are located on chromosomes, but nobody seriously takes this fact into account while discussing molecular evolution. However, structural and numerical chromosome changes are especially important for speciation, being powerful mechanisms for establishing reproductive barriers. The huge amount of data accumulated by cytogeneticists should be integrated into our picture of molecular evolution. At critical points of phylogenetic trees, combined cytogenetic and protein sequence analysis could help to solve problems of classification and phylogeny.

Examples for each of the abovementioned possibilities have been selected in order to explain the principle. We shall start with an attempt to convert the time scale into a replication scale.

2. Generation Time and Replication Cycles in Different Species

With some unimportant exceptions, replication is connected with cell division. Therefore, estimation of the number of replication cycles means assessment of cell divisions per time unit. In dividing bacteria, for example, up to three division cycles can be observed within an hour. Obviously, the division rate in higher organisms must be much slower, but there are almost no actual data available. In most cases, even the question of how many division cycles in the germ line correspond to one generation has not been investigated. This lack of knowledge is not caused by insuperable obstacles. With respect to *Homo sapiens*, a relatively well-founded estimate can be made on the basis of available evidence. (For details and literature, see Vogel and Rathenberg, 1975.)

What do we know about germ cell development? Between day 33 and day 37 of embryonic development, human germ cells leave the yolk sac to colonize the gonadal ridges. From here on, the two sexes must be considered separately. With the exception of meiosis, oogenesis is completed at the time of birth. The maximum number of germ cells (6.8×10^6) is reached during the fifth month of pregnancy (Fig. 1). It then declines to about 2×10^6 at birth because of cell degeneration. On the reasonable assumption that proliferation occurs by dichotomous divisions, this means about 21–23 division cycles. To get the total number of divisions per generation, the two meiotic divisions must be added. The number of divisions per generation is independent of the mother's age at fertilization.

In the male, the estimation is much more complicated. Spermatogonial stem cells are present up to the age of puberty. Until recently, nobody had counted them, although it was certainly possible. We have tried estimates

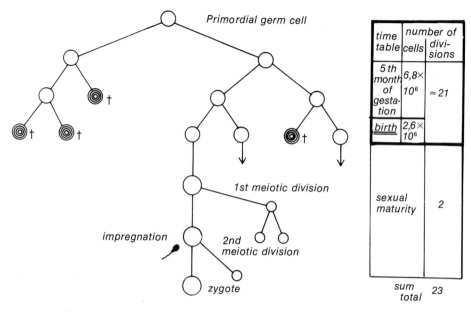

Fig. 1. Diagram of human oogenesis (from Vogel and Rathenberg, 1975).

starting from three completely different approaches, with identical results (Table 1). These stem cells, numbering approximately $5.7–6.2 \times 10^8$ per testicle, can be formed by approximately 29–30 dichotomous divisions. In order to estimate the additional divisions required for spermatogenesis in later life, we need to know more about the cell kinetics in spermatogenesis. In man, this information is available in the anatomical literature, mainly from studies with [³H]thymidine-labeling experiments *in vivo*. The emerging picture is seen in Fig. 2, and the number of divisions in relation to the age of the father at reproduction is seen in Table 2. This estimate oversimplifies the problem. The most important additional feature for a better understanding of the paternal age

Table 1.   Three Methods to Calculate the Number of Stem Cells per Testicle in Spermatogenesis[a]

Source of information	Number of cells/testicle	Dichotomous divisions
1. Volumetric data	$5.7–6.4 \times 10^8$	
2. Tubule length and cells per tubule diameter	$5.7–6.2 \times 10^8$	≈ 29 per testicle
3. Maximum daily sperm production	6.2×10^8	

[a]For details, see Vogel and Rathenberg (1975).

effect found in human mutation rates is the cell kinetic aspect of the well-known decline of spermatogenesis with age. Incredibly, this aspect has simply never been examined. From an evolutionary point of view, it may be of less importance because aging animals usually die early and reproduction occurs at a biologically young age.

There are two main differences between spermatogenesis and oogenesis: (1) the absolute number of divisions is more than tenfold larger during spermatogenesis, and (2) it increases with increasing age at the time of reproduction. These differences are compatible with statistical results in humans as well as in mice. In humans, we observe a strong increase in mutation rate with paternal age, but no independent maternal age effect. A higher mutation rate in male than in female germ cells was demonstrated in mice (Searle, 1972); the evidence for humans is more ambiguous.

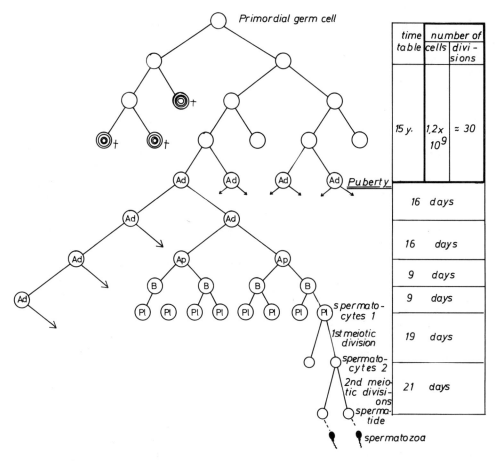

time table	number of cells	divisions
15 y.	1,2x 10⁹	≈ 30
16 days		
16 days		
9 days		
9 days		
19 days		
21 days		

Fig. 2. Diagram of human spermatogenesis with number of cells and cell divisions. From Vogel and Rathenberg (1975).

Table 2. Number of Cell Divisions in Spermatogenesis from Embryonic Development to Meiosis

From embryonic age to puberty (13 years):	≈ 30
Ad-type spermatogonia (1 division per cycle = 16 days):	≈ 23 per year
Proliferation + maturation:	6
Total (at the age of 28):	≈ 380 divisions
(at the age of 35):	≈ 540 divisions

For the sake of simplicity, let us take the average human (male) generation time as 25 years. There would then be about 300 division cycles per generation in the male and about 23 cycles in the female. For other species, especially for those which can serve as models for our own ancestors in evolution, such calculations have not been carried out. They should, however, be possible in principle. Figure 3 shows an estimation for the male mouse. At the end of the reproductive period (about 1 year), a male germ cell has undergone about 61–86 divisions. The average generation time can be estimated very crudely to be about 4 months. This gives us an average number of divisions of 28–34 per generation. In Fig. 3, humans and mice are compared (assuming about 30 cycles per generation in the mouse). Generation time is different by a factor of 75, and the time scale as measured by division cycles differs by a factor of about 7.5. This means that, all other conditions being equal, the rate of evolution for mammals with generation times similar to that of mice could be expected to be about 7.5 times higher than for manlike primates (or other species with longer generation times). Humans and mice also permit a comparison of mutation rates leading to well-defined dominant phenotypes (Tables 3A and 3B). Here, for the most frequently observed dominant mutations, the rates estimated in mice were almost one order of magnitude below the corresponding figures for humans. To evaluate these data, a number of possible biases must be critically examined; in the mouse, for example, mutant phenotypes which are readily recognized in humans would frequently go unnoticed. Taken at face value, however, the data are very similar when measured on the division scale. They are hardly compatible with a generation scale on the one hand and with a simple time scale on the other.

In order to arrive at a biologically meaningful scale for measuring the rate of evolution, similar data for germ cell kinetics as well as generation time would be needed for other species. A systematic search of the literature yields some data on division rates of stem spermatogonia and on the number of divisions during sperm maturation, but data on the number of stem spermatogonia are usually lacking (Table 4). Therefore, at this time, there is no reasonable estimate for the average number of cell divisions per generation for any other animal. It is to be hoped that formulation of the problem and demonstration that its solution is, in principle, feasible may induce some zoologists to make similar calculations on species with which they are familiar.

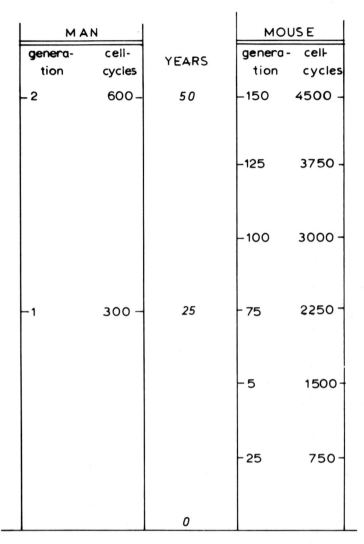

MAN		YEARS	MOUSE	
genera-tion	cell-cycles		genera-tion	cell-cycles
2	600	50	150	4500
			125	3750
			100	3000
1	300	25	75	2250
			5	1500
			25	750
		0		

Fig. 3. Estimated average number of cell division cycles per time unit in man and mouse.

3. Direction of the Mutation

The second oversimplification which is usually made concerns the direction of mutations. In a crude way, all amino acid substitutions which are possible on the basis of the genetic code can be regarded as equivalent. One refinement would be to calculate expectations in the code (see below). Another refinement concerns the base replacements. They can be subdivided in two groups: transitions and transversions. This means that, for all the four DNA

Table 3A. Selected "Classical" Mutation Rates for Human Genes

Trait	$\geq 10^{-4}$		$<10^{-4}\text{--}10^{-5}$		$<10^{-5}\text{--}10^{-6}$		$<10^{-6}$	
Trait	Trait	Number of series[a]	Trait	Number of series[a]	Trait	Number of series	Trait	Number of series
	Neurofibromatosis	1	Achondroplasia	3	Aniridia	2	von Hippel-Lindau's disease	1
			Osteogenesis imperfecta	2	Dystrophia myotonica	2		
					Retinoblastoma	5		
			Polyposis intestini	1	Apert's syndrome	2		
			Polycystic kidney disease	1	Tuberous sclerosis	2		
			Hemophilia A	5	Marfan's syndrome	1		
			Duchenne-type muscular dystrophy	10	Diaphyseal aclasis	1		
			Incontinentia pigmenti	1	Hemophilia B	2		
					OFD Syndrome	1		

[a]Number of estimates from different populations.

Table 3B. Spontaneous Dominant Visible Mutations in Mice: Both Sexes[a]

Tested gametes	Mutations	Frequency
369,944	3	8.1×10^{-6} per gamete
10,453,062 or		
14,021,464[b]	54	0.4×10^{-6} per locus[c]

[a]From Searle (1972).
[b]The lower figure is for color mutations, the higher is for other types.
[c]Mutations occured at 12 loci.

bases together, four transitions and eight transversions are possible (Fig. 4). Almost all specific molecular models suggested so far for base replacements, including the first mechanism proposed by Watson and Crick (1953), lead to transitions. The same is true for the chemical manipulation of single bases, e.g., with nitrite or base analogues. Additionally, every transversion involves one replication cycle in which base pairing cannot proceed properly. We do not know the reasons for naturally occurring base replacements. Yet it is not unreasonable to ask whether all transitions and transversions occur with identical probabilities, or whether certain events are more likely.

In 1965, we reported a higher incidence of transitions in certain human hemoglobin variants. Later on, the problem was examined from different points of view by many authors (Vogel, 1972). The following results seem to be undisputed:

1. In hemoglobin variants observed in present-day human populations, cytosine–thymine transitions (C↔T) in the DNA code corresponding to

Table 4. Some Comparative Data on Spermatogenesis

	Number of premeiotic (spermatogonial) divisions[a]	Daily sperm production
Mammals		
Man	4	1.5×10^8
Monkey	5	
Mouse	5–6	
Rat	5–6	
Birds		
Cock	1–3	2.5×10^9
Drake	3	$<1.2 \times 10^9$
Quail	3	
Turkey		1.12×10^9
Fish		
Guppy	10–14	
Trout	6	
Shark (dogfish)	13	

[a]From stem spermatogonia to spermatocytes.

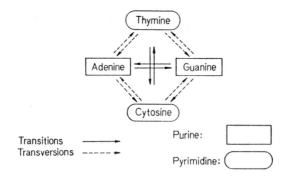

Fig. 4. Possible transitions and transversions.

transitions (A↔G) in the mRNA code are more frequently observed than would be expected with random substitution.

2. The same deviation from randomness can also be observed if homologous proteins from different species, which are separated by a number of different amino acid substitutions caused by one or a few mutations, are compared with each other.

3. The effect is not caused by unequal representation of the four bases in the cistrons concerned.

The mutations leading to differences in homologous proteins date back many, many generations, and are a very small sample of all mutations which have occurred during this time. The situation is much better for the present-day hemoglobin variants because they are of much more recent origin. Unfortunately, even most of these are not the immediate result of new mutations, but have been maintained in the population for an unknown number of generations. Therefore, it is still questionable whether the difference found is caused by selection, or whether it reflects a true nonrandomness of the mutation process. In order to elucidate this question, we have carried out some additional investigations:

1. For the recent hemoglobin variants, we repeated the examinations using additional material (see Lehmann and Huntsman, 1974).

2. For the proteins screened in evolution, we tested whether the observed nonrandomness could be explained by selection in favor of substitution by similar amino acids. Here, the dissimilarity (D) index of Sneath (1966) was used as a measure.*

Table 5 shows the results of the new analysis for hemoglobin variants, and, as can be seen, the effect is again present. In order to test the hypothesis of

---

*The Sneath index may not be the best measure to use here. It is possible that, for example, the conformational index used by Goodman (personal communication) would give more clear-cut results.

Table 5. Types of Transitions and Transversions Leading to Human Hemoglobin Variants[a]

Transitions			Transversions		
	All mutations in α-, β-, γ-, δ-cistrons	Only mutations in positions without specified function (α-, β-chain)		All mutations in α-, β-, γ-, δ-cistrons	Only mutations in positions without specified function (α-, β-chain)
A → G	11	0	A → T	6	2
G → A	8	0	T → A	6	5
T → C	28	15	A → C	9	2
C → T	42	26	C → A	10	4
			G → T	14	10
			T → G	15	8
			G → C	12	5
			C → G	15	9
			Unspecified	22	0
Total	89	41	Total	109	45

[a]Based on data from Lehmann and Huntsman (1974).

Table 6. Involvement of First and Second Bases in the Mutation Process Leading to Human Hemoglobin Variants[a]

Cistron	Base		
	First	Second	First and second
α-chain	30	28	58
β-chain	46	60	106
γ-chain	5	4	9
δ-chain	1	3	4
Total	82	95	177 $\chi_1^2 = 0.9548$

[a]Based on data from Lehmann and Huntsman (1974).

selection, a separate analysis was made for those substitutions for which selective influence is at least not obvious. The analysis was confined to amino acid sites without obvious, known functional restrictions (e.g., contact sites with other chains, internal surface of the heme pocket). The effect does not disappear. It may be asked what other factor apart from the selection could bias the result. Involvement of first and second bases seems to be random (Table 6). The most obvious bias is the preferential detection of polar substitutions, because electrophoresis is the technique usually employed to detect variation. Therefore, expectations for relative frequencies of polar and nonpolar substitutions were calculated for the hemoglobin α- and β-chains. It turns out that, of all possible transversions, 49% are expected to be polar, whereas for transitions only 28% are expected to be polar (Tables 7 and 8). Therefore, the deviation from randomness found is opposite to the expectation if the deviation were caused by this special bias. Within the group of transitions, however, the different expectations for (C↔T) transitions on the one hand and (A↔G) transitions on the other are in the expected direction (C↔T 40% polar, A↔G 14% polar). Therefore, it appears that the higher incidence of transitions in general as compared with transversions is real, whereas the preponderance of C↔T as compared with A↔G transitions is caused at least partially by the preferential ascertainment of polar substitutions. Selection for function and bias by preferential discovery of polar substitutions are, of course, only the most obvious causes for differences between mutation and its ascertainment. A definite answer cannot be given until a sufficient number of really new mutants

Table 7. Expectations for the Percentage of Polar Amino Acid Replacements in the Hemoglobin α- and β-Chains[a]

All transitions:	28%
C↔T:	40.4%
A↔G:	13.7%
All transversions:	48.6%

[a]From Vogel (1972, Table 7).

Table 8. Expected and Observed Polar (Charge Differences) and
Nonpolar (No Charge Differences) Amino Acid Replacements in
Human Hemoglobin α- and β-Chains (Recent Amino Acid
Variants)

| | Polar | | Nonpolar, |
	Expected	Observed	observed
Transitions			
Total		56 (72.7%)	21
C↔T	48.7	51 (86.4%)	8
A↔G	9.5	5 (27.8%)	13
Transversions		143 (79.0%)	38

from random populations of newborns become available. Neel (see Neel *et al.*,
1973) is now pressing hard for such studies in different human populations.

In looking at the amino acid differences between homologous proteins, we
have repeated our previously published (Vogel, 1972) analysis using additional
material and introducing the index of dissimilarity developed by Sneath (1966).
In all cases, the observed frequencies of single amino acid substitutions were
compared with expectations, the calculation of which has been described
before (see also Table 9). Figure 5 shows the following results: (1) There is a relatively strong negative relationship between dissimilarity of substitutions and the
observed/expected (O/E) ratio. (2) As suspected by Fitch, Zuckerkandl, and
others, the C↔T transitions are on the average more similar than, for example,
the A↔G transitions. (3) A regression analysis carried out for transitions and
transversions separately (Fig. 5) shows that a for $D = 0$ is much higher for transitions. Taken at face value, this result would also point toward a higher mutation
rate for transitions. However, the slope of the regression line, b, is steeper in
transitions. It is therefore possible that the true regression of O/E on D is not

Table 9. Example of Calculation of Expectations for Base
Substitutions[a]

Amino acid:	Ala	
DNA codons:	CGA, CGG, CGT, GGC	
First base:	Giving codon for:	Probability:
C → A	Ser	1/3
→ G	Pro	1/3
→ T	Thr	1/3
Second base:		
G → A	Val	1/3
→ T	Asp (depending on	1/6
or	third base	
	Glu in Ala codon)	1/6
→ C	Gly	1/3

[a]For a more formal description of the methods, see Vogel (1972).

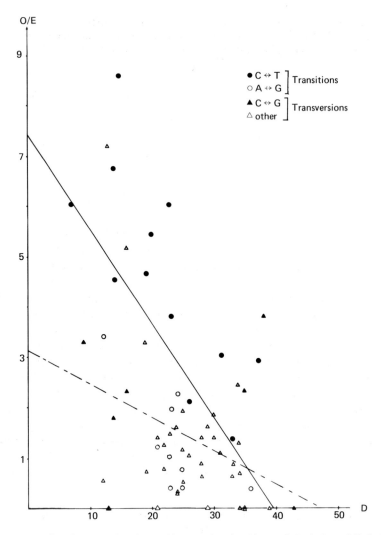

Fig. 5.   Comparison between the observed/expected ratio (O/E) and the index of dissimilarity (D) for a number of related proteins. For the proteins which were included in this analysis, see Vogel (1972).

linear, and there is, indeed, no *a priori* reason why it should be linear. Therefore, the argument is not altogether convincing.

Nevertheless, it is obvious that a trend toward transitions among the new hemoglobin variants cannot be explained away by any of the logically deducible biases. The tendency is also brought out by a quite independent source of evidence that we obtained on homologous nucleotide sequences of tRNA in which all transitions were also more frequent than expected under the assumption of randomness (Table 10).

Table 10. Comparison of Transitions (T) and Transversions (TV) in tRNA Substitutions[a]

	A	C	G	U	
A					$\chi^2 = 3.36$
C	5				$P(T/TV) = 0.07$
G	10	5			$P(T>TV) = 0.035$
U	8	10	5		
—	0	0	1	2	

[a]Data from Dayhoff (1972).

The method used here for analysis of amino acid substitutions in homologous proteins can also be used to examine the "neutral" hypothesis quantitatively. Regression analysis allows an estimate of the fraction of substitutions which are not observed because they would be too dissimilar. Hence it permits a quantitative estimate for selection. This estimate is certainly too low, because it covers only selection for similarity. This type of selection is of obvious importance in the "steady state" in which a gene has to maintain its function. Ohno (1970) is correct when he stresses the conservative character of selection. Another type of selection concerns the acquisition of new functions. This occurs, for example, when a duplication has been established and a cistron has become free to take up a new function (Goodman, this volume).

Nevertheless, pursuing this type of reasoning will help in better understanding the relative importance of neutral substitutions vs. selection, especially when different proteins are compared. However, the method has one inescapable flaw: It is not really known which of the possible codons are actually present in the genes concerned. We have to assume equal probabilities for all possible codons.

4. Chromosome Mutations in Evolution

Recent years have brought an abundance of literature on the topic of chromosome mutations in evolution, most of which has been reviewed by White (1973). Many helpful methods of chromosome analysis, the so-called banding techniques, have been developed and have allowed us to draw more precise conclusions. It is a real pity that there is so little coordination between this work and work on molecular evolution, which could lead to a better understanding of genetic determination in higher organisms in general and help to solve problems of phylogeny, especially for taxonomically closely related species for which evidence from protein sequences alone or together with the "classical" taxonomy gives ambiguous results. Besides, it seems that chromosome data should permit certain conclusions on the mode of speciation which cannot be drawn from protein sequences. Inclusion of such independent evidence may be a much better research strategy than further refinements of mathematical methods for the construction of phylogenetic trees, which may be subject to the law of diminishing return.

The principle can be described and its application explained using the examples of man and his closest relatives, the great apes:

1. The principle: In the human chromosome complement, every chromosome arm can now be identified using the banding techniques (Fig. 6). In the present-day human population, structural variants are being observed which usually go unnoticed if they are balanced but may lead to the formation of unbalanced, and frequently lethal, variants in the zygotes of the next generation (one example in Fig. 6).

2. In essence, the same structural "aberrations" separate the karyotypes of the great apes from the human karyotype. It turns out first that the apes have $2n = 48$ chromosomes, whereas *Homo* has only $2n = 46$ chromosomes. Analysis of banding patterns has recently permitted identification of the differences between the species (see Table 11; for all details, see Dutrillaux, 1975). Different types of rearrangements can be seen, the most important ones being the following:

   a. Man is different from all the apes in his chromosome No. 2, which corresponds to two D chromosomes in the apes. When this difference was discovered, it was thought to be caused by a centric fusion, i.e., a fusion of centromeres combined with a loss of the short arms of the two acrocentric chromosomes. This rearrangement is also the most frequently observed one in the present-day human population. After more careful analysis of the banding patterns, however, the basic event turned out to be a telomeric fusion; i.e., the telomeres are not lost but are fused end to end. This event results in a chromosome with two centromeres. One of these seems to be functionally suppressed; suppression of a centromere has also been observed in rare cases of chromosome pathology. (For details of the cytogenetic arguments, see Dutrillaux, 1975.)

   b. The most frequently observed type of chromosomal rearrangement is the pericentric inversion: a chromosome segment which contains the centromere is broken out of the chromosome (two breaks are necessary) and reintegrated upside down. In human chromosome pathology, pericentric inversions have also been observed occasionally, albeit much more rarely than, for example, centric fusions (Fig. 6).

   c. Paracentric inversions and other more complicated rearrangements are to be seen in single cases.

   d. There are differences in heterochromatic regions, for example, the so-called T bands, which are the inverse of the better-known Q and G bands.

Dutrillaux (1975) has tried to construct a phylogenetic tree, combining the evidence from the different chromosomes. In doing so, he encountered a

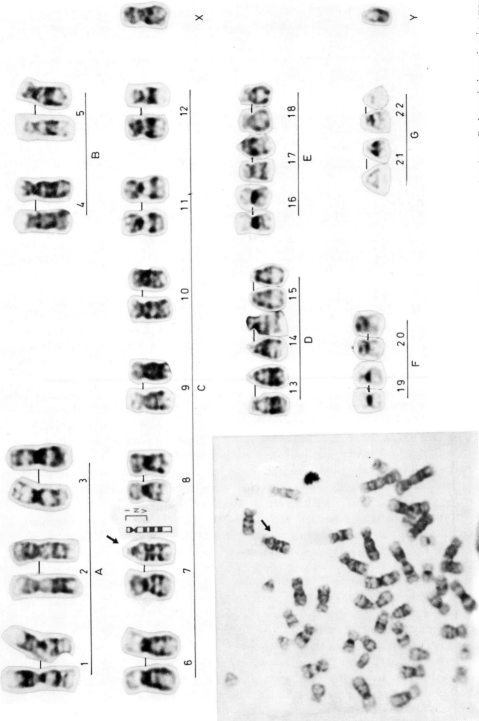

Fig. 6. Karyotype of a phenotypically normal male, who came to examination because of his wife's recurrent abortions. Pericentric inversion in one chromosome No. 7, G bands. Courtesy of T. M. Schroeder, Institut fur Humangenetik, Heidelberg.

Table 11. Comparison of Structural Differences[a] between the Human Karyotype and the Karyotype of the Great Apes[b]

Comparison	Pericentric inversions	Paracentric inversions	Telomeric fusions	Other
H. sapiens–P. troglodytes	6	0	1	
H. sapiens–P. paniscus	6	(1)	1	
H. sapiens–G. gorilla	8	2	1	1 translocation, 1 translation
H. sapiens–P. pygmaeus	7	3	1	3 translations, 1 complex change
P. paniscus–P. troglodytes	0	(1)	0	
P. paniscus–G. gorilla	6	2(+1)	0	1 translocation, 1 translation
P. paniscus–P. pygmaeus	9	3(+1)	0	3 translations, 1 complex change
P. troglodytes–G. gorilla	6	2	0	1 translocation, 1 translation
P. troglodytes–P. pygmaeus	9	3	0	3 translations, 1 complex change
G. gorilla–P. pygmaeus	10	3	0	1 translocation, 2 translations, 1 complex change

[a]Additions of bands or heterochromatic material are not taken into account.
[b]Condensed from Dutrillaux (1975).

contradiction: In two chromosomes (5 and 12) *Homo* is identical to *Pongo*, and in a third one (17) *Homo* is more similar to *Pongo* than to *Gorilla* or *Pan*. This could suggest a common ancestry of *Gorilla* and *Pan* with an earlier separation of *Homo* from the common ancestor. This hypothesis, however, is not compatible with other data, which suggest a sequence *Gorilla* → *Pan* → *Homo* (inversion in No. 7 and in 2p). Dutrillaux (1975), in discussing the different possibilities to overcome this difficulty (Fig. 7), favors the hypothesis that *Pan* separated at a later time from *Homo* but that there was a period (or periods) of hybridization with *Gorilla* which led to the sharing of 5, 12, and 17 (and possibly 4) (Fig. 8). This hypothesis, of course, implies a new principle not contained in the usual thinking on speciation, which always implies establishment of a reproductive barrier. However, the other possibilities, especially independent fixation of the same rearrangements in different species, seem to be, at least at first glance, still more unlikely. There is an interesting difference between the evidence from chromosome and protein data in that for the protein data the possibility of parallel evolution can never be excluded with confidence since identical amino acid replacements could occur in different branches. The following questions thus arise: (1) What exactly are the formal differences between protein and chromosome data? (2) Is the "unorthodox" hybridization hypothesis really the most likely explanation? (3) Can the hybridization hypothesis be used to solve some contradictions in the protein data, possibly not so much for amino acid replacements as for "Braunitzer gaps" (deletions)?

There is one important difference between amino acid replacements and chromosome arrangements, at least for their most frequent type, the pericentric inversions. In the heterozygous state, pericentric inversions lead to disturbances during meiosis, and hence to a selective (reproductive) disadvantage. This disadvantage disappears in the homozygote. The situation is similar to that encountered for Rh incompatibility and analyzed by Haldane (1942): there is an unstable equilibrium which makes it very difficult for a mutation to establish itself in a population as long as it has not reached a frequency of one-half of all alleles (or chromosomes). For all practical purposes, its establishment can be considered impossible except in a very special situation where in a population which is subdivided into many small, highly homogamous groups, a new species with a deviating chromosome complement (e.g., Fig. 6) may be started by one or a few individuals. Interestingly enough, the very same situation—homozygosity of a pericentric inversion in a human produced by an incestuous mating—has been observed once (Betz *et al.*, 1974).

Another problem posed by this analysis is the question of why those

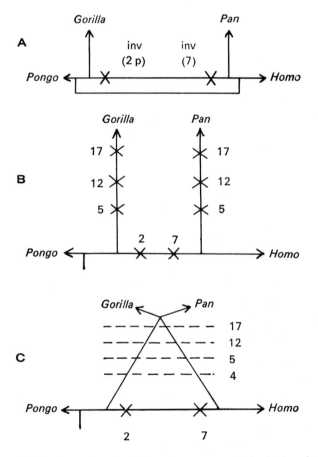

Fig. 7. Three possibilities for explanation of the apparent contradiction in the relationship between humans and the great apes. A: Heterozygosity for Inv (2p) and Inv (7) for a very long period during which separation of *Gorilla* and *Pan* from the common ancestor occurred. B: Independent double occurrence of three rare rearrangements. C: Repeated hybridization between the lines of *Gorilla* and *Pan*. From Dutrillaux (1975).

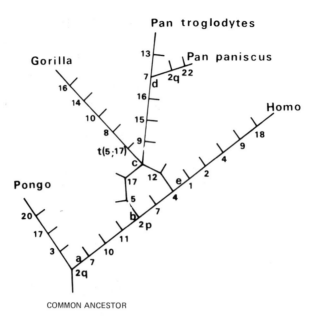

Fig. 8. Proposed phylogenetic tree of the human species and the great apes. Every number indicates a chromosome rearrangement which became homozygous. The succession in which the mutations have occurred between two phyla is arbitrary. From Dutrillaux (1975).

chromosome rearrangements which are observed most frequently in human populations, namely centric fusions and reciprocal translocations, have had so little importance in speciation. No convincing answer seems to be at hand.

A further problem is posed by those chromosome parts which contain heterochromatin, especially the R and T bands. Here, some evidence seems to point toward a *de novo* synthesis of relatively large stretches of DNA in special phases of evolution, especially at telomeric and centromeric areas. These results need urgently to be integrated with those found by Jones (this volume) with the *in situ* hybridization technique.

These relatively brief suggestions may serve to show in which aspects data from chromosome evolution may help to complement the picture of evolution gained from molecular data. First, they may help to solve questions of the relationships between closely related species, for which molecular data alone or together with "classical" morphological evidence give ambiguous results. Second, they may induce us to look at problems from a different aspect and to consider new solutions.

Cytogeneticists working on problems of evolution frequently pick the species to be examined for external reasons, e.g., availability in a zoo. The polypeptide sequencers' reasons for selection of species may not be much more sophisticated. However, the time has now come to establish cooperation and to exchange data and especially theoretical concepts.

Three aspects of mutation may be summarized in a very cursory way: time scale, direction of point mutations, and the additional information to be gained from chromosome mutations. It was not so much our intention to present new results which could alter our picture of evolution profoundly as it was to indicate possible new directions of research which could help to create a more realistic picture of evolution. Surprisingly, much material is available which has been collected from quite different points of view. Most of us are eager to make discoveries in the laboratory. However, considering the vast amount of scientific work which is being done all over the world, it should not be too surprising if some of the most interesting discoveries could be made in the literature of other fields.

5. References

Betz, A., Turleau, C., and de Grouchy, J., 1974, Hétérozygotie et homozygotie pour une inversion péricentrique du 3 humain, *Ann. Genet.* **17**:77.

Clarke, B., 1973, The effect of mutation on population size, *Nature (London)* **242**:196–197.

Dayhoff, M. O., 1972, *Atlas of Protein Sequence and Structure*, Vol. 5, National Biomedical Research Foundation, Baltimore.

Drake, J. W. (ed.), 1973, Proceedings of an International Workshop on the Genetic Control of Mutation, *Genetics,* Suppl. 73.

Dutrillaux, B., 1975, Sur la nature et l'origine des chromosomes humains, *Monographies des Annales de Génétique.*

Fitch, W. M., 1972, Are human hemoglobin variants distributed randomly among the positions? *Haematol. Bluttransfus.*

Haldane, J. B. S., 1942, Selection against heterozygotes in man, *Ann. Eugen. (London)* **11**:333.

Kimura, M., 1968, Evolutionary rate at the molecular level, *Nature (London)* **217**:624–626.

Lehmann, H., and Huntsman, R. G., 1974, *Man's Haemoglobins: Including the Haemoglobinopathies and Their Investigation,* rev. ed., North-Holland, Amsterdam.

Neel, J. V., Tiffany, T. O., and Anderson, N. G., 1973, Approaches to monitoring human populations for mutation rates and genetic disease, in: *Chemical Mutagens* (A. Hollaender, ed.), pp. 105–150, Plenum, New York.

Ohno, S., 1970, *Evolution and Gene Duplication*, Springer-Verlag, New York.

Searle, A. G., 1972, Spontaneous frequencies of point mutations in mice, *Humangenetik* **16**:33–38.

Sneath, P. H. A., 1966, Relations between chemical structure and biological activity in peptides, *J. Theor. Biol.* **12**:157–195.

Vogel, F., 1972, Non-randomness of base replacement in point mutation, *J. Mol. Evol.* **1**:334–367.

Vogel, F., and Rathenberg, R., 1975, Spontaneous mutation in man, in: *Advances in Human Genetics,* Vol. 5, Plenum, New York.

Vogel, F., and Röhrborn, G., 1965, Mutationsvorgänge bei der Entstehung von Hämoglobinvarianten, *Humangenetik* **1**:635–650.

Watson, J. D., and Crick, F. H. C., 1953, The structure of DNA, *Cold Spring Harbor Symp. Quant. Biol.* **18**:123–132.

White, W. J. D., 1973, *Animal Cytology and Evolution,* Cambridge University Press, Cambridge.

Zuckerkandl, E., Derancourt, J., and Vogel, H., 1971, Mutational trends and random process in the evolution of informational macromolecules, *J. Mol. Biol.* **59**:473–490.

The Fossil Record of Primate Phylogeny

3

ELWYN L. SIMONS

1. Introduction

Phylogeny is the study of the real genealogical relationship of taxa. It is not fully knowable. Phylogenetic trees combine information from fossil forms, from living species, from dating, from anatomy, and from chemistry, but none of these fields can stand alone in reconstructing phylogeny. Neither dendrograms grouping extrapolated dates of dichotomies with present-day species nor phylogenetic trees including only fossil forms placed on noncommittal side branches are adequate. Molecular biology can be seen as one field contributing to the understanding of phylogeny, but in no sense should it be viewed as overthrowing or supplanting the evidence from other disciplines. Dendrograms with branch-point dates extrapolated from presently living species can contribute to understanding phylogeny in some cases, but such charts are devoid of historical content, that is, the names, the nature, and the way of life of the fossil forms themselves. In addition, many splitting-time dendrograms, for primates at least, bear little relation to the record left by extinct members of this order. A definite problem exists in resolving these two major sorts of approaches.

Fossil primate species and genera are much more abundant than those now living. Each month more and more accurate geochemically based absolute dates for fossil primates are published. There are now hundreds of such dates in the literature. These dates are of themselves a major challenge to the need for molecular anthropologists to concern themselves with dating. The dating

ELWYN L. SIMONS • Department of Geology and Geophysics, Peabody Museum, Yale University, New Haven, Connecticut 06520.

and detailing of primate history are basically paleontological problems. Nevertheless, the biochemical differences which separate living species can bring useful evidence to bear on speculations about the relatedness of one group of primates to another, particularly when the fossil record itself does not provide any information. Such spacings, devoid of any time implications, have some value of their own.

Biochemical spacing data are in fact only a variation on the classical techniques of morphological spacing. Instead of denoting macroscopic structural differences from related animals (observed in the gross appearance of an animal's phenotype), immunochemists and those studying molecular structure measure one particular set of microscopic differences. Biochemical spacings tend to confirm morphological spacings. Great apes are a case in point. The African apes have no fossil record in the Pliocene or Pleistocene. However, it has been shown (Goodman, 1962) that in various biochemical systems these animals *(Pan, Gorilla)* are closer to humans than is the Asian orangutan *(Pongo),* an animal whose ancestry is also without a well-documented fossil record. Both Darwin *(Descent of Man,* 1871) and T. H. Huxley *(Man's Place in Nature,* 1863) suggest that the African apes are closer to humans than are the Asian. Thus this conclusion, based in the nineteenth century on gross anatomy, was further bolstered by biochemists working with blood precipitation as far back as the early part of this century (Nuttall, 1904). Since the 1950s at least, biochemical evidence has been incorporated into the thinking of paleontologists concerned with spacing in the order Primates. Even though DNA encodes the entirety of an organism's heredity, there is no sure method at present for converting this molecular record into a clock that keeps reliable time. For all these reasons, it seems to me that the framework of primate history will always have to be based on the fossils themselves and on their ecological, temporal, and geographic contexts, not on conjectured or extrapolated branch-point times.

2. Some Problems of Paleontology and of the Origin of Higher Groups

A few students may suppose that, because of a high degree of exactitude in the method of determining potassium/argon dates, branch points in primate phylogeny have been (or can be) determined with precision. However, as Walker makes clear in his contribution to this volume, fossil evidence as to time of dichotomies tends to make paleontologists estimate such times as of younger occurrence than they actually are. This is because evidence that the branching has happened has to be clear-cut, convincing to others, and demonstrable on morphological grounds. It also has to be preceded by fossil evidence of a common ancestral species for complete proof that a given taxon did not exist earlier. These requirements are seldom met. Moreover, the sort of behavioral and habitat changes in species populations that are thought to bring about allopatric species differentiation might not from the first be reflected in changes in the bones and teeth. Therefore, two already diverged populations sampled by the paleontologist can be interpreted as one. First appearances in the fossil record often come later than their real divergence times, because of deficiencies in the sedimentary record in relevant geographic regions. Such appearances do not indicate origin times. For example, this is the problem in determining the time of origin of such genera as *Theropithecus (Simopithecus), Parapapio (Papio* ancestry), *Australopithecus,* and *Macaca* in Africa. First occurrences of species of all these genera are usually given as being at 4 or 5 m.y. (million years) ago. This should not be taken as showing that they arose then. There are almost no primate-yielding sites in subsaharan Africa that date between 4 and 12 m.y. Origin times for these genera are unknown.

Beyond such considerations, there is another, provisional element in paleontology that derives from difficulties in solving two central issues: (1) who and what are the ancestors? and (2) where are the ancestors?

These two sets of problems are both questions requiring considered judgment and assessment of complex sets of facts. Experts within the paleontological field are bound to differ individually in assessing the same data. The first question is one of taxonomic ranking and it in turn involves a dual judgment: whether the find is of a direct ancestor or is only representative of an ancestral group. Ranking an animal as belonging in a species, genus, family, or yet higher category is an act indicating that the group represented by the particular taxon concerned has come into existence. Nevertheless, all taxonomic rankings above the species level have an arbitrary element.

The second set of questions deals with where living and fossil forms occur. This is essentially the subject of primate biogeography. This subject itself is a new one that, so far, has been characterized by a high level of uncritical speculation. In spite of the fact that much new and exciting evidence can be brought to bear on problems of primate distribution, well-known gaps in the fossil record make some of the details of past distributions unknowable. Such a presently unsolvable supposition relating to earliest primates is the hypothetical placement by McKenna (1966) of the order Primates in Africa in the Cretaceous Period. In Africa no Cretaceous (or even subsequent Paleocene) mammalian fauna is known. Therefore, his hypothesis or prediction has no testable meaning. When such fauna is found in fossil form (if ever), it will either show that Primates were in Africa or show that they were not there. If the fossils are found in Africa, McKenna will be said to have predicted their occurrence there. If they are not found there, his view will not presumably be discounted. Such a suppositional placement of a group where there is no evidence for (or need for) their presence is confusing. Others then repeat such speculations, giving them a life of their own. This has, in fact, occurred with the point of posited African Cretaceous primate distribution in the recent studies of Walker (1972) and Hoffstetter (1974).

Primates are extremely abundant and diversified in early Paleocene times in Holarctica. The pattern indicates an ancient primate history there. When the group can be documented first in Africa and South America in early

Oligocene times, the degree of diversity these animals show is far less, suggesting a much more recent arrival time in those continents. Had Primates been in Africa since the Cretaceous Period, they would have, in all likelihood, given rise to several different primate-derived orders, just as an early (apparently condylarth) stock there produced the related orders Embrithopoda, Hyracoidea, Sirenia, and Proboscidea. There is no need to place Primates in Africa before the Eocene. The 20 m.y. time duration of the latter epoch provides quite enough time for the lemurs to have reached Africa via a route across Africa from southern Eurasia on their way to Madagascar, if this was indeed their route into that island. Another theory is that they reached Madagascar via the Indian plate. It should be pointed out that at present there is no fossil evidence from any African Eocene, Oligocene, and Miocene sites that lemurs were ever in Africa, not even a single suspected tooth! A mandibular fragment supposed to be a prosimian and recently described by Sudré (1975) from the Eocene of Algeria would be hard to sustain as directly related to lemurs.

2.1. What Are Ancestors?

The question of what qualifies a fossil as an ancestral form is an important one. At one level, without pedigree, no individual fossil can actually be confirmed as an ancestor of anything. Whether it ever sired offspring is unknown and unknowable. For example, the sole skull so far discovered of an Oligocene ape (the only skull known from the entire first half of catarrhine evolution), that of *Aegyptopithecus zeuxis* from the Fayum badlands of Egypt, is subadult. (This is judged by the small amount of wear on the cheek-tooth crowns and the fact that the large canines had a little farther to erupt.) This individual, probably a male, may never have mated, and if this is so he would not be anything's ancestor. At the other end of the spectrum from such an exclusion from ancestry is the position taken by Le Gros Clark (1959). In his view, paleoprimatologists are really looking for knowledge of the broad level of anatomical and ecological organization that a particular grade in the history of a group represents. Thus a skeleton of *Australopithecus* (such as that found in the Hadar region of Ethiopia November 24, 1974, and nicknamed "Lucy" by the Cleveland paleontological group) provides much that scientists want to know about a particular stage or grade in hominoid evolution. This information is supplied whether or not the individual (or in fact any other *Australopithecus* individual now known) is the ancestor of any later group. The deductive reasoning which leads to such a conclusion is simply put: since *Australopithecus* does represent an anatomical intermediate between *Homo* and *Pan* it is most logical to assume that in very broad terms it represents a better approximation to an earlier stage in the divergence of the phylogeny of man than do members of *Homo*—defined so as to exclude *Australopithecus*. In this context, such finds produce inferential evidence about ancestral stages without necessarily being direct ancestors.

2.2. Direct Ancestors and Adaptive Radiations

Another issue which often sidetracks interpretation of the phyletic meaning of known fossils is that dealt with by Walker in his chapter of this volume. This is the question of whether such fossils as scientists find have a high or a low probability of being actual ancestors. Nevertheless, paleontologists are just as interested in fossils that are a little (or even quite a lot) off the main line as they are in those on the direct line to later varieties. Paleontology is basically a historical discipline and should, I believe, take notice of the core of what history is about: the recounting of the names, the conditions, the events of the past. In paleontology, understanding of past species also becomes more clear as more and more is found out about them. It is commonplace that the discovery of previously unknown parts of an extinct species can change an assessment of its taxonomic position. In the case of poorly known species, suspended judgment is frequent.

It is difficult to prove whether any fossil population samples actual ancestors or not. It is also a point of debate whether the first stages in the history of a particular taxon have any significant chance of recovery (see Eldridge and Gould, 1972; Walker, this volume). Nevertheless, the demonstrated adaptive radiation of a group does have an important meaning for phylogeny. Such significance stands independent of the seemingly always controversial attempts at proof that any one particular species is in a line of direct ancestry. It can often be shown that a divergence has occurred without having to prove that any particular fossil is a definite ancestor of a specific modern species or group. In other words, finding several fossil species of a group, all at essentially the same time level, *does* demonstrate that the particular taxonomic grade of organization has arisen, regardless of whether or not any of the known species found are ancestors of later species.

Another commonplace of paleontology is that when a group has radiated into a variety of forms it almost always implies the existence of a more extensive radiation than can be proven from actual fossils. This is because no taxonomic group and no continent are so well known paleontologically that scientists can believe they have documented the majority of the species which once existed there. Such a case can be exemplified by the early Miocene pongids and hylobatids of East Africa. Although there are over 1000 specimens of these apes, they all come from about a dozen sites in Kenya and Uganda. The total geographic area exposed at these sites can be estimated as about 5 km.^{2} This is about 1/3,000,000th part of the present-day surface area of Africa. Africa must have been about the same total size in the Miocene, so that the probability that known fossil collections, all from these two East African countries, have adequately sampled all the species and genera of African Miocene apes is extremely small. Two principal conclusions follow from this sort of patchy occurrence of fossils: (1) It is unwise to try to discount even partial evidence of a radiation of a group of mammals. (2) It is just as unwise to insist that known fossils are always, or even often, actual ancestors. I follow Le Gros Clark's

suggestion that relevant fossils should be regarded as being broadly representative of ancestral types. On the other hand, when fossils come from the right
time and place it is equally difficult to prove that some are not on direct
ancestral lines. From the foregoing derives the scholarly tendency to produce
phylogenetic trees which never place actual fossils on the main line. Some of the
varieties of such trees have even been named: "hat-rack," "candelabra," etc.
This subject is whimsically discussed by Howells (1947). Le Gros Clark (1959)
has also well summarized and illustrated the manner in which actual fossils are
typically placed apart from and outside main lines of descent. Obviously in such
a tradition a scholar would be out of step ever to insist, categorically, that known
fossils *have to be* on direct ancestral lines and therefore *prove* branch-point
times because of their geochemical datings. Faced with this dilemma, scholars
should concentrate their attention on the evidence that particular taxonomic
groups have arisen, and not on the attempt to show that given fossil species are
direct ancestors.

2.3. Does Biochemical Evidence Necessitate Change in Higher Groupings?

The determination of those scholars riding the current wave of Hennigian
thinking about taxonomy is likely to serve one nonuseful purpose: that of
making it appear that a single agreed-on kind of taxonomy can, or must, be
developed. The practice of ranking and assorting higher taxonomic categories
is more an art than a science and pursued for its own sake is a sterile one.
Although taxonomy has established rules of practice given by the International
Commission on Nomenclature, these rules have not made and probably cannot
make for total agreement. No matter what it is called, each school of classification, whether vertical or horizontal, gradist or cladist, Simpsonian or Hennigian, has its own merits and defects. This point was made long ago by W. K.
Gregory (1949). The traditional system which was incorporated for mammals
by Simpson (1945) has the advantage of comprehensiveness and is well established. The intrusion of radical new changes now into mammalian classification
of the sort advocated by extreme cladists or Hennigians is not worth the effort it
would take to establish them. As recently outlined (Simons, 1974a), it seems that
the transfer and rearrangement of mammalian higher categories represent an
increasingly puerile field for scientific inquiry. For the order Primates, the mass
of accumulated taxonomic data is already so great, and most of it has been so
long available, that new discoveries requiring major taxonomic changes in this
order are unlikely. The details of primate phylogeny are in part poorly known,
but these may soon be further clarified by new fossil finds. There is no such
thing as a correct classification. Even if scientists could agree on a phylogeny,
reasonable men would differ about its meaning for classification. Playing the
game of rearranging higher categories in Primates is to substitute intellectual
constructs for real findings about primate biology. Molecular biologists can

usefully contribute to primate classification insofar as it relates to identification of present-day species and their various degrees of chemical difference. Such spacings do not change materially the classification of Primates, nor does it seem that they will provide much real need to do so in future. At best, the molecular evidence provides one of several lines of evidence that is relevant to primate classification. Just as evolutionary shortcuts in the ontogenetic development of given species have obscured detailed knowledge of stages in their phylogeny, the biochemical encoding by which animals and plants are reproduced cannot be read out as a phylogeny, nor as any more than an outline of classification.

2.4. The Fossil Record

Recently, Romero-Herrera *et al.* (1973) discussed the fossil evidence for eight splitting times, or dichotomies, in the fossil record of the order Primates. In general, the discussions they give are well founded, but in most cases can now be expanded or restated. In addition, there are three or four other radiations evidenced by primate fossils. These will also be taken up here in sequence. For convenience of reference to their work on molecular evolution of myoglobin, the numbers Romero-Herrera *et al.* assigned to the various splitting times will be reproduced after each of my headings below, *viz.*, R-H Nos. 3, 4, 9–15. Their discussions lay the basis for what follows and should be read in connection with it.

2.5. The Dichotomy between Primates and Other Placentals   (R-H Nos. 3, 4)

Evidence from the fossil record as to the branch point of primates with other placentals is the most vague and inferential of all. Origin of the order Primates has usually been discussed in reference to the species *Purgatorius ceratops,* oldest asserted primate (see Van Valen and Sloan, 1965). The status of this animal and the related *Purgatorius unio,* i.e., their taxonomic placement in an order, is currently under review (see discussion in Clemens, 1974). As I have recently pointed out (Simons, 1975), the known diversity of North American middle Paleocene plesiadapiform primates (60–63 m.y. old) implies that they would not have had a common ancestry later than the end of the Maestrictian. This is the time of first *Purgatorius* and the last stage of the Cretaceous Period. Thus whether *Purgatorius* occurs in the Maestrictian and whether it is primate or not are both partly extraneous issues. Recent research reported on the oldest primate skull, that of a mid-Paleocene member of the primate infraorder Plesiadapiformes, *Palaechthon nacimienti,* by Kay and Cartmill (1974) shows that this mid-Paleocene radiation (that of the plesiadapiform primates) was of animals that may have been almost at the insectivoran level of organization. It is probable that ancestors of all the living primates were already separate in the middle Paleocene from the plesiadapiforms, thus suggesting a splitting time of these two branches much earlier than 65 m.y. Should plesiadapiform and tarsiiform primates have a special relatedness, as recently suggested by Gingerich (1975), then the dichotomy between the lemurs and the ancestors of present-day higher primates (Anthropoidea) would perhaps fall even earlier.

Apart from these issues is the new evidence about the time of radiation of the most ancient placental orders during the Cretaceous Period. A few orders such as Perissodactyla and Artiodactyla have been proven by fossils to have had their basic branching within the Tertiary. Previous discussions, based largely on North American fossils, make Primates one of the most anciently differentiated orders, along with an early ungulate and an archaic carnivore stock (e.g., see Van Valen and Sloan, 1965; Clemens, 1974). Recently, McKenna (1975) and Kielan-Jaworowska (1975*a,b*) have studied newly found placental skulls from the Gobi Desert of Mongolia. These come from fossil sites located in the Djadokhta (and somewhat younger) Barun Goyot formations, both of late Cretaceous age. Faunal correlations are not completely confirmed, but it appears that the age of all these placental skulls, about 50 in number, is more than 80 m.y. Considering their antiquity, they show more diversification than most students of Tertiary mammals had previously expected. Even that far back in time, McKenna (1976) has identified distinct dichotomies between rabbitlike and elephant-shrew-like forms. The implication is that the diversification of the main branches (cohorts and orders) of placental mammals was then well under way. Thus it is possible that the particular species which was to give rise to the order Primates could have diverged from other placentals prior to 80 m.y. but probably not before 100 m.y. These deductions are inferential, but it is evident that the date of 60 m.y. still sometimes given for the age of the basal primate radiation is wrong and might be no more than two-thirds the actual number.

2.6. The Radiation of Primates of Modern Aspect (R-H No. 9)

The "second primate radiation" discussed by Radinsky (1975) or the "radiation of primates of modern aspect" discussed by Simons (1963, 1972) is actually an unresolved trichotomy rather than a dichotomy in that it involves the branching of Lemuriformes, Tarsiiformes, and basal Anthropoidea. Unfortunately, neither phenotype anatomy (including gross morphology of living or fossil Tarsiidae) nor manner of placentation has clearly settled whether tarsiers are closer to Anthropoidea than other Prosimii or simply share with Anthropoidea several features evolved in parallel because they also share a pronounced commitment to visually oriented food-getting. Parallelisms evoked by the visual orientation could explain the partial orbital closure, reduction in olfactory dependence, and concomitant loss of a glandular rhinarium as well as the presence of a retinal macula, all features *Tarsius* shares with Anthropoidea.

Almost certainly this is a case where molecular anthropology could contribute to understanding primate phylogeny. There are conflicting opinions as to whether details of placentation of tarsiers and Anthropoidea are as similar as some assert. Alternatively, they are thought similar because (1) they are shared derived characters, (2) they have evolved in parallel, or (3) they are characters of common heritage from preprimate stages. These conflicts, I believe, eliminate type of placentation from a central role in determining whether the major divisions among living primates are Haplorhini/Strepsirhini or Prosimii/Anthropoidea (e.g., see the varying opinions expressed in the recent summaries of Luckett, 1974; Starck, 1974; Gingerich, 1974). Lemurs and lorises have a glandular rhinarium, while in tarsiers and Anthropoidea the external nose is dry and the upper lip not cleft. This character complex of the soft anatomy was the entire basis of Pocock's (1918) grouping of *Tarsius* and Anthropoidea in Haplorhini, with lemurs and lorises in a correlative group, Strepsirhini. These terms were used by almost no one until revived by Hill (1953). More recently, with a wave of cladism in progress, the terms have also been circulated by Martin (1972) and Luckett (1974). A majority of scientists have, however, disregarded this classification because (as stated already) the reorganization of the visual and olfactory senses in *Tarsius* together with snout reduction gives the similarity no phylogenetic importance, *per se,* but strongly suggests a parallelism. Moreover, nowhere in their anatomy do tarsiers resemble, even in the broadest sense, the ancestors of Anthropoidea (Starck, 1974).

The radiation of primates of modern aspect might even be older than that of the plesiadapiform primates. First dated occurrences in this radiation at about 56 m.y. already include adapids, tarsiids, and omomyids of the genera *Agerina, Pelycodus, Protoadapis, Anemorphysis, Tetonius, Pseudotetonius, Loveina, Omomys,* and *Teilhardina.* By the end of the first third of the Eocene, several additional genera appear. Probably not all of these can be directly derived from the above genera and thus they imply the existence of several earlier and separate branches. None of these "modern aspect" primates can with any probability be derived from the plesiadapiform radiation. Thus their ancestors might go back 10–15 m.y. separately before reaching a common stock species. Where the ancestors of modern aspect primates lived earlier is presently unknown.

2.7. The Platyrrhine-Catarrhine Dichotomy (R-H No. 11)

Interestingly, the best evidence as to the time of the platyrrhine-catarrhine split is not that given by Romero-Herrera *et al.* (1973). It is based on neither soft anatomy, fossils, nor molecules but on biogeographic considerations. These have never been fully outlined and to do so will occupy considerable space in what follows. This must be regarded as a very short review of a much larger and important body of scientific literature. This dichotomy is the oldest among

Primates that can be enclosed in relatively narrow temporal brackets. Indications are that the split occurred between 59 and 49 m.y. ago.

There is general agreement that the Ceboidea (New World monkeys) or platyrrhines and the combined Cercopithecoidea (Old World monkeys) plus Hominoidea (apes and man) or catarrhine primates had their basal radiations in South America and Africa, respectively (e.g., see Simons, 1963, 1972; Von Koenigswald, 1969). Primate arrivals in these southern continents have been suggested as being in the Eocene (40–50 m.y.) for one or both groups by various authors, a few of the most recent of which are Simons (1972), Walker (1972), Hoffstetter (1974), and Orlosky and Swindler (1975). If the arrival of each in the southern continents was about 45 m.y., then their latest common ancestry was likely much earlier, say around 55 m.y. The reasoning behind such a conclusion is as follows.

Primates are diversified in the Fayum deposits of Egypt. There are four hominoid genera, each with a single described species, and two cercopithecoid genera, each with two described species. Other species are presently in the process of description (Simons, ms). The Fayum fauna comes from beds overlying later Eocene marine deposits (Qasr el Sagha Formation), and the Jebel el Qatrani Formation containing this fauma is capped by a basalt dated at 25 or 26 m.y. Faunal correlations of the primate-bearing levels are discussed in Simons (1968) and Coryndon and Savage (1973). Species of at least four creodont genera and of two artiodactyl genera which also occur in Europe are known from the Fayum. One Fayum specimen of *Apterodon* (a creodont described first in Europe) comes from the aforementioned Qasr el Sagha Formation (Simons and Gingerich, 1976). This occurrence, taken together with the other genera in common, proves a filter migration route between Africa and Eurasia in the Eocene. Outside Africa, the five genera concerned are typical of late Eocene and early Oligocene deposits. Thus the provisional correlation of the part of the Fayum deposits containing primates is with the North American early Oligocene Chadronian provincial age. Time duration of the latter is 38 m.y. to about 32 m.y. ago: for Chadronian dates, see Savage (1975). All phenotype and immunochemical as well as other evidence is in agreement in showing that Cercopithecoidea and Hominoidea are closer than either is to Ceboidea. However, by Fayumian times they have differentiated to the level of two families, six genera, and ten-plus species, but of course not all of these are synchronous. These belong to both parapithecid monkeys and dryopithecine apes (pongids) (see Simons, 1974b).

There is general agreement that these catarrhines radiated in Africa (Simons, 1972). As itemized above, they were clearly diversified in Fayumian times. A common stock species (single species ancestor) would probably not have existed later than 5–10 m.y. earlier and the common ancestral catarrhine-platyrrhine species a comparable period earlier than that. In fact, a considerable time lapse between the catarrhine-platyrrhine and the cercopithecoid-hominoid divergences is requisite in order to account for the significantly closer ID distances and other molecular similarities of the latter two superfamilies when contrasted to Ceboidea. Independent introduction of parapithecids and dryopithecine ancestors into Africa would not substantially alter the above deductions but is less attractive because only one other mammalian species need have reached Africa in the middle Eocene: the ancestor of phiomyid rodents.

The nature of platyrrhine introduction into South America appears to be similar to that of the catarrhine arrival in Africa. That is, the introduction is suspected of having happened in the middle Eocene. Proof that it was definitely accomplished by early Oligocene times exists in Bolivia, where a primate *Branisella* is present. Like the Fayumian provincial age, the Deseadan of South America appears to be a Chadronian temporal equivalent. However, it is well known that correlation of provincial ages between continents is uncertain. There is some evidence that a Chadronian equivalency is a latest correlation for Fayumian and Deseadan. This is suggested by the fact that most of the type species of the six immigrant genera to the Fayum are Eocene in age. An earlier Deseadan is also a possibility (Bryan Patterson, personal communication).

Platyrrhine-catarrhine segregation and deployment must be seen as occurring as three separate events: (1) the segregation of the two basic stocks into the two separate hemispheres concerned, (2) the deployment of catarrhines into Africa, and (3) the movement of the ceboids into South America. These events will be dealt with in sequence.

2.8. Geographic Segregation of Platyrrhines and Catarrhines

It is generally agreed that early primates were all animals of the tropical forest. Even up to the present date no ceboids have left this environment. Moreover, of all the primates, and throughout the whole of the Tertiary, only *Homo sapiens* has crossed the northern land bridge (Alaskan-Siberian) between the two hemispheres. Early primates do not distribute through colder climates, but through tropical or warm-temperate forests. As I will attempt to demonstrate in a later section, they also do not distribute across broad water barriers. The nearly complete disappearance of the previously abundant prosimians of the North American and European Eocene by early Oligocene times is generally thought due to southward migration resulting from climatic deterioration into regions where their fossils have not been found (see Simons, 1963).

If basal catarrhines and platyrrhines were restricted to tropical forests and could cross only narrow spaces of open ocean, how can their distribution be accounted for? The species ancestral to both must have exhibited one of three distributions: (1) an eastern hemisphere distribution, (2) a western hemisphere distribution, or (3) a distribution in both hemispheres. In either of the first two cases, after the dichotomy one of the stocks would have had to cross into the other hemisphere. With the final rifting in the North Atlantic seaway (dated after Wasatchian/Sparnacian times) at about 49 m.y. (see McKenna, 1972), this dispersal would have had to be across the open ocean. Since the Atlantic split from south to north and the spreading was initiated about 110 m.y. in the South

Atlantic (see Ladd *et al.*, 1973), this dispersal would have to be across an open ocean at least two-thirds the present minimum distance, i.e., 2000 rather than 3000 km, by 40–50 m.y. ago. If rafted by the South Equatorial Current, distances would be 3000 km (length of current then, 40 m.y.) and about 4500 km today.

An important recent discovery is that the ancestral platyrrhines and catarrhines did not have to migrate nor to leave forested environment at all to reach a distribution in the two hemispheres. This is because of what is now known of the remarkable faunal community between North America and Western Europe (only) in the Wasatchian/Sparnacian provincial age of the early Eocene (duration 56–50 m.y.; see Savage, 1975, for summary). As Simpson originally suggested (1947), the faunal community between the European early Eocene and that of Western North America is so great as to indicate a single geographic region at that time. This is here termed Euronearctica. Until the Oligocene Epoch, Western Europe was separated from Asia by a seaway (Ouralian Sea). Recent tabulations of mine (Simons, 1969) as well as fuller discussions by McKenna (1972) and Russell (1975) and by others have demonstrated that in the early Eocene Euronearctica was tropical continuous land in the north and possessed over 50% of mammalian genera in common, including the primates *Phenacolemur, Plesiadapis, Pelycodus, Chiromyoides,* and *Teilhardina.* The last of these, *Teilhardina belgica,* has even been stated by Quinet (1966) as possibly being the earliest and ancestral species for Anthropoidea. More remarkable perhaps is the fact that in the summer of 1974 scientists working for the University of Wyoming discovered a new North American species of *Teilhardina* in the Wasatch deposits of the Bighorn Basin of Wyoming (Bown, 1976). In view of this find, one can now consider, for the first time, whether or not these two species could actually be the forebears of catarrhines and platyrrhines, respectively. But even if not, it is my view that in the Wasatchian/Sparnacian of Euronearctica we have the time and place where the platyrrhine-catarrhine dichotomy took place. The approximate time would be 56 m.y. or 20 m.y. earlier that the molecular clock date reiterated by Sarich and Cronin in this volume of 36 ± 3 m.y.

In summary then, the Oligocene/Recent distribution of New and Old World higher primates can be taken as an exact case of what Hallam (1973) recently termed "disjunct endemism," which he defines as "a type of regionally restricted distribution of a group of fossil organisms in which two or more component parts are separated by a major physical barrier and hence not readily explicable in terms of present-day geography. In the case of continental or shelf organisms such a barrier could be a zone of deep ocean." As McKenna (1972) has discussed, a whole fauna can be carried by a drifting continent, transporting animals in "Noah's Ark" fashion into a new zoogeographic region. It was in all probability by this means that elements of the extensively diversified North American Wasatchian primate fauna were isolated in the eastern hemisphere following the North Atlantic rifting of Euronearctica at about 49 m.y. ago; these almost certainly included the catarrhine ancestor, and thus a catarrhine-platyrrhine split point of at least 49 m.y. is suggested. A split-point date of

36 m.y. for the catarrhine-platyrrhine dichotomy requires rafting of platyrrhine ancestors from east to west in the early Oligocene. All suggestions of such
rafting have been of a casual and uncritical character. They arose originally
(Lavocat, 1969) to attempt an explanation of the distribution of hystricomorph
rodents, not primates. Moreover, this theory was immediately opposed for
good reason by Wood and Patterson (1970), who concluded: "we believe that,
even in the earliest Tertiary, South America and Africa were far enough from
each other so that trans-Atlantic migration did not occur, and close enough to
North America and Eurasia respectively so that their mammalian faunas were
ultimately derived from the adjacent northern continents." Wood (1974) again
analyzes the problem of rafting rodents and concludes "the possibility of a
trans-Atlantic crossing still seems to me to be as improbable as it ever did." The
question of the facility, or lack of it, by which primates could colonize a new
continent by rafting across open ocean has never been seriously analyzed. Cox
(1973) states the general prerequisites for mammals as follows: "To show that it
was feasible for any group to spread from one area to another at a particular
time it is necessary to prove several things:1) that the group was then in
existence; 2) that there was continuous land, or at least only minor sea crossings
(which would probably be almost undetectable geologically) between the two
areas; 3) that there was no impassable climatic barrier at the time in question."
Hoffstetter (1974) is the only author to state a mechanism for accomplishing a
South Atlantic rafting of primates: that the raft be carried across by the South
Equatorial Current, or its early Tertiary equivalent. Hoffstetter's speculation
was soon employed uncritically as a justification of a late platyrrhine-catarrhine
split (Washburn, 1973). There are many important objections to primate long-
distance rafting. Since the improbability of such rafting is directly related to the
improbability of a catarrhine-platyrrhine split in the early Oligocene, evidence
about rafting will be outlined below.

2.9. Primate Colonization: Rafting

Rafting of mammals as a possible explanation of zoogeographic distributions was proposed by Wallace (1876). He remarked that floating islands "are
sometime seen drifting a hundred miles from the mouth of the Ganges with
living trees erect upon them" and that islands are also seen off the mouth of the
Amazon, having been swept down it by floods. Nevertheless, little documentation of such floating islands, or "rafts" as they have come to be called, has been
accumulated since his time. It seems, according to Spruce (1908), that the
islands in the Amazon are "sudd" islands, consisting of tangled masses of grass,
often very large but unsuitable for long-distance animal transport. Wallace also
posited that the rafts would have to retain standing trees which could act as sails
and was uncertain whether currents could act as means of transport. Neither of
two comprehensive books about plant dispersal across oceans, Guppy (1917) or
Ridley (1930), mentions floating islands. Ridley does point out the arrival alive

in the Cocoa Islands, 700 miles west of Java, of clumps of sugar cane, lalang grass, bamboo, and a crocodile. All these organisms are highly resistant to salt water.

Wallace had in mind rafting of mammals across the ocean in the East Indian Archipelago and elsewhere because, in his view, hardly any nonaquatic mammal was capable of swimming across a channel more than 20 or so miles wide. Therefore, floating transport seemed to be all that was left to explain such distributions across island arcs. There are at least four major lines of evidence, poorly understood in Wallace's time, which relate to this problem. The first two are sets of questions about raft formation and conditions of survival on them. The last two are about geological changes that may affect distribution routes across water.

1. How do arboreal mammals such as primates and rodents meet the stresses of rafting, such as lack of food and water and the extremes of temperature and exposure? Have such rafts ever been seen with primates on them? Could any such floating islands provide food, water, and shelter for more than a few days?

2. Are such rafts ever formed by the smaller rivers of islands in an archipelago? If not, can they be used as part of the explanation of "island hopping"? King (1962) observed raft formation in the Rio Tortuguero, Costa Rica, and references most of the literature on possible rafting. The mats or rafts he observed were low-lying tangles of grass, branches, and water hyacinths (12–24 inches in height) which passed out of sight about 4 miles offshore and their fate beyond this point remained unknown to King. King concluded that "These rafts could not transport large land vertebrates whose weight would tend to break the rafts apart," and from his description it would appear that these rafts would be unsuitable for primate transport.

3. Tectonically active island arcs. According to newer geological data, such chains of islands are constantly forming, sinking, and coalescing near plate margins. Rare contacts between islands might allow a few forms, only, to cross on dry land or even through forests. Linear distance between islands would fluctuate. Such an island arc has been in existence in middle America since earliest Tertiary (see recent review by Weyl, 1974). Although there are two geologically active areas in the South Atlantic, the Walvis Ridge and the Rio Grand Rise, there is no evidence that they ever comprised an island arc that could have served as an alternate route (see Hekinian, 1972; Pastouret and Goslin, 1974; Barker et al., 1974; Perchnielson et al., 1975).

4. Pleistocene sea-level lowering. Whether primates ever have rafted across seaways can be tested by looking at the distribution of Pleistocene/Recent primate species and all other suspected cases of rafting. These include the introduction of primates into Africa and South America as already mentioned, into Madagascar (see Walker, 1972),

into the islands surrounding the two southern continents, and into those on the Sunda Shelf in Southeast Asia.

One important point noted by Wallace (1876) and numerous observers since should be kept in mind. A great many primate species, particularly platyrrhine monkeys, have species ranges limited by rivers. During flooding of lowland rivers, trees are often swept from one bank to the other, but it seems that when undermined trees begin to wobble the primates leave them. Otherwise, rivers could not so frequently be found to be barriers to distribution of primates.

The 100-m isobath is generally accepted as a tentative position of sea level at the maximum of the last glaciation (Flint, personal communication). Of course, this figure is an estimate. With local fluctuations, owing to emergence and submergence of coastline, some areas now just below the 100-m isobath could at some time formerly have been above it. When this Pleistocene sea-level lowering is taken into account, it becomes clear that of all the present-day island living primates none needs to have rafted a significant distance to be where it is now. That is, none farther than about 50 km or less. Raven and Axelrod (1974) cite the Jamaican monkey *Xenothrix* as a possible case of rafting about 600 km from South America. Nevertheless, neither the discoverer nor the two describers of this primate could eliminate its having been a human introduction. Consequently, *Xenothrix* has to be shelved as an unproven case of rafting. Primates on Madagascar have also been given as an example of relatively long-distance rafting, the present width of the Mozambique Channel being about 400 km. Nevertheless, Walker (1972) has shown that the geology of the area would allow for the channel having been much narrower, perhaps nonexistent, at some time in the early Tertiary. The earlier in time that lemurs reached the area the more likely this would have been. Also, Simpson (1940) probably overstressed the point that the character of the Malagasy fauna indicates sweepstakes introduction into that island. Walker (personal communication) and I agree that the variety of mammals on the mainland of Africa in the early Tertiary was atypical in being limited to fewer major groups than in most other continents. Therefore, even across dry land only an assemblage lacking in diversification could have reached Madagascar in the early Tertiary.

2.10. Primate Colonization: Island Hopping

It seemed plausible to Simpson (1950, 1953) and indeed most later writers on the subject of island hopping that primates and rodents, being arboreal, might easily be transported by rafts. Even so, no one investigated this probability in detail. However, nature has provided several control experiments. These include evidence of the distribution of mammals across archipelagos in the Pleistocene and Recent, and the nonpresence of mammals on isolated oceanic islands such as the Hawaiian chain and Tristan da Cunha. These evidences show how far primates and rodents can and cannot be distributed across water. Two points emerge clearly from such studies: (1) Rodents distribute across water much better than primates. (2) Neither rodents nor primates have been rafted great distances to isolated oceanic islands. Recent work suggests a 70 m.y. age for the Hawaiian Islands but no mammals ever reached them. Their present distance from the nearest mainland of 3500 km is only about 500 km greater than the esimated mid-Tertiary length given above for the South Equatorial Current.

By "island hopping" Simpson clearly meant transport across only short distances. Furthermore, in the late Paleocene or earliest Eocene a stock of the South American notoungulate order, represented by the Clarkforkian Wasatchian mammal *Arctostylops steini*, reached Wyoming. It can only have come from South America. Woodring (1954), Haffer (1970), and most recently Weyl (1974) have all discussed this island arc which linked North and South America before the isthmus arose in the late Miocene or Pliocene. The distribution of *Arctostylops* proves the existence of an early Tertiary route out of South America through the isthmus region and establishes one that would not have to involve long-distance rafting. (That is because *Arctostylops* is a hoofed ungulate, incapable of clinging to floating trees.) As stated above, the distribution of several late Eocene and early Oligocene genera which occur both in Europe and in the Fayum, North Africa, shows that a similar route between Europe and Africa once existed. Presence of these migration routes (involving at most only short water crossings) explains the Eocene introduction of rodents and primates into both southern continents. In consequence of the foregoing, a fair conclusion can be drawn: no explanation involving transport across wide reaches of ocean is tenable in accounting for the distribution of any primate.

2.11. The Probability of Primate Colonization

As stated, the probability that primates can colonize by rafting across oceans seems much lower than would be the case for certain other mammals, for instance, the murid rodents. A case in point would be the distribution of these rodents across the East Indian Archipelago through New Guinea and into Australia (Simpson, 1961). In the same area and under the same circumstances, primates reached only the first island (Celebes; i.e., macaque and tarsier). During the various Pleistocene sea-level low stands, continental Asia extended out as Sundaland to include islands as far as Borneo; the channel between this peninsula and the Celebes need not have been more than 25 km wide. In fact, there may have been an overland route through the Philippines into the Celebes. Primates never reached Australia-New Guinea, although no necessary crossings would have exceeded 50 km.

The physiological and behavioral objections to primate rafting over long distances are great. The most crucial objection would be dehydration due to lack of water, accelerated by heat, sun, and lack of shelter. Exposed to salt spray, vegetation would wither and cease to provide a food or water source, thus upsetting the water/salt balance. Small monkeys do not normally utilize tree holes and thus would be exposed during the day to heat and wind stresses. Platyrrhines are known to be sensitive to heat stress and are easily susceptible to sun stroke. Lack of food and water, salt imbalance, and the aforementioned environmental stresses would render most small primates comatose or unconscious in 4–6 days. Primates would likely jump off a raft when it was forming, of, if not, the probable isolated rafting group would be only a few individuals. Mother-son and sister-brother incest avoidance occurs in marmosets and cebids and might further impede colonization. Also, because they are social animals, small group size and the characteristic slow birth rate might lead to abnormal behavior. Finally, the sudden adjustment to a new environment long-distance rafting necessarily requires (new predators, new foods) argues against a successful colonization. Rafting is the reverse of gradual overland migration, which allows for long-term accumulation of adjustments to environmental differences. As stated, colonists would probably be exposed to new food, predators, diseases, and parasites and there might be different seasonality as well. Many primates will not eat unfamiliar plants and, in others, reproduction is governed by seasonality.

The greater the time lapse after splitting of two continents such as Africa and South America, the greater the environmental differences between them. If a splitting time of 36 m.y. produced the New World monkeys, their rafting across the South Atlantic would have been at some time after that date. If the rafting transit time were calculated on the basis of the South Atlantic being two-thirds its present width, evidence exists as to the time posited for a crossing in that current. Scheltema (1971) has studied distribution of pelagic marine larvae in the South Equatorial Current with float cards. His data show that minimum crossing time at the present day in a current approximately 4500 km long is 60 days, maximum crossing time is 154 days. Geophysical evidence is that at around 36 m.y. ago the South Atlantic was at least 70–80% its present width. Even if it could be shown that the ocean (and current) was two-thirds or even a half its present dimensions at that time, minimum crossing times would be, respectively, 40 and 30 days, both impossible for small primate survival. I reject faster rates of raft movement due to wind since no evidence exists that such raft-sailing occurs. Moreover, these calculations omit another temporal factor in the rafting model entirely. This is the matter of days or perhaps even weeks that such a raft might spend in random drift between the mouth of the river of its formation and an ocean current. Because of these objections, I conclude that it is more probable that the 36 m.y. ID divergence date of platyrrhines and catarrhines is wrong than that these animals could have rafted the South Atlantic subsequent to it.

2.12. The Later Primate Divergences

Considerations of space do not allow for a full summary of all the voluminous literature on the later primate divergences. Just as is the case of the splitting times already discussed, eight more splits, itemized below, also indicate divergence times 5–10 m.y. earlier than those calculated with molecular clocks. Discounting one or another of these divergence times does little to weaken the whole body of evidence from fossils. The radiation of apes with thick-enameled teeth ended with *Gigantopithecus* and is therefore extinct and not subject to molecular studies. Four of the following itemized divergences are discussed by Romero-Herrera *et al.* (1973) as indicated.

1. The hominoid-cercopithecoid split. (R-H No. 11)
2. The marmoset-cebid dichotomy. (R-H No. 12)
3. The Miocene lorisid radiation.
4. The *Papio-Macaca-Theropithecus* diversification.
5. The hylobatid-pongid divergence. (R-H No. 14)
6. The pongine-panine separation and the Miocene apes.
7. The radiation of thick-enameled apes (*Gigantopithecus* and others).
8. The divergence of hominids from apes. (R-H No. 15)

2.12.1. The Hominoid-Cercopithecoid Split (R-H No. 11)

As Romero-Herrera *et al.* (1973) indicate, the hominoid-cercopithecoid split was most probably accomplished prior to the occurrence of the Fayum hominoids *Propliopithecus* and *Aegyptopithecus* and parapithecid cercopithecoids *Apidium* and *Parapithecus*. Since these Fayum primates have already diverged to the family level, the split is prior to *their* own date. As discussed, the age of the Fayum is uncertain. The latest possible date, but improbable, is 27 m.y.; the earliest possible date, equally improbable, is 39 m.y. With reservations, this could be expressed as 33 ± 6 m.y. Delson (1975) has stated doubts that parapithecids were ancestral cercopithecoids but nevertheless places them in the cercopithecoid radiation.* We both agree that Fayum primates document that hominoids and cercopithecoids had split by then. Neither the Fayum hominoids nor cercopithecoids resemble platyrrhine monkeys in any significant way (except for the expected shared primitive features). This comment is made notwithstanding the statement of Hoffstetter (1974) that there is an "astonishing resemblance" between the petrosal of *Apidium* figured by Gingerich (1973) and certain platyrrhines (not specified by him). The resemblance of this petrosal to those of modern cercopithecoids is just as great (even if not astonishing). The radiations which the Fayum primates constitute do not have to do with the ceboid radiation, which must have branched off much earlier. On the assumption that the ancestral catarrhine introduction into Africa was in the form of a single species, I would suspect the differentiation of the Fayum parapithecids and pongids to the level of six genera of two different families to

---

*His views are not founded on extensive study of the relevant material, and some of his conclusions are implausible. For instance, without adequate discussion he places *Oligopithecus* at the base of the cercopithecoids, not with hominoids, where it more reasonably belongs. Simons (1974c and ms.) sets forth from direct study why parapithecids are both monkeys and cercopithecoids.

have taken at least 5–10 m.y. Thus 33 ± 6 plus 5–10 m.y. gives 27 + 5 to 39 + 10. Such age determinations are whimsically called guesstimates, but, taken for what they are worth, this would put the hominoid-cercopithecoid split somewhere between 32 and 48 m.y. Delson's chart (1975:195) puts it at 34 m.y.

2.12.2. The Marmoset-Cebid Dichotomy (R-H No. 12)

There is not much evidence about the radiation of the South American monkeys. In general, the various Miocene species exhibit a radiation well under way and are surprisingly modern (See Hershkovitz, 1970, 1974). Several genera from the late Miocene of Colombia are known *(Stirtonia, Cebupithecia, Neosaimiri)* as well as older forms from elsewhere in South America. Of these, only *Neosaimiri* bears significant resemblance to a modern genus. I would emphasize the possibility mentioned by Hershkovitz (1970) that *Neosaimiri* is enough like modern *Saimiri* to imply that ancestors of squirrel monkeys were already a distinct group at the time when it lived, and reject on other grounds the alternate idea that all ancestral cebids looked like *Saimiri*. The fossil is from the La Venta fauna, Honda Formation, near Huila, Colombia. This fauna is thought to belong to the late Miocene Friasian provincial age (see Patterson and Pascual, 1972). Radiometric dates are badly needed in South America, but Patterson and Pascual cite one date from the Santa Cruz (Santacruzian is the faunal age that comes immediately before Friasian) as 21.6 m.y. In order to be correlated as pre-Pliocene (old style), the end of the Friasian would be about 12–15 m.y. Therefore, *Neosaimiri* may indicate that at about 18 m.y. ago ancestors of some, at least, of the present-day cebid genera had split from ancestors of the others. The marmoset-cebid dichotomy must have been much earlier.

2.12.3. The Miocene Lorisid Radiation

Walker (1974) provides the latest review of the Miocene lorisids of Kenya and Uganda. He recognizes six species in the total sample: two of a lorisine genus *Mioeuoticus* and four species of galagine type. In his view, the four latter species should probably all be referred to genus *Progalago* of which *Komba* would then be either a synonym or subgenus, somewhat as modern *Euoticus* should be sunk in *Galago*. One of the species of *Mioeuoticus*, *M. bishopi*, is based on an excellent palate originally described by L. S. B. Leakey, and the other species, as yet undescribed, is exemplified by a skull from Rusinga (originally referred to *Progalago dorae* by Le Gros Clark, 1956). The bulk of the Miocene material is of the four galagine species. It includes many mandibles, a partial skull, and much postcranial material. Walker (1974) draws two major conclusions relevant to splitting times in Lorisidae, as follows: "The new specimens of fossil lorisids from Kenya help to show that both the Lorisinae and Galaginae were established by lower Miocene times. . . . There is a strong possibility that the ancestors of at least some modern *Galago* species are present in the Miocene assemblage." The East African Miocene lorisids come from six major sites. One of these, Moroto I, has not been radiometrically dated. Of the remaining five sites, four have been dated and the fifth, Mfwanganu, correlates with Rusinga, which has been dated. Bishop *et al.* (1969) discuss accuracy of dating at all of the four dated sites, which are (1) Rusinga, Kenya, 19.0–19.3 m.y., (2) Songhor, Kenya, 19.7–19.9 m.y., (3) Koru, Kenya, 19.5–19.6 m.y., and (4) Napak, Uganda, 17.8 ± 0.4–19.0 ± 0.2. If the radiation of species within the genus *Galago* had already begun at around 20 m.y., then the divergence of the major groupings of lorisids must have been yet earlier. That this 20-m.y.-old lorisid radiation can be explained away as an early parallel radiation, not relevant to a later 10-m.y.-old basal branching of lorisids, is unlikely for geographic reasons. If it is a coincidentally parallel radiation, why is it documented in exactly the same region where the later undoubted lorisids occur? Interestingly, the same matching of a radiation with the area where the undoubted members of a group later occur applies in all the other cases of paleontologically determined branch points among primates.

2.12.4. The Papio-Macaca-Theropithecus Diversification

The latest discussion of the history of the *Papio-Macaca-Theropithecus* group is by Delson (1975). Specimens he tentatively refers to as ?*Macaca* sp. come from a fauna of Turolian provincial age at Marceau, Algeria (probable age 7.5 m.y.), and from Wadi Natrun, Northern Egypt, where a population of some 20 fossil monkeys is identified by him as cf. *Macaca libyca* and correlated at about 6 m.y. Elsewhere he indicates that undoubtedly referred European and North African Plio-Pleistocene *Macaca* spp. go back in unbroken sequence to at least 5 m.y. Similarly, he traces the *Papio (Chaeropithecus)* spp. line (savannah/hamadryas baboons) back to 4 m.y. in the Kanapoi Beds, Turkana, Kenya (see Patterson *et al.,* 1970). The earliest evidence for *Theropithecus* is likewise 4 m.y. at Lothagam, Kenya. Delson (1975) remarks: "As Simons has noted . . . the older fossils document a longer separation of *Theropithecus* from Papionini than would be expected from Sarich's (1970) finding of no protein differences between *Theropithecus* and some *Papio* species; this indicates that, if Sarich is in error on this point (by a factor of greater than four), some of his other estimated divergence times might be equally inaccurate."

The real problem in documenting the time of the papionin radiation is the one mentioned already in the introduction to this chapter. All these genera go back as far as an adequate fossil record goes in Africa. Between 4 or 5 m.y. and 13 m.y. in Africa, none of the known sites (which are very few in number) contains any primates related to living African monkeys. None of the earliest occurrence dates for papionins indicates anything about the antiquity of the genera they date, but all would likely have originated before the first dated fossils.

2.12.5. The Hylobatid-Pongid Divergence (R-H No. 14)

Schlosser (1911) originally suggested *Propliopithecus* from the Fayum as a relative of the gibbons. After many other rankings in the intervening 65 years, a relationship to *Pliopithecus* and to the gibbons still remains possible (Simons, 1972; Simons and Fleagle, 1973). *Aeolopithecus* (Simons, 1965) has also been shown to have had gibbonlike features, its mandible implying a broad short palate with large incisors; it is quite unlike its contemporary *Aegyptopithecus* in jaw and tooth proportions (Simons, 1972). *Aegyptopithecus* is clearly like the later dryopithecines of the Miocene of East Africa (subgenus *Proconsul*) and could well be in or near the dryopithecine ancestry (Simons, 1965, 1967, 1969, 1972, 1974a,b). Although skull teeth and jaws are about gibbon-size, *Aegyptopithecus* does not resemble cranially or dentally any of the Miocene gibbonlike apes, including "*Dendropithecus,*" *Limnopithecus, Pliopithecus,* and what is probably another hylobatid genus from Napak in Uganda (Fleagle, 1975). Instead, dentally it looks like *Dryopithecus* (subgenus *Proconsul*). Thus, on grounds of both size and cranial and dental anatomy, *Aegyptopithecus* is in or near the line of the great apes, and either *Propliopithecus* or *Aeolopithecus* could represent the gibbon line. If so, the hylobatid-pongid divergence was more than 27 m.y. ago.

More complete data on when great apes and lesser apes split comes from four separate sources of evidence from the Miocene (Table 1). Each of four different species shows definite anatomical resemblance to gibbons, not only in teeth but also in details of facial or cranial anatomy, and in two of the species, which preserve postcranial remains, hylobatid resemblances are to be seen in these bones, also.

None of these species has to be ancestral to modern *Hylobates* and *Symphalangus* to indicate that hylobatids had split off from pongids by 18–20 m.y. ago because each can on its own merits be ranked in Hylobatidae and each separately shows the same probability: that hylobatids had then diverged. The relegation of all these similarities to the rank of parallelisms seems unlikely. The new form from Uganda has now been described in detail by Fleagle (1975). In

Table 1. Gibbonlike Species of the Miocene

Name	Locality(s)	Potassium/argon age of locality
"*Dendropithecus*" *macinnesi*[a]	Rusinga, Kenya	19.0–19.3 m.y.
	Songhor, Kenya	19.7–19.9 m.y.
Limnopithecus legetet[a]	Koru, Kenya	19.5–19.6 m.y.
	Mt. Elgon, Uganda	21.9 (overlies fossils)
New (?) genus and species (Uganda)[a]	Napak, Uganda	17.8 ± 0.4, 19.0 ± 0.2
Pliopithecus vindobonensis[b]	Neudorf an der March, Czechoslovakia	> 15 ± < 19 ±

[a]From Bishop *et al.* (1969).
[b]From Cicha *et al.* (1972).

this analysis, he is suitably cautious about phyletic ranking but reaffirms Simons and Fleagle (1973) that there "is nothing that would bar it from the ancestry of extant gibbons." Nor can it be said just any small ape of this sort would look dentally and cranially like a gibbon; as stated already, *Aegyptopithecus* is known from a complete skull which falls in the gibbon size range, yet it does not look like a gibbon.

2.12.6. The Pongine-Panine Separation and the Miocene Apes

The Miocene great apes from Kenya and Uganda were first monographed by Le Gros Clark and Leakey (1951) and reviewed extensively by Pilbeam (1969) and Simons (1972). Original studies on postcranial elements were published by Napier and Davis (1959) and by Walker and Rose (1968). Recently, Andrews (1970, 1974) has published on two new species. All these apes are considered to belong to genus *Dryopithecus* (this, in turn, contains two subgenera, *Proconsul* Hopwood and *Rangwapithecus* Andrews). To date, five valid species have been described: *Dryopithecus africanus, D. nyanzae, D. africanus, D. gordoni,* and *D. vancouveringi.* The type specimen (only) of a sixth possible species is that of "*Sivapithecus africanus*" (Le Gros Clark and Leakey, 1951). This upper dentition may represent a distinct species or may be referable to *D. nyanzae.*

A considerable amount has been written about these apes, which range in size of comparable parts from about the size of a female pigmy chimpanzee to the size of a female gorilla. This can be summarized here as follows: No one who has done original work on these fossils believes that they represent a radiation of animals which resemble apes only in the teeth. Almost every part of a skull, face, or postcranial bone shows unmistakable resemblance to the modern great apes. Equally, all these specimens, which (together with parts of dentitions) total hundreds, show primitive features unlike those of the modern apes. This is a typical situation to be observed with all fossil groups of any antiquity. Each individual species exhibits a blend of characteristics resembling both earlier and later forms. In this case, among primitive or generalized features are characteristics which resemble those of monkeys or sometimes even prosimians. That does not make this radiation one of some kind of primitive monkeys, "dental apes," or "pseudo-apes." Were such the case, then all the ape resemblances would become inexplicable since such features do not typify any monkeys. The advanced or "shared derived characteristics," the apelike features, are what are phyletically important for determining relatedness.

As has already been discussed at length Simons (1969, 1972), I believe Pilbeam (1969) was correct in demonstrating that the *Dryopithecus major* palate from Moroto, Uganda, dated as older than 14 m.y. but probably 18–19 m.y., bears extensive, special resemblance to modern *Gorilla gorilla.* If the species it belonged to lay in or close to the ancestry of the gorilla, then it would be plausible to suppose that the chimp-gorilla line had branched by the time of its existence. It follows that the line to man had also branched by then, unless man is closer to one of the African apes than to the other.

How much earlier it was that the line of *Pongo* branched off is uncertain, but one point about the time of migration of its forebears from Africa is clear. As is adequately discussed by Walker elsewhere in this volume, the collective evidence suggests that the optimal time for forested interconnections between Africa and Eurasia is around 17–19 m.y. ago (see also Berggren and Van Couvering, 1974). This is the time of the Neogene paleotemperature curve maxima and is the most probable time that forest apes such as *Dryopithecus fontani* and *D. laietanus* could have reached Europe. The latter species, which occurs in the Vallesian provincial age and earlier deposits of Spain (age ± 13.0 to ± 9.5 m.y.), is especially like *Pan* and might represent a remnant of the ancestral chimpanzee stock that had migrated through coastal forests and across Gibraltar from West Africa. Separately from other evidence this could indicate that by the latter dates chimpanzees had split off from other hominoids.

At Pasalar, Turkey, two Miocene hominoids occur, one closely resembling both *D. major* and *D. indicus.* The age of Paşalar is thought to be between 11.5 and 19.8 m.y., and closer to the latter. Either the ancestors of living great apes and man had begun their divergences in Africa before about 18 m.y. with the orangutan representing an outmigrant through the early/mid-Miocene forest corridors, or (2) the ancestors of the four living higher hominoid genera branched in Eurasia and ancestors of *Pan, Gorilla,* and *Homo,* respectively, reentered Africa with the extensive East African/Siwalik faunal exchange that was essentially complete (including *Ramapithecus*) at the time of deposition of the Ft. Ternan site in Kenya, whose age is greater than 12.5 m.y. but less than 1.40 m.y. After about this period, no forest forms are known to have entered Africa and consequently it is improbable that the three divergences under consideration could have taken place in Eurasia after that date. Washburn (1973) suggests a model similar to the second possibility above but involving a late split in Eurasia between African apes and hominoids. It should be noted that such a late divergence would require that the ancestors of the living African pongids were not only terrestrial but also open-country forms. This seems unlikely.

2.12.7. The Radiation of Thick-Enameled Apes (Gigantopithecus and Others)

Many recent discoveries in Europe and Asia document a radiation of medium to large apes with thick tooth enamel. These discoveries are from Greece, Turkey, Hungary, Pakistan, and India. My statements are provisional because many of these forms are yet to be described, but their discoverers have kindly acquainted me with the material (de Bonis for the Macedonian species, Tekkaya for the Turkish forms, and Kretzoi for those from Hungary). Most of these large and medium-sized, thick-enameled hominoids are best assigned to *Dryopithecus indicus* (see Simons and Pilbeam, 1971), to *Dryopithecus macedoniensis* (see De Bonis, 1974), and to *Ramapithecus*. The two species of *Dryopithecus* (as presently understood) are both thought to be close to the ancestry of *Gigantopithecus*. This line may well be closer than the African apes are to that of *Ramapithecus* and subsequent hominids—in fact, part of the same late Miocene radiation. This would go far to explain the long-mooted resemblances of *Gigantopithecus* to hominids. The many newly discovered occurrences of apes and hominids belonging to this radiation date its diversification to about 14 ± 4 m.y. This is a date, at least in the younger part of its range, compatible with oldest dates of *Ramapithecus*.

2.12.8. The Divergence of Hominids from Apes (R-H No. 15)

Two recent papers (Simons 1975; Simons and Pilbeam, 1975) stress that the probability that *Ramapithecus* is a hominid has been strengthened since the late 1960s rather than weakened, and the data presented by Andrews and Walker (1974) or by Greenfield (1974) have not challenged the case. As stated already, in 1974 important new material turned up in Turkey, Hungary, and Greece. These finds extend and confirm the anatomical features of resemblance to the Plio-Pleistocene hominids which were the basis of the original reference of *Ramapithecus* to Hominidae.

As Walker in his contribution to this book independently points out, advances in understanding both *Ramapithecus* and *Australopithecus* since my work with Pilbeam in the 1960s show that neither typically has a rounded or hemicircular dental arcade arrangement as was then thought. Both are alike in their arcade arrangement, narrow at front with cheek-tooth rows diverging posteriorly. *Ramapithecus* comes from at least seven main sites with dated sediments or dated faunal correlations (Table 2).

3. Conclusion

From the foregoing discussions, it is clear that none of the evidence that is to be taken from the fossil record for individual branch points in the order

Table 2. Main Sites for *Ramopithecus*

Site	Date[a]
Haritalyangar, India	~ 9 m.y.
Chinji, Pakistan	~12 m.y.
Pyrgos, Greece	younger than 11.5 m.y.
Rudabanya, Hungary	about 11 m.y.
Pasalar, Turkey	16 to 14 m.y.
Ft. Ternan, Kenya	older than 12.5 m.y.
Maboko Island, Kenya	older than 12.5 m.y.

[a]These dates summate as ± 14 to ± 9—inclusive time range of earliest recognized hominids.

Primates yields dates that are as precise as we would like. Nevertheless, the fossil evidence, when considered together with the paleogeographic, summates to suggest very strongly that the branch points suggested by Sarich and Cronin in their contribution to this volume are too young. Either the "clock" is incorrectly calibrated or it doesn't keep proper time.

4. References

Andrews, P., 1970, Two new fossil primates from the Lower Miocene of Kenya, *Nature (London)* **228:**537–540.

Andrews, P., 1974, New species of *Dryopithecus* from Kenya, *Nature (London)* **249:**188–190.

Andrews, P., and Walker, A. C., 1974, A new look at *Ramapithecus* specimens from Fort Ternan, Kenya, *Nature (London)* **244:**313–314.

Barker, P. F., *et al.,* 1974, Southwestern Atlantic Leg 36, *Geotimes,* November, pp. 16–18.

Berggren, W. A., and Van Couvering, J. A., 1974, The Late Neogene, *Palaeogeogr. Palaeoclimatol. Palaeoecol.* **16:**1–216.

Bishop, W. W., Miller, J. A., and Fitch, F. J., 1969, New potassium-argon age determinations relevant to the Miocene fossil mammal sequence in East Africa, *Am. J. Sci.* **267:**669–699.

Bown, T. M., 1976, Affinities of *Teilhardina* (Primates, Omomyidae) with description of a new species from North America, *Folia Primatol.* **25:**62–72.

Cicha, I., Fahlbusch, V., and Fejfar, O., 1972, Die biostratigraphische Korrelation einiger jungtertiärer Wirbeltierfaunen Mitteleuropas, *Neues Jahrb. Geol. Palaeontol. Abh.* **140:**129–145.

Clemens, W. A., 1974, *Purgatorius,* an early paromomyid primate (Mammalia), *Science* **184:**903–905.

Coryndon, S. C., and Savage, R. J. G., 1973, The origin and affinities of African mammal faunas, in: *Organisms and Continents through Time* (N. F. Hughes, ed.), pp. 121–135, Vol. 12 of *Special Papers in Paleontology,* The Paleontological Association, London.

Cox, C. B., 1973, Systematics and plate tectonics in the spread of marsupials, in: *Organisms and Continents through Time* (N. F. Hughes, ed.), pp. 113–120, Vol. 12 of *Special Papers in Paleontology,* The Paleontological Association, London.

De Bonis, L., Bouvrain, G., and Melentis, J., 1974, Première découverte d'un primate hominoide dans le Miocène supérior de Macédoine (Grèce), *Acad. Sci. Ser. D* **278:**3063–3066.

Delson, E., 1975, Evolutionary history of Cercopithecoidae, in: *Approaches to Primate Paleobiology: Contributions to Primatology,* Vol. 5 (F. S. Szalay, ed.), pp. 167–217, Karger, Basel.

Eldridge, N., and Gould, S. J., 1972, Speciation and punctuated equilibria: An alternative to phyletic gradualism, in: *Models in Paleobiology* (T. Schopf, ed.), pp. 82–115, Freeman and Cooper, San Francisco.

Fleagle, J. G., 1975, A small gibbon-like hominoid from the Miocene of Uganda, *Folia Primatol.* **24:**1–15.

Gingerich, P. D., 1973, Anatomy of the temporal bone in the Oligocene anthropoid *Apidium* and the origin of Anthropoidea, *Folia Primatol.* **19:**329–337.

Gingerich, P. D., 1974, Cranial anatomy and evolution of early Tertiary Plesiadapidae (Mammalia, Primates), Ph.D. thesis, Yale University, pp. 1–370.

Gingerich, P. D., 1975, Systematic position of *Plesiadapis, Nature (London)* **253:**111–113.

Goodman, M., 1962, Immunochemistry of the primates and primate evolution, *Ann. N.Y. Acad. Sci.* **102:**219–234.

Greenfield, L. M., 1974, Taxonomic reassessment of two *Ramapithecus* specimens, *Folia Primatol.* **22:**97–115.

Gregory, W. K., 1949, The bearing of the Australopithecinae upon the problem of man's place in nature, *Am. J. Phys. Anthropol.* **7:**485–512.

Guppy, H. B., 1917, *Plants, Seeds, and Currents in the West Indies and Azores,* Williams and Norgate, London.

Haffer, J., 1970, Geologic-climatic history and zoogeographic significance of the Uraba Region in northwestern Columbia, *Caldasia* **50:**603–636.

Hallam, A., 1973, Distributional patterns in contemporary terrestrial and marine animals, in: *Organisms and Continents through Time* (N. F. Hughes, ed.), pp. 93–106, Vol. 12 of *Special Papers in Paleontology,* The Paleontological Association, London.

Hekinian, R., 1972, Volcanics from the Walvis Ridge, *Nature (London)* **239:**91–93.

Hershkovitz, P., 1970, Notes on Tertiary platyrrhine monkeys and description of a new genus from the late Miocene of Colombia, *Folia Primatol.* **12:**1–37.

Hershkovitz, P., 1974, A new genus of late Oligocene monkey (Cebidae, Platyrrhini) with notes on postorbital closure and platyrrhine evolution, *Folia primatol.* **21:**1–35.

Hill, W. C. O., 1953, *Primates: Comparative Anatomy and Taxonomy, Strepsirhini,* Edinburgh Univ. Press, Edinburgh.

Hoffstetter, R., 1974, Phylogeny and geographical deployment of the Primates, *J. Hum. Evol.* **3:**327–350.

Howells, W. W., 1947, *Mankind So Far,* Doubleday, Garden City, N.Y.

Kay, R. F., and Cartmill, M., 1974, The skull of *Palaechthon nacimienti, Nature (London)* **252:**37–38.

Kielan-Jaworowska, Z., 1975a, Preliminary description of two new eutherian genera from the late Cretaceous of Mongolia, *Palaeontol. Pol.* **33:**5–16.

Kielan-Jaworowska, Z., 1975b, Evolution of the therian mammals in the late Cretaceous of Asia. Part 1. Deltatheridiidae, *Palaeontol. Pol.* **33:**103–132.

King, W., 1962, The occurrence of rafts for dispersal of land animals in the West Indies, *J. Fla. Acad. Sci.* **25(1):**45–52.

Ladd, W. J., Dickson, G. O., and Pitmann, W. C., III, 1973, The age of the South Atlantic, in: *The Ocean Basins and Margins,* Vol. 1 (A. E. M. Nairn and F. G. Stehli, eds.), pp. 555–573, Plenum Press, New York.

Lavocat, R., 1969, La systematique des rongeurs histricomorphes et la derive des continents, *C. R. Acad. Sci. Ser. D* **269:**1496–1497.

Le Gros Clark, W. E., 1956, A Miocene lemuroid skull from East Africa, *Brit. Mus. Nat. Hist. Fossil Mamm. Afr.* **9:**1–6.

Le Gros Clark, W. E., 1959 and later eds., *The Antecedents of Man: An Introduction to the Evolution of the Primates,* pp. 1–374, Edinburgh University Press, Edinburgh.

Le Gros Clark, W. E., and Leakey, L. S. B., 1951, The Miocene Hominoidea of East Africa, *Brit. Mus. Nat. Hist. Fossil Mamm. Afr.* **1:**1–117.

Luckett, W. P., 1974, The phylogenetic relationship of the prosimian primates: Evidence from the morphogenesis of the placenta and foetal membranes, in: *Prosimian Biology* (R. D. M. Martin, G. A. Doyle, and A. C. Walker, eds.), pp. 475–488, Duckworth, Glouster.

Martin, R. D. M., 1972, Adaptive radiation and behavior of the Malagasy lemurs, *Philos. Trans. R. Soc. London, Ser. B* **264:**295–352.

McKenna, M. C., 1966, Paleontology and the origin of the primates, *Folia Primatol.* **4:**1–25.

McKenna, M. C., 1972, Eocene final separation of the Eurasian and Greenland-North American landmasses, *24th Int. Geol. Congr. Sect. 7,* pp. 275–281.

McKenna, M. C., 1975, Toward a phylogenetic classification of the mammalia, in: *Phylogeny of the Primates* (W. P. Luckett and F. S. Szalay, eds.), pp. 21–46, Plenum Press, New York.

Napier, J. R., and Davis, P. R., 1959, The fore-limb skeleton and associated remains of *Proconsul africanus, Brit. Mus. Nat. Hist. Fossil Mamm. Afr.* **16:**1–69.

Nuttall, G. H. F., 1904, *Blood Immunity and Blood Relationships,* pp. 1–444, Cambridge University Press, Cambridge.

Orlosky, F. J., and Swindler, D. R., 1975, Origins of New World monkeys, *J. Hum. Evol.* **4:**77–83.

Pastouret, L., and Goslin, J., 1974, Middle Cretaceous sediments from the eastern part of Walvis Ridge, *Nature (London)* **248:**495–496.

Patterson, B., and Pascual, R., 1972, The fossil mammal fauna of South America, in: *Evolution, Mammals and Southern Continents* (A. Keast, F. C. Erk, and B. Glass, eds.), pp. 247–310, State of New York University Press, Albany, N.Y.

Patterson, B., Behrensmeyer, A. K., and Dill, W. D., 1970, Geology and fauna of a new Pliocene locality in northwestern Kenya, *Nature (London)* **226:**918–921.

Perchnielson, K., *et al.,* 1975, Leg 39 examines facies changes in South Atlantic, *Geotimes,* March, pp. 26–28.

Pilbeam, D. R., 1969, Tertiary Pongidae of East Africa: Evolutionary relationships and taxonomy, *Peabody Mus. Bull.* **31:**1–185.

Pocock, R. I., 1918, On the external characters of lemurs and *Tarsius, Proc. Zool. Soc. London,* pp. 19–53.

Quinet, G. E., 1966, *Teilhardina belgica,* ancêtre des Anthropoidea de l'ancien monde, *Bull. Inst. R. Sci. Nat. Belg.* **42:**1–14.

Radinsky, L., 1976, Later mammalian radiations, in: *Evolution of Brain and Behavior in Vertebrates* (R. B. Masterson *et al.,* eds.), Lawrence Erlbaum Assoc., Hillsdale, N.J. (in press).

Raven, P. H., and Axelrod, D. I., 1974, Angiosperm biogeography and past continental movements, *Ann. M. Bot. Gard.* **61:**539–673.

Ridley, H. N., 1930, *The Dispersal of Plants Throughout the World,* L. Reeve, Ashford, Kent.

Romero-Herrera, A. E., Lehmann, H., Joysey, K. A., and Friday, A. E., 1973, Molecular evolution of myoglobin and the fossil record: A phylogenetic synthesis, *Nature (London)* **246:**389–395.

Russell, D. E., 1975, Paleoecology of the Paleocene-Eocene transition in Europe, in: *Approaches to Primate Paleobiology* (F. S. Szalay, ed.), pp. 28–61, Vol. 12 of *Contributions to Primatology,* Karger, Basel.

Sarich, V. M., 1970, Primate systematics with special reference to Old World monkeys: A protein perspective, in: *Old World Monkeys* (J. R. and P. H. Napier, eds.), pp. 175–226, Academic Press, New York.

Savage, D. E., 1975, Genozoic—The primate episode, in: *Approaches to Primate Paleobiology* (F. S. Szalay, ed.), pp. 2–27, Vol. 5 of *Contributions to Primatology,* Karger, Basel.

Scheltema, R. S., 1971, Larval dispersal as a means of genetic exchange between geographically separated populations of shallow-water benthic marine gastropods, *Biol. Bull.* **140:**284–322.

Schlosser, 1911, Beiträge zur Kenntniss der Oligozänen Landsäugetiere aus dem Fayûm (Ägypten), *Beitrage Pal. Geol. Österr.-Ung.* **24:**51–167.

Simons, E. L., 1963, A critical reappraisal of Tertiary Primates, in: *Evolutionary and Genetic Biology of Primates,* Vol. 1 (J. Buettner-Janusch, ed.), pp. 65–129, Academic Press, New York.

Simons, E. L., 1965, New fossil apes from Egypt and the initial differentiation of Hominoidea, *Nature (London)* **205:**135–139.

Simons, E. L., 1967, The earliest apes, *Sci. Am.* **217(6):**28–35.

Simons, E. L., 1968, Early Cenozoic mammalian faunas Fayum province, Egypt, Part I, *Peabody Mus. Bull.* **28:**1–21.

Simons, E. L., 1969, The origin and radiation of the Primates, *Ann. N.Y. Acad. Sci.* **167:**319–331.

Simons, E. L., 1972, *Primate Evolution: An Introduction to Man's Place in Nature,* pp. 1–322, Macmillan, New York.

Simons, E. L., 1974a, Notes on Early Tertiary prosimians, in: *Prosimian Biology* (R. D. M. Martin, G. A. Doyle, and A. C. Walker, eds.), pp. 415–433, Duckworth, Glouster.

Simons, E. L., 1974b, The relationship of *Aegyptopithecus* to other primates, *Ann. Geol. Surv. Egypt* **4:**149–156.

Simons, E. L., 1974c, *Parapithecus grangeri* (Parapithecidae, Old World Higher Primates): New species from the Oligocene of Egypt and the initial differentiation of Cercopithecoidea, *Postilla,* Yale University, Peabody Museum.

Simons, E. L., 1975, Diversity among the early hominids: A vertebrate paleontologist's viewpoint, in: *African Hominidae of the Pleistocene* (C. J. Jolly, ed.), Duckworth, London.

Simons, E. L., manuscript, Review of the Primates of the Oligocene deposits of the Fayum Province, Egypt.

Simons, E. L., and Fleagle, J. G., 1973, The history of extinct gibbon-like primates, in: *Gibbon and Siamang,* Vol. 2 (D. M. Rumbaugh, ed.), pp. 121–148, Karger, Basel.

Simons, E. L., and Gingerich, P. D., 1976, A new species of *Apterodon* (Mammalia, Credonta) from the Upper Eocene Qasr El-Sagha Formation of Egypt, *Postilla,* Yale University, Peabody Museum.

Simons, E. L., and Pilbeam, D. R., 1971, A gorilla-sized ape from the Miocene of India, *Science* **173**:23–27.

Simons, E. L., and Pilbeam, D. R., 1975, Anthropoidea. Part III. *Ramapithecus*, in: *Evolution of Mammals in Africa* (V. J. Maglio, ed.), Princeton University Press, Princeton, N.J.

Simons, E. L., Pilbeam, D. R., and Wood, A. E., 1968, Early Cenozoic mammalian faunas, Fayum Province, Egypt. *Bull. Peabody Mus. Nat. Hist.* **28**:1–105.

Simpson, G. G., 1940, Mammals and land bridges, *J. Wash. Acad. Sci.* **30**:137–163.

Simpson, G. G., 1945, The principles of classification and a classification of mammals, *Bull. Amer. Mus. Nat. Hist.* **85**:1–350.

Simpson, G. G., 1947, Holarctic mammalian faunas and continental relationships during the Cenozoic, *Bull. Geol. Soc. Am.* **58**:613–688.

Simpson, G. G., 1950, History of the fauna of Latin America, *Am. Sci.* **38**:361–389.

Simpson, G. G., 1953, Evolution and geography, 64 pp., Condon Lecture, Oregon, State System of Higher Education.

Simpson, G. G., 1961, Historical zoogeography of Australian mammals, *Evolution* **15**:431–446.

Spruce, R., 1908, *Notes of a Botanist on the Amazon and Andes,* 2 vols., 542 pp., Macmillan, London.

Starck, D., 1974, Stellung der Hominiden im Rahmen der Säugetiere, in: *Die Evolution der Organismen,* Vol. III (G. Heberer, eds.), pp. 1–131, Gustav Fischer, Stuttgart.

Sudré, M. J., 1975, Un prosimien du Paléogène ancien du Sahara nordoccidental: *Azibius trerke* n. g. n. sp., *C. R. Acad. Sci. Ser. D* **280**:1539–1542.

Van Valen, L., and Sloan, R. E., 1965, The earliest primates, *Science* **150**:743–745.

Von Koenigswald, G. H. R., 1969, Miocene Cercopithecoidea and Oreopithecoidea from the Miocene of East Africa, *Fossil Vertebr. Afr.* **1**:39–52.

Walker, A. C., 1972, The dissemination and segregation of early Primates in relation to continental configuration, in: *Calibration of Hominoid Evolution* (W. W. Bishop and J. A. Miller, eds.), pp. 195–218, Scottish Academic Press, Edinburgh.

Walker, A. C., 1974, A review of the Miocene Lorisidae of East Africa, in: *Prosimian Biology* (R. D. M. Martin, G. A. Doyle, and A. C. Walker, eds.), pp. 435–447, Duckworth, Glouster.

Walker, A. C., and Rose, M. D., 1968, Fossil hominoid vertebra from the Miocene of Uganda, *Nature (London)* **217**:980–981.

Wallace, A. R., 1876, *The Geographical Distribution of Animals,* 2 vols., 607 pp., Macmillan, London.

Washburn, S. L., 1973, The evolution game, *J. Hum. Evol.* **2**:557–561.

Weyl, R., 1974, Die paläogeographische Entwicklung Mittelamerikas, *Zentralb. Geol. Palaeontal. Teil 1* **5/6**:432–466.

Wood, A. E., 1974, The evolution of the Old World and New World hystricomorphs, *Symp. Zool. Soc. London* **34**:21–60.

Wood, A. E., and Patterson, B., 1970, Relationships among hystricognathous and hystricomorphous rodents, *Mammalia* **34**:628–639.

Woodring, W. P., 1954, Caribbean land and sea through the ages, *Bull. Geol. Soc. Am.* **65**:719–732.

Splitting Times among Hominoids Deduced from the Fossil Record

4

ALAN WALKER

1. Introduction

This chapter is concerned with some of the difficulties encountered when trying to establish divergence times from the fossil record. I have not presented much of the evidence for the divergence dates postulated here, but I have given reference to new discoveries made since the publication of Simons' 1972 review. The problems to be faced include relevant evolutionary models, the nature of the sedimentary and fossil records, ways of determining temporal sequences and actual dates of fossils, and the implications of paleogeographic reconstructions. The deductions that follow from these considerations are, at the very best, the nearest approximations to paleontological history that the record allows, or, at worst, just educated guesses. New finds often offer more than new interpretations of the material, and it is a peculiarity and one of the fascinations of paleontological inquiry that we are continually reassessing our notions with each successive new discovery.

Before dealing with the major difficulties, I must point out at the beginning that a paleontologist would never recognize a divergence if he had the remains of all the individuals at his disposal, since populations that had just undergone division would, for all intents and purposes, probably be identical in the parts of the individuals that would be fossilized. Only after some period of

ALAN WALKER • Department of Anatomy, Harvard Medical School, Boston, Massachusetts 02115.

differentiation would the divergent populations be separable morphologically. The period of differentiation would, of course, be variable, from long to short; but inevitably a paleontologist would err on the side of estimating a divergence time to be younger than it really was.

2. The Sedimentary Record

A glance at geological maps of any of the continental areas will show the twofold nature of the inconsistent representation of geological events. Much of the continental surface is composed of ancient rocks of cratons and either is not covered by more recent sediment because it never was or was once covered and is now denuded. When sediments of the age we are interested in are found, extensive deposits may represent relatively rapid geological events. In other areas only a short interval is represented by surface exposures, the older horizons being covered and inaccessible. All too often a record that represents only a short period of geological time is preserved in an area. In Africa, for example, fossiliferous continental Oligocene sediments are found in a small area of Egypt and nowhere else. For land vertebrates, then, we have to be content with a record that is patchy in time and space.

3. Ecological Sampling in the Sediments

The types of plant and animal communities that may be preserved in sediments will be controlled by the geological circumstances in the sedimentary basin at the particular time. Several major habitats may be sampled in any one deposit, however, although the most frequently sampled mammalian habitats are lakeshore, deltaic, floodplain, and those next to fluvial environments. This does not mean that other habitats will never be sampled, but just that the chances will be greater for the preservation of parts of the communities from some than others. In almost every case, the remains of plants and animals must undergo sedimentary burial in order to become fossilized. There is a perennial myth that concerns fossil primates. The myth runs: primates mostly live in forests, forest soils are notoriously acid, and hence primate bones are quickly dissolved and are rare in the record. This is based on a misconception. Soils are not sediments, they are the products of chemical erosion. Fossils are very rarely found in soils proper anywhere, but when soils are reworked into a sediment then bones may become fossilized. It may be that forests generally have poor runoff and that sedimentation is less there, but I would think that the chances of a sedimentary basin being present are just as high near forests as near open country. This leads us to the next problem, the sampling of species from particular habitats, with a consideration of the chances of their survival as fossils.

4. Species Sampling

Species sampling is affected by such diverse factors as population density, generation periods, mortality rates, scavenging patterns, sediment particle size, and water flow rates. Inevitably, we are left with a fossil record of any one species that is only a tiny fraction of the original populations and is often fragmentary in that a species might be represented only by jaws and teeth or fragments of the skeleton. Only a very few reasonably complete skeletons of fossil primates have survived the various taphonomic processes. Given all the processes that can obliterate the record of communities during their life, death, burial, and exhumation, it is not surprising that sometimes we find ourselves in the position of having to compare unlike assemblages of bones. Taxonomic judgments are based best, as all judgments, on the most complete evidence, but often whole populations are known from only a few teeth and jaws or a few isolated limb bones. On the whole, the most complete specimens and the most complete assemblages of specimens engender the greatest confidence. All too often in primate paleontology small fragments have too great an emphasis placed upon them, despite the fact that nomenclatural rules provide for a noncommittal referral of such material which would lead to less confusion.

Our samples of populations, like the samples of sedimentary events, tend to be patchy. Sometimes many individuals may be sampled, as with the Miocene hominoid and lorisid primates from East Africa, but often only a few individuals are known from the possible temporal and geographic ranges. Another small point that might be borne in mind here is that it can sometimes happen that when we search museums for comparative samples of living species the numbers obtainable are often less than those of the fossil sample we wish to compare them with!

It might be thought that, on the whole, the chances of sampling a peripheral population of a species are much less than sampling the main species' population (Eldredge and Gould, 1972), and although this is probably true for marine invertebrates I have reason to think that this might not always be so for terrestrial vertebrates. I will return to this point later.

5. Choice of Evolutionary Model

There are alternative evolutionary models with which to interpret the fossil record, but recently two new models have been proposed that, if workable, might enable us to reevaluate our present notions of mammalian phylogeny. Eldredge and Gould (1972) have suggested that phyletic gradualism, whereby a species changes through time by a long and imperceptibly graded series, might not be the way in which most, if not all, of evolution takes place. They propose instead a model of punctuated equilibrium, by which stable species are relatively long-lived and new ones, when they appear, do so suddenly (as far as the geological record is concerned) and are themselves recognizable as new stable forms. Their examples are taken from the record of marine invertebrate evolution, for which the record of populations through time is excellent, because the organisms have large populations. Sedimentation under marine conditions is also more widespread and continuous than, say, under continental ones. These authors list the following basic premises of the gradualism model:

1. New species arise by the transformation of an ancestral population into its modified descendents.
2. The change is even and slow.
3. The change involves usually all of the ancestral population.
4. The change takes place over all or most of the species' range.

By this manner, the fossil record should show a long sequence of intermediate forms, and morphological "breaks" in the record are due to an imperfect record.

By using Mayr's (1963) concepts of allopatric speciation, these authors' punctuated equilibrium model calls for

1. New species to arise by lineage splitting.
2. Rapid development of the new species.
3. Only a small part of the ancestral species' population being involved.
4. This population to be a peripheral one in an isolated area at the periphery of the ancestral species' range.

This would lead to the fossil record showing (in any one area) sharp morphological breaks between the two related forms. Thus many breaks in the fossil record could be real and not due to imperfections of the record. Following the implications of the model, they point out that successive species in a given lineage *could* appear as a more or less gradual series if there is an overall morphological trend in the origin of successive species, but, even if that is so, the species that comprise the lineage would themselves be stable over considerable periods of time.

Examples of the gradualist approach to primate evolution are that of Brace (1967) for the later stages of human evolution and that of Eckhardt (1972), although, in fact, most students have assumed a gradualist position without stating so explicitly. These two authors, using the gradualist approach to the full, show that small changes in morphology, given the expected variability in any population by age, sex, and geographic range, can enable the fossils to be fitted into members of gradually evolving lineages. Eckhardt (1972), calculating amounts of change between putative chimpanzee ancestors and living chimpanzees, comes to the conclusion that there need have been only one evolving line of early apes before the middle Pliocene. He shows that the metric changes per generation need only be minute—and this appears easily accomplishable at first glance. But what is easily accomplishable in evolutionary terms? I find it just as difficult to decide whether a change in tooth width of $0.03\mu m$ per generation is reasonable as I do to conceptualize the difference between a fossil being 23 million and 43 million years old. The numbers are, by my everyday human terms of reference, *unreal,* and measured rates of change of *Drosophila* wing length do not increase my understanding, at least as far as changes in ape teeth over millions of years.

If, however, the gradualist model is not the correct one to use in all cases, then variability in populations through time might not be more than the observed variability in populations of related organisms at any one time, and major changes in morphotypes would be expected if new species had developed. Whichever model may prove the correct one in any one case, using the gradualist model to the full it would be possible, given the enormity of geological time, to derive practically any modern species from practically any fossil one. Using the punctuated equilibrium model, we should be able to observe the stasis in each recorded species and develop our phylogenetic schemes accordingly. The evidence from the marine invertebrate record is swinging in favor of the Eldredge and Gould model in many cases, but the evidence from the land vertebrates is still equivocal (Gingerich, 1974).

In a recent stimulating contribution, Hayami and Ozawa (1975) postulated the following four models for the often seen gaps in the fossil record:

1. Discontinuity by lack of sedimentation or preservation or by typological treatment of fossils.
2. Discontinuity caused by unequal amounts of morphological transformation.
3. Actual discontinuity as following the punctuated equilibrium model.
4. Actual discontinuity of morphology by phenotypic substitution within a single lineage.

This last, modeled in their account, is based on theoretical transient non-sex-associated polymorphism, where, depending on the level of selection pressure, the substitution of one phenotype by another could be either noted or missed in the fossil record. This is, in effect, a population genetics model of the punctuated equilibrium one, but not necessarily with allopatric speciation.

Stanley (1975) suggested some tests that would critically evaluate evolutionary models by using the fossil record as it is now known, and concluded that major changes in evolution are unlikely to be brought about by phyletic gradualism. He made a clear distinction between speciation events on the one hand and gradual changes brought about by natural selection on the other, and regarded natural selection as providing only the raw material and fine adjustment of large-scale evolution. Harper (1975), after an analysis in which the alternative hypotheses were formally stated, cautions against such tests, which may only be pointing out that extreme versions of any evolutionary hypotheses are unlikely.

In summary, as Hayami and Ozawa point out, the cause of an observed paleontological discontinuity should be deduced for each case on its own evidence. Whether lack of record, punctuated equilibrium, gradualism, random genetic drift, recurrent mutation, or migration models can best account for known facts is not likely to be decided by an arbitrary choice of model.

I would like to return, for a moment, to the problem of sampling main or peripheral populations, since, whatever the causes of observed change in species, allopatric speciation is most likely the means by which new mammalian species arise. If things were left to chance, then presumably the main populations would most often, perhaps always, be the ones that would be sampled. However, habitat boundaries are defined, by and large, by features that are geomorphologically determined, such as rivers, lakes, and mountains. It is also often the case that these boundaries coincide, therefore, with areas of active sedimentation. Over the main parts of a species' range, conditions are often stable, uniform in habitat, and without many sedimentary basins. On the contrary, then, it is quite likely that we are most often given the record of peripheral populations (at least as far as land mammals are concerned) and not main ones. Peripheral populations in Mayr's sense need not be actually at the periphery of a species' range. For instance, mountains in the middle of a forest or desert could easily provide quite different habitats and sedimentary opportunities for both the development of peripheral populations and their sedimentary burial. Three important examples from the African fossil record illustrate this. The Fayum sediments are deltaic and floodplain ones, laid down close to the ancient North African (Tethyan) shoreline. The Kenya-Uganda Miocene sites are small sedimentary patches associated with large central volcanoes that were very close to the main continental watershed of those times, and there is evidence that they were near the eastern limits of the Miocene rainforest belt (Andrews and Van Couvering, 1975). The rift valley sites show many different sedimentary environments in a series of sequential basins associated with rift faulting. As now, the rift valleys would then have been major habitat boundaries. These few examples do suggest that peripheral populations have been sampled more than main ones. What this says of the fossil record is that part of the multiplicity of species seen by paleontologists might be due to paleosampling of marginal habitats and sequential isolated peripheral populations. The usual fate of such peripheral populations is extinction rather than speciation, and hence most if not all of them would not be expected to be ancestral to later species. Such populations do indicate, however, the nature of the main (ancestral) species from which they were derived. It would be quite possible, for instance, to take the large species of *Dryopithecus* from the postulated East African peripheral populations and derive (by cladistic or other methods) an idea of the larger species that would have been present over the area of the Miocene forest belt. This postulated species, for which little or no fossil evidence would exist, might be a better model for the species from which *Ramapithecus* might have arisen if, indeed, species of that genus arose from the populations of the larger *Dryopithecus*.

6. Dating of the Record

There are problems concerned with the dating of the fossils under consideration. For the time periods we are considering, the fossils themselves are never dated, but rather the sediments that contain them. The rocks containing the fossils, or rocks above and below the fossiliferous sediments, are dated and by a variety of methods if possible. By far the most useful method for this period has proved to be the potassium/argon method (Fitch, 1972; Miller, 1972) whereby volcanic rocks may be dated to their time of last fusion, usually the time of volcanic discharge. By using the step-heating method, a certain amount also can be learned of the rock's subsequent history. Apart from problems associated with the rock itself and the technicalities of the method, the geological context is all important, and, unless the geological relationships between the dated samples and the sediments are clearly understood, false conclusions can follow. I would like to point out that there really is no such thing as a wrong potassium/argon date. Every date gives information about geological episodes, but whether the episode is the one in which the paleontologist is interested is an entirely different matter.

Except in those cases where the rocks containing the fossils can be dated, the best that can be hoped for is that the sediments with their contained fossils can be bracketed between two dates on overlying and underlying volcanic rocks. We have many examples of sites that have been dated with fair accuracy, but we also have many more that have either only an upper or lower limiting date or no radioisotopic date at all.

Fission-track dating (Fleischer and Hart, 1972) is another method based on unstable isotope decay; it has different sources of error than those for the potassium/argon method. Again, it can be used to date the time of origin or last heating of certain volcanic materials. With one or more key radioisotopic dates in a sedimentary sequence, it is also possible to use remnant paleomagnetism in the deposits to make a paleomagnetic column that can be compared with the determined worldwide time scale (Cox, 1972). This can enable certain sequences to be allotted to their correct time span, but for reasons of the experimental error in potassium/argon dating this can only be used for the time period less than 4.5 m.y. before the present (Dalrymple, 1972). Paleomagnetic studies on their own cannot give age determinations.

There have been attempts to devise other methods for the period that concerns us, but none comes close to radioisotopic methods in accuracy, objectivity, and lack of ambiguity. The recent attempt by Partridge (1973) to assess the ages of opening of the South African cave sites by geomorphological means is a case in which the validity of the method was not checked before its use on the specific problem and which has been under attack on theoretical and practical grounds. This is leading to the acceptance of the derived "dates" by those with no geological expertise, and it is unfortunate that many anthropologists will grasp at this particularly flimsy straw because it concerns some very important fossil evidence for hominid evolution.

Sometimes we have to rely on faunal correlations. In one or two instances, there are faunal events that now have been so well established and that have been such a widespread phenomenon that they can be taken as faunal datum points in a temporal sequence. The occurrence of the horse *Hipparion primigenium* appears to have been an almost simultaneous event across Eurasia and

Africa at about 11.5 m.y. ago and the presence of this species can be taken to mean that the associated faunas are younger than that date (Berggren and Van Couvering, 1974). The first murid rodent to appear in the paleo-Mediterranean basin *(Progonomys)* is another such example. Both macro- and micro-faunas can be used, therefore, an important point when remembering that taphonomic processes can sort bones into samples containing extremes of size, but it must be remembered that species of small mammals are usually more restricted geographically. On the face of it, however, faunal correlations leave much to be desired. There are problems ranging from the identification of faunal elements to habitat sampling. There is, as well, the peculiar instance where a worker interested in knowing how old his hominoid primates are will use correlations based on other vertebrate groups, excluding the hominoids from the analysis, and, at the same time and with the same information, an artiodactyl specialist, say, will exclude those and include the hominoids. There is no way, seemingly, to avoid this circularity, since the fauna as a whole should be the best indicator of similarity and contemporaneity. Those sites that are in close proximity are best correlated by fauna than more distant ones, but it is often the long-distance correlations that would be most helpful.

7. The Problem of Higher Categories

For the purposes of this volume, attention will focus on the differentiation of higher categories. This presents real problems for the paleontologist, since whichever evolutionary model turns out to be the most pertinent in a particular case, the development of characters that typify higher categories is not likely to be so rapid that the development of one new species will show them. When a paleontologist is asked to estimate a divergence date for hominids and pongids, is he really being asked to say when, once upon a time, a pongid mother gave birth to a hominid infant? If it is true that hominids developed from pongids, then the first hominids will be very pongidlike, and how to categorize intermediate species within higher categories becomes a matter of resolving the conceptual conflict between horizontal and vertical classifications. All we can hope to determine is the time of the first record of the species that show unequivocal hominid characteristics. This will not be the splitting time, of course, but a much later event. If it could be demonstrated further from which pongid species these early hominids evolved, it would still remain to establish from which ancestral species the living pongids were descended in order to deduce the last common ancestor and hence a divergence date for modern hominids and pongids. If the higher categories are not monophyletic, then the problems that face paleontologists trying to establish divergence dates will be increased enormously. In my view, there is good evidence that all currently recognized families within the order have a monophyletic origin.

8. Species Dispersal

The seasonal migration of many birds or the wildebeest in Tanzania is a special activity that is the result of a population having two suitable home ranges separated by some area of unsuitable habitat. This is a special adaptation to habitats that change seasonally, and is not the usual way in which a species could spread from one part of the world to another. The method of species dispersal is a function of population pressure and territoriality. A species will spread to fill all available suitable habitat by territorial incrementation (Walker, 1972). Species do not get up and move across unsuitable habitats in order to spread their range. If the available habitat spreads or shrinks, then the same forces of population pressure and territoriality will act and the population will either expand into the newly available habitat or shrink with the decreasing habitat, leaving peripheral populations at the same time. Climate is one major determinant of habitats, and climatic deterioration rather than amelioration is most likely to isolate populations and thereby increase possibilities for speciation. It might be the case that a habitat will contract on one front and expand on another, and the populations dependent on it appear to spread in the direction of habitat shift, but it is habitats that shift and the populations that follow. In practical terms, habitat shift is so slow that the normal processes of territorial incrementation easily will keep pace. In the case of a newly arrived species such as *Hipparion primigenium,* a grazing horse, its relatively rapid spread across Eurasia into Africa at about 11.5 m.y. ago is sufficient grounds for believing that the suitable grassland habitat was available and that the complexities of the intraspecific competitive processes did not lead to stabilization of grazing faunas until after the first overwhelming invasion. The simultaneous appearance of the first murid rodents into the paleo-Mediterranean area must be a correlated faunal event. The impact of new (and especially invasive) species upon the habitats is not to be underestimated, and *Hipparion* might have had a similar environmental impact as goats have made in Africa, but not so great as the impact of *Homo sapiens* on the habitats which that species has invaded.

The paleogeographic and paleohabitat reconstructions that are gradually being built up for the areas critical for hominoid evolution will enable us to establish the probabilities of various species' dispersal over successive periods of geologic time. The summaries of Van Couvering (1972) and Van Couvering and Van Couvering (1975a,b) give the broad picture, although the details of tectonic microplates are still being sorted out.

The main faunal events taken from Van Couvering (Berggren and Van Couvering, 1974) are these:

30–17.5 m.y.—Hominoid faunas limited to Africa by Tethyan sea.

17.5 m.y.—Proboscideans disperse from Africa into Eurasia.

17.5 m.y.—*Pliopithecus antiquus* appears in Europe (dispersal via Turkish-Iranian region).

15.0–13.0 m.y.—*Dryopithecus* (and related) species disperse into Eurasia.

11.5 m.y.—*Hipparion primigenium* disperses from America over Eurasia
and into Africa; first murid rodents appear simultaneously.

10.0–9.5 m.y.—"Pikermi" steppe faunas spread over Eurasia.

6.0–5.5 m.y.—Microtine (arvicoline) rodents appear in Europe.

3.5–3.0 m.y.—Villafranchian faunas established.

These faunal changes correlate well with what can be inferred from
paleoclimatic studies, with essentially wet and warm conditions prevailing over
the whole area until about 10.0 m.y. ago, when relatively cool, dry-summer
climates began to predominate in temperate regions—the initiation of the
progressive cooling process that culminates in today's miserable cyclical glacial
climate in the temperate Old World.

As far as the hominoids are concerned, it is clear that no major ecological
barriers to territorial spread from Africa to Eurasia were developed until about
10.5 m.y. ago. There are some discrepancies, however, and one that is mentioned by Van Couvering and Van Couvering (1975*b*) concerns the higher
primates. All the evidence suggests that the smaller hominoids (e.g., *Pliopithecus*) entered the European forests quite some time earlier than the large
Dryopithecus species. It might be, as the Van Couverings suggest, that the latter
made use of the northward tectonic doming and rifting trend toward Turkey
that would perhaps have provided forested environments by the effect of
altitude. The smaller hominoids could have maintained their populations in
more ephemeral wooded environments that were probably alternately developing and breaking up across the northern African and southern European zone,
whereas the large forms might not have had the bioenergetic flexibility to
spread except through longer-lived forest belts. In any case, as the Van Couverings mention, it is highly likely that only light vegetation covered the greater
part of the area north of the African forest belt, for the configurations of
continents and oceans that direct climate today were established by the mid-
Tertiary collision of Africa and Eurasia.

9. Hominoid Divergence Dates

I have shown the dates for single important fossils or the ranges for species
groups in Fig. 1. I have used all the available evidence for these time placements
and have given "best guess" relationships as I see them. I am not documenting
the sources here, but advise the interested reader to refer to Berggren and Van
Couvering (1974) not only for information but also to gain an impression of the
sort of fine time control that can be gained by use of combined methods of
analysis. One point is immediately obvious from this compilation. The origin of
the hominoid primates was in Africa. This is not the case of the fossil record in
Eurasia being incomplete, for there are well-documented faunas that have
absolutely no hominoids in them.

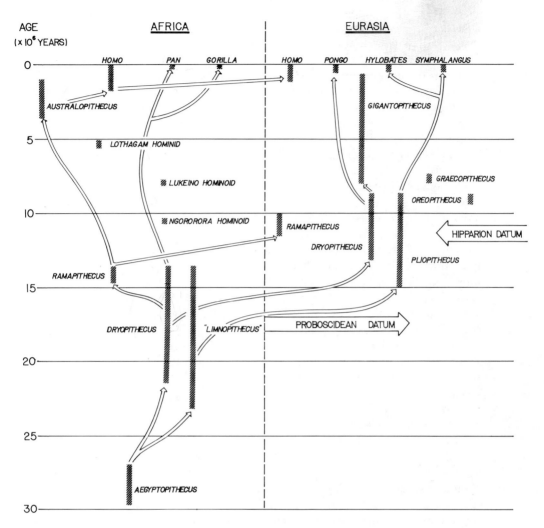

Fig. 1. Diagram to indicate the temporal and geographic ranges of hominoid primate species groups. The open arrows are personal "best guess" indications of probable relationships.

It seems to me that the relationships between the *Dryopithecus* and "*Limnopithecus*" species of East Africa could be quite complex and that allometric considerations might show that small *Dryopithecus* species could look very like similar-sized "*Limnopithecus*" species (Andrews, 1974). Andrews (1970) suggested that *Aegyptopithecus* was also present in the East African Miocene. Whether or not this notion is correct is not particularly important for our purposes, but it does illustrate the point that there are many resemblances between the Fayum early hominoids and the species that were part of the

Dryopithecus/"Limnopithecus" radiation. The simplest interpretation is possibly that a radiation of hominoids of diverse sizes developed out of an ancestral form such as *Aegyptopithecus zeuxis* (Simons, 1972). Some of the species in this radiation demonstrate remarkable morphological stability over several million years. Species from this radiation spread to Eurasia about 15 m.y. ago, some developing their own characteristics, but all disappearing from the fossil record about 10 m.y. ago as steppe faunas became established over the areas that have sedimentary records.

The earliest *Ramapithecus* specimens are from 14-m.y.-old deposits in East Africa. It is clearly a species that developed from one of the *Dryopithecus* species, probably one of the larger ones (*D. nyanzae* sized). Andrews and I (1973) were able to reconstruct the shape of the dental arcades of this species. I have been amazed at the response to this little exercise, for although we were careful to try not to make any taxonomic judgments at the familial level, many writers have considered that our work showed that *R. wickeri* could not be a hominid or a hominid ancestor. This can only mean that for many people a hominid ancestor must look like a human in its arcade shape, a position that would leave *Australopithecus* species out of our ancestry. I repeat again that "these peculiar and unique gnathic features very clearly place *Ramapithecus wickeri* apart from contemporary species of *Dryopithecus.*" What I see in the Ft. Ternan *Ramapithecus* is a species that in its dental adaptations has diverged from its ancestral *Dryopithecus* pattern in a way that no other hominoid has and that makes it, as far as the evidence will allow, the best candidate we have for the ancestry of the Hominidae. It is a fascinating point that this species appears in the record when the first antelopes, ostriches, and long-legged giraffes appear in the African record and at the earliest fossil site to record the breakup of the Miocene rainforest belt (Andrews and Walker, 1975).

Ramapithecus appears in the Eurasian record at a later time as *R. punjabicus*, again demonstrating, in the parts we know, remarkable stability. Several more complete specimens of *Ramapithecus* have now been reported from Hungary (Kretzoi, 1975) and Turkey (Tekkaya, 1974) and they allow more unambiguous statements to be made about the morphology of the jaws and teeth. With the benefit of these new specimens, I would now support the notion that the *Ramapithecus* specimens are sampled from one stable species that had a widespread distribution throughout the Old World and a duration of at least 5 m.y. *Ramapithecus punjabicus* is a small hominoid with relatively large posterior teeth and small anterior teeth. The incisors are primitively small but the canines are reduced in size from the ancestral condition. The dental arcades diverge slightly posteriorly and the teeth are set in a thick, shallow mandible that has double mandibular tori and a strong postincisive plane. The zygomatic processes of the maxillae are wide and flat, the pyriform aperture is small, and the premaxillary region is short. Despite this being a small species, the biomechanical implications are that relatively powerful muscles were producing high loading and shearing forces on the cheek teeth and that the canine teeth did not prevent lateral movements of the anterior part of the jaw near to the centric occlusal posture. In the overall morphology of the preserved parts it resembles a small version of the robust *Australopithecus* specimens from East and South Africa.

The newly reported *Graecopithecus freybergi* (Von Koenigswald, 1972) is remarkable in that it has a mandibular morphology that is very close to that deduced by Andrews and me for *R. wickeri* and yet has even larger cheek teeth relative to its mandible size. There is little doubt in my mind that if, say, the right body of the *G. freybergi* mandible had been found in an African deposit of the same age it would have been placed in the genus *Australopithecus*. The faunal associations of this specimen are not, as with earlier *Ramapithecus,* indicative of woodland or forest edge habitats, but are the classical steppe fauna of Eurasia. However, the type and only specimen is heavily damaged and does not lend itself to unequivocal determination. *Gigantopithecus* species seem to me to have clear affinities with pongids and no special relationships with hominids other than a common *Dryopithecus* ancestry perhaps. New later *Dryopithecus* specimens from Greece (De Bonis, 1974, 1975) add support to this supposition.

I see no good evidence for the genus *Australopithecus* outside Africa. I do see evidence, however, for the development of the genus *Homo* in Africa and a dispersal of man to Eurasia at about 2 m.y. ago. I strongly suspect that the impetus for this dispersal was culture, enabling early *Homo* to break out of a habitat-bound range by becoming a species that could take advantage of other species' adaptations to habitats that were previously not available to it.

As far as this dated evidence fits in with paleogeographic and climatic evidence, I conclude that there have been two basic dispersals of hominoids from Africa. One—which involved dispersal through warm, wet, forested or densely wooded environments about 15 m.y. ago and by which a variety of hominoids, from small to large, entered Eurasia—was initially determined by tectonic events. The other—by which members of the genus *Homo* dispersed into Eurasia—was determined by the development in our own genus of culture.

As far as the great apes are concerned, I conclude that the last common ancestor of the orangutan, on the one hand, and the African apes, on the other, can have been no younger than 15 m.y. Remembering my remarks on peripheral species populations, and by just taking the fossil record of chimpanzees and gorillas, we could postulate a divergence date for them at just before the first historical record of the discovery of those species. However, recalling that morphological stasis is common in many species and that the African forest belt has remained stable since before the early Miocene, it is quite possible that they diverged during some period of tectonic upheaval, the most reasonable one being in the middle Miocene. Within the *Dryopithecus* species of East Africa there are some that seem to show features reminiscent of some living apes, but I agree with Andrews (1974) that it would be inadvisable to place too much emphasis on these resemblances.

The origin of the gibbons must surely be found among the smaller hominoids of the Miocene, and some species do show very striking resemblances to living gibbons. The Napak palatal specimen (Fleagle, 1975) is a case in point, and nearly all original work on the *"Limnopithecus"* and *Pliopithecus* remains reaches the conclusion that they are plausible ancestral forms (Simons and Fleagle, 1973). Recently, Fleagle (1975) has pointed out that these similarities could be due to allometry, retention of primitive hominoid characters, or true phylogenetic relationship, and clearly more studies are needed to settle the issue. Again, taking all factors into account, a divergence date of later than 15 m.y. seems out of the question and in all probability the gibbons developed from an early Miocene hominoid stock.

The critical issue of the origin of the hominids appears to me to be better documented than, say, that of the gibbons. Using the *Ramapithecus* evidence, which I feel to be the best case to be made, a divergence date would again be 15 m.y. at the latest. The newer evidence from the East African Plio-Pleistocene sites shows us that hominid dental, cranial, and particularly postcranial patterns have been established for a considerable time. It is important to realize in this respect that the postcranial adaptations of *Australopithecus* are by no means intermediate between apes and men, but rather they are hyperhuman. It is unfortunate that the deposits that are known to cover the critically important time period in Africa have so far yielded only two isolated teeth, but we can have reasonable hopes that future expeditions to those areas will recover hominoid fossils that will help to resolve the major issues in hominoid phylogeny. In closing, however, I cannot help pointing out that as the records of fossil groups go the hominoid story is one of the more complete.

10. References

Andrews, P. J., 1970, Two new fossil primates from the lower Miocene of Kenya, *Nature (London)* **228**:537.

Andrews, P. J., 1974, New species of *Dryopithecus* from Kenya, *Nature (London)* **249**:188.

Andrews, P. J., and Van Couvering, J. A. H., 1975, Palaeoenvironments in the East African Miocene, *Contrib. Primatol.* **5**:62.

Andrews, P. J., and Walker, A., 1975, A new look at the *Ramapithecus* specimens from Fort Ternan, Kenya, in: *Perspectives on Human Evolution*, Vol. 3 (G. L. Isaac, ed.), pp. 279–304, Benjamin, Menlo Park, Calif.

Berggren, W. A., and Van Couvering, J. A., 1974, The late Neogene, *Palaeogeogr. Palaeoclimatol. Palaeoecol.* **16**:1.

Brace, C. L., 1967, *The Stages of Human Evolution*, Prentice-Hall, Englewood Cliffs, N.J.

Cox, A., 1972, Geomagnetic reversals—Their frequency, their origin and some problems of correlation, in: *Calibration of Hominoid Evolution* (W. W. Bishop and J. A. Miller, eds.,), pp. 93–106, Scottish Academic Press, Edinburgh.

Dalrymple, G. B., 1972, Potassium-argon dating of geomagnetic reversals and North American glaciations, in: *Calibration of Hominoid Evolution* (W. W. Bishop and J. A. Miller, eds.), pp. 107–134, Scottish Academic Press, Edinburgh.

De Bonis, L., 1974, Première découverte d'un primate hominoide dans le Miocène supérieur de Macédoine (Grèce), *C. R. Acad. Sci. Ser. D* **278**:3063.

De Bonis, L., 1975, Nouveaux restes de primate hominoides dans le Vallésian de Macédoine (Grèce), *C. R. Acad. Sci. Ser. D* **281**:379.

Eckhardt, R. B., 1972, Population genetics and human origins, *Sci Am.* **226**:94.

Eldridge, N., and Gould, S. J., 1972, Punctuated equilibria: An alternative to phyletic gradualism, in: *Models in Paleobiology* (T. J. M. Schopf, ed.), pp. 82–115, Freeman, San Francisco.

Fitch, F. J., 1972, Selection of suitable material for dating and the assessment of geological error in potassium-argon age determination, in: *Calibration of Hominoid Evolution* (W. W. Bishop and J. A. Miller, eds.), pp. 77–92, Scottish Academic Press, Edinburgh.

Fleagle, J. G., 1975, A small gibbon-like hominoid from the Miocene of Uganda, *Folia Primatol.* **24**:1.

Fleischer, R. L., and Hart, H. R., 1972, Fission track dating: Techniques and problems, in: *Calibration of Hominoid Evolution* (W. W. Bishop and J. A. Miller, eds.), pp. 135–170, Scottish Academic Press, Edinburgh.

Gingerich, P. D., 1974, Stratigraphic record of early Eocene *Hyopsodus* and the geometry of mammalian phylogeny, *Nature (London)* **248**:107.

Harper, C. W., 1975, Origin of species in geologic time: alternatives to the Eldredge-Gould model, *Science* **190**:47.

Hayami, I., and Ozawa, T., 1975, Evolutionary models of lineage zones, *Lethaia* **8**:2.

Kretzoi, M., 1975, New ramapithecines and *Pliopithecus* from the lower Pliocene of Rudabanya in north-eastern Hungary, *Nature (London)* **257**:578.

Mayr, E., 1963, *Animal Species and Evolution,* Harvard University Press, Cambridge, Mass.

Miller, J. A., 1972, Dating Pliocene and Pleistocene strata using the potassium-argon and argon-40/argon 39 methods, in: *Calibration of Hominoid Evolution* (W. W. Bishop and J. A. Miller, eds.), pp. 63–76, Scottish Academic Press, Edinburgh.

Partridge, T. C., 1973, Geomorphological dating of cave openings at Makapansgat, Sterkfontein, Swartkrans and Taung, *Nature (London)* **246**:75.

Simons, E. L., 1972, *Primate Evolution,* Macmillan, New York.

Simons, E. L., and Fleagle, J. G., 1973, The history of extinct gibbon-like primates, in: *Gibbon and Siamang,* Vol. 2 (D. M. Rumbaugh, ed.), pp. 121–148, Karger, Basel.

Stanley, S. M., 1975. A theory of evolution above the species level. *Proc. Nat. Acad. Sci. U.S.A.* **72**:646.

Tekkaya, I., 1974, A new species of Tortonian anthropoid (Primates, Mammalia) from Anatolia, *Bull. Min. Res. Exp. Inst. Turkey* **83**:148.

Van Couvering, J. A., 1972, Radiometric calibration of European Neogene, in: *Calibration of Hominoid Evolution* (W. W. Bishop and J. A. Miller, eds.), pp. 247–272, Scottish Academic Press, Edinburgh.

Van Couvering, J. A., and Van Couvering, J. H., 1975a, African isolation and the Tethys seaway, in: *VIth Regional Conference on Neogene Stratigraphy, Bratislava,* pp. 360–367.

Van Couvering, J. A., and Van Couvering, J. H., 1975b, Geological setting and faunal analysis of the African early Miocene, in: *Perceptions on Human Evolution,* Vol. 3 (G. L. Isaac, ed.), pp. 155–207, Benjamin, Menlo Park, Calif.

Von Koenigswald, G. H. R., 1972, Ein Unterkiefer eines fossilen Hominoiden aus dem Unterpliozan Griechlands, *K. Ned. Akad. Wet. Versl. Gewone Vergad. Afd. Natuurkd.* **75**:385–394.

Walker, A., 1972, The dissemination and segregation of early primates in relation to continental configuration, in: *Calibration of Hominoid Evolution* (W. W. Bishop and J. A. Miller, eds.), pp. 195–218, Scottish Academic Press, Edinburgh.

Walker, A., and Andrews, P. J., 1973, Reconstruction of the dental arcades of *Ramapithecus wickeri,* *Nature (London)* **244**:313–314.

Molecular Evolution as Interpreted by Mathematical Models

II

Information Theory, Molecular Evolution, and the Concepts of von Neumann

5

LILA L. GATLIN

1. Introduction

In his theory of self-reproducing automata, von Neumann (1966) begins with the concept of complexity as a given primitive notion. Von Neumann, with creative original work to his credit in quantum theory, theory of operators, ergodic theory, and many other areas of pure mathematics, was well versed in the art of beginning with an axiom, definition, or primitive notion as *given* and deducing its consequences rigorously. He did not hesitate to begin with the axiomatic observation that today's organisms are phylogenetically derived from ancestral forms which were much less *complex.*

He noted, however, that for nonliving machines an opposite tendency exists. If machine A manufactures B, another machine, A is always·more complex than B. The automaton A must contain not only a complete description of B but also a program specifying its own behavior during the synthesizing process. von Neumann expressed this situation by saying that the "complication" or "productive potentiality" was "degenerate."

However, he drew the conclusion that once the system achieves a certain

LILA L. GATLIN • Genetics Department, University of California, Davis, California 95616.

level of organization or complexity, which we may regard as a kind of theoretical complexity barrier, then complication is no longer degenerative and the production of a more complex from a less complex system is, in principle, possible.

If we adopt von Neumann's approach and regard complexity as an intuitive first principle, it follows that increasing information is required to specify increasing complexity, and this information must be created during the evolutionary process by the coincidental synchronization of changes in the genome with changes in the environment.

The concept of information creation is a study in itself; here we will be concerned primarily with the complexity barrier concept of von Neumann, which implies a dichotomy in informational properties of more complex vs. less complex organisms. Experimental evidence of this dichotomy can now be offered.

This difference should reside in the total DNA of the organism, not merely in the protein-coding sequences. King and Wilson (1975) have offered evidence that the difference in evolutionary (behavioral and anatomical) complexity between man and chimpanzee is probably not due to differences in their proteins since these differ less than those of sibling species. They regard the evolutionary difference as arising from the regulation of protein production.

There are two main types of views as to how the information for the regulation of the more complex system is acquired and stored: the simple automation view, and the complex, multicomponent automation view. The difference between these two views is not merely quantitative. Von Neumann regarded complex systems as qualitatively different from simple ones.

Under the simple automation view, proteins regulate the synthesis of other proteins through simple feedback loops, creating a kind of blind, automatically progressing automaton which is read off like a tape. Once the initial conditions are chosen which set the rigid, robotlike process in motion, no further information is required. Indeed, under the simple automation view, the difference between man and chimpanzee could well reside in minute differences in regulatory proteins undetectable by the methods of King and Wilson (1975).

However, the complex automation view essentially assumes that information regulating the behavior of the automaton is contained outside the blueprints for proteins themselves, i.e., in the non-protein-coding DNA. This is in accord with von Neumann's concept that nondegenerative complication requires a flexible, multicomponent system with a self-correcting behavioral program. The most likely candidates for the components of the behavioral or regulatory program are the repetitive sequences of highly redundant or rapidly reassociating DNA interspersed throughout the main DNA (Britten and Kohne, 1968; see also Jones, this volume).

We will adopt the complex automation view since it is most consistent with von Neumann's concepts and proceed to show that a dichotomy in informational complexity exists between vertebrates and nonvertebrates.

2. Theory

Classical information theory was content to use entropy or "information" as a measure of the technical capacity of a sequence of symbols to convey meaning, with specific admonitions not to confuse "information" and meaning. Present-day thought is much bolder. Bongard (1970) has defined "useful" information, and Gatlin (1972) has shown that the redundancy, which Shannon (1949) regarded as a scalar, has in reality a complex n-dimensional vector structure. Hutten (1973) has suggested that this redundancy structure may serve as a measure of meaning or complexity in the von Neumann sense, thus allowing us to introduce a value element lacking in classical theory. Any change in redundancy structure represents a change in meaning under this measure.

We deal with a finite sequence of symbols from an information source with a finite alphabet

$$A_1 = \{x_i : i = 1 \cdots a\} \tag{1}$$

Where a is the number of letters in the alphabet. With each letter x_i there is associated a probability of emission, $0 \leq p_i \leq 1$; $\Sigma_{i=1}^{a} p_i = 1$.

The entropy of the source is

$$H_1 = - \sum_{i=1}^{a} p_i \log p_i \tag{2}$$

The entropy function takes on its maximum value, $\log a$, if and only if all the p_i are equal (Khinchin, 1957). The function is unique for this property.

The maximum entropy state or the most random state is characterized by events which are both equiprobable and independent. These two fundamental properties can be separated as follows.

If the sequence of symbols diverges from the maximum entropy state due only to a divergence from equiprobability of the single letters in the sequence, this divergence is given by

$$D_1 = \log a - H_1 \tag{3}$$

If we wish to analyze the divergence from independence of the symbols in the sequence we must define a space of n-tuples

$$S_n = \{x_{i_\sigma} : \sigma = 1 \cdots n : i = 1 \cdots a\} \tag{4}$$

If the letters in the n-tuple are independent of each other, the probability of the n-tuple, p_n, is given by

$$p_n = p_{i_1} p_{i_2} \cdots p_{i_n} \tag{5}$$

However, if the occurrence of a given letter, x_i, in the sequence depends on the m letters preceding it in the sequence, we are dealing with the output of a Markov source with a memory of m.

The probability of the n-tuple for a Markov source with memory m is given by

$$p_n^{(m)} = p_{i_1} p_{i_1 i_2} p_{i_1 i_2 i_3} \cdots p_{i_1 \ldots i_{(m+1)}} \cdots p_{i_{(n-m)} \ldots i_{(n-1)i_n}} \tag{6}$$

where there are always n probabilities and $m + 1$ subscripts maximal.

The entropy of S_n for a Markov source of memory m is

$$H_n^{(m)} = -\sum_{i=1}^{a} p_n^{(m)} \log p_n^{(m)} \tag{7}$$

From Gatlin (1972) this is equivalent to

$$H_n^{(m)} = H_1 + H_M^{(1)} + H_M^{(2)} + \cdots + H_M^{(m-1)} + (n - m) H_M^{(m)} \tag{8}$$

where

$$H_M^{(m)} = -\sum_{i_\sigma=1}^{a} p_{i_1} p_{i_1 i_2} p_{i_1 i_2 i_3} \cdots p_{i_1 \cdots i_{(m+1)}} \log p_{i_1 \cdots i_{(m+1)}} \tag{9}$$

If $m \ll n$, a condition we can always impose simply by considering long n-tuples,

$$\lim_{n \to \infty} \frac{H_n^{(m)}}{n} = H_M^{(m)} \tag{10}$$

Following the general definition of D_2 (Gatlin, 1972), let us define

$$D_{m+1} = H_M^{(m-1)} - H_M^{(m)} \tag{11}$$

If we allow $H_M^{(0)}$ to be equivalent notation for H_1, equation (11) becomes a general definition of the increments of divergence from independence of the single letters in the sequence. The total divergence from the maximum entropy state is related to Shannon's redundancy R by the relation

$$R \log a = D_1 + D_2 + D_3 + \cdots + D_{m+1} \tag{12}$$

Any change in $H_M^{(m)}$ for any m changes the structure of the redundancy. Since $m + 1$ components are required to specify the structure of R completely, R is an $m + 1$-dimensional vector quantity.

3. Concepts

The theory may be interpreted for the nonspecialist as follows. The DNA nearest-neighbor data (Josse et al., 1961; Swartz et al., 1962) are the only extensive experimental DNA data measuring the parameters necessary for calculating the information functions. Therefore, in applications to DNA we deal only with D_1 and D_2. D_1 measures the departure from randomness due only to the divergence from equiprobability and D_2 that due only to the divergence from independence of the single letters in the sequence. D_2 may also be regarded as the divergence from equiprobability of the doublet sequences, beyond that fixed by D_1. The redundancy

$$R = \frac{1}{\log a} (D_1 + D_2) \qquad (13)$$

measures the total amount of departure from randomness, and the structure or organization of this redundancy may be characterized by the index

$$\text{RD2} = \frac{D_2}{D_1 + D_2} \qquad (14)$$

which measures what fraction of the total organization comes from D_2. RD2 is especially important as a measure of the efficiency of information processing.

This concept arises from the second theorem of information theory, which states essentially that under certain conditions it is possible to reduce error in the transmission of messages through noisy channels without loss of message rate if the message has been efficiently encoded at its source. Efficient encoding involves an optimal blend of the canonically conjugate variables, variety vs. reliability.

As D_1 increases, potential message variety is rapidly reduced. In the limit, maximal D_1 characterizes the monotone, the absolute zero of information theory with no message variety.

The nature of the process of approach to the monotone is easy to grasp intuitively. If we have a jar full of letters from the English alphabet, and we remove or drastically reduce the frequency of one letter, we have greatly limited the number of different words which can be formed. As we remove more and more letters of a given kind, in the limit very few words can be formed from one kind of letter, and with a fixed word length only one word can be formed. A more quantitative formulation of this process is given in Gatlin (1972:94–95).

As D_2 is increased, potential message variety also declines, but at a slower rate. However, the increase of both D_1 and D_2 contributes to the reliability and fidelity of the message as measured by the total R value.

It is therefore clear that the most efficient encoding for a given value of R is that with the smallest D_1 and highest D_2, i.e., highest RD2. This achieves maximum potential message variety without undue loss of fidelity, which fulfills the second theorem's promise.

4. Real DNA Data

Figure 1 is a plot of RD2 vs. R for all the available vertebrate and bacterial DNA nearest-neighbor data. The vertebrates form a relatively localized cluster bounded by $R \cong 0.02$–0.04 and RD2 $\cong 0.6$–0.8. Not a single bacterium

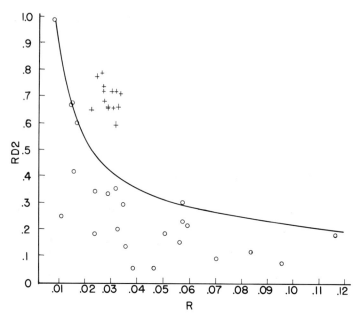

Fig. 1. RD2 v. R for all available vertebrate and bacterial DNA nearest-neighbor data. All bacterial data are compiled in Russell *et al.* (1973) and all vertebrate data are compiled in Gatlin (1972).

appears in this cluster, and all other available nonvertebrate data, which include data for insects, plants, protozoans, and invertebrates, also fall within the bacterial domain. I have tabulated and summarized these data (Gatlin, 1972) as follows:

> Vertebrates have achieved their higher R values by holding D_1 relatively constant and increasing D_2, whereas the lower organisms which achieve R values in the vertebrate range or higher do so primarily by increasing D_1. The mechanism is fundamentally different. If this is the case, we would expect to find the vertebrate D_2 values higher in general than those for lower organisms. This is what we observe.

At that time no bacterium with D_2 values in the vertebrate range of $D_2 \cong$ 0.029–0.048 bit had been observed. In 1973, Russel *et al.* reported the nearest-neighbor frequencies of rodent mycoplasma DNA, for which $D_2 = 0.044$ bit. However, they also reported that their preparation "appeared to be prone to degradation." Since D_2 tends to increase with decreasing length (see Fig. 29 of Gatlin, 1972), the high D_2 value could be simply an artifact of the degradation. There is also one other bacterium in their study, *Rhodopseudomonas capsulata*, a photosynthetic bacterium, which has $D_2 = 0.036$ bit, a value that overlaps the vertebrate range. No discussion appears as to the stability of this DNA preparation.

Redundancy structure as calculated from the DNA nearest-neighbor data is a two-dimensional vector quantity and requires two numbers for its complete description. Figure 1 shows clearly that the two bacterial DNAs described above are nowhere near the vertebrate domain in a plot of RD2 vs. R. For rodent mycoplasma, RD2 = 0.186, R = 0.117; for *Rhodopseudomonas capsulata*, RD2 = 0.309, R = 0.058.

The relatively low RD2 values signify reduced potential message variety. Another type of DNA with a high R−low RD2 pattern is satellite DNA. For example, for mouse satellite, RD2 = 0.421, R = 0.0539, and for the poly(AT) of the crab (Sueoka, 1960), RD2 = 0.490, R = 0.817. We might regard this type of DNA in a very general sense as a simple but accurate genetic message. Vertebrate DNA, on the other hand, with its moderately high R and very high RD2 retains sufficient fidelity but at a high level of complexity.

Thus we may interpret Fig. 1 as an expression of von Neumann's concept of a complexity barrier. Vertebrates represent organisms which have surmounted the barrier and achieved the level of nondegenerate productive potentiality.

ACKNOWLEDGMENT

I thank the Department of Applied Science, University of California, Davis, for use of the CDC 3400.

5. References

Bongard, N., 1970, *Pattern Recognition* (J. K. Hawkins, ed.), Spartan Books, New York.
Britten, R. J., and Kohne, D. E., 1968, *Science* **161**:529.
Gatlin, L. L., 1972, *Information Theory and the Living System*, Columbia University Press, New York.
Hutten, E. H., 1973, *Ann. Ist. Super. Sanita* **9**:335.
Josse, J., Kaiser, A. D., and Kornberg, A., 1961, *J. Biol. Chem.* **236**:864.
Khinchin, A. I., 1957, *Mathematical Foundations of Information Theory*, Dover, New York.
King, M. C., and Wilson, A. C., 1975, *Science* **188**:107.
Russell, G. J., McGoech, D. J., Elton, R. A., and Subak-Sharpe, J. H., 1973, *J. Mol. Evol.* **2**:277.
Shannon, C. E., 1949, *The Mathematical Theory of Communication*, University of Illinois Press, Urbana, Ill.
Sueoka, M., 1960, *J. Mol. Biol.* **3**:31.
Swartz, M. N., Trautner, T. A., and Kornberg. A., 1962, *J. Biol. Chem.* **237**:1961.
von Neumann, J., 1966, *Theory of Self-reproducing Automata* (A. W. Burkes ed.), University of Illinois Press, Urbana, Ill.

Random and Nonrandom Processes in the Molecular Evolution of Higher Organisms

6

RICHARD HOLMQUIST

1. Introduction

I will discuss the usefulness of approaching an understanding of evolution through a conceptual analysis of the random and nonrandom processes which occur at the molecular level of proteins and nucleic acids, rather than volunteer an exposition of mathematical methodology available elsewhere in the literature. A qualitatively incorrect concept, however mathematically transformed and quantified, remains biologically uninformative. A correct concept, even though imperfectly quantified, is at least useful; the quantitation can be improved as additional data and insight permit. With cautious optimism, Vogel *et al.* (this volume) state: "the prospects may not be so poor, provided that we do not expect to develop a final and altogether perfect concept. Our model should be refined step by step. After the first few steps, the picture admittedly may still be oversimplified, but at least the most obvious flaws of the old one are corrected." On the other hand, the known facts are not likely to be accounted for by an arbitrary choice of model. The evolutionary model presented here embodies both selective (i.e., deterministic) and probabilistic evolutionary mechanisms; it reasonably accounts, both qualitatively and quantitatively, for both the observed random and nonrandom and the Darwinian and selectively

RICHARD HOLMQUIST • Space Sciences Laboratory, University of California, Berkeley, California 94720.

neutral patterns of protein and nucleic acid variation. The values for evolutionary divergence between species which were calculated from this model when it was first published in 1972 (Holmquist *et al.*, 1972; Jukes and Holmquist, 1972*a*) were at that time 2–4 times higher than the values then considered correct. As more sequence data have become available, estimates of evolutionary divergence based on the alternative model of parsimony (Goodman *et al.*, 1974; Goodman, this volume; Moore *et al.*, 1976; Holmquist *et al.*, 1976) have been steadily approaching the values we then calculated. It is thus a predictive model that can be tested through observation and refined in the sense given by Vogel and his colleagues.

Before describing the particular model just referred to, I will discuss, in some depth, certain basic evolutionary principles that are model independent. This discussion will emphasize how unrecognized or uncritically accepted assumptions may hinder attempts to discover our evolutionary past.

2. Rationale for a Probabilistic Approach

So many modalities of biological expression are the clear result of selection that with reason one asks "Why study random processes at all?" First, certain common mutational mechanisms are known to be probabilistic in nature: radioactive decay is a Poisson process; the known thermal fluctuations inherent to all chemical reactions, including those causing mutagenesis, are described by the statistics of Brownian motion and the Boltzmann statistics of classical thermodynamics. DNA copying errors induced by the normal tautomerism of the nucleotides are one example of the latter type of phenomenon. The probabilistic description of these processes is not due to our ignorance of the mechanisms, but is physically inherent in them. A probabilistic description is thus not always to be equated to a lack of knowledge, as implied by Crowson (1975), but for some processes is an essential component of that knowledge. Second, randomness does not imply selective neutrality or lack of biological function: Mendelian genetics is based on the independent segregation of chromosomes during meiosis in sexually reproducing organisms. Most biologists would accept that this random pattern of segregation is both functional and the result of positive Darwinian selection. In a well-defined informational sense, randomization of certain biological patterns optimizes an organism's or a population's potential to cope with environmental change (Gatlin, 1972; Holmquist, 1975; Holmquist and Moise, 1975). Third, a random process is a maximally inefficient mechanism with respect to the number of steps it takes to get between two defined biological states in the absence of directing constraints. Thus not only does a random process correspond to a Darwinian process of trial and error, but for some parameters of evolutionary importance, such as the degree of evolutionary divergence between two species (as measured by estimating the number of single-step nucleotide replacements separating homologous nucleic acids from those species), a random process provides a reasonable upper limit to that divergence. As the most parsimonious estimates provide a reasonable lower limit for the divergence, the two methods together conveniently bracket the numerical magnitude of the divergence. Fourth, a random process provides a stable frame-independent standard from which one can measure deviations to determine both the nature and quantitative magnitude of those biological effects which are due to natural selection. Fifth, a partially probabilistic description avoids the biologically unreal oversimplification and/or overdeterminism of methods based on mathematical convenience or ease of computation. The rigorous method of maximum entropy inference developed by Jaynes (1957) permits the experimental data themselves to determine the probabilities (and their uncertainties) that we assign to various events. These estimates are the least biased possible on the given information in that they are maximally noncommittal with regard to missing information—i. e., in assigning the probabilities, positive weight is assigned to every situation not absolutely excluded by the data themselves. The effects of Darwinian selection are naturally accommodated by assigning determined events a probability of unity. The Jaynesian formalism guarantees that we do not prematurely exclude evolutionary possible solutions until the data are of such quality to warrant their exclusion. This is a more objective approach than forcing the data to accommodate themselves to *ad hoc* assumptions such as the additive hypothesis, the assumption of parsimony, the assumption of selective neutrality, or any other pattern of variation which neglects or distorts the effects of natural selection. The evolutionary biases inherent in such assumptions are discussed in Section 3.4. Sixth, there is evidence that the processes of speciation and extinction are themselves stochastic in nature (Gould *et al.*, 1975; Raup and Gould, 1974; Raup *et al.*, 1973).

3. Phylogenetic Reconstructions

3.1. Separation of Temporal and Spatial Aspects of Evolution

A minimal goal of evolutionary studies is to approximate, from the extant data, the cladistic relationships between species, that is, the historical order in which these species branched off from their common ancestors and the times at which they did so. To achieve this goal, and independent of whether or not a particular evolutionary phenomenon resulted from a random or nonrandom process, it is necessary at the outset to make a clear distinction between temporal and spatial evolutionary events. Fitch and Langley (this volume) discuss the former. The variation of biological structures in space—specifically, the distribution of the number and type of nucleotide fixations along the gene, or the number and type of amino acid replacements along the protein chain—is the concern of this chapter. The evolution of biological function is a consequence of both temporal and spatial processes; these are, of course, interdependent. Phylogenetic trees are a pictorial representation of the temporal history of a set of contemporary species. However, they are often reconstructed from spatial data alone: sets of homologous amino acid sequences determined on proteins isolated from the contemporary species. Extrapolation of these purely spatial data to the temporal domain is normally made by assuming *ad hoc* that evolution is in some sense temporally predictable—parsimoniously or stochastically regular, for example. A preferable method is to integrate the temporal information from the fossil record with the spatial information from the molecular record (Romero-Herrera *et al.*, 1973). In this regard, paleontological information such as the approximate species branch-point dates provided in the chapters by Simons and Walker (both in this volume) is invaluable.

3.2. Separation of Topological and Numerical Aspects in Phylogenetic Reconstructions

The branching order which depicts the species genealogy in a phylogenetic tree is a topological result, and although numerical methods (Sneath and Sokal, 1973; Jardine and Sibson, 1971) may be of use in obtaining this topology the problem is conceptually separate from that of determining the numerical distances between nodes of the tree. In the global version of the parsimony method (Moore, this volume), the two aspects are mutually dependent. One might also take the topology from the fossil record, and then apply the parsimony or other method to determine the internodal distances (Romero-Herrera *et al.*, 1973). Again on the assumption that shared structural features reflect descent from a common ancestor rather than convergent evolution or accidental similarity, the topological branching order could be determined by some version of the parsimony principle, as this principle minimizes the effect of the latter parallelisms. One could then, not at all inconsistently, recognize that a proper measure of the genetic distances between species must take into account such parallelisms and use an entirely different and nonparsimonious method of calculating the numerical values of the internodal separations. This is in fact the procedure used by Moore (1976) in constructing his estimates of the internodal separations, the augmented distances (AD).

3.3. The Importance of the Triangle Inequality for Phylogenetic Reconstructions

A minimal and mandatory topological constraint that all phylogenetic reconstructions must satisfy if they are to have biologically meaningful cladistic (historical) content is the triangle inequality (Lipschutz, 1965), which states simply that the separation between any two contemporary species must be less than or equal to the sum of the separation of each from any ancestral sequence (see Fig. 1). This inequality is in fact the only real constraint that is known to be

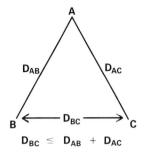

Fig. 1. The triangle inequality: $D_{BC} \leq D_{AB} + D_{AC}$. The more restrictive additive hypothesis states that $D_{BC} = D_{AB} + D_{AC}$. If the
input distance is greater than the output distance, i.e., $D_{BC} >
D_{AB} + D_{AC}$, an error has been made in the phylogenetic recon-
struction. The statement that $D_{BC} < D_{AB} + D_{AC}$ is simply the
mathematical formulization of the biological fact that evolutionary
parallelisms and back mutations exist. All distances must be in the
same units.

true with certainty independent of any of the other details of the evolutionary
process. If for any phylogenetic tree the cladistic distances do not satisfy all the
$N(N - 1)/2$ triangle inequalities for that tree (N is the number of species), and/
or the distances are not all positive, then the tree is somewhere incorrect either
in its topological branching arrangement or in the numerical estimates of the
internodal separations (genetic distances) between ancestors, or both. Violations
of the triangle inequality are a major cause of incorrectly reconstructed cladograms. The biological impossibility of the results they depict has not hindered
their publication, however. Such errors are particularly frequent among cladograms inferred from the additive hypothesis (see Section 3.4). Violations
are easily spotted in practice: if for any tree both the input and output pairwise
distances between species are shown (the output distance is the sum of the
reconstructed distances along the branches leading to the two contemporary
species from their common ancestor), wherever an output distance is less than
an input distance the triangle inequality is violated (Fig. 1). Methods of phylogenetic reconstruction which guarantee that the triangle inequality will be satisfied
have been published by Farris (1972) and by Beyer *et al.* (1974). These methods
are applicable to all types of similarity matrices of pairwise species differences
whether their elements be amino acid differences (Zuckerkandl and Pauling,
1962), minimum base differences (Jukes, 1963), nucleotide differences from
homologous messenger RNA pairs, random evolutionary hits (Holmquist *et al.*,
1972; Jukes and Holmquist, 1972*a*), the chemical similarity distances defined by
Sneath (1966), or the immunological distances of Sarich and Wilson (1967) and
of Goodman and Moore (1971).

3.4. The Dangers in Ad Hoc Assumptions and Oversimplification

The additive hypothesis states that the evolutionary distance separating
two contemporary species is equal to the sum of their distances from their
common ancestor (Fig. 1). Moore *et al.* (1973) give an excellent discussion of the
utility and limitations of this hypothesis, which is an exceedingly restrictive
special case of the triangle inequality; the hypothesis can lead to biologically
correct phylogenetic trees only if the distance measure used to estimate the

separation between species includes all parallelisms and back mutations. The existence of parallelisms in evolution is well documented at both the morphological and molecular level. With respect to myoglobin sequences, Romero-Herrera *et al.* (this volume) comment: "it is evident that the unguided use of similarity as a basis for phylogeny is likely to lead to some false conclusions." Beard and Goodman (this volume) give further examples of parallelisms in the β-hemoglobin chains during the descent of *Tarsius bancanus, Nycticebus,* and the anthropoids from a common ancestor. The distance measures most commonly used in conjunction with the additive hypothesis are immunological distances (ID), amino acid distances (AAD), minimum base differences (MBD), nucleotide differences between homologous messenger RNAs, random evolutionary hits (REH), and augmented maximum parsimony distances (AD). With the exception of the latter two, these measures either ignore or grossly underestimate parallelisms. Some suggested corrections (Dickerson, 1971; Dayhoff *et al.,* 1972) are not at all adequate because they neglect the fact that different codons in the structural gene for a given protein fix mutations at different species-dependent rates. The version of the additive hypothesis that is most often used for constructing phylogenetic trees from similarity matrices of the above distance measures is the algorithm of Fitch and Margoliash (1967). In that algorithm, neglected parallelisms can and do lead to some implausible cladistic reconstructions. For the cytochrome *c* phylogeny reconstructed from 20 contemporary species by those authors, out of the 190 reconstructed distances in their Table 3, 75 are biological impossibilities because they violate the triangle inequality; these errors are distributed approximately uniformly throughout the phylogenetic tree. Perhaps the greater danger in unrestrained use of the additive hypothesis with nonmetric distance measures or those that neglect parallelisms is that one thereby excludes or distorts those cladograms where parallelisms are permitted, and it is precisely these that more correctly represent the evolutionary history.

From the above discussion, one might anticipate, erroneously, that the pairwise evolutionary distances between species would be additive provided that they were adequately corrected for parallelisms. The measure of genetic distance developed in our own laboratory—random evolutionary hits (REH)—probably comes as close to an adequate correction for parallelisms as any measure now in use. It is in agreement (Holmquist *et al.,* 1976; Moore *et al.,* 1976) with a second estimate, the augmented distance (AD), of parallelism obtained by an entirely independent method, the augmented maximum parsimony approach of Moore (1976). As REH and AD correct approximately maximally and minimally, respectively, for the extent of parallelism present, the agreement of the two measures implies that corrections derived by other methods will be of the same quantitative magnitude. Unfortunately, however, even if an estimate has been corrected perfectly for parallelisms, it would still not be true that the elements of the difference matrix formed from that measure would form an additive set. The reason is that the inherent probabilistic component of evolution has a finite scatter. If the evolutionary process could be often repeated, the average REH or AD values derived from independent evolutionary trials should be additive. But, in fact, contemporary terrestrial life is the result of a single unique evolutionary history. Each protein or nucleic acid sequence is, like speciation, a unique evolutionary event—a single spin of the roulette wheel, so to speak. The probability that for a set of real data the single measured values correspond to the average values is vanishingly small. The most satisfactory way to cope with this problem may be to assume that proteins with very different biological functions and structures approximate independent evolutionary trials. By averaging genetic distance measures (corrected for parallelisms) over a sufficiently large number of proteins, an additive set of differences might be approached. However, one does not have this recourse in constructing a phylogenetic tree from data for a single protein such as cytochrome *c*. Here the best that may be done is to use the method of Beyer *et al.* (1974) on the matrix of, say, REH values and to choose the objective function in that method so as to minimize the deviations from additivity in the final tree. A possibly equally satisfactory, but not equivalent, alternative is to derive a correction (taking care to justify it both biologically and theoretically) to the original matrix of, for example, REH values such that the corrected values form a perfectly additive set. The procedure of Moore (1973) could then be used to construct the cladogram.

A different form of biologically unreal mathematical oversimplification and overdeterminism is illustrated by the global form of the parsimony principle (Moore, this volume) in which the "best" phylogenetic reconstructions are defined as those trees which minimize the *total* number of nucleotide replacements required to go from the most ancestral sequence to the set of homologous contemporary sequences. Romero-Herrera *et al.* (this volume) note that when this algorithm is applied to the myoglobin sequences one might reject the common ancestry of the baboon *(Papio anubis)* and macaque *(Macaca fascicularis),* even though their known biology as documented by the fossil record and the fact that the two myoglobins differ from each other in only one residue out of 153 makes their common ancestry almost certain. But from the fact that whale and camel cytochromes *c* are identical (Goldstone and Smith, 1966; Sokolovsky and Moldovan, 1972) is one also to conclude they too share a recent common ancestry? The question is an open one, as here an adequate fossil record of intermediate forms is lacking.

4. The Value of Free Exploration

I have tried to point out in the above sections that though all of the methods currently being used to analyze molecular data have some rational basis, some are based on assumptions that are sometimes arbitrary and sometimes demonstrably wrong. These assumptions have significant quantitative and qualitative effects on the evolutionary inferences we draw from the experimental data. It is therefore necessary not only that our analytical methodology be based on sound biological and theoretical principles but also that in the absence of data to the contrary we do not limit our biological perspective by unjustifiably restrictive assumptions or tricks of mathematical convenience which may exclude that unique cladogram which *is* our evolutionary history.

5. A Quantitative Interactive Model* for Evolution

I suggest and review here an evolutionary perspective in which the selective and probabilistic aspects of evolution are viewed as interacting complements. Neither aspect is dominant; both are important, although in any particular case the balance may be toward one or the other. The existence of a probabilistic component does not necessarily imply that when we observe a random pattern of nucleotide or amino acid substitution, or of variation in evolutionary rate, that that pattern arose by a process of pure chance. Such an interpretation may be the simplest and sometimes the correct explanation of the observed randomness. But this pattern could also have arisen by Darwinian selection for randomness (Section 2) or could be the result of the superposition of many complex independent selective events, just as height in human populations is distributed in a Gaussian manner but is in fact the result of many complex polygenic interactions. Similarly, when we observe a nonrandom pattern, although the correct explanation may be a selective one, in a percentage of such cases the observed pattern will represent no more than an event in the tails of some random distribution.

5.1. Definition of a Random Variable

A random variable is a function (for example, genetic divergence) defined on a sample space. The sample space is a set of objects (as the $N(N - 1)/2$ aligned homologous pairs of myoglobin sequences from N biological species), determined by the nature of the problem we are investigating, for each one of which there exists a unique rule (count the number of amino acid differences between the aligned pair) which associates a number X (in this case, the number of amino acid differences), the function, to that object. The number X may or may not be different for different objects. The number of objects associated with the numerical value x_j divided by the total number of objects is the probability that the function X has the value x_j. The totality of such probabili-

---

*I originally termed this model the "stochastic model of evolution" because of its probabilistic formulation and mathematical methodology. It is usually referred to as such in the literature. However, that name unfortunately suggests that the model ignores Darwinian natural selection when in fact it is an integral part of the model. It is hoped the new choice of words will more accurately convey the real conceptual content of the model, at least for the purpose of this chapter.

ties, one for each x_j, is the probability distribution of the random variable X. Knowing the probability distribution can give us insight into the mechanisms (here biological) that gave rise to the set of objects. If the functions X are time dependent, $X = X(t)$, the mechanism(s) giving rise to these random variables is called a stochastic process. If we know the mathematical form of $X(t)$, either from experiment or from theory, then we know the complete evolutionary history of the random variable. Partial differentiation of $X(t)$ with respect to time then provides additional information (in this example, the first partial derivative gives us the "mutation rate," i.e., in more correct terminology the rate at which amino acid replacements have been fixed by natural selection). Thus formulated, the theory of stochastic processes achieves both analytical power and generality. No special assumptions about the random variables being equiprobable, about the uniformity of evolutionary rates, or about the existence (or lack thereof) of selective neutrality have been made. If an evolutionary hypothesis can be formulated in terms of such processes, the entire analytical armory of a well-developed and still active branch of mathematics is at one's disposal. The text by the late William Feller (1968) is an admirable introduction to the necessary mathematical techniques at a level which permits one to successfully tackle the complications which arise in real, as opposed to ideal, sets of data.

It is important to note that the concepts of a random variable and of a stochastic process say nothing about the relative contributions of chance and necessity. The concepts simply associate a certain time-dependent probability with each evolutionary event. So long as we determine these probabilities from the data, no *a priori* assumptions are involved.

5.2. The Model

I now outline an interactive quantitative model of molecular evolution with the purpose of suggesting its theoretical and biological reasonableness as well as its practical utility. It is a model that "works." There do not appear to be any conceptual or computational obstacles which stand in the way of refining it to whatever degree of realism required by the data. It is the only model, to my knowledge, which possesses predictive as well as retrodictive value. These predictions can be verified against the experimental data (which consist of protein amino acid sequences or messenger RNA nucleotide sequences) and thus form a check on the current status of the model's realism. Discrepancies thus uncovered provide guidelines for further refinement.

5.2.1. The Role of Darwinian Selection

The observable effects of natural selection are so pervasive that any model which does not include these as an integral part of its basic formalism need not be seriously considered. It does not do to tack these selective constraints on "at the end" as some sort of perturbation to an otherwise pure (but useless) probabilistic model. Some examples of such effects (taken mostly from this volume for ease of reference) are as follows: the amino acid substitutions in the variable region of the human κ-immunoglobulin chains are not uniformly distributed over the protein but are linked (Hilschmann *et al.*, this volume); the accumulation of a charge altering amino acid substitution in the papionine α-hemoglobins is often accompanied by a second compensating charge altering substitution elsewhere in the reverse direction so as to maintain overall electrical charge balance (Hewett-Emmett *et al.*, this volume); unequal rates of substitutions along different lineages of the same protein or along the same lineage in different proteins are the norm, not the exception (Goodman, this volume; Jukes and Holmquist, 1972*b;* Fitch and Langley, this volume; Matsuda, this volume; Romero-Herrera *et al.*, this volume; Tashian *et al.*, this volume; Moore *et al.*, 1976; Holmquist *et al.*, 1976); the probability that a change in the primary structure of the DNA will occur depends on the efficiency of the enzyme DNA polymerase, while the probability (much lower) that this change will occur, become viable as a mutation, and be fixed by selection depends on the efficiency of the repair processes as well (Vogel *et al.*, this volume). Such selective effects are allowed for as follows. For each protein, for a given time period of genetic divergence away from that structure, we define at each amino acid site i a mutability μ_i equal to the total number of nucleotide replacements which the codon corresponding to that site has sustained, i.e., fixed. The total number of nucleotide replacements for the structural gene we designate REH, a mnemonic for the words "random evolutionary hits," and it is this that we use as a direct measure of the genetic divergence between homologous nucleic acids. Thus

$$\text{REH} = \sum_{i=1}^{T} \mu_i = \sum_{i=1}^{n} \bar{\mu}_i T_i = \mu_2 T_2 \qquad (1)$$

where T is the total number of residues in the protein. As each μ_i is determined by the totality of selective and probabilistic interactions during this divergence, the end result of natural selection on the numerical value of REH is fully accounted for through equation (1). Further, since REH is the total number of nucleotide replacements, parallelisms and back mutations are included in this measure of genetic divergence. In the expression between the second and third equal signs, the amino acid sites are classified into n mutability classes, the ith class having an average mutability $\bar{\mu}_i$ and containing T_i sites. In the expression to the right of the last equal sign, the amino acid sites are divided into only two classes. $T - T_2$ of the sites are assigned zero mutability, and the remainder an average mutability μ_2. As there exists a range of mutabilities among the amino acid sites in a protein because of the different functional importance of amino acids in different positions, on proceeding from left to right in equation (1) the various expressions represent successively poorer approximations to the true situation. However, because all these expressions are equal to each other, even the poorest approximation yields as accurate an estimate of REH as does the exact expression (between the first and second equal signs) provided that μ_2 and T_2 are correctly chosen. Because of this fact, and the overwhelming simplifications that having to consider only one class of nonzero mutability brings to the computational effort, the immediate practical problem facing an investigator wishing to determine the total number of nucleotide replacements REH which separate the structural genes coding for two homologous proteins is to have a method for obtaining, by an analysis of the amino acid sequences of these proteins, numerical values for μ_2 and T_2 in the expression REH $= \mu_2 T_2$.

5.2.2. The Mechanics of Evolutionary Divergence at the Macromolecular Level

The following three additional requirements enable us to solve for μ_2 and T_2 uniquely: over the variable part of the structural gene (those T_2 codon sites having nonzero mutability), the accepted point mutations in the DNA are distributed in a known or estimable manner; the four individual nucleotides in these T_2 codon sites occur with known or estimable frequency and order; and a given nucleotide is likely to mutate to any one of the other three with known or estimable probabilities. These requirements define the dynamic, i.e., predictive, accuracy of the model. The required distributions can be obtained experimentally from mRNA nucleotide sequences, if available; they can be estimated with less but still sufficient accuracy from ancestral mRNA sequences inferred from the principle of maximum parsimony (Moore *et al.*, 1973); or they can be assumed with an accuracy corresponding to the quality of our guess. Our model is now completely defined. It suffices, given μ_2 and T_2, and an initial nucleotide sequence, to predict for a given period of genetic divergence the expected number of observable base differences (i.e., directly countable nucleotide replacements between the initial and final sequence) of each of the ten [4(3)/2 + 4] possible types, or in the case of diverging proteins the expected number of observable amino acid replacements of each of the 210 [20(19)/2 + 20] possible types. Conversely, from the observed pattern of nucleotide or amino acid replacements between two aligned homologues, that value of μ_2 and T_2 can be found (by a least-squares or other suitable method) which best fits the observed pattern. The total genetic divergence is then REH $= \mu_2 T_2$. The model is amenable to computer simulation, but it is more efficient, cheaper,* and more biologically informative to derive the necessary analytical expressions because then one can see at a glance what effect varying a certain parameter has.

Up to this point, we have introduced no asumptions of any sort. We have only rephrased, with somewhat more precision than is customary, and in a form amenable to further mathematical formalization, known biological facts.

5.2.3. Some Assumptions—Not Ad Hoc

We now turn to some practical considerations. The distributions required to implement the above analysis cannot always be obtained empirically because

---

*The cost of determining all 1770 REH values for the pairwise separations from the 60 homologous protein sequences (of about 100 residues each) from 60 species is $4–50 depending on the speed of the computer one has available. The corresponding cost by the method of parsimony is several hundred dollars.

too few homologous mRNAs have been sequenced. With the exception of the analysis of Goodman *et al.* (1974) for the vertebrate globin chains and the analysis of Romero-Herrera *et al.* (1973) for myoglobin, individuals using the parsimony method generally limit themselves to stating the total divergence and do not list all the nodal ancestral nucleotide sequences. In this state of ignorance, one is forced to guess at the required distributions. However, this guess need not be arbitrary. The best guess possible given the lack of information is provided by Jaynes' principle of maximum entropy inference (Jaynes, 1957; Tribus, 1962; Christenson, 1963; Tribus *et al.*, 1966; MacQueen and Marschak, 1975) from information theory: bias is minimized by making all outcomes equiprobable. Accordingly, we adopt the following three assumptions for the positive reason that they bias our results least for those situations where detailed empirical information about the distributions is lacking. The accepted point mutations in the variable part of the structural gene are spatially random; the four individual nucleotides in the variable part of the structural gene occur with equal frequency and in random order; and, last, a given nucleotide in this region is equally likely to mutate to any one of the other three.

Having accepted these assumptions as a working basis, we ask how far they bias our results and how far they deviate from a known situation where we do have some empirical information. The assumption of a stochastically uniform (i.e., Poisson) distribution of accepted point mutations along the variable part of the structural gene appears to describe the observed pattern of amino acid substitution in proteins adequately (Jukes and Holmquist, 1972*a*); it biases the numerical magnitude of REH in the direction of being too low (Table 10 in Jukes and Holmquist, 1972*a*).

When tested against known protein sequence data, the assumption that the four nucleotides occur with equal frequency appears (Table 1) to be reasonable when all gene positions are considered. A more stringent test of this assumption can be made against messenger RNAs whose sequences have been determined experimentally as here it is not necessary to assume that all codons for a given amino acid are used equally frequently. Such a test is shown (Table 2) for the A-protein gene of bacteriophage MS2. The observed frequency of occurrence of G:A:U:C in this gene is 26.3:23.4:24.2:26.0, whereas the frequency expected on the assumption of equiprobability is 25.0:25.0:25.0:25.0. The agreement is good. The assumption that the four bases occur in random order can be examined by looking at the three codon positions separately (Tables 1 and 2). The assumption of random order seems reasonable for the A-protein gene from bacteriophage MS2 and for the immunoglobulin genes and less reasonable for the other structural genes. A similar result was obtained when we examined (Holmquist *et al.*, 1973) a total of 3049 nucleotide loci from 43 transfer RNA genes G:A:U:C::31.3:19.9:23.2:25.6, a base composition not far from that expected. However, since the base compositions of the helical and nonhelical regions of transfer RNA were quite different, the assumption of random order must be considered an approximation.

A direct experimental measure of the relative frequency with which any

Table 1. Frequencies of Occurrence of the Four Nucleotides as Deduced[a] from Amino Acid Sequence Data

		All proteins[b]	All globins[c]	Myoglobin[d]	Cytochrome c[e]	Immunoglobulins[f]
First codon position	G	37 (10)	38	37	33	28
	A	27 (5)	23	27	34	27
	U	18 (5)	17	14	16	23
	C	18 (7)	21	22	17	21
Second codon position	G	19 (8)	14	13	18	21
	A	33 (7)	33	41	39	28
	U	24 (5)	29	28	20	22
	C	24 (5)	24	18	23	29
Third codon position	G	25	25	27	26	23
	A	24	25	27	26	24
	U	26	25	23	24	26
	C	26	25	23	24	26
All codon positions	G	27	26	26	26	24
	A	28	27	32	33	26
	U	22	24	21	20	24
	C	23	23	21	21	26

[a]On the assumption that all codons for a given amino acid are used equally frequently.
[b]The value given is the average from 78 distinct protein families comprised of 169 individual sequences and represents the totality of published information at the time of this writing. The individual sequences used are listed in Holmquist and Moise (1975). The value in parentheses is the standard deviation for the population of 78 families.
[c]The value given is the average from 19 α-hemoglobins, 17 β-hemoglobins, 1 δ-hemoglobin, 8 myoglobins, 2 lamprey globins, 1 mollusc globin, and 2 plant leghemoglobins. A complete listing can be found in Holmquist *et al.* (1976).
[d]The value given is the average for 20 myoglobins. A complete listing can be found in Holmquist *et al.* (1976).
[e]The value given is the average for 57 cytochromes *c*. A complete listing can be found in Moore *et al.* (1976).
[f]The value given is the average for 18 immunoglobulins.

base mutates to any other is provided by the clinically known human hemoglobin variants. The data are in Table 3. The measured frequencies are not all equal, but they do appear symmetrical: the probability of nucleotide i mutating to nucleotide j appears equal to the probability of nucleotide j mutating to nucleotide i. What is important for evolution, however, is not mutations *per se* but mutations that are fixed by natural selection, and these transition probabilities are in general neither equal nor symmetrical (Table 4). The deviations from the equiprobable transition frequencies average 40%, and the maximum deviation is about a factor of 2 from them.

Calculations in which known deviations from a base ratio of 1:1:1:1 and in which nonequiprobable transition probabilities are used show that the deviations from the assumptions of equiprobability which are found in real data sets may cause an under- or overestimation of REH by about 5–15% in any particular case (Holmquist, 1972a; Holmquist, 1973).

Table 2. Frequencies[a] of Occurrence of the Four Nucleotides as Experimentally Determined from the Nucleotide Sequence of Bacteriophage MS2 A-Protein Gene

First codon position	G	31.6
	A	24.2
	U	22.9
	C	21.4
Second codon position	G	20.1
	A	25.4
	U	27.0
	C	27.5
Third codon position	G	27.2
	A	20.6
	U	22.9
	C	29.3
All codon positions	G	26.3
	A	23.4
	U	24.2
	C	26.0

[a]The frequencies were calculated from the tabulations in Fig. 3 of Fiers *et al.* (1975).

In both Tables 3 and 4, the data shown are averages for the first two codon positions. The transition probabilities for these two positions considered separately differ from each other as well as from the average value shown. In Table 4, the results are also averaged along six lineages of evolutionary descent. If the lineages are examined separately, even for the same codon position, the transition probabilities differ for different lineages. Each of these particularities is further evidence of natural selection at the macromolecular level. If each of these complex particularities were to be included in a more realistic model, one would need a separate model not only for each protein but also for each evolutionary lineage. Theory would be reduced to description devoid of predictive (hence testable) content. As a more useful alternative, the principle of maximum entropy inference has permitted us to derive a set of assumptions adequate with respect to the purpose at hand, that of estimating the evolutionary divergence between species in such a manner that its numerical magnitude is realistic, and such that the predictive content and general applicability of the theory to many protein families and lineages are not lost. Any agreement of the prediction from this model with experiment is due to the correctness of the concepts, not to an ever-increasing *a posteriori* number of parameters designed to induce a better fit with that data.

Table 3. Transition Frequencies for Single Amino Acid Mutations in Proteins as Related to Base Changes B → B′ in mRNA[a,b]

B \ B′	A	G	C	U
A	—	16.4	7.8	4.1
G	18.3	—	6.8	7.3
C	7.8	5.9	—	7.8
U	3.6	5.0	9.1	—

[a]If all mutations are equiprobable, then the expected frequency for each cell is 8.3.

[b]A total of 219 observed mutations is included, 175 from human hemoglobin variants and 44 from other proteins. No nitrous acid induced mutations are included. The numbers in the cells are percentages of the total. Calculated from data in Jukes (1975).

5.2.4. Classification of the Experimental Data

One additional facet demands attention before we can obtain useful numerical results. Because of the very few residues in a typical protein, even the longer ones containing but 500 or so amino acids, classification of the experimentally observed amino acid replacements between two homologues into 210 types would result in many classes being empty, with a consequent decrease in the reliability of any statistical estimates made. It is necessary to reduce the number

Table 4. Transition Frequencies for Single Amino Acid Fixations in Six Lineages of Globin Chains as Related to Base Changes B → B′ in mRNA[a,b]

B \ B′	A	G	C	U
A	—	11.1	12.3	4.3
G	17.5	—	11.5	6.8
C	9.6	8.4	—	4.8
U	4.6	3.9	5.1	—

[a]If all mutations are equiprobable, then the expected frequency for each cell is 8.3.

[b]A total of 867 fixations along six lines of descent in the globin chains is included. These fixations were inferred from ancestral sequences derived by the maximum parsimony method. The numbers in the cells are percentages of total. Calculated from data in Goodman *et al.* (1974).

of classes so that the number of replacements which falls in most is appreciable. This can be done in any way that works as the theory can predict the expected number of replacements in each class for any arbitrary division of the classes. We have classified the amino acid replacements between any two homologous proteins into the four classes most commonly used by workers in the field of molecular evolution, namely those of the minimal zero-, one-, two-, and three-base types. Examples of these types are, respectively, Ala/GCA to Ala/GCG, Pro/CCC to Arg/AGA, Gly/GGA to Thr/ACU, and Lys/AAG to Phe/UUU, because the first three of these amino acid replacements can be explained with fewer base replacements in the corresponding codons as GCA to GCA, CCA to CGA, and GGA to ACA, respectively. As illustration, in the comparison of the α-hemoglobin chains of the cebus monkey and kangaroo 112, 23, 6, and 0 amino acid replacements of each of the four types have occurred, respectively. Even with this reduction in the number of classes from 210 to 4, one can see from the illustration, which is fairly typical, that the number of replacements falling into each class is not statistically overwhelming—one would prefer more. It is for this reason, as well as for the ones given earlier, that the simplest formulation in equation (1) has been used: there are simply not enough data available to estimate much more than the two parameters μ_2 and T_2 accurately.

5.2.5. The Essential and Peripheral Structure of the Model

The formal structure of the interactive evolutionary model was complete with Section 5.2.2. The assumptions of equiprobability adopted in Section 5.2.3 and the particular data classification scheme adopted in Section 5.2.4 can be replaced by any biologically reasonable alternative of the user's choice. I have tried to make clear the rationale behind adopting these particular assumptions and this particular classification scheme with the full realization that other assumptions and classification schemes might better apply to a particular set of protein or nucleic acid sequence data. Provided that the deviations from the assumptions are not too large, and they are not, the form of the equations rather than their precise numerical parametric values governs the dynamics of evolution. If evolutionary scatter, both that caused by natural selection and that caused by stochastic fluctuation, is comparable (and it is on both counts) in magnitude or larger than the deviations from equiprobability, such deviations will affect measurable parameters only slightly. This is the reason our model works at all and why when the entire gene is free to fix mutations our results and those estimated by the PAM method of Dayhoff and her colleagues are nearly identical (Holmquist, 1972a, 1973) even though the latter group does not assume equiprobability of events and makes a valid effort to reason out the correct probabilities. Where the PAM method errs, however, is in ignoring the fact that natural selection does more than alter the transition probabilities of one nucleotide (or amino acid) to another, and more than alter the base (or amino acid) composition. By far the more quantitatively significant effect of natural selection is to decide how many and which particular residues in a protein can vary at all. This can be determined only by looking at *each* protein pair residue by residue and seeing how many sites are unvaried and how many amino acid replacements are of each of the 190 possible types (or as we more simply have done of each of the minimal one-, two-, or three-base types). It does not suffice to use matrices of average values to do this, and that is why, except when the entire gene is free to fix mutations, the PAM values are not correct (usually too low). It is also why our own equations (Holmquist, 1972*b,c,d*), or Table 1 or Fig. 1 in Holmquist (1972*a*), cannot be directly used to calculate REH values. One must use the more accurate methodology outlined here, which takes into account these selective effects in as simple a manner as has biological utility.

5.2.6. Final Numerical Calculation of REH, μ_2, and T_2

We are now in a position to estimate numerical values for μ_2 and T_2 for any aligned homologous pair of proteins. The essential idea is that the higher the ratio of amino acid replacements of the minimal two- and three-base types to those of the one-base type, the greater is μ_2 (the total number—countable plus superimposed—of fixed nucleotide replacements per variable codon). A complete analytical exposition of the theory and numerical examples from real data sets are in the original papers (Holmquist *et al.*, 1972; Jukes and Holmquist, 1972*a*). Once μ_2 and T_2 have been estimated from the pair of homologous proteins, the number of countable (observable) differences between the corresponding pair of messenger RNAs can be predicted *a priori* from theory alone. Once these messengers become available and are sequenced, this prediction can be tested and will form a critical test of the theory and indicate those really necessary refinements that would improve it.

Thus viewed, the evolutionary divergence of proteins and nucleic acids has been a chance phenomenon operating *within* selective constraints. Its philosophical basis is that by and large the process has been the inefficient one of trial and error tempered, strongly or weakly, by natural selection. Its mathematical structure is rooted in the statistical theory of random variables, branching processes, and Markov chains, constrained by the boundary conditions of natural selection.

6. The Method of Parsimony

Our attention now turns to a very different method, that of maximum parsimony, for estimating evolutionary divergence. Its philosophical basis is that the process of divergence has been parsimonious and efficient. Its mathematical structure is rooted in set theory and topology. Its operational methodology is deterministic rather than probabilistic.

6.1. Evidence against Biological Evolution as an Efficient Minimum Path Process

The principle of parsimony has been the object of many misinterpretations. It does not assert that genetic divergence has followed that pathway requiring the fewest nucleotide substitutions. Flamboyant pronouncements that "without a general belief that evolution has followed the shortest pathways there is no constraint on the wildest of postulated pathways" (Sneath, 1974) are not only conceptually misleading but also incorrect. There is no experimental evidence whatsoever in support of such a belief, and the experimental evidence against this dynamic interpretation of the principle is very strong. Nucleic acid hybridization experiments (Li *et al.*, 1973), the high rate at which amino acid replacements have occurred (Kimura, 1968), the increase in the proportion of experimentally observed amino acid substitutions of the minimal two-base type with increasing taxonomic separation (Jukes and Holmquist, 1972a), the common occurrence of parallelisms (Romero-Herrera *et al.*, this volume), and the conformance of the amino acid compositions of proteins as a whole (King and Jukes, 1969), and to a major extent as individual proteins (Holmquist and Moise, 1975), to the genetic code table frequencies—all are evidence that the minimum path required by a dynamic interpretation of the principle of parsimony has not been followed. Dynamically, the principle violates causality: the reconstructed ancestral sequences of some past time minus t are conditional upon events (the contemporary protein and nucleic acid sequences) at the present time. From the known list of mutagens, it is impossible to devise a mutagenic mechanism as selective as a dynamic interpretation requires. Finally, the Darwinian process of trial-and-error survival of the fittest by natural selection is incompatible with always choosing the minimum path as the latter permits neither trial nor error. In summary, as an *a priori*, i.e., predictive, explanation or guideline for evolutionary divergence, the principle simply fails.

6.2. The Distinction between a Mathematically Unique Phylogenetic Solution and an Evolutionarily Correct One

In the method of parsimony as one extrapolates from the set of contemporary messenger RNA sequences backward toward the more ancestral sequences, common sense dictates that the more ancestral the sequence the less its probability of being correct; yet in the maximum parsimony method it is the most ancestral sequence that is least ambiguously determined (Moore, this volume). The point is not minor, as the input data themselves, the set of mRNA sequences, are usually inferred from the amino acid sequence data. For N protein sequences each 100 residues long, the probability that all N inferred contemporary mRNA sequences are correct is almost zero [$(1/4)^{100N}$] because of the degeneracy of the third nucleotide of the coding triplet. Stated another way, one-third of the nucleotide sequence has three-fourths of a chance of being wrong, so that on the average a quarter of each inferred contemporary sequence is incorrect, and the situation must be worse for the more ancestral sequences.

At the first and second codon positions, the above type of ambiguity is much less, but here estimates of the extent of parallelism (Section 3.4) indicate that the method of maximum parsimony is capturing on the average only 60% of the fixations that have actually occurred. With 40% of the nucleotide replacements unaccounted for, it is not possible to place much confidence in the accuracy of the ancestral sequences.

In summary, the near mathematical uniqueness of a cladogram inferred by the parsimony method does not imply its biological or evolutionary correctness. Such a cladogram is a useful but blurred reflection of the actual evolutionary history.

6.3. The Method of Parsimony as a Valid Tool for Evolutionary Reconstruction

The preceding two sections are meant to provide perspective for the unjustified firmness with which conclusions inferred by the parsimony method are at times enunciated and uncritically accepted.

Correctly interpreted, *a priori* the method of parsimony is dynamically neutral, and the results from the phylogenies reconstructed from sets of contemporary homologous protein or nucleic acid sequences by this method in fact give us information on the dynamic pathway actually followed during genetic divergence. The correct algorithms for this reconstruction have been given by Fitch (1971) for nucleic acids and by Moore *et al.* (1973) for proteins. The accuracy of the information retrieved from these ancestral reconstructions depends on the topological quality of the set of contemporary protein or nucleic acid sequences from which they were inferred: the more numerous the number of species represented along *each* branch of the evolutionary family tree, the greater the accuracy. This property of the denseness of the maximum parsimony evolutionary trees bears importantly on the reliability of the biological conclusions we draw from them. The total nucleotide replacements separating two sequences are underestimated in the sparse regions of the tree. Recently, Moore (1976) has invented an ingenious set theoretic algorithm which corrects (augments) these divergence estimates so that the augmented total nucleotide replacements correspond to what would have been obtained were the evolutionary tree everywhere equally as dense as its densest region. Moore's algorithm is an important conceptual advance because, for the first time, it endows the method of parsimony with a predictive property: as future sequences are added to the sparse regions of the tree, the unaugmented distances should approach the augmented distances, and the latter should converge upon themselves. Both predictions are experimentally testable. Nothing can be done to improve the accuracy of the inferred ancestral nodal sequences (Section 6.2) in sparse regions of the tree other than to sequence more proteins in those regions. As an *Aufbau-Prinzip*, the method of maximum parsimony guarantees that provided the branching order of the cladogram is correct* the phylogenetic reconstructions (ancestral sequences and the number of nucleotide replacements separating these sequences) will approach, without necessarily ever reaching, the correct historical record. The method sets a lower bound for genetic divergence and gives us much other useful information as well.

A parallel to paleontology may be drawn: the method of maximum parsimony tells us how to dig through the contemporary strata of sequence data in small-enough steps so that we do not miss any ancestral forms that left behind fossil remains. There is a difference, however, between the paleontological results and those from molecular evolution. The fossils of the former are real; those from the latter are inferred. And in neither case do the methods used provide us with a satisfactory explanation of the results of our excavations.

7. The Logical Consistency of the Interactive Model and the Method of Parsimony

The message we want to leave is that despite the contradictory philosophical motivations which led to the development of the interactive model and the method of maximum parsimony, the two approaches are mutually complementary. Their logical structures are both compatible and consistent. Perhaps the most useful immediate contribution the latter approach could make would be a complete listing of the nodal ancestral nucleotide sequences for those phylogenies so far published, with an accompanying tabulation at each codon position for each of the frequencies of occurrence of each of the four nucleotides, and a matrix of the frequencies with which each of the 16 transitions between nucleotides occurs.

8. Convergence of Independent Quantitative Estimates of Genetic Divergence and Evolutionary Parallelisms

During this last year, correlation of our data with those of M. Goodman and G. W. Moore demonstrated that the estimates of genetic divergence derived from the method of maximum parsimony (augmented distance, AD) and from the interactive model (REH values) converge (Holmquist *et al.*, 1976; Moore *et al.*, 1976). As these two estimates are, respectively, approximate

---

*It should be noted that the global form of the method of parsimony (Moore, this volume; see also the last paragraph of Section 3.4 in this chapter) determines the branching order of the cladogram. There is no guarantee that this branching order is correct. It may be preferable in some cases to take the branching order from an independent source (such as the fossil record) and then apply the method of parsimony, even though the total number of mutations in the resulting cladogram may exceed the most parsimonious global solution.

minimal and maximal estimates of this divergence, their convergence establishes that the process of genetic divergence is mechanistically probabilistic, essentially, although not exactly, along the lines sketched in the first section of this chapter. Before now, the only careful estimate of the extent of parallelisms and back mutations in molecular evolution was given by the excess of REH estimates over the unaugmented internodal separations taken from maximum parsimony trees. This excess was large: on the average, REH was about 50% greater than the unaugmented distances, and, not uncommonly, the REH value was two- to four-fold larger than the unaugmented distances for the more distantly separated species. Some biologists and evolutionary theorists found it difficult to believe that back mutations and parallelism could be so extensive. This in turn cast doubt on the REH measure itself as a proper estimate of genetic divergence. Moore's augmentation algorithm (1976) provides a new and independent estimate of the extent of parallelisms which is in agreement with the REH estimates. The experimental work of Romero-Herrerra *et al.* (this volume) demonstrates that such extensive parallelism occurs in practice.

The conceptual underpinnings and mathemetical methodologies of the interactive model and of models based on some version of parsimony are so disjoint that the quantitative agreement of the evolutionary estimates provided by the two approaches comes as a pleasant surprise. Perhaps the most important practical consequence of the agreement is that as the two models were in part designed to place approximate upper and lower bounds, respectively, on the magnitude of genetic divergence, the convergence of these two bounds implies that any numerical estimate of genetic divergence made by some other adequate evolutionary model cannot be too disparate* from those given by the REH and AD values. This in turn implies that the present estimates of genetic divergence are free of at least the grosser effects of the particular assumptions of any individual model, and increases their biological credibility.

9. The Nonequivalence of Selective Neutrality and Randomness

There is one important question, fiercely debated by both its protagonists and its antagonists, each drawing on the same molecular data for support, that is *in principle* not capable of being answered by those data alone: Does the random fixation of selectively neutral or very slightly deleterious mutants occur far more frequently in evolution than positive Darwinian selection of definitely advantageous mutants (Kimura and Ohta, 1974; Ayala, 1974; Ohta, 1975)? The arguments for selective neutrality experimentally rest in part on the observed

---

*This comment is not meant to imply that the present values of REH and AD are absolute final estimates of genetic divergence. AD will increase as additional experimental sequence information becomes available, and REH will increase when full account is taken of the differential ability of each nucleotide site to fix mutations. If the explanation of the convergence given in Sections 6, 7, and 8 is correct, then the limiting values of REH and AD will still be such that their ratio is on the average unity.

random distribution of some measurable evolutionary character. But on the simple observation that randomness can be selected *for,* in the Darwinian sense, these arguments lose all their force. A familiar example is the advantages accruing to sexual vs. asexual organisms by the randomization of their chromosome complement (Section 2). The molecular data can tell us nothing about whether the observed random distribution arose by a stochastic or a selective mechanism. Additionally, a random distribution can arise from the superposition of individually nonrandom distributions. By way of illustration, Kimura and Ohta point to the approximate constancy of the rate at which amino acid substitutions have been fixed as evidence for selective neutrality on the basis that this is what one would expect if the mutations were distributed in time in a Poisson manner. Although others in this volume (Goodman, Fitch and Langley, Matsuda, Tashian *et al.,* and Romero-Herrera *et al.*) and elsewhere (Jukes and Holmquist, 1972*b;* Romero-Herrera *et al.,* 1973; Moore *et al.,* 1976; Holmquist *et al.,* 1976) might dispute this constancy, even if true it is relatively easy to devise selective schemes (Van Valen, 1973) which also give rise to a constant rate. Conversely, the arguments for positive Darwinian selection experimentally rest on the observed nonrandom distribution of measurable evolutionary characters. For at least some characters, there is clear quantitative bias toward randomness and there is no way to prove *a posteriori* that this randomness did not arise by a stochastic process. For example, in Table 4 in Jukes (1971) some amino acid loci in the globin chains have been occupied by as many as ten different amino acid types having little chemical similarity.

10. Speciation, Morphological Change, and Gene Structure

What can and cannot be learned from comparison of molecular sequence data from structural genes? The data so far have not given us useful information about the process of speciation. Man and chimpanzee; donkey and zebra; cow, pig, and sheep; camel and whale; chicken and turkey; and cauliflower and grape—the cytochromes *c* for every species listed within any one of these six groups are identical.* Although composite comparisons which utilize many protein families, rather than one, may reduce the number of such identities, it is noteworthy that for the man/chimpanzee divergence the protein sequences are identical not only for cytochrome *c* but also for α-hemoglobin and β-hemoglobin. At the opposite extreme are the immunoglobins, where for the variable region of the κ-chain, the proteins isolated from different individuals from the same species (mouse) differ at 42% of their amino acid loci (Dayhoff, 1972*b*).† Thus speciation can be relatively uncoupled to genetic divergence arising from

---

*The first reference to the peculiar identity of the camel and whale occurs in a manuscript dated 1604, by William Shakespeare: *Hamlet,* Act III, Scene ii.

†This large divergence suggests the two mouse genes are related by gene duplication and subsequent divergence rather than by direct divergent descent from a common ancestral gene.

point mutations in the DNA coding for protein. This may be of adaptive advantage, for it gives organisms two independent ways rather than a single way of coping with environmental changes. With respect to one of these, that of point mutation, the molecular data can give us detailed information about the dynamics of that process. It has been shown already that this process is one of chance as well as necessity (Sections 5–9, and Langley and Fitch, 1974), that it is nonuniform in time (Ohta and Kimura, 1971; Jukes and Holmquist, 1972*b;* Prager *et al.,* 1974; Goodman *et al.,* 1974; Matsuda, this volume; and Fitch and Langley, this volume), and that there are measurable evolutionary invariants, stable over time, and relatively independent of both speciation and biological function. Two of these invariants are Shannon's redundancy R (Gatlin, 1974) and the extent of nonrandomness with respect to amino acid composition in proteins (Holmquist and Moise, 1975).

The intriguing species-specific patterns found by Jones (this volume) for satellite DNA may or may not prove to be related to speciation and/or adaptive morphological change.

Wilson *et al.* (1974) have suggested that gene rearrangements may be more important than point mutations as sources for evolutionary changes in anatomy and way of life. Their evidence is that whereas the rate of protein evolution in mammals has been roughly equal to that in frogs, the rate of change in chromosome number has been much faster in the former than in the latter. In parallel to this observation, they earlier noted that the rate of change in the number of species that could hybridize was much more rapid among mammals than among frogs (Wilson *et al.,* 1974*a*). Attributing the inability to form interspecies hybrids to an incompatibility in the parental systems which regulate the expression of genes, they conclude that gene rearrangement may provide an important means of achieving new patterns of regulation.

However, the correlations they observed need not be causal. An alternative explanation of their results is that prior to the differentiation of an organism into a new morphological form a particular gene rearrangement may be adaptively deleterious whereas after that differentiation it may be adaptively allowed.

Gene expression is regulated through a complex hierarchy of biological control systems. From what is known about these systems (D'Azzo and Houpis, 1966), it is unlikely that improved regulation can result from gene rearrangements which cause a large change in the chromosome structure of existing functional control regions of the nucleic acid. A drastic alteration in the arrangement of the components of a complex control system usually leads to a loss of previous function. Rather, in these complex systems, improved response to an environmental change is achieved by quantitative changes in the control components already present. If the environmental disturbance is large enough, it may be necessary to add additional control loops to the control circuits already present. Only for the largest nontransient disturbances is it necessary to design a radically new system. Whether or not morphological adaptations among mammals represent radical change, the most likely form of regulatory gene rearrangement would be a duplication of the entire regulatory gene, or a substantial portion thereof, because the old control system must continue to function while the new is being built. Elaboration of the duplicated gene would then lead to the new pattern of regulation. Kolata (1975) summarizes some of the more recent experimental data with respect to gene regulation. Stebbins (1969) notes:

> The kinds of adaptive responses that a particular population will evolve at any time depend to a large extent upon the nature of those adaptations that it already possesses. Once a particular level of organizational complexity has been achieved, mutations that elaborate upon this complexity have a much greater chance of success than do mutations tending to destroy it, simplify it, or start a trend towards lower levels of complexity.

Wilson's dismissal of the possible importance of point mutations with reference to changes in morphology and hybridizability seems premature and inconsistent with some of his own findings: although the rate of protein evolution in both frogs and mammals is similar and uncorrelated with either the rate of morphological change or the rate at which the ability to form interspecific hybrids has been lost in the two classes, in birds the correlation exists with respect to mammals but not with respect to frogs. The rate of protein evolution in birds is one-third that in mammals (Prager *et al.,* 1974); the rate of chromosomal evolution and the rate of loss of interspecific hybridization potential in birds are also slower (Prager and Wilson, 1975) than in mammals, and in fact comparable to the slow rates of these phenomena in frogs. Some morphological changes of adaptive significance, such as changes in bone length, are known to be under the control of polypeptide hormones which exert their effect by interacting with protein or protein-associated receptor sites on the cell surface. This interaction triggers a succession of enzymatically controlled biochemical events which lead to the morphological change. Point mutations in one or more of these proteins could alter receptor or enzymatic properties sufficiently to cause significant morphological change. African pigmies have normal plasma growth hormone levels, but their tissues are subnormally responsive to this hormone (Ganong, 1973). Here there is no obvious need to invoke gene rearrangement to explain the observed morphological change. The amino acid sequences of the polypeptide hormones themselves differ from species to species, and these differences are the known result of point mutations in the structural (not regulatory) genes which code for them (Dayhoff, 1972*a*). Tomkins (1975) gives further discussion of the evolution of hormonal systems. The probable importance of the acidic nonhistone nuclear proteins to gene regulation is experimentally documented in Cameron and Jeter's (1974) recent book. A likely reason that the study of protein structure has so far failed to give us much useful information with respect to adaptive anatomical changes or interspecies hybridizability is that the right proteins have not yet been looked at, not that proteins and the point mutations which occur in them are unimportant for adaptive change.

Whatever the cause of morphological or adaptive change, it is well to recall R. A. Fisher's postulate that the probability of individual mutations contributing to evolution is inversely correlated with the intensity of the effect on the developing phenotype. The available data are sufficiently scant that to accept any of the current explanations is unwise except as a working hypothesis for suggesting more definitive experiments.

Finally, it is only by the editing action of natural selection on the phenotypes which result from any proposed molecular mechanism(s) that evolutionary effective change occurs. Van Valen (1971, 1974) discusses the complexity of this interaction with respect to the origin of higher taxa and the origin of the orders of mammals.

11. Conclusion

In conclusion, we return to the title theme—the extent to which the processes discussed in the preceding sections are operative in higher organisms, and, with respect to this symposium, particularly within the order Primates. Were an observer from another galaxy to arrive on Earth and be handed the *Atlas of Protein Sequence and Structure,* with labels and interpretive matter removed, but the protein sequence data themselves intact, he would, even though the sequences had been scrambled among but not within themselves, be able to classify these data into families of biological function, and within a given family to tell that the different sequences therein belonged to different species. Assuming that galaxy had had its Darwin, he could even deduce a phylogenetic tree for these species depicting their approximate evolutionary order of branching, placing by that galaxy's tradition the most advanced species on the right. He could not, however, tell whether that most singlulary advanced form was a bacterium or a man. Random and nonrandom processes appear to operate at all levels of organismal complexity.

Those characteristics which distinguish man and the other primates from the rest of life lie in the hierarchical interrelationships of their organ systems. Although these relationships are ultimately to be understood in terms of molecular structure, the data so far available contribute minimally toward that goal. Such data are, however, perhaps one necessary small first step.

12. Summary

Temporal, spatial, topological, and numerical aspects of phylogenetic reconstructions have been examined. Mathematical oversimplifications and certain commonly used assumptions affect these reconstructions significantly with respect to their ability to accurately depict our evolutionary past. The importance of Darwinian selection has been stressed. A quantitative evolutionary model has been proposed in which the selective and stochastic components of biological evolution are viewed as interacting rather than mutually exclusive. The principle of parsimony is dynamically untenable as an evolutionary hypothesis. At the same time it is a valid *Aufbau-Prinzip* for examining our evolutionary origins. The interactive and parsimonious models of molecular evolution are logically consistent complements. The observed random distribution of some measurable evolutionary character does not suffice to establish its selective neutrality and at the same time makes impossible a proof that such a distribution arose by positive Darwinian selection. The relationship of gene structure to speciation and morphological change has been considered.

ACKNOWLEDGMENT

This work was supported by grant NGR 05-003-460 from the National Aeronautics and Space Administration.

13. References

Ayala, F., 1974, Evolution: Natural selection or random walk? *Am. Sci* **62**:692–701.

Beyer, W., Stein, M., Smith, T., and Ulam, S., 1974, A molecular sequence metric and evolutionary trees, *Math. Biosci.* **19**:9–25.

Cameron, I., and Jeter, S. (eds.), 1974, *Acidic Proteins of the Nucleus*, Academic Press, New York.

Christenson, R., 1963, in: *Induction and the Evolution of Language*, Arthur D. Little, Inc., Cambridge, Mass.

Crowson, R., 1975, Anti-Darwinism among the molecular biologists, *Nature (London)* **254**:464.

Dayhoff, M., 1972a, Hormones, active peptides, and toxins, *Atlas of Protein Sequence and Structure* **5**:D173.

Dayhoff, M., 1972b, *Atlas of Protein Sequence and Structure* **5**:D234, Matrix 24.

Dayhoff, M., Eck, R., and Park, C., 1972, A model of evolutionary change in proteins, *Atlas of Protein Sequence and Structure* **5**:89–99.

D'Azzo, J., and Houpis, C., 1966, *Feedback Control System Analysis and Synthesis*, 2nd ed., Chapter 15, McGraw-Hill, New York.

Dickerson, R., 1971, The structure of cytochrome *c* and the rates of molecular evolution, *J. Mol. Evol.* **1**:26–45.

Farris, J., 1972, Estimating phylogenetic trees from distance matrices, *Am. Nat.* **106**:645–668.

Feller, W., 1968, *An Introduction to Probability Theory and Its Applications*, Vol. 1, Wiley, New York.

Fiers, W., Contreras, R., Duerinck, F., Haegman, G., Merregaert, J., Min Jou, W., Raeymakers, A., Volckaert, G., Ysebaert, M., Van de Kerckhove, J., Nolf, F., and Van Montagu, M., 1975, A-protein gene of bacteriophage MS2, *Nature (London)* **256**:273–278.

Fitch, W., 1971, Toward defining the course of evolution: minimum change for a specific tree topology, *Syst. Zool.* **20**:406–416.

Fitch, W., and Margoliash, 1967, Construction of phylogenetic tress, *Science* **155**:279–284.

Ganong, W., 1973, *Review of Medical Physiology*, 6th ed., p. 307, Lange Medical Publications, Los Altos, Calif.

Gatlin, L., 1972, *Information Theory and the Living System*, Columbia University Press, New York.

Gatlin, L., 1974, Conservation of Shannon's redundancy for proteins, *J. Mol. Evol.* **3**:182–208.

Goldstone, A., and Smith, E., 1966, Amino acid sequence of whale heart cytochrome *c*, *J. Biol. Chem.* **241**:4480–4486.

Goodman, M., and Moore, G., 1971, Immunodiffusion systematics of the primates. I. The Catarrhini, *Syst. Zool.* **20**:19–62.

Goodman, M., Moore, G., Barnabas, J., and Matsuda, G., 1974, The phylogeny of human globin genes investigated by the maximum parsimony method, *J. Mol. Evol.* **3**:1–48.

Gould, S., Raup, D., Schopf, T., and Simberloff, D., 1975, in: Research news (G. B. Kolata, reviewer), Paleobiology: Random events over geological time, *Science* **189**:625–626, 660.

Holmquist, R., 1972a, Empirical support for a stochastic model of evolution, *J. Mol. Evol.* **1**:211–222.

Holmquist, R., 1972b, Theoretical foundations for a quantitative approach to paleogenetics. Part I. DNA, *J. Mol. Evol.* **1**:115–133.

Holmquist, R., 1972c, Theoretical foundations for a quantitative approach to paleogenetics. Part II. Proteins, *J. Mol. Evol.* **1**:134–149.

Holmquist, R., 1972d, Theoretical foundations of paleogenetics, in: *Proceedings of the Sixth Berkeley Symposium on Mathematical Statistics and Probability: Darwinian, Neo-Darwinian, and Non-Darwinian Evolution,* Vol. 5 (L. LeCam, J. Neyman, and E. Scott, eds.), pp. 315–350, University of California Press, Berkeley.

Holmquist, R., 1973, The stochastic model and deviations from randomness in eukaryotic tRNAs: Comparison with the PAM approach, *J. Mol. Evol.* **2**:145–148.

Holmquist, R., 1975, Deviations from compositional randomness in eukaryotic and prokaryotic proteins: The hypothesis of selective-stochastic stability and a principle of charge conservation, *J. Mol. Evol.* **4**:277–306.

Holmquist, R., and Moise, H., 1975, Compositional non-randomness: A quantitatively conserved evolutionary invariant, *J. Mol. Evol.* **6**:1–14.

Holmquist, R., Cantor, C., and Jukes, T., 1972, Improved procedures for comparing homologous sequences in molecules of proteins and nucleic acids, *J. Mol. Biol.* **64**:145–161.

Holmquist, R., Jukes, T., and Pangburn, S., 1973, Evolution of transfer RNA, *J. Mol. Biol.* **78**:91–116.

Holmquist, R., Jukes, T., Moise, H., Goodman, M., and Moore, G., 1976, The evolution of the globin family genes: Concordance of stochastic and augmented maximum parsimony genetic distances for alpha hemoglobin, beta hemoglobin, and myoglobin phylogenies, *J. Mol. Biol.* **105**:39–74.

Jardine, H., and Sibson, R., 1971, *Mathematical Taxonomy,* Interscience, New York.

Jaynes, E., 1957, Information theory and statistical mechanics, *Phys. Rev.* **106**:620–630, **108**:171–190.

Jukes, T., 1963, Some recent advances in studies of the transcription of the genetic message, *Adv. Biol. Med. Phys.* **9**:1–41.

Jukes, T., 1971, Comparison of the polypeptide chains of globins, *J. Mol. Evol.* **1**:46–62.

Jukes, T., 1975, Mutations in proteins and base changes in codons, *Biochem. Biophys. Res. Commun.* **66**:1–8.

Jukes, T., and Holmquist, R., 1972a, Estimation of evolutionary changes in certain homologous polypeptide chains, *J. Mol. Biol.* **64**:163–179.

Jukes, T., and Holmquist, R., 1972b, Evolutionary clock: Non-constancy of rate in different species, *Science* **177**:530–532.

Kimura, M., 1968, Evolutionary rate at the molecular level, *Nature (London)* **217**:624–626.

Kimura, M., and Ohta, T., 1974, On some principles governing molecular evolution, *Proc. Natl. Acad. Sci. U.S.A.* **71**:2848–2852.

King, J., and Jukes, T., 1969, Non-Darwinian evolution: Random fixation of selectively neutral mutations, *Science* **164**:788–798.

Kolata, G. (reviewer), 1975, Evolution of DNA: Changes in gene regulation, *Science* **189**:446–447.

Langley, C., and Fitch, W., 1974, An examination of the constancy of the rate of molecular evolution, *J. Mol. Evol* **3**:161–177.

Li, S., Denney, R., and Yanofsky, C., 1973, Nucleotide sequence divergence in the α-chain structural genes of tryptophan synthetase from *Escherichia coli, Salmonella typhimurium,* and *Aerobactor aerogenes, Proc. Natl. Acad. Sci. USA* **70**:1112–1116.

Lipschutz, S., 1965, *Theory and Problems of General Topology,* McGraw-Hill, New York.

MacQueen, J., and Marschak, J., 1975, Partial knowledge, entropy, and estimation, *Proc. Natl. Acad. Sci. USA* **72**:3819–3824.

McLaughlin, P., and Dayhoff, M., 1972, Evolution of species and proteins: a time scale, *Atlas of Protein Sequence and Structure* **5**:47–52.

Moore, G., 1973, An iterative approach from the standpoint of the additive hypothesis to the dendrogram problem posed by molecular data sets, *J. Theor. Biol.* **38**:423–457.

Moore, G., 1976, Proof of the populous path algorithm for missing mutations in parsimony trees, *J. Theor. Biol.* (in press).

Moore, G., Barnabas, J., and Goodman, M., 1973, A method for constructing maximum parsimony ancestral amino acid sequences on a given network, *J. Theor. Biol.* **38:**459–485.

Moore, G., Goodman, M., Callahan, C., Holmquist, R., and Moise, H. 1976, Stochastic vs. augmented maximum parsimony method for estimation of superimposed mutations in the divergent evolution of protein sequences—Methods tested on cytochrome *c* amino acid sequences, *J. Mol. Biol.* **105:**15–38.

Ohta, T., 1975, Statistical analyses of *Drosophila* and human protein polymorphisms, *Proc. Natl. Acad. Sci. USA* **72:**3194–3196.

Ohta, T., and Kimura, M., 1971, On the constancy of the evolutionary rate of cistrons, *J. Mol. Evol.* **1:**18–25.

Prager, E., and Wilson, A., 1975, Slow evolutionary loss of the potential for interspecific hybridization in birds: A manifestation of slow regulatory evolution, *Proc. Natl. Acad. Sci. USA* **72:**200–204.

Prager, E., Brush, A., Nolan, R., Nakanishi, M., and Wilson, A., 1974, Slow evolution of transferrin and albumin in birds according to microcomplement fixation analysis, *J. Mol. Evol.* **3:**243–262.

Raup, D., and Gould, S., 1974, Stochastic simulation and evolution of morphology—towards a nomothetic paleontology, *Syst. Zool.* **23:**305–322.

Raup, D., Gould, S., Schopf, T., and Simberloff, D., 1973, Stochastic models of phylogeny and the evolution of diversity, *J. Geol.* **81:**525–542.

Romero-Herrera, A., Lehmann, H., Joysey, K., and Friday, A., 1973, Molecular evolution of myoglobin and the fossil record: A phylogenetic synthesis, *Nature (London)* **246:**389–395.

Sarich, V., and Wilson, A., 1967, Immunological time scale for hominid evolution, *Science* **158:**1200–1203.

Sneath, P., 1966, Relations between chemical structure and biological activity in peptides, *J. Theor. Biol.* **12:**157–195.

Sneath, P., 1974, Phylogeny of micro-organisms, *Symp. Soc. Gen. Microbiol.* **24:**1–39.

Sneath, P., and Sokal, R., 1973, *Numerical Taxonomy,* Freeman, San Francisco.

Sokolovsky, M., and Moldovan, M., 1972, Primary structure of cytochrome *c* from the camel, *Camelus dromedarius, Biochemistry* **11:**145–149.

Stebbins, G., 1969, *The Basis of Progressive Evolution,* pp. 29, 124, University of North Carolina Press, Chapel Hill, N.C.

Tomkins, G., 1975, The metabolic code, *Science* **189:**760–763.

Tribus, M., 1962, The use of the maximum entropy estimate in the estimation of reliability, in: *Recent Developments in Information and Decision Processes* (R. E. Marshall and Paul Grey, eds.), Macmillan, New York.

Tribus, M., Shannon, P., and Evans, R., 1966, Why thermodynamics is a logical consequence of information theory, *AIChE J.* **12:**244–248.

Van Valen, L., 1971, Adaptive zones and the orders of mammals, *Evolution* **25:**420–428.

Van Valen, L., 1973, A new evolutionary law, *Evol. Theory* **1:**1–30.

Van Valen, L., 1974, A natural model for the origin of some higher taxa, *J. Herpetol.* **8:**109–121.

Wilson, A., Maxson, L., and Sarich, V., 1974*a,* Two types of molecular evolution: Evidence from studies of interspecific hybridization, *Proc. Natl. Acad. Sci. USA* **71:**2843–2847.

Wilson, A., Sarich, V., and Maxson, L., 1974*b,* The importance of gene rearrangement in evolution: Evidence from studies on rates of chromosomal, protein, and anatomical evolution, *Proc. Natl. Acad. Sci. USA* **71:**3028–3030.

Zuckerkandl, E., and Pauling, L., 1962, Molecular disease, evolution, and genic heterogeneity, in: *Horizons in Biochemistry* (M. Kasha and B. Pullman, eds.), pp. 189–225, Academic Press, New York.

Proof for the Maximum Parsimony ("Red King") Algorithm

7

G. WILLIAM MOORE

Alice had never been in a court of justice before, but she had read about them in books, and she was quite pleased to find that she knew the name of nearly everything there. "That's the judge," she said to herself, "because of his great wig."

1. Principles of Tree Construction

1.1 Introduction

In reconstructing hypothetical messenger RNA (mRNA) sequences which were contained in the ancestors of present-day species, we find ourselves in much the same predicament as Alice, sitting in a courtroom for the first time. Alice, who has never seen a judge before, has to infer that the man sitting before her with the "great wig" is a judge. He *looks* like a judge, so he must *be* a judge. In examining mRNA sequences from contemporary species (obtained by inference from amino acid sequences and the genetic code), we assume that when two sequences *look* alike they must *be* alike, in the sense of sharing a common ancestry. Similarity does not always imply common ancestry, but we assume that that reconstruction of hypothetical ancestors which maximizes the similarity due to common ancestry (and thus minimizes similarity due to

G. WILLIAM MOORE • Department of Anatomy, Wayne State University School of Medicine, Detroit, Michigan 48201. Present address: Department of Pathology, Johns Hopkins Hospital, Baltimore, Maryland 21205.

parallel and back mutations) is the best reconstruction. This is known as the *maximum parsimony hypothesis.* Since the judge is really the King of Hearts, or Red King, we call this the Red King hypothesis (see Van Valen, 1974). In this chapter, I shall review the highlights of the proof for the current computer algorithm for reconstructing hypothetical mRNA sequence ancestors consistent with contemporary amino acid sequences and the Red King hypothesis.

1.2. Trees and Networks

A *tree* is a branching arrangement consisting of points connected by lines (Fig. 1). A tree has a single most ancestral point, or *root.* Each ancestral point gives rise to exactly two descendants. Points which have no further descendants are called *exterior points,* and correspond to contemporary species. All nonroot, nonexterior points are called *interior points.* It is customary to number points on the tree in ascending order starting with exterior points and ending with the root.

In many applications, it is useful to work with the *network* corresponding to a given tree. The network is obtained by removing the root and disregarding the ancestral directionality of the branching arrangement (Fig. 2).

1.3. Evolutionary Hypotheses

Each computer reconstruction algorithm is based, more or less rigorously, on an assumed pattern of evolutionary development, or *evolutionary hypothesis.* Evolutionary hypotheses which have been tested by our group on molecular data include the maximum homology hypothesis (Red King hypothesis), the uniform rate hypothesis, and the uniform convergence (uniform homoplasy) hypothesis. The *maximum homology hypothesis* states that the best reconstruction

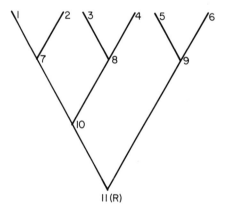

Fig. 1. A tree is a branching arrangement consisting of points connected by lines, with a single most ancestral point, or root (*R*). This tree has six exterior points (1–6), four interior points (7–10), and one root (11).

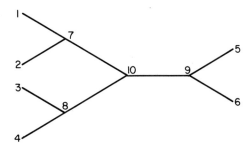

Fig. 2. This network is obtained by removing the root and disregarding the ancestral directionality of the branching arrangement in Fig. 1. There are six exterior points (1–6) and four interior points (7–10) in this network.

of ancestors and the evolutionary tree is that which maximizes identity at homologous positions on the amino acid sequence due to common ancestry. Maximizing identity due to common ancestry is equivalent to minimizing mutational steps, the so-called minimum evolution (Edwards and Cavalli-Sforza, 1963) or maximum parsimony approach (Farris, 1970). Empirically, we find that the most successful way to analyze molecular data is by the maximum parsimony approach. Later in this chapter we discuss a mathematical proof linking this approach to a computer algorithm introduced by Fitch (1971). The *Camin-Sokal (1965) hypothesis* is a special case of the maximum homology hypothesis. As this hypothesis does not permit back mutations to prior character states, it is unrealistic with codon sequence data.

The *uniform rate hypothesis* is the hypothesis that any two lines of descent in an evolutionary tree diverge at a constant rate with respect to one another. One consequence of this is that more ancestrally separated lines have a greater separation of mutational steps. This paradigm permits use of the unweighted pair-group (Sokal and Michener, 1958), complete linkage (Sørenson, 1948), and single linkage (Sneath, 1957) algorithms, although the first of these is most stable in the face of minor variations in the data (Moore, 1971).

The *uniform convergence hypothesis* assumes that events of convergence are distributed in a sufficiently uniform manner throughout the tree that they do not falsely obscure any real, regional differences in rates of evolutionary change. One consequence of this is the additive hypothesis (Cavalli-Sforza and Edwards, 1967), which states that there is a value corresponding to each link on the tree, and the distance between each pair of contemporary species is directly proportional to the sum of the values for the links connecting that pair of species. The Moore *et al.* (1973*a*) algorithm and the Wagner tree algorithm (Farris, 1970) are examples of numerous algorithms which employ the uniform convergence hypothesis.

1.4. Maximum Parsimony Principle

According to the *maximum parsimony principle* (Red King hypothesis), the best reconstruction of ancestral mRNA sequences is that which requires the fewest nucleotide changes (or mutations) in the evolutionary tree to account for contemporary amino acid sequences. In this context, the *little maximum parsimony problem* is the problem of how to reconstruct mRNA sequence ancestors, given a prior knowledge of the contemporary amino acid sequences and the structure or topology of the evolutionary tree. The *big maximum parsimony problem* is the problem of ancestor reconstruction from contemporary sequences without knowledge of the tree topology; i.e., the tree topology of minimum nucleotide difference must be found as part of the solution procedure. The little problem was solved by Fitch in 1971, and proofs were discovered and published independently by Hartigan (1973), Sankoff (1973), and Moore *et al.* (1973). The big problem belongs to a class of very difficult problems from graph theory known as "nondeterministic polynomial complete" (NPC) problems (Kolata, 1974). Other NPC problems include circuit board optimization problems, minimal telephone wire problems, and a problem related to the famous four-color map problem in mathematics. It is safe to say that a definitive solution to the big maximum parsimony problem is not forthcoming in the near future.

In this chapter, I shall discuss the thought processes which were used in proving the solution to the little maximum parsimony problem. First, we shall review the Fitch algorithm as it applies to the reconstruction of amino acid ancestors. Then we shall discuss general proof techniques—the *method of contradiction* and the *method of induction*. We shall develop a specific tool known as the *method of partial sums*. Finally, we shall show how performance of the Fitch procedure guarantees the retention of all possible maximum parsimony ancestors.

2. The Fitch Algorithm

2.1. Fitch Procedure for Codons

In the Fitch procedure for codons, each contemporary species (exterior point) in the network corresponds to a known amino acid. The algorithm uses the *codons* corresponding to this amino acid, where each codon has been appended with the *accumulation number*, 0. For example (Fig. 3), species 1 with the amino acid tyrosine has two codons: UAC^0 and UAU^0. The accumulation number indicates how much length we have accumulated on the network so far. When we first start the algorithm (at the exterior points), we have not accumulated any length yet.

We construct ancestral codons for a particular interior point (ancestor) in the network (say, point 10, Fig. 4) by working stepwise from the outside inward. Points 1 and 2 are used to construct point 7, points 3 and 4 to construct point 8, ..., and finally points 7, 8, and 9 to construct point 10 (Fig. 4).

A pair of points used to construct a new point, say 1 and 2 to construct 7 (Fig. 3), are combined according to the *union rule* for codons. To combine

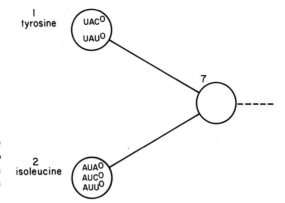

Fig. 3. The Fitch algorithm uses the codons corresponding to each amino acid, where each codon has been appended with accumulation number, 0.

bubble 1 and bubble 2 to obtain the contents of bubble 7, we procceed as follows: Take *any member of bubble 1* (say UAC^0) and *any member of bubble 2* (say AUA^0) and produce all codons of the form

$$\begin{pmatrix} U \\ A \end{pmatrix} \begin{pmatrix} A \\ U \end{pmatrix} \begin{pmatrix} C \\ A \end{pmatrix}$$

i.e., with the first nucleotide U or A, second nucleotide A or U, and third nucleotide C or A. These are AAA, AAC, AUA, AUC, UAA, UAC, UUA, and UUC. The accumulation number for codon AAA in bubble 7, for example, is

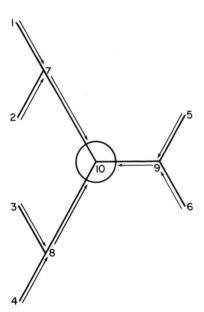

Fig. 4. Stepwise solution for the contents of point 10 by the Fitch procedure.

the amount of length accumulated on the network so far. Codon UAC has accumulated 0 length up to bubble 1 (an exterior point), but the distance from UAC to AAA is an additional two mutations. Codon AUA^0 has accumulated 0 length up to bubble 2, but the distance from AUA to AAA is an additional one mutation. Therefore, codon AAA has accumulated a total of $0 + 0 + 2 + 1 = 3$ mutations by the time we reach bubble 7. The other codons in bubble 7 are calculated in a similar manner, using all combinations of codons, one codon from bubble 1 and one codon from bubble 2. In such calculations, the same codon is sometimes found more than once for a bubble; then only the codon of least accumulation number is saved.

2.2. Worked-Out Example

Let me warn the reader that performing this calculation to completion by hand is quite tedious. For anyone bold enough to try, it is easier if one uses a smaller, less complicated "genetic code." The following is intended purely for illustrative purposes. All of our research calculations are of course performed with the true genetic code. Consider a hypothetical genetic code with three nucleotides (A, B, C), two-letter codons, and five amino acids (V, W, X, Y, Z), according to the scheme

		Second letter		
		A	B	C
	A	W	X	V
First letter	B	W	Z	Y
	C	Z	X	V

Further suppose we have the network in Fig. 4 with input data (exterior amino acids) as illustrated in Fig. 5. To obtain the contents of bubble 9, for example, we must consider every possible pair of codons such that one member of the pair belongs to W and the other to Z. There are four such pairs:

$$AA^0 \times BB^0$$
$$AA^0 \times CA^0$$
$$BA^0 \times BB^0$$
$$BA^0 \times CA^0$$

Each pair is expanded by the union rule. For example, the pair $AA^0 \times BB^0$ has a choice of A or B in the first codon and A or B in the second codon. The set of all codons which can be constructed in this fashion is

$$AA$$
$$AB$$
$$BA$$
$$BB$$

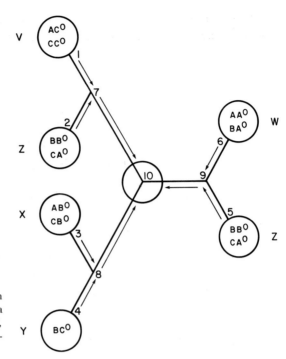

Fig. 5. The network of Fig. 4. Each exterior point (1–6) is assigned a hypothetical amino acid (V, W, X, Y, Z) belonging to our simplified "genetic code."

The accumulation number for AB, for example, includes 0 from contributory codon AA^0, 0 from contributory codon BB^0, 1 as the distance from AA^0 to AB, and 1 as the distance from BB^0 to AB. The total is $0 + 0 + 1 + 1 = 2$. Therefore, bubble 9, prior to revision (see below), contains

$$AA^2$$
$$AB^2$$
$$BA^2$$
$$BB^2$$

For the pair $AA^0 \times CA^0$, we obtain

$$AA^1$$
$$CA^1$$

For $BA^0 \times BB^0$,

$$BA^1$$
$$BB^1$$

For $BA^0 \times CA^0$,

$$BA^1$$
$$CA^1$$

Finally, repeat codons are discarded (only a codon with least accumulation number is saved). The revised bubble 9 thus contains

$$AA^1$$
$$AB^2$$
$$BA^1$$
$$BB^1$$
$$CA^1$$

The remaining revised bubbles are as follows:

Bubble 7: AA^2, AB^2, AC^2, BB^2, BC^2, CA^1, CB^2, CC^1
Bubble 8: AB^2, AC^2, BB^2, BC^2, CB^2, CC^2
Bubble 10: AA^6, AB^6, AC^6, BA^7, BB^5, BC^6, CA^5, CB^6, CC^5

The contents of bubble 10 (end point of the calculation) tell us that the minimal length for the network is 5 (= minimal accumulation number in bubble 10), and every parsimonious solution for this network must contain BB or CA or CC at point 10.

To solve for point 9, we must start over again, this time using bubble 9 as the end point of our calculation (Fig. 6). Actually, we do not have to start from

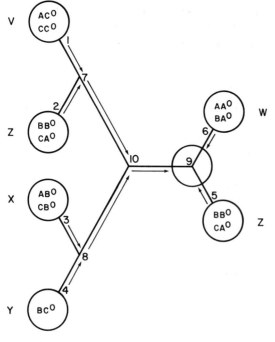

Fig. 6. The network of Fig. 2, containing the input data from Fig. 5. This time, bubble 9 is used as the end point of calculation.

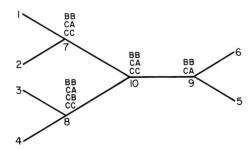

Fig. 7. Possible codon solutions at each of the
interior points (7–10).

scratch, because bubbles 7 and 8 are calculated the same way as before. This
time we have

> Bubble 7: AA^2, AB^2, AC^2, BB^2, BC^2, CA^1, CB^2, CC^1
> Bubble 8: AB^2, AC^2, BB^2, BC^2, CB^2, CC^2
> Bubble 10: AA^6, AB^4, AC^4, BA^6, BB^4, BC^4, CA^6, CB^4, CC^3
> Bubble 9: AA^6, AB^6, AC^7, BA^6, BB^5, BC^6, CA^5, CB^7, CC^6

Therefore, the minimal length for the network is 5 (no surprise) and every
parsimonious solution for this network must contain BB or CA at point 9.
    Using bubble 8 as end point, we obtain

> Bubble 7: AA^2, AB^2, AC^2, BA^2, BB^2, BC^2, CA^2, CB^2, CC^2
> Bubble 9: AA^1, AB^2, BA^1, BB^1, CA^1
> Bubble 10: AA^3, AB^4, AC^4, BA^3, BB^3, BC^4, CA^2, CB^4, CC^3
> Bubble 8: AA^6, AB^6, AC^6, BA^6, BB^5, BC^6, CA^5, CB^5, CC^5

Using bubble 7 as end point, we obtain

> Bubble 8: AB^2, AC^2, BB^2, BC^2, CB^2, CC^2
> Bubble 9: AA^1, AB^2, BA^1, BB^1, CA^1
> Bubble 10: AA^4, AB^4, AC^4, BA^4, BB^3, BC^4, CA^4, CB^4, CC^4
> Bubble 7: AA^6, AB^6, AC^6, BA^7, BB^5, BC^6, CA^5, CB^6, CC^5

Figure 7 shows the possible codons at each of the interior points. Figure 8
shows the collection of all parsimonious solutions (five all told).

3. Mathematical Methods

3.1. Method of Contradiction

    The *method of contradiction* is a very basic approach in any discussion of
mathematical proof. If we wish to prove that a statement S is true, then we

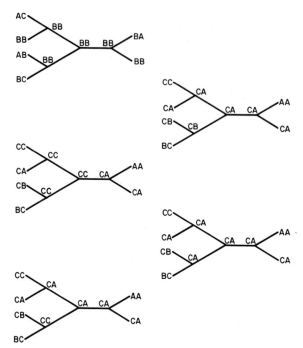

Fig. 8. The five parsimonious solutions for the problem in Fig. 5.

consider the exact negation of statement S, namely "not-S." If we can show that "not-S" leads to a contradiction, then "not-S" cannot be true, and by elimination S must be true. This completes the proof of statement S. Recall that a prime number is any number divisible without a remainder only by itself or by 1. Suppose we wish to prove the statement "Some odd numbers are not prime." The *exact negation* of this statement is "All odd numbers are prime." This would mean that 1 is prime (true), 3 is prime (true), 5 is prime (true), 7 is prime (true), 9 is prime (false),. . . . Since the negation of the statement is false, the statement itself must be true.

3.2. Method of Induction

In the *method of induction*, the statements to be proved are arranged *in order:* the 1st, the 2nd, the 3rd, . . . , the nth, etc. First, you prove that the first statement is true. Second, you prove that *if* the kth statement is true *then* the $(k+1)$th statement is true. Since the 1st statement is true (you proved it), the 2nd must be true. Since the 2nd is true, therefore the 3rd must be, and thus every statement in the ordering is true. For example, suppose we wish to prove that the sum of all the whole numbers from 1 to n equals $[(n+1)(n)]/2$. That is, we wish to prove that

$$1 + 2 + 3 + \cdots + n = \frac{(n + 1)(n)}{2}$$

This proposition is certainly true for the first few n that we try out.

For $n = 1$,
$$1 = \frac{(1 + 1)(1)}{2}$$

For $n = 2$,
$$1 + 2 = \frac{(2 + 1)(2)}{2}$$

For $n = 3$,
$$1 + 2 + 3 = \frac{(3 + 1)(3)}{2}$$

Suppose the proposition is true for $n = k$, that is, suppose for $n = k$,

$$1 + \cdots + k = \frac{(k + 1)(k)}{2}$$

We wish to show the proposition is true for $n = k + 1$. This is a simple algebraic maneuver:

$$[1 + \cdots + (k + 1)] = [(1 + \cdots + k) + (k + 1)]$$
$$= \frac{(k + 1)(1)}{2} + (k + 1)$$
$$= \frac{(k + 2)(k + 1)}{2}$$

3.3. Contradiction and Induction

We can combine the contradiction and induction approaches for a collection of statements we wish to show are true. The statements are arranged in order: the 1st, the 2nd, ..., the nth, etc. Suppose that one of the statements in this collection is false. Pick the *first* such false statement in the collection. State the *exact negation* of that statement, and show that this negation leads to a contradiction. Since the negation is false, the statement is true, and it could not be the first false statement in the collection. Therefore, all statements in the collection are true. Paraphrased, the method of contradiction-induction relies on the principle that if you are going to make a mistake, you have to make it the first time. If you do not make it the first time, then you never make it.

Our proof of the Fitch theorem employs this device. On a given network, all points are numbered in ascending order from the outside inward (Fig. 4). The outside points (contemporary species) must have the right codon choices, because these codons are given data. The Fitch procedure shows how to construct ancestral codon choices from the outside inward in a stepwise fashion. If the Fitch procedure is in error, there must be a first ancestor (interior) point at which it drops a required codon. Let us call this point z. We examine this first ancestor, z, and proceed to demonstrate that the Fitch procedure did *not* drop a required codon. Therefore, the Fitch procedure never drops a required codon.

3.4. *Method of Partial Sums*

The *method of partial sums* employs the simple fact that if $a \geq d$, $b \geq e$, and $c \geq f$, then $a + b + c \geq d + e + f$. In examining the ancestral point, z, on the evolutionary network (Fig. 9), we are always examining a point with three neighbors, say w, x, and y. We are trying to demonstrate that the Fitch procedure never drops a codon which belongs to a solution of least length, L. But since point z radiates off in three directions, it is fair to separate length L into three components, $L = L_w + L_x + L_y$. For a solution obtained by the Fitch procedure, append the superscript F: $L^F = L^F_w + L^F_x + L^F_y$. For a solution obtained by any other means, append the superscript G: $L^G = L^G_w + L^G_x + L^G_y$. If a solution obtained by the Fitch procedure is indeed parsimonious, then $L^G \geq L^F$. The method of partial sums assures us that if each component of length is minimized by the Fitch procedure, i.e., if $L^G_w \geq L^F_w$, $L^G_x \geq L^F_x$, and $L^G_y \geq L^F_y$, then $L^G \geq L^F$. In fact, it suffices to examine a single direction, because we would examine the other directions in an analogous fashion. Arbitrarily, let us choose to examine direction x.

Remember, the objective of our algorithm is to find all codons for bubble z which would appear in at least one parsimonious solution. If we can find any codon which belongs to a parsimonious solution but which was dropped by the Fitch algorithm, then we are in trouble. Suppose we have such a codon. Then the length which that codon generates for the network must be smaller than the length generated by any Fitch codon. Partial sums assures us that the length must be smaller in either direction x, direction w, or direction y. Suppose this non-Fitch codon generates a smaller length in the x direction. (The proof for the w or z directions would be the same.)

In examination of direction x, we can consider nucleotide positions 1, 2, and 3, one by one. Again by the method of partial sums, if $L^G_{x1} \geq L^F_{x1}$, $L^G_{x2} \geq$

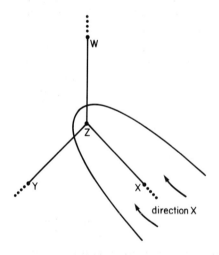

Fig. 9. Each ancestral (interior) point z always has three neighbors: w, x, and y. They radiate out from point z in three directions, which form three distinct components of the network.

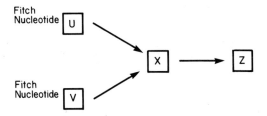

Fig. 10. Suppose it is possible to find any nucleotide at point z and a non-Fitch nucleotide at point x with a shorter overall distance than a Fitch nucleotide at point x.

L^F_{x2}, and $L^G_{x3} \geq L^F_{x3}$, then $L^G_x \geq L^F_x$ (because $L^G_x = L^G_{x1} + L^G_{x2} + L^G_{x3}$ and $L^F_x = L^F_{x1} + L^F_{x2} + L^F_{x3}$). It suffices to examine a single position, j, because the proof for the other two positions would be analogous.

3.5. Telescoping the Problem

Let us take stock of what we have done so far. We started with an algorithm which works from the outermost points of a network stepwise to the innermost points. Using the methods of contradiction and induction, we agree that we need only to examine the first network point, z, on our inward sweep where we drop a required codon. In examining point z, we need only to look at one of the three directions radiating from point z, namely direction x. In examining direction x, we need only to look at the jth nucleotide position. If we can show that $L^G_{xj} \geq L^F_{xj}$, then we have not dropped a required codon at this step, and the Fitch procedure is vindicated. Observe that we have taken a relatively large problem (i.e., proof of the Fitch algorithm for an entire network) and *telescoped* it down to the far more manageable problem at a single point (i.e., point z) arising from a single direction (i.e., direction x) with respect to a single nucleotide position (i.e., position j). This telescoped region of the network is called a *nexus*.

3.6. Solution of the Nexus Problem

Suppose that the devil proposes a nucleotide (*any* nucleotide) at position j of point z which belongs to some parsimonious solution, but which has been dropped by the Fitch algorithm. Furthermore, the devil informs us that the parsimony criterion is violated for the first time at point x, coming from direction x (i.e., via points u and v, Fig. 10). This means that points u and v, which we have constructed by the Fitch procedure have *not* yet violated parsimony. There are two alternatives: either $u = v$ or $u \neq v$. If $u = v$, then let $u = v = A$. If $u \neq v$, then let $u = A$ and $u = B$ (Fig. 11). Remember, z could be anything: it could be equal to u, equal to v, or unequal to both. Furthermore, the devil knows what z is, but we do not because we are constructing *stepwise from the x direction*, and we have not reached point z yet. Obviously, we cannot

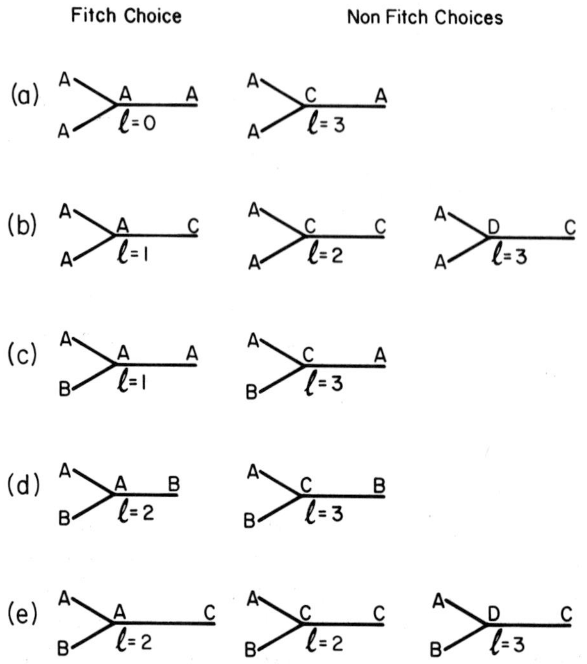

Fitch Choice   Non Fitch Choices

Fig. 11. For every possible
nexus, a Fitch nucleotide
choice at point x results in at
least as short a length as any
nonFitch choice at point x.
(a,b) All possible z choices for
$u = v = A$ and Fitch $x = A$.
(c,d,e) All possible choices for
$u = A$, $v = B$, and Fitch $x = A$.
The proof for $u = A$, $v = B$,
and Fitch $x = B$ is analogous to
(c), (d), and (e).

use any knowledge of z to help us in constructing the contents of point x. We shall prove that, no matter what nucleotide the devil chose at point z position j, the Fitch procedure will not fail to find a parsimonious nucleotide for point x.

By the *union rule*, if $u = v = A$, then $x = A$, but if $u = A$ and $v = B$, then x can equal either A or B. Figure 11 demonstrates that for every possible nexus a Fitch nucleotide choice at point x results in at least as short a length as any non-Fitch choice at point x. Figure 11(a,b) shows all possible z choices for $u = v = A$ and Fitch $x = A$. Figure 11(c,d,e) shows all possible z choices for $u = A$, $v = B$, and Fitch $x = A$. The proof for $u = A$, $v = B$, and Fitch $x = B$ is analogous to Fig. 11(c,d,e).

4. Other Parsimony Procedures

4.1. Branch Swapping

Since there is no known method for solving the big maximum parsimony problem (i.e., finding the branching arrangement of least length) short of brute force examination of all networks, investigators often resort to "local" search procedures. For a network containing three exterior points, called a 3-network, there is only one possible branching arrangement. A 4-network has three possible branching arrangements. A 5-network has 15, a 6-network has 105, a 7-network has 945, an 8-network has 10,395, and a 20-network has over 10^{20} arrangements. If you could solve networks at 1 per millisecond, the 20-network problem would consume almost 10^{10} years.

Our approach is a "branch swap" procedure which performs all possible "nearest-neighbor single-step changes" (NNSSCs) on a given topology. The NNSSC topology which maximally decreases the mutation count of the initial topology is then used as the starting point for a new round of NNSSC trials. When a complete round of NNSSC trials fails to discover any new topologies of lower mutation count, then the procedure is terminated. Execution of this iterative procedure to a termination point in no way guarantees that an overall ("global") minimum has been reached; the result is better thought of as a "local minimum," i.e., a local depression in a landscape of topologies whose deepest valley may still be undiscovered. The uncertainties (and expense) of our calculation would multiply if we demanded an error estimate as well. In our opinion, the small gain in information would not justify the outlay of computer time.

In the branch swap procedure, we begin with any link EF within the tree topology whose end points E and F (Fig. 12) are *interior points* (i.e., hypothetical ancestors, not contemporary species). The two branches leading into point E are labeled A and B; the two branches leading to point F are labeled C and D. In branch swap alternate I, branches B and C are exchanged; in branch swap alternate II, branches B and D are exchanged. No other swaps are possible with respect to link EF. In a complete round of branch swaps, all possible links EF are subjected to all possible swaps. In a topology containing n contemporary species, $2(n-3)$ new topologies are examined.

The cost of branch swapping is relatively cheap in comparison to the

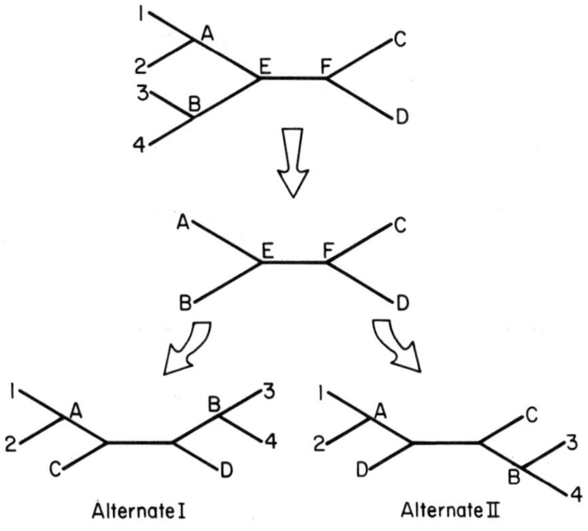

Fig. 12. In a hypothetical tree, consider any link EF whose end points E and F are interior points. In alternate I, branches B and C are exchanged; in alternate II, branches B and D are exchanged. All possible links EF are subjected to all possible swaps.

number of solutions examined. In a tree with n contemporary species and m nontrivial amino acid sequence positions, the cost for a single length calculation is roughly proportional to $n \times m$, depending on the ambiguities inherent in each position. (For example, positions containing numerous Leu, Arg, Ser residues, with an ambiguity of six codons apiece, are relatively more expensive.) A complete round of branch swaps which explores an additional $2(n - 3)$ topologies costs somewhat more than twice as much as the initial length calculation. The trick is as follows. In obtaining the length for the initial tree, it suffices to examine the lowest accumulation number attached to any codon in the final bubble of the stepwise calculation. Each step (and thus each unit of cost) leading up to the final bubble is represented by a single arrow over each link in Fig. 4. The arrow always points toward the final bubble. If we wished to evaluate *every* interior point in the topology, we would merely require that over each link we have calculated an arrow going in both directions. This costs twice as much as the calculations for a single bubble. Since every final bubble has now been evaluated separately, we can lift out any network as in Fig. 13(a) with the assurance that the calculation has been completed up to points A, B, C, and D. It is relatively trivial to complete the calculation for the alternate topologies illustrated in Fig. 13(b,c).

4.2. Alignments

In examining amino acid sequences spanning a broad range of evolutionary diversity, it is often apparent that a gap is present in one group of sequences which is filled up with amino acids in the remaining sequences (Fig. 14). After positioning certain portions of the protein, it is often desirable to position the remainder of the gap consistent with the parsimony principle. In our work, we

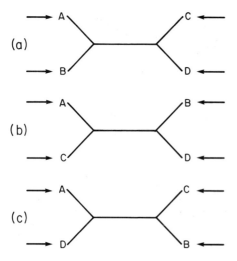

(a)

(b)

(c)

Fig. 13. (a) If an arrow going in both directions has been calculated for every link in the initial topology, then any subnetwork lifted out of the topology has been evaluated up to points A, B, C, and D. The remaining calculations for subnetworks (b) and (c) are thus easy to complete.

Fig. 14. A single gap of two codons on a prototype sequence of six codons can be positioned in five ways.

examine all possible gap positions of the *ancestral* mRNA with the gap against the *ancestral* mRNA without the gap. The result is that we can choose a parsimonious gap position consistent with the given evolutionary branching arrangement. Our current approach employs a sequence of back-and-forth maneuvers—first optimize the network by branch swapping, then optimize the gap, then reoptimize the network, etc. As before, no foolproof search algorithm is known for this problem, short of brute force.

5. Assumptions in Tree Construction

5.1. Unweighted Pair Group

It is important to realize that different methods of tree construction do not necessarily lead to the same result with a given set of data. The fact that several methods currently popular in molecular studies of the primates—e.g., the unweighted pair-group method, the "additive" approach (Sarich and Cronin, this volume), and the maximum parsimony approach—often yield comparable results suggests an unexpected internal consistency in these data. When one considers the highly divergent assumptions on which these methods are based, occasional discrepancies are hardly surprising.

The unweighted pair-group method of Sokal and Michener (1958) requires that evolutionary rates be more or less constant throughout the tree.

①	②	③	④	⑤	⑥
A	C	A	A	A	A
C	A	A	A	A	A
C	A	A	A	A	A
C	A	A	A	A	A
A	A	C	A	A	A
A	A	A	C	A	A
A	A	A	C	A	A
A	A	A	C	A	A
A	A	A	C	A	A
A	A	C	C	A	C

Fig. 15. Hypothetical mRNA sequences for unweighted pair group calculations in subsequent figures. Sequences 5 and 6 are parsimonious ancestors for Fig. 17(b).

Among the several popular pair-group methods, the unweighted pair-group method can probably best tolerate minor deviations from the uniform rate hypothesis (Moore, 1971). However, large discrepancies in evolutionary rate are not tolerated. For example, if we begin with the nucleotide sequences in Fig. 15, convert them to matrix form (Fig. 16), and solve for the unweighted pair group tree (Fig. 17a), the result we obtain does not agree with the maximum parsimony tree (Fig. 17b).

5.2. Additive Hypothesis

Sarich and Cronin (this volume) make the point that two taxonomic groups within an order (lemurs and hominoids within primates) which are roughly equally distant from a known outside species (cow) have therefore diverged at an equal rate from the common ancestor of primates and cow, subject to certain tests of internal consistency within the data. It is important to make the point

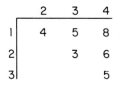

	2	3	4
1	4	5	8
2		3	6
3			5

Fig. 16. Matrix of differences for sequences in Fig. 15.

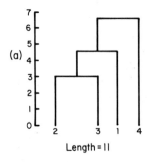

(a) Length = 11

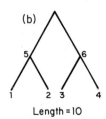

(b) Length = 10

Fig. 17. (a) Unweighted pair-group solution for the matrix in Fig. 16. (b) Parsimonious solution for sequences in Fig. 15. Sequence ancestors are given in Fig. 15.

	w	x	y	z
v	10	11	11	11
w		1	3	3
x			2	2
y				2

Fig. 18. Matrix of observed direct distances for the sequences in Fig. 19. When the actual distances are calculated through the most parsimonious ancestors, it is seen that the direct distances are not consistent with the additive hypothesis.

that this claim, plus consistency tests, is meaningful only when the so-called additive hypothesis of Cavalli-Sforza and Edwards (1967) is satisfied. The additive hypothesis states that the distance between two contemporary species equals or is directly proportional to the sum of distances over the links connecting the species. This follows from the assumption of "random homoplasy," or events of convergence distributed proportionately throughout the tree. If convergence is *not* so distributed throughout the tree, then Sarich and Cronin's analysis no longer holds. We can illustrate this with an example. Starting with the matrix in Fig. 18, in which species v is taken as the "outside species," it appears that species x, y, and z have each diverged to the same extent from species v and that species w has diverged to almost the same extent. Figure 19 shows a set of sequence data which conform to this matrix. However, observe that when the maximum parsimony tree is found for these data the actual distances over the tree do not parallel the observed, direct distances. Evolutionary rates have markedly decelerated in the lineages to species w and x. Admittedly, the uniform convergence hypothesis is not satisfied for this hypothetical data set, and the data were intentionally constructed with certain pathological properties. Whether nature is equally capable of such pathologies is unanswerable in our present state of ignorance. The purpose of this exercise is to show that Sarich and Cronin's analysis is only as good as their assumptions, and that

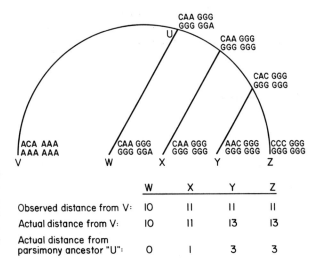

Fig. 19. In the maximum parsimony solution consistent with the matrix of Fig. 18, observed and actual distances do not match. Species w and x appear to have "decelerated."

	W	X	Y	Z
Observed distance from V:	10	11	11	11
Actual distance from V:	10	11	13	13
Actual distance from parsimony ancestor "U":	0	1	3	3

data sets in which there is inequality of convergence sufficient to negate the premises of the analysis are not inconceivable.

If some lineages indeed accumulate more mutations in a fixed period of time than other lineages, confusion could result when an inappropriate analysis is employed. We feel that the Red King hypothesis makes the best use of existing information. The additive hypothesis may inflate some evolutionary rates and underestimate others. We call this the mushroom fallacy, in honor of the magic mushroom from *Alice's Adventures in Wonderland:*

> In a minute or two the Caterpillar took the hookah out of its mouth and yawned once or twice, and shook itself. Then it got down off the mushroom, and crawled away into the grass, merely remarking as it went, "One side will make you grow taller, and the other side will make you grow shorter."

6. Summary

The maximum parsimony hypothesis is gaining increasing importance in studying the protein evolution of animal groups such as the primates. According to maximum parsimony, the best reconstruction of ancestral messenger RNA sequences is that which requires the fewest nucleotide substitutions in hypothetical ancestral sequences to account for contemporary amino acid sequences. The highlights of the proof for the current computer algorithm for maximum parsimony ancestral reconstruction have been reviewed herein.

ACKNOWLEDGMENT

I wish to thank Dr. Morris Goodman for constructive criticism of the manuscript and Miss Elaine Krobock for secretarial assistance. This work was supported in part by NSF grant GB36157.

7. References

Camin, J. H., and Sokal, R. R., 1965, A method for deducing branching sequences in phylogeny, *Evolution* **19**:311–326.
Cavalli-Sforza, L. L., and Edwards, A. W. F., 1967, Phylogenetic analysis: Models and estimation procedures, *Evolution* **21**:550–570.
Edwards, A. F. W., and Cavalli-Sforza, L. L., 1963, The reconstruction of evolution, *Ann. Hum. Genet.* **27**:104–105.
Farris, J. S., 1970, Methods for computing Wagner trees, *Syst. Zool.* **19**:83–92.
Farris, J. S., 1973, Probability model for inferring evolutionary trees, *Syst. Zool.* **22**:250–256.
Felsenstein, J., 1973, Maximum likelihood estimation of evolutionary trees from continuous characters, *Am. J. Hum. Genet.* **25**:471–492.

Fitch, W. M., 1971, Toward defining the course of evolution: Minimum change for a specific tree topology, *Syst. Zool.* **20:**406–416.

Hartigan, J. A., 1973, Minimum mutation fits to a given tree, *Biometrics* **29:**53–65.

Kolata, G. B., 1974, Analysis of algorithms: Coping with hard problems, *Science* **186:**520–521.

Moore, G. W., 1971, A mathematical model for the construction of cladograms, *Institute of Statistics Mimeograph Series. No. 731,* North Carolina State University, Raleigh, N.C.

Moore, G. W., Goodman, M., and Barnabas, J., 1973a, An iterative approach from the standpoint of the additive hypothesis to the dendrogram problem posed by molecular data sets, *J. Theor. Biol.* **38:**423–457.

Moore, G. W., Barnabas, J., and Goodman, M., 1973b, A method for constructing maximum parsimony ancestral amino acid sequences on a given network, *J. Theor. Biol.* **38:**459–485.

Sankoff, D., 1973, *Publication Centre de Recherches Mathematiques Technical Report No. 262,* University of Montreal, Montreal.

Sneath, P. H. A., 1957, The application of computers to taxonomy, *J.Gen. Microbiol.* **17:**201–226.

Sokal, R. R., and Michener, C. D., 1958, A statistical method for evaluating systematic relationships, *Univ. Kan. Sci. Bull.* **38:**1409–1438.

Sørenson, T., 1948, Method of establishing groups of equal amplitude in plant sociology based on similarity of species content and its application to analyses of the vegetation on Danish commons, *Biol. Skr.* **5:**1–34.

Van Valen, L., 1974, Molecular evolution as predicted by natural selection, *J. Mol. Evol.* **3:**89–101.

Primate Phylogeny and the Molecular Clock Controversy

III

Molecular Systematics of the Primates

8

VINCENT M. SARICH and JOHN E. CRONIN

1. Introduction

A large body of comparative macromolecular data bearing on primate systematics is now available. There does not exist, however, any consensus as to just what that bearing is or what it should be. We hope that this chapter can contribute toward such a consensus.

Protein and nucleic acid data have a limited but vital role in the development of an understanding of the systematics of a group. They have the potential of providing, independent of any other information, the cladistic and temporal dimensions of the phylogeny of that group. We do not think it overstates the case to argue that until we have the cladistics we, for all practical purposes, really have nothing at all. The knowledge of the cladistics is the necessary framework upon which we can properly assess the meaning of the evidence gleaned from other areas of research. Once the cladistics have been worked out, and if there is a significant body of fossil data to integrate into our understanding of the group, we can attempt to provide a time estimate for each node or branch point. Just how much the molecules can contribute to dating is, of course, a most controversial matter and one which will be a major concern of ours here.

On the lineages so defined are then placed such anatomical, behavioral, physiological, and molecular events as we can infer to have occurred along them. Finally, we attempt to answer the basic evolutionary questions of the selective "whys" and "hows" for each of these events. Thus the cladistic and

VINCENT M. SARICH and JOHN E. CRONIN • Departments of Anthropology and Biochemistry, University of California, Berkeley, California 94707.

temporal dimensions of the phylogeny become the framework upon which an understanding of what happened within it is to be structured.

We believe that the potential of proteins and nucleic acids to provide such cladistic and temporal information is inherently greater than that of more traditional sources of comparative data—and, for the primates at least, much of that potential has already been realized. Certainly one could not hold a conference on the molecular systematics of any other vertebrate order. The reason for this claim of molecular superiority in providing information in the limited realm of cladistics and time is the empirical observation that the molecular approach effectively circumvents the major problems present in more traditional approaches. The success of any evolutionary reconstructive effort is in large part determined by its success in discriminating between "primitive" and "advanced" features; that is, in deciding whether a particular character state is ancestral or derived. The problem is that one can neither really decide this about a character state without knowing about the cladistics nor develop the cladistics without deciding about the character states. The inevitable danger of circularity here cannot be ignored, but the fact that there can be only a single history for any group of organisms means that ultimately all the anatomy, behavior, and physiology must be fitted into it through some sort of successive approximation procedure.

These problems are effectively circumvented at the molecular level for several reasons. The first vital feature is the fact that the unit of change at the molecular level is known—a single base pair or amino acid substitution. Then the same molecules are present in different species; thus the differences between the albumins of *Homo* and *Pan* can be measured in precisely the same units as between those of *Canis* and *Felis*. Finally, the analytic rationale of the molecular approach is the observation that nucleic acid and protein evolution produce, in the main, differentiation. In other words, at the protein and nucleic acid levels we are working with systems that evolve in such a manner as to generally approach or approximate the ideal situation of continual divergence. The differences among modern species are then measurable in the same units along a common scale and are patently derived characters. To the extent that these observations are valid, one should be able to simply count the amino acid or nucleotide sequence differences among extant species and apportion these along a unique, derived phylogeny. Conversely, as will be shown, one measure of the quality of our molecular data sets is the ease with which they are so apportionable. Proteins and nucleic acids are finite structures, and thus some parallelisms and convergences must occur, either by chance or by natural selection. To the extent that they are randomly distributed, however, the unique apportionment process ought to produce a cladogram congruent with the actual history of the taxa being studied.

Thus not only should the cladistics be derivable from the molecules, but also those cladistics then come to represent a working hypothesis to be tested in terms of the degree to which they facilitate an understanding of the physiology, behavior, and anatomy. Some few discrepancies among the pictures given by different molecules will inevitably develop; and many more disagreements have already appeared between certain molecular phylogenies and those derived using more traditional approaches. To resolve these issues, it will prove wise to remember that no comparative technique can produce *the* ultimate, actual history. The actual history or phylogeny does exist and is in that sense real. Any comparative data set and the resulting cladogram are also real enough, but can, for various reasons, only approximate the actual history. They are, nonetheless, all that we have to work with today. This same logic applies to any body of contemporary comparative data and, in a slightly altered form, to such information as is available from the fossil record. The fossil record, it should be noted here, is invaluable in telling us what happened and in what sequence; but it has certainly been much less important than contemporary comparative anatomy in providing phylogenies and taxonomies. Thus the desired actual history can have no operational reality save that given it through our efforts at reconstruction. Nor, obviously, can its reality be tested except in terms of the facilitation of our understanding of all the available comparative data.

2. Molecular Approaches to Systematics

2.1. Overview

Currently four techniques are available for macromolecular comparisons: direct amino acid sequencing, electrophoresis, nucleic acid hybridization, and immunology. We now know that within the limits set by their inherent resolving powers the cladistics and relative genetic distances provided within the same group are generally congruent. Thus the choice as to which should be used can be made objectively in terms of the task at hand.

The three "indirect" techniques (electrophoresis, immunology, hybridization) are probably broadly comparable as to cost (time, effort, money) per unit of information (placement of a lineage) obtained. It is also probable that each can be applied without undue strain by a nonbiochemist systematist to a particular problem which interests him. It is certain that amino acid sequencing costs at least a hundred times as much per unit of information obtained, is not at all easily performed by the nonspecialist, and is not going to become routinely applicable in the foreseeable future. Unfortunately, sequences have become some sort of Holy Grail to far too many individuals, with the "indirect" methods either relegated to a second-class status or just ignored. The expression of this sentiment should not be taken to imply that sequences as such are unimportant. To those interested in the processes of molecular evolution, relationships among classes of proteins, and comparative and functional biochemistry in general, they are of fundamental importance. On the other hand, their contribution to systematics generally and to knowledge of primate evolution specifically has been at best supplemental or confirmatory. There is probably not a single case for the primates where sequence data have told us something that we did not already know about the phylogeny involved. Having said this, we must hasten to point out that the psychological impact of sequence data as "direct" evidence is out of all proportion to their intrinsic value—which gets into the difficult area of the difference between knowing something and convincing the rest of the community of interest that you know it.

2.2. Immunology

We, of course, along with Goodman and his collaborators, have concentrated on immunology. This is admittedly an "indirect" approach, and one of the major objective problems delaying full utilization of immunological data in systematics has been the certain knowledge that each measured immunological distance can only approximate the "actual" sequence difference between the two protein species being measured. The actual sequences are, of course, not known—if they were, there would be no point in doing the immunology. However, one can estimate internally the level of immunological "noise" present and so provide confidence limits on the cladistic conclusions drawn. We emphasize "internal," for we want to avoid the very strong temptation to test our conclusions in terms of what is "known" of the phylogenies we are trying to work out.

The first of these internal tests stems from the observation that while the number of sequence differences between two protein species A and B is fixed, the immunological distance between them measured with antisera to A will generally not be equal to that measured with antisera to B. This lack of agreement we term "nonreciprocity" and feel that it is, in the main, most readily attributable to the fact that the sets of surface differences between the protein of the immunized animal (in our work, usually the rabbit) and those of species A and B will generally vary. That is, the proteins of species A and B, being different from one another, will share different portions of their sequences with the homologous rabbit protein. This being the case, the rabbits will produce populations of antibodies with differing specificities to the two protein species. An illustration of this effect is provided by Reichlin's (1974) immunological study of human hemoglobin A variants. Rabbits were immunized with "normal" hemoglobin A and various mutant hemoglobins. The resulting antisera were then used to see which hemoglobins they could distinguish from the particular hemoglobin used as the antigen. The results followed a logical pattern which could be understood in terms of the amino acid sequences of the three hemoglobins involved (the immunogen A, the rabbit B, and the one to be tested C). Thus if C differed from A at a site where A and B possessed the same amino acid, then A and C reacted equally well with antisera to A. If, on the other hand, A and C differed at a site where B also differed from A, then generally A and C could be distinguished immunologically. We therefore cannot view the rabbit as an unbiased participant in the immunological approach, and it appears highly probable that much of the bias so introduced is seen experimentally as "nonreciprocity" (T. White, thesis, Berkeley, 1976).

We define "nonreciprocity" as

$$\text{percent nonreciprocity} = 100 \left(\frac{\text{anti-}A \text{ with } B - \text{anti-}B \text{ with } A}{\text{anti-}A \text{ with } B + \text{anti-}B \text{ with } A} \right)$$

and find that it is generally in the range of 4–10% for most of our large albumin and transferrin data sets involving, by now, antisera to some 150 albumins and 100 transferrins. If the nonreciprocities were randomly distributed, nothing more could be done and this level of "noise" would simply have to be accepted. It has now become quite clear, however, that there is a significant nonrandom element in the distribution of nonreciprocities, and this suggests that it should be possible to reduce the noise level by the removal of this nonrandom element. Our solution is illustrated with reference to the New World monkey albumin data set, which has the appalling (for albumin) nonreciprocity level of 18% to start with (Table 1). As seen in that table, a straightforward correction procedure reduces the 18% to a much more reasonable 5.7%. We have found that all of our data matrices are improved by this correction procedure to a consistent nonreciprocity level of 4–6%, which can then be accepted as a reasonable estimate of the irreducible noise level associated with the immunological approach. Thus, for use in predicting degrees of sequence difference or in doing cladistic analyses, this random noise level can be taken into account in determining the probable error of any conclusions drawn. The conversion of the data sets into cladograms can then follow.

2.3. Cladistic Analysis

The conversion of an immunological distance data matrix into a cladogram follows directly once the context within which the analysis is to be done is chosen. This is illustrated in Fig. 1, where it is evident that setting A through D as a clade relative to E within an additive context reproduces directly the cladogram with which we began. It should also be noted that no assumptions concerning rate equalities or inequalities are made prior to the analysis; on the contrary, we discern them by means of the analysis.

Clearly, in the real world it is not always possible to choose an outside reference species such as E with any degree of confidence; in that case, the analysis proceeds by finding those taxa which give a consistent apportioning of the measured distance between the pair at issue. We then conclude that those which do are probably cladistically outside this pair. For example, in Fig. 1, C, D, and E each see B as 9 units farther away than A; it then becomes a reasonable working hypothesis that A and B form a clade relative to C, D, and E. This procedure will generate from the data of Fig. 1 the original cladogram, and the only uncertainty will be the placement of the nodes along a lineage

Table 1. Albumin Immunological Distances among the New World Monkeys[a]

Antigens	Antisera								Row sum	
	Ao	AA	Cc	PC	Sm	Ce	Sg	Ca		
Aotus	0	30	26	20	32	29	32	27	196	Raw data
Ateles-Alouatta	35	0	21	26	35	31	47	48	243	
Callicebus	36	35	0	26	42	29	50	41	259	
Pithecia-Cacajao	26	33	18	0	34	24	45	32	212	
Saimiri	40	39	34	41	0	31	60	46	291	
Cebus	37	39	30	31	38	0	59	56	290	
Saguinus	36	26	16	20	27	17	0	29	171	
Callithrix	29	42	22	30	40	37	29	0	229	
Column sum	239	244	167	194	248	198	322	279		
Row: column	0.82	1.00	1.55	1.09	1.17	1.46	0.53	0.82		
Aotus	0	29	38	21	35	39	18	23	203	Corrected matrix
Ateles-Alouatta	30	0	30	27	39	41	26	41	234	
Callicebus	31	34	0	27	46	39	27	35	239	
Pithecia-Cacajao	22	32	26	0	38	32	25	27	202	
Saimiri	34	37	49	43	0	41	33	39	276	
Cebus	31	37	43	32	42	0	32	48	265	
Saguinus	30	25	23	21	30	23	0	25	177	
Callithrix	24	40	32	31	44	49	16	0	236	
Column sum	202	234	241	202	274	264	177	238		

[a]To achieve the corrected matrix, each value in a particular raw data column is multiplied by the corresponding row-to-column ratio starting with the most discrepant—in this case, 0.53 for all the anti-*Saguinus* values. Corrected row sums are then recalculated, a new set of correction factors is calculated from these, the next most discrepant column is corrected using its row-to-column ratio, and the process is repeated until the row and column sums agree as closely as possible.

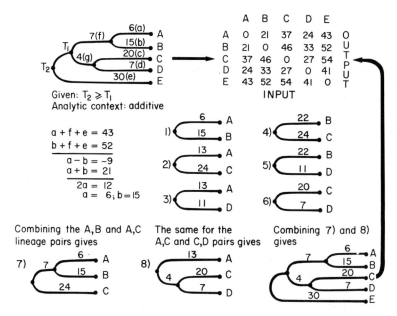

Fig. 1. Cladistic analysis exemplified in an ideal additive context. The analysis is presented in detail to document that the solution achieved is direct, unique, and "correct." This type of effort implies that the reliability of any cladistic solution using real data can be tested in terms of the degree to which they approximate the additive ideal—assuming, of course, that homoplasies are randomly distributed.

leading to E. Thus, in general, a cladistic analysis of an additive data set where an outside reference species cannot be fixed will leave ambiguous the placement of one primary lineage relative to the others. In the example, then, the node from which the lineage leading to E stems can be placed anywhere along the lineages linking nodes A, B and C, D.

The above represents the ideal, and any analysis of an actual data set must take cognizance of the uncertainties introduced by parallelisms, convergences, back mutations, and particularly the already discussed vagaries of immunology. Thus perfect additivity becomes impossible and any cladistic solutions become approximations. Accepting this, criteria for selecting among the many possible solutions need to be chosen. The obvious primary criterion is to minimize the total difference between the output and input data matrices. However, no practical procedure can give us all possible approximate solutions to which this criterion can be applied. This seeming impasse is readily bypassed if we accept the known immunological "noise" level as an indicator of when a solution is "good enough," that is, when the input-output difference decreases to a level compatible with perfect additivity blurred by the degree of immunological "noise" indicated by the reciprocity tests. Although a theoretical analysis of just how much blurring is to be expected has not yet been carried out, we do know that our best current trees show average differences between elements of the input and output data matrices that approximate the nonreciprocity levels associated with the input data. Thus it would appear that the additivity ideal is closely approached after the immunological "noise" is allowed for. These findings, illustrated in the various phylogenies of this chapter, at once strongly support the additivity hypothesis and indicate the excellent resolving power inherent in the use of immunological techniques.

As an example of the approach used, the analysis of two small real data sets involving the Madagascar genera *Lemur, Lepilemur, Propithecus,* and *Avahi* is presented in Fig. 2. Clearly, space limitations and the patience of the reader preclude the presentation of the various analyses leading to the phylogenies in Figs. 3–7, but we have included the input-output differences as indicators of their reliabilities. It should be noted here that the approach is basically that given by Fitch and Margoliash (1967) in their original analysis of the cytochrome *c* data and differs from it in only two important respects. First, we consider lineages with negative numbers along them as unrealistic and do not allow them either in the course of the analysis or in the final cladogram. Second, we do not use all available outside reference species to apportion amounts of change along lineages within a group. For example, the *Homo-Hylobates* albumin immunological distance is 14 units, and this rather small value makes it very unlikely that reference species farther removed than the Old World monkeys (~ 30–35 units) will contribute very much in the way of information as to how that 14 units is to be distributed along the *Homo* and *Hylobates* lineages.

	ALB				Tf			
	L	*Lep*	*P,A*	Rate	*L*	*Lep*	*P,A*	Rate
Lemur	0	47	64	+4	0	70	63	0
Lepilemur	67	0	61	−9	74	0	76	0
Propithecus } *Avahi*	57	57	0	0	70	70	0	+5

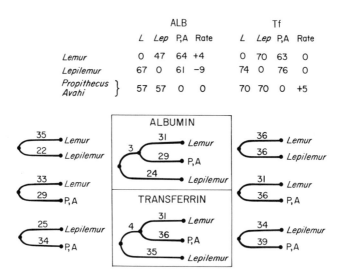

Fig. 2. Cladistic analysis exemplified in a small real data system. Again, the logic is as in Fig. 1. The rate column in each case refers to the relative distances of the four lemur molecules from their anthropoid counterparts. For example, the +4 for *Lemur* albumin means that it is, on the average, 4 immunological distance units farther from the various anthropoid albumins than are the albumins of *Propithecus* and *Avahi*.

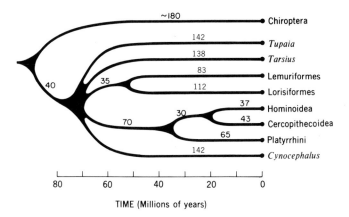

Fig. 3. Albumin plus transferrin phylogeny of Primates and related mammals. The elements of the input and output data half-matrices differ by an average of 3%. The nonreciprocity of the corrected input data matrices is 4%. Note that *Tupaia* and the flying lemur, *Cynocephalus,* are clearly members of the clade which includes the other primates—whatever their final taxonomic allocation might be. The bats are included to indicate the position of the "nonungulate" placental radiation, which also involves the edentates, carnivores, rodents, and insectivores. The calibration of the time scale is discussed in the text.

Thus, once clear species clusters are seen in the data, each is collapsed into a single lineage and one is no longer faced, for example, with trying to fit simultaneously some 15 catarrhine and 8 platyrrhine lineages relative to one another. The analysis of the primate data then proceeds by collapsing the original total data set into six major groups (Anthropoidea, Lemuriformes, Lorisiformes, *Cynocephalus, Tupaia,* and *Tarsius*) and using other placental mammal reference species such as various bats, carnivores, and edentates to place these six relative to one another (Fig. 3). Then the lemurs, lorises, flying lemur, and tree shrews become reference species for placing the catarrhines and platyrrhines, and each of the latter two for placing the major groups in the other (Fig. 3–7). Proceeding in this fashion, we can place, with appreciable internal confidence, almost all extant primate taxa.

We again emphasize that the general question of how closely these cladograms approximate what actually happened cannot be answered directly, that is, by checking it against what actually happened. We do not know what actually happened; if we did, there would be little, if any, point in doing the molecular studies. We are then left with three indirect checks on the reliability of the cladograms. In the first of these, we compare the input data matrices (actual data) with the output matrices (those obtained by summing the distances on the cladograms). The result of that comparison is provided with each cladogram.

Second, we can compare the cladograms given for the same taxa by different molecules. One example of the congruence of the albumin and transferrin pictures is given in Fig. 2, and this pattern exemplifies the rule and not the exception. As yet, no statistically significant example of noncongruence

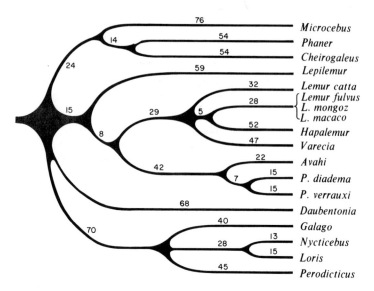

Fig. 4. Albumin plus transferrin phylogeny of Lemuriformes and Lorisiformes. The elements of the input and output data half-matrices differ, on the average, by 3 %. The nonreciprocity of the corrected input matrices is 5 %. It should be noted that the Malagasy forms do not make up a clade relative to the lorises and galagos and thus most probably represent a relict group of African Eocene primates left in Madagascar after the formation of the Mozambique Channel. Note also the first firm positioning of *Daubentonia*, the early divergences of the *Lepilemur* and *Lemur variegatus* lineages, and the close association of *Hapalemur* and the remaining *Lemur* species. A detailed presentation of the data involved and significance of those results will follow (Sarich, Cronin, and Rumpler, in preparation).

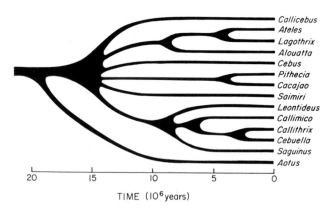

Fig. 5. Albumin plus transferrin phylogeny of the New World monkeys. In this case, the amounts of change along the lineages are not included as the transferrin data are not as extensive as for the albumins. The elements of the input and output data half-matrices differ, on the average, by 3–4 %. The nonreciprocity of the corrected input matrices is 5 %. We feel uneasy as to the placement of *Aotus* outside the clade including all other New World monkeys, but the albumin data in particular require it (Sarich, 1970). Both the albumin and transferrin data indicate that *Callimico* is clearly a callithricid. The complete data sets and extensive discussion are presented in Cronin and Sarich (1975).

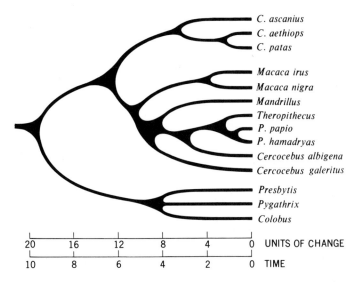

Fig. 6. Albumin plus transferrin phylogeny of the Old World monkeys. The units of change scale indicates the average number of albumin plus transferrin changes which accumulated along that lineage; for example, 8 along an average colobine lineage since the beginning of their radiation. We emphasize that the total number of changes is small relative to the number of lineages involved, and therefore the resolution is somewhat reduced. The detailed cladistics of this complex group, insofar as we have taken them, will be presented elsewhere (Cronin and Sarich, in preparation). In addition, a good deal more remains to be done—in particular, with the genus *Cercopithecus* and the Colobinae. Note especially that *Mandrillus* does not associate closely with *Papio* and that the mangabeys have a dual origin. The *albigena–aterrimus* lineage seems to stem from an early *Papio-Theropithecus* stock, while the other mangabeys are part of an older lineage stemming from the original 42-chromosome group radiation. The elements of the input and output data half-matrices differ by an average of 1–2 units.

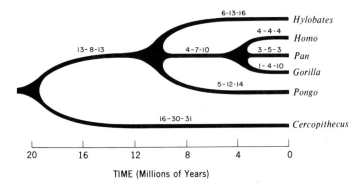

Fig. 7. Albumin, transferrin, and DNA phylogeny of the Hominoidea. The numbers on the lineages are, in order, the number of albumin and transferrin units of change, and the $\Delta T_m \times 10$ for that lineage. In this case, the input and output data matrix elements agree to 1–2 units for the immunological data, and to 0.2 C for the ΔT_m. The placements of the eastern gorilla and the pigmy chimpanzee on the basis of the serum protein electrophoretic data are discussed in the text.

has been seen in comparisons of our primate, carnivore, and marsupial cladograms for the two molecules. We fully expect that further work will turn up
some incongruities, but the level of cladistic accord to date is remarkably high.

Finally we can assess the degree to which these cladograms facilitate our
understanding of the other comparative data. We will deal with the hominoid
aspects of this test after a consideration of molecular clocks.

3. Molecular Clocks and a Time Scale for Primate Evolution

The concept of molecular clocks stems from observations of situations such
as that diagrammed in Fig. 3, where the number of albumin plus transferrin
units of change along the several primate lineages from their origin to the
present fall in a narrow range. Nonetheless, almost everyone who has commented on our applications of the molecular clock concept to studies of primate
evolution seems to be under the impression that we have somehow imposed the
assumption of regularity upon our analysis of the molecular evidence. Thus we
again emphasize that the regularity hypothesis follows from the data, and that
much of the orientation of our work has been specifically to the question of
testing rates of molecular evolution—and not to making *a priori* assumptions
about them. So we now know that while regularity of protein and nucleic acid
change is the rule, our rate analyses have pointed up numerous (in absolute
terms) exceptions. Among the primates, *Aotus* albumin (slow), the common
anthropoid albumin lineage (fast), *Phaner* albumin (fast), *Phaner* transferrin
(slow), and the common lorisiform transferrin lineage (fast) are fairly clear
exceptions to a purely time-dependent model of change. The *Phaner* case
points up an important possible interdependence of rates of albumin and
transferrin evolution which implies that many of the errors in either clock may
well be corrected when the summed data are used. Although this interdependence does not always hold, it has now shown up in far too many cases to be
ignored. For example, among the bats, *Rousettus* has the most changed albumin
and least changed transferrin; *Phaner* shows the same phenomenon for the
primates; *Canis* albumin is fast and the transferrin slow; *Ursus* albumin is slow
and the transferrin fast and the same is true for *Felis;* and, finally, among the
marsupials, *Caluromys* has the least changed albumin and the most changed
transferrin. We emphasize again, however, that one cannot assume that this will
hold in every case (it certainly does not for *Aotus*). The question of whether
change along a particular lineage or in a particular protein is regular or
irregular is not something about which assumptions can be made—it is a
question which needs to be tested directly if further use of the data, be it in
formulating molecular clocks or in considering mechanisms of molecular evolution, is to be made.

The questions of the existence of molecular clocks and of their precision
can only be answered empirically. Their utility is dependent on the ability to
carry out the requisite rate tests and also on the nature of the relationships

being studied. We are particularly fortunate in regard to this latter point with the primates, where the taxonomic diversity gives us a practically continuous range of genetic distance among the existing forms. For example, *Homo* and *Pan* are very similar at the molecular level, as are *Elephas* and *Loxodonta*. However, because *Homo* and *Pan* are members of a diverse group of extant taxa, one can answer directly the question of whether their molecular similarity is due to a very recent common ancestry or a relatively ancient divergence coupled with little change after it. For the elephants, on the other hand, no close relatives exist to provide rate test reference species. Thus, no matter how many molecules might be compared in the two genera, the question is impossible to answer without reference to the fossil record. Such reference in the *Homo-Pan* case has, of course, so thoroughly misled so many for so long as to the significance of the molecular data that great caution should be exercised in the future in this area.

The calculation of specific divergence times among the various primate lineages does not, then, depend on any assumptions as to the existence of albumin, transferrin, or DNA clocks—these must in each case be documented by the appropriate rate tests. Neither does it depend on a demonstration of equivalence of amounts of change along lineages of equal time depth for a particular molecule, nor on a demonstration that all proteins and nucleic acids are evolving in a time-dependent fashion. We have already indicated numerous exceptions in any attempts at such demonstrations. The question, then, is not whether there are exceptions—but whether they are there in sufficient numbers to preclude seeing the rule. From the very large body of available comparative protein and nucleic acid data, there can no longer be much doubt that there is a very strong time dependence in the accumulation of amino acid and nucleotide substitutions. Fitch and Langley calculate (this volume) that the variance of amounts of change along lineages of equal time depth is only twice that expected if the molecular clock kept perfect stochastic time.

Nonetheless, this cannot be assumed to be necessarily true for primate albumins, transferrins, and DNAs. We allocate the amounts of change along the various lineages involved, note which of these are similar enough to be compatible with the clock model, and then use the mean amount of change along those that are as a measure of their age. This gives us relative ages; the calculation of absolute ages requires that the clock be calibrated.

Any such calibration is of course dependent on the availability of well-documented divergence times—and these must derive from judicious considerations of the evidence from the fossil record, zoogeography, the anatomy of living forms, and the cladistic framework within which all of these data sets are to be interpreted. For the mammals, at least, the abundance of data available from all of these areas places marked constraints on the range of dates which might be given to each of a known sequence of divergence events. Thus once we have sets of lineages which show sufficiently similar amounts of change in some protein over the same periods of time (that is, once the presence of a molecular clock is documented) the setting of the clock is not much of a problem.

Our original calibration of the albumin clock, based primarily on the early primate data, set 100 albumin immunological distance units between two taxa equal to a most probable divergence time of 60 million years (m.y.) ago (Sarich, 1968). A large body of later work on carnivores (Sarich, 1969), iguanid lizards (Gorman et al., 1971), ranid frogs (Wallace et al., 1972), and marsupials and hylid frogs (Maxson et al., 1975; Maxson and Wilson, 1975) has produced data supportive of that original calibration. In addition to these published data, we have recently been involved in extensive studies in which we have attempted to place the primates in an overall mammalian context, thereby placing as many constraints as possible on the cladistic framework within which we must then operate. Although much remains to be done in this area, particularly on many of the details of interordinal relationships among the mammals, we now know enough to be able to follow the amounts of change at the albumin locus over a representative sample of lineages encompassing the range of modern mammals. This analysis provides us with a series of divergence events which occur subsequent to the separation of the placental and marsupial lineages and predate the adaptive radiation leading to the modern primates. Thus, if we take the approximately 105 units of change which has occurred along an average mammalian lineage since that marsupial-placental separation, about 55 occurred subsequent to the primate adaptive radiation, 20 between it and that involving bats, rodents, carnivores, edentates, and insectivores, perhaps another 10–15 (here we do not have much reliable data) along the lineage going back to the original placental radiation, with the remainder belonging to the common placental lineage. This, then, is a sample of the context within which we are constrained to operate. It is not so much a matter of choosing some single known date and calculating all others from it (although one good date would make this feasible) but rather of seeing that the dates involved are interdependent. That is, we cannot readily alter one without altering the others—or without making the observed amount of change regularities coincidental. This latter option can of course be exercised only rarely without making a mockery of the rules of the scientific game, but limitations on involving it increase in severity as the complexity of cladistic framework within which we are operating increases. Thus, if one for any reason wishes to question a specific molecular clock date, he cannot avoid considering what altering that date will do to the rest of the picture.

Our dates for the basic divergence events are given in Fig. 3. On the specific issue of the beginning of primate adaptive radiation, we have indicated 58 units of albumin change along an average primate lineage from that time to the present—corresponding to a time of $58 \times 60/50$ or 70 m.y. On this scale the marsupial-placental albumin immunological distance is about 210 units, is set at about 125 m.y. ago (earliest Cretaceous). This accords well with current paleontological opinion on this point (Lillegraven, 1974). Similarly, the indicated dates for the antiquity of the basic placental adaptive radiations—90–100 m.y. ago—are in accord with the growing realization that these radiations must significantly predate the latest Cretaceous or earliest Paleocene where they have usually been placed (see Simons, this volume). Thus we feel that starting the primate radiation at about 70 m.y. ago is reasonable in terms of the basic criteria given above; that is, it fits without strain in the total mammalian picture.

Once this date is set, other dates within the Primates can be calculated using increasing amounts of molecular information as we come to more recent separations. A check of the data in Fig. 3 indicates that, since the beginnings of the primitive adaptive radiation, their various lineages have accumulated very similar amounts of albumin plus transferrin change. Since the amount of change along the common anthropoid lineage (70 units) is about equal to that along the various anthropoid lineages since the catarrhine-platyrrhine divergence, we conclude that divergence must have occurred about one-half as long ago as the beginning of the primate adaptive radiation, that is, about 35 m.y. ago. Hoffstetter (1974) has recently indicated the reasonableness of this figure, although reserving judgment on some of our other intraanthropoid dates. Once this point is fixed, most others within the anthropoids follow directly, especially after one notes the quantitative agreement among the various available intraanthropoid genetic distance estimates (Table 2). The addition of the DNA data is particularly useful here as it substantially increases the effective number of changes seen along each lineage and contributes a new dimension to the comparisons. No significant rate variations involving the major lineages are apparent and the divergence times can thus be calculated as fractions of the catarrhine-platyrrhine time. If the several available systems tell us that the hominoid-cercopithecoid genetic distance is 0.56 ± 0.04 (Table 2) of the catarrhine-platyrrhine distance, and no significant rate differences are apparent along those lineages, then there is little option but to conclude that the origin of the Old World monkey lineage must be seen as having occurred about 35 × (0.56 ± 0.04) or 20 ± 2 m.y. ago. Similarly, an examination of the data in Fig. 6, where the transferrin, albumin, and DNA distances are apportioned along the various lineages involved, tells us that about as much change occurred along the common hominoid lineage as along the hylobatine, pongine, and hominine lineages since the hominoid adaptive radiation began and somewhat

Table 2. Relative Molecular Differences among Various Primate Groups Referred to a Catarrhine-Platyrrhine Distance of 1.0

Groups compared	Albumin[a]	Transferrin[b]	DNA[c]	Mixed immunology[d]
Homo-Pan	0.12	0.13	0.15	0.123
Homo-Pongo-Hylobates	0.25	0.30	0.33	0.293
Hominoidea-Cercopithecoidea	0.58	0.53	0.61	0.533
Catarrhini-Platyrrhini	1.0	1.0	1.0	1.0
Anthropoidea-Prosimii	2.1	1.8	2.6	1.67

[a]From Sarich (1970) and subsequent work in this laboratory. The number is immunological distance between the groups divided by the catarrhine-platyrrhine albumin immunological distance.
[b]Data from this laboratory and expressed as in *a*.
[c]Calculated from the data in Hoyer *et al.* (1972) and Kohne *et al.* (1972) by dividing the ΔT_m for the groups by the corresponding catarrhine-platyrrhine value.
[d]From Goodman (1975).

more along the common hominine lineage than along the gorilla, chimpanzee, and hominid lineages since their radiation. Thus we see no reason in these data to doubt that the adaptive radiation leading to the modern hominoids began 10–12 m.y. ago and that leading to the African apes and man about 4 m.y. ago.

Now, of course, there has been an almost unanimous rejection of these dates for various reasons, and this rejection will be commented on shortly. However, one should be reminded here that a cladistic network as complex as that linking the various extant primate taxa, plus the observed general regularity of change is their molecules, severely circumscribes the amount of "give" in these dates. If, for example, the Old and New World monkeys did not exist and we had only the hominoids and prosimians (a situation analogous to the Camelidae *vis-à-vis* the artiodactyls), our rate tests would lose a great deal of sensitivity, and much more doubt would exist as to any calculated dates. This is not the case for the primates generally and the anthropoids in particular. If one, for whatever reason, decides that a date of 10 m.y. ago for the gorilla-chimpanzee-hominid radiation is a more realistic figure, the very simple picture just outlined becomes more complex. The reason is that the existence of gibbons, orangutans, plus Old and New World monkeys cannot be ignored. To stretch a lineage or set of lineages in time is to contract others, which ultimately means that an observed regularity is seen as being a chance combination of grossly irregular rates of change. This is of course possible in the abstract, but it seems to us that to choose such an "explanation" of the hominid data is clearly anthropocentric—an implicit statement that somehow the rules of the scientific game just do not apply to the study of man.

Inevitably some proteins will show accelerated change in some lineages and deceleration in others, and it may well be that hominoid globins are an example of the latter phenomenon. As we have already pointed out, if one looks at enough proteins in enough lineages one can readily accumulate a number of such cases. Each of these may be of appreciable intrinsic interest, but there can be no justification for calculating times of divergence in such cases without correcting for any *observed* departures from regularity. For example, there is only one catarrhine-platyrrhine divergence time, but it would be ludicrous to attempt a calculation of it from the *Homo-Aotus* albumin immunological distance without allowing for the fact that the *Homo* lineage has accumulated about 35 units of change as compared to 10 for *Aotus* since the divergence of the two genera. Calibrations based on observations of one rate of change clearly cannot apply if the rate changes, yet some objections to our dates have been raised along precisely such lines (Goodman, 1975). Even in those cases, where molecules other than albumin, transferrin, and DNA are being studied, one still finds along the common hominine lineage subsequent to the divergence of the *Hylobates* and *Pongo* lineages two fibrinopeptide (Wooding and Doolittle, 1972), two carbonic anhydrase (Tashian *et al.*, 1972), and two β-globin (Boyer *et al.*, 1972) substitutions—a total which must represent a significant (5–10 m.y.) period of time. The numbers are getting very small here, however, and considerations of the statistical significance of the results, of the possibility of sequence determination errors, and of the effects of generation length and population size on the length of time required to complete a substitution become of increasing significance. It is for these reasons that the relative paucity of sequence differences among human, chimpanzee, and gorilla globins and fibrinopeptides does not present any real challenge to the conclusions drawn from the albumin, transferrin, and DNA data. Finally, it should be noted that sufficient time has elapsed since the three hominine lineages began their separate existences to allow three globin substitutions in the gorilla lineage (one α-globin, one β-globin, one myoglobin), one along the hominid lineage (γ-globin), and one along the chimpanzee lineage (myoglobin).

4. Objections to the Albumin-Transferrin-DNA Clock

The objections to the albumin-transferrin-DNA clock have been recently dealt with at some length (Sarich, 1973) and will only be outlined here. As pointed out in that article, almost all of the objections were demonstrably wrong in terms of the published data available at the time they were made. All of them have been part of the tautology that a separation of the African apes and man at least 15 m.y. ago, as would be required by the acceptance of *Ramapithecus* as a hominid, meant clearly that there was something wrong with a molecular dating of this divergence at 4–5 m.y. ago. We now know, of course, that the hominid status of *Ramapithecus* is, to say the least, open to serious question (for example, Greenfield, 1974), but that has nothing to do with the objections themselves—which should have been tested and not simply asserted as a tautology.

Basically there have been two kinds of explanations for the molecular similarities of man and the African apes at the molecular level other than that of a very recent divergence. The first is the suggestion of a deceleration of molecular evolution in the higher primates (Goodman, 1961, *et seq.*; Kohne *et al.*, 1972; Lovejoy *et al.*, 1972). It is interesting to note that this hypothesis has never been treated by its authors as such, that is, as subject to testing without reference to the fossil record. If there had been significantly more or less change in the albumins, transferrins, or DNAs of the hominoids as compared to those of the Old World monkeys, New World monkeys, or prosimians, then rate tests using appropriate outside reference points have always been available, and none of these have ever shown such a deceleration. Indeed, Weigle had shown as long ago as 1961 that human albumin was just as different from that of *Bos* as were the albumins of rat, mouse, and dog (that is, they had changed equally from the ancestral condition). Thus the appearance of the deceleration hypothesis and data disproving it were almost simultaneous, but these latter data and those appearing subsequently have consistently been ignored by workers outside of our Berkeley group.

The second, and more significant, kind of objection questions the scale used in relating immunological differences to sequence differences or times of separation (Read and Lestrel, 1970; Read, 1975). A commentary on the 1975

article (which has just appeared at the time this is being written) is in preparation, but basically the problem remains one of not looking at the implications of
the proposed "solution." In other words, any solution to the problem of the
seeming impasse between our molecular time scale for primate evolution and
that currently favored by most primate paleontologists should be considered as
a hypothesis subject to testing. One of us (Sarich) had the opportunity of
reviewing the Read article when it was first submitted in 1973 and (among other
things) discussed in extensive detail that such testing did not support the Read
hypothesis. Unfortunately, those criticisms have been ignored by Read.

This question of scaling has long been a concern of ours, as is evident from
a reading of the Sarich dissertation (1967:68–78). One can readily choose a set
of dates for particular primate divergences which would be acceptable to
Simons (this volume) and Walker (this volume) and fit along a smooth curve
linking the immunological distances involved. Once that is done, any number of
equations generating such a curve can be written and a wide variety of possible
variables can be used in the "derivation" of one or another of them. For
example, in Sarich (1967:74) we find the equation:

$$\log \text{I.D.} = kT^{(1 + \alpha T/80)}$$

which will for $\alpha = 0.4$ (and I.D. = 10 for $T = 60$) generate "reasonable"
divergence dates among the primates. For example: *Homo-Pan*, 13 m.y.; *Homo-*

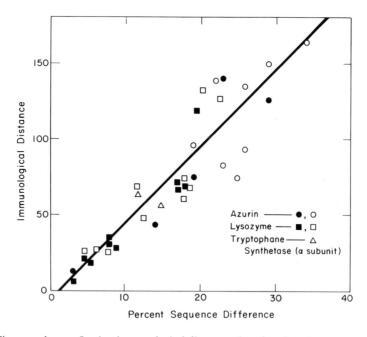

Fig. 8. Microcomplement fixation immunological distances plotted against the percent amino acid
sequence differences for lysozymes azurins, and tryptophan synthetases. The points are averages
of reciprocal measurements. From Champion *et al.* (1975).

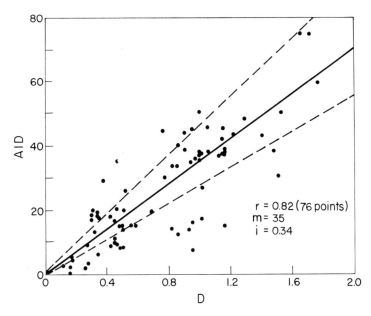

Fig. 9. Microcomplement fixation immunological distances plotted against the Nei electrophoretic distances obtained for the same species pairs. The comparisons involve *Anolis, Dipodomys, Sigmodon, Peromyscus,* and *Thomomys.* The dashed lines indicate the standard errors of the Nei distances. The deviant cluster of 7 points involves the *D. ordii* comparisons (Johnson and Selander, 1971). The *Anolis* electrophoretic data were obtained from G. C. Gorman of UCLA, the *Sigmodon* and *Peromyscus* values come from the work of R. K. Selander and co-workers at the University of Texas, and the *Thomomys* data are from S. Y. Yang and J. L. Patton of this campus. All the immunological comparisons were carried out in this laboratory.

Pongo-Hylobates, 20 m.y.; *Homo-Macaca,* 35 m.y.; catarrhine-platyrrhine, 48 m.y.; anthropoid-prosimian, 63 m.y. It is also easy to justify this equation as allowing for an immunological bias of the rabbit stemming from the fact that its albumin must also share a good deal of sequence homology with those of the primates.

The above equation is but one of many that will have the desired property. Thus we need not deal with each independently (or "derive" each one), but only with the consequences of the possession of a characteristic common to all. This characteristic is one which compensates for the desired spreading of times of divergence at the lower end of the scale by compressing them for large immunological distances. Thus in our example the marsupial-placental divergence would occur at 76 m.y., a figure at least 40 m.y. low.

Immunological distances are linearly proportional to the number of amino acid sequence differences between two lysozymes, azurins, or myoglobins (Fig. 8). We can of course only infer that this also holds for albumin, but it does seem a reasonable inference. Albumin immunological distances are highly correlated ($r = 0.82$) with Nei electrophoretic distances for the same species pairs (Fig. 9), and with the number of nucleotide replacements based on DNA hybridization

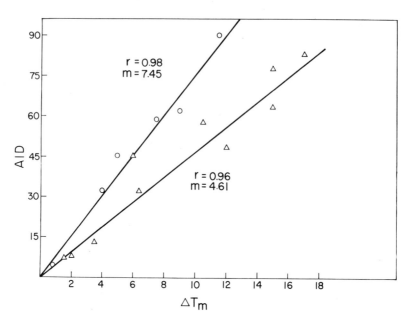

Fig. 10. Albumin immunological distances plotted against the DNA hybridization ΔT_m values for various comparisons involving rodents, primates, carnivores, and artiodactyls (Kohne *et al.*, 1972; Hoyer *et al.*, 1972; Benveniste and Todaro, 1974; Rice, unpublished observations). The distribution is clearly bimodal as differing hybridization criteria were used by the Carnegie and NIH workers. Thus where the two groups carried out the same comparisons the NIH values were always lower. All immunological comparisons were carried out in this laboratory.

data (Fig. 10, $r = 0.97$). Thus there appears to be no empirically justifiable reason for using anything other than the simple immunological distance parameter to estimate the number of sequence differences with which one is dealing. In addition, of course, these comparisons are highly supportive of the albumin clock, as it is difficult to see how differing rates of electrophoretic and DNA change would somehow be providentially associated in the same species with similarly differing rates of albumin immunological change.

The Read hypothesis of our use of an incorrect scale also requires that other nonimmunological measures of genetic distance among the species involved ought to reflect the desired times of divergence. The data in Table 2 indicate clearly that this is not the case, with the DNA differences quantitatively paralleling those obtained through immunological comparisons of the same species albumins and transferrins.

Finally we should note that unless immunological distances were linearly related to sequence differences one could not apportion them into additive cladograms as in Figs. 2–7. We conclude, then, that neither the deceleration hypothesis nor one suggesting incorrect scaling has any current validity. In the absence of any other interpretations of the albumin, transferrin, and DNA data, we still opt for the time scales of Figs. 3–7 and a man-chimpanzee-gorilla radiation beginning in Africa about 4 m.y. ago.

5. *Electrophoresis*

In recent years, by far the most popular technique of comparative bio-chemistry has been electrophoresis. Nonetheless, relatively few such compari-sons among the primates have been carried out, and those, with the exception of the *Homo-Pan* comparison of King and Wilson (1975), have involved only macaques and baboons. Although our efforts in this area have not yet been extensive, some recent observations do indicate an as yet untapped potential of the electrophoretic approach in resolving questions of relationship at those lower taxon levels where neither immunology nor DNA hybridization provides adequate resolving power.

We have already documented in Fig. 9 a very good correlation between albumin immunological distances and Nei electrophoretic distances for the same pairs of species, with the albumin immunological distance (AID) for a D of 1.0 averaging some 35 units. However, the work of King and Wilson (1975) suggested that for the *Homo-Pan* comparison involving, in the main, blood protein loci, it was inappropriate to treat the various loci as part of a single distribution. That is, although the D value for the "intracellular" loci was 0.3, eight of ten serum proteins had different mobilities in the two species ($D = 1.6$). It has now become evident to us that this single datum is entirely representative for all the mammals thus far tested.

What we have done is to carry out vertical slab polyacrylamide gel electro-phoreses of numerous sera in a continuous system (0.1 M Tris, 0.05 M glycine) at an acrylamide concentration of 7%. Under these conditions, the system will resolve, in reasonably fresh sera, some 20–25 bands—thus providing, in a single experiment, a significant amount of information. Some 5–10 bands can be easily added also surveying the serum and red cell esterases. In addition, if liver samples are available, a further 10–15 esterase bands can be seen. As an indicator of genetic similarity in such comparisons, we use simply the ratio of the number of bands that have the same mobility to the total number of bands in the sample showing the smaller number. This then means that at poly-morphic loci two individuals will be counted as different only when they are homozygous for different alleles. Thus allele frequency differences are in effect ignored, and single individuals become representative of populations. For example, if two alleles at a single locus are present at frequencies of 0.5, the probability that two individuals drawn from the same population will contribute a genetic distance component at that locus is only 1 in 8. Thus even in highly polymorphic populations it will be rare to find two individuals differing by more than one band of the 20–25 usually seen. Within populations, then, coefficients of genetic similarity (S) will be $\geqslant 0.95$, and $D = (-\ln S)$ will be $\leqslant 0.05$.

If we now compare albumin immunological distances and serum protein electrophoretic D values, we find that the slope of Fig. 9 is completely inappro-priate. In fact, pairs of taxa separated by a serum protein D value of 1 show an average AID of 4–5 units. If we also take into account that the D values of Fig. 9

typically include loci such as albumin and transferrin, then the difference in rates of electrophoretic differentiation (using AID as the constant) must be close to ten fold. Thus it would appear that rates of protein evolution are distributed rather tightly around two modes—one characteristic for what might be termed the "intracellular" or slowly evolving group (for example, LDH, MDH, GOT, IDH) and the other for the "extracellular-secreted" or rapidly evolving group (plasma proteins, esterases).

The practical effect of these observations is that we now have available an inexpensive and readily applied tool for probing the closest of evolutionary relationships. If we accept the equivalence of 4–5 AID units and a serum protein D value of 1 (where 8 or 9 of the 20–25 bands show the same mobility), then this degree of genetic distance corresponds to 2–3 m.y. of separation. This implies that we will begin seeing fixations of different alleles in diverging lineages within 0.25 m.y., and, of course, we retain the capability of assessing even shorter divergence times through the use of allele frequency differences.

Our application of this approach to specific problems in primate systematics has been limited, but it has allowed the resolution of certain issues. Among the Hominoidea, useful information has been obtained among and within the genera *Pan*, *Gorilla*, and *Homo*. Consistent with all other molecular data on this trio, we find that each of the intergeneric D values is in the range 1.6 ± 0.1; that is, they are equidistant from one another. More important, however, are the first reliable measures of genetic distance within *Pan* and *Gorilla*. For the *troglodytes-paniscus* comparison, the D value is 0.8 (10 of 21 bands line up), and for the eastern and western gorillas it is 0.7 (8 of 16 bands line up). Thus we would date these intergeneric divergences as having occured about half as long ago as the original hominine radiation; i.e., around 2 m.y. ago.

Among the Old World monkeys, we have concentrated our efforts on *Papio* and forms closely related to it (we do not include the drill and mandrill in *Papio*). We cannot differentiate *P. papio* and *P. anubis* ($D = 0$), but all other "species" in the genus can be discriminated from one another, with the D values ranging from 0.35 for the *cynocephalus-anubis* comparison to 0.5 for those involving *ursinus* and *hamadryas*. The *Papio-Theropithecus* distance is 0.7, the *albigena-aterrimus* group of mangabeys is at 1.2, and finally the distances to *Macaca*, *Mandrillus*, and the other *Cercocebus* species are around 2.

Thus all the data we currently have on the *Theropithecus-Papio* association (albumin and transferrin immunology, serum protein electrophoresis) remain consistent with the original albumin picture of a very close relationship, and force us to continue to seriously question the suggestions of a divergence time between the two genera of 5 m.y. or more. Again, as with the earlier hominoid evolution discussion, we cannot look at the *Theropithecus-Papio* question except in the total cercopithecine context. That view (Fig. 6) tells us that the sequence of well-resolved divergence events starts with the separation of *Cercopithecus*, then we have the basic 42-chomosome group radiation, then the *albigena-aterrimus* mangabey lineage splits off, and finally the *Theropithecus-Papio* divergence occurs. We also know that, relative to other anthropoid groups, albumin and transferrin evolution in the Old World monkeys generally, and in *Theropithecus* and *Papio* particularly, has not slowed down. Thus, if the current fossil-based picture accurately depicts the actual history of the Old World monkeys, we would have to accept that the molecules are somehow misleading us. If it were just the albumins, as in 1970, such a deception would have to be considered a very real possibility, but the addition of other molecular perspectives has lengthened those odds very considerably. As long as we cannot find a molecular difference commensurate with 5 m.y. or so of separate existence for the *Papio* and *Theropithecus* lineages, there has to remain doubt that they have been separated for that long. Currently, we are very doubtful indeed, and we suggest that the definitive remaining study is to compare unique-sequence DNAs among the Old World monkeys using hybridization techniques. If that study produces data supportive of the current molecular picture, then a serious reevaluation of the inferential value of certain kinds of paleontological evidence is in order; if it does not, then we would have to place some serious limitations on the degrees of confidence with which our molecularly based conclusions could be put forward.

6. Implications for Anatomy and the Fossil Record

This section will only briefly outline some of the arguments made elsewhere. The Miocene hominoids remain "dental" apes with typically monkeylike upper limbs (McHenry, 1975). If this is the case, then it is possible to view the cercopithecoid molar pattern, which is certainly a derived character, as evolving from the primitive Oligocene and Miocene hominoid pattern. Beyond this one would almost have to view the adaptive success of the Old World monkeys as rooted in their behavior and social organization, which immediately gets us into an uncomfortably speculative realm. For the hominoids we have to conclude that the widespread Miocene dryopithecines left but a single surviving lineage. A small, Asian representative of this lineage became uniquely successful through the development of the locomotor-feeding adaptation termed brachiation. The subsequent adaptive radiation from this lineage left it the only survivor of the many apes present throughout the tropical and subtropical Miocene forests of the Old World. The products of such an adaptive radiation show similarities in basic pattern, reflecting the attainment of a new grade of organization, but differences in detail, representing the various lines comprising the adaptive radiation. The unity of the modern Hominoidea, then, is based on the relatively short period of time during which a major adaptation was being evolved, and their diversity on the relatively long periods of time that each line has been evolving independently of the others (taken with minor editing from Sarich, 1968).

The suggested Asian origin of the modern hominoids is an answer to Simons' objection (1969) as to having "highly arboreal apes wander hundreds of miles out of Africa across the Pontian steppes of Eurasia in search of tropical rain forests. . . ." Brachiation is without doubt an arboreal adaptation, and an Asian origin allows an *in situ* radiation into three extant lineages of which two are still Asian and highly arboreal. The third today consists of three basically terrestrial forms, and it is therefore not unreasonable to posit that the common ancestor of the three was also terrestrial. We then take this terrestrial (knuckle-walking?) species across the Pontian steppes *to* Africa where the adaptive radiation to *Australopithecus, Pan,* and *Gorilla* takes place. This then requires the African immigrant to be "presumably not particularly unlike a small chimpanzee" (Sarich, 1968) as it is difficult to conceive of a situation where the transition from pongid to hominid would entail a size reduction.

7. Organismal Evolution and Taxonomy

The argument to this point has been direct, quantitative, and objective. However, the use of molecular data in solving problems concerning the cladistic and temporal dimensions of particular phylogenies leaves open numerous questions as to their "deeper" significance. Unfortunately, at least two of these do not lend themselves very readily to objective analysis and their consideration has tended to involve a good deal of advocacy and emotionalism. First, there is the matter of "Darwinian" vs. "non-Darwinian" evolution at the protein and nucleic acid level, and then the taxonomic relevance, if any, of the molecular data.

We emphasize first that these two areas of controversy have nothing to do either with the cladistic utility of molecular data or with molecular clocks. The latter form, as already discussed, testable hypotheses and are dealt with accordingly. It is true that most people find it somehow easier to accept the existence of molecular clocks in the context of neutral mutations, but such an association is unfortunately extremely counterproductive. We do not have to "understand" a phenomenon in terms of basic causation in order to make use of it—a classic case being gravitational attraction. It has been possible since Newton to describe gravity with the elegance of mathematical precision and since Cavendish to measure its magnitude, but even today we are hardly farther along than they were in understanding the hows and whys of gravity. We wonder if somehow weather satellites or lunar explorations should have been postponed until we understood the ultimate whys and hows of our capabilities in orbiting satellites or putting men on the moon or bringing them back. The questions of why the albumins, transferrins, DNAs, and globins vary as they do among species are perhaps equally unanswerable at this time, but it is difficult to see how this lack has anything to do with the use of the data.

The taxonomic situation is equally muddled, but at least potentially amenable to immediate resolution. Here we need but decide what information is to be retrievable from the taxonomy of a group and the form of that taxonomy follows directly. The considerable progress recently made toward understanding the relationship between molecular and organismal evolution should lead to a rapprochement between the molecular and organismal schools of thought in this area. The most important thing to appreciate is that there is generally no relationship between molecular and organismal evolution. We know of no examples indicating a correlation between rates of molecular and organismal change, nor do we think that there is much point in continuing a search for them at random. The problem will not even become potentially soluble until the time-dependent majority of protein and nucleic acid changes can be specifically isolated from that minority which are not. Again, current impotence in this area need not hamper our progress in others.

Several authors have recently suggested that in order to understand organismal evolution one needs to focus attention on the control of gene expression rather than on the amino acid sequences of proteins coded for by structural genes (Davidson and Britten, 1973; Ohno, 1973; Soulé, 1973; Wallace and Kass, 1974). Geneticists have long appreciated the importance of gene association and order along chromosomes, which must be an integral part of such control (Ford, 1965; Dobzhansky, 1970). Molecular evidence consistent with this hypothesis has recently appeared (Wilson *et al.*, 1974*a,b;* Prager and Wilson, 1975; Wilson, 1975). Those articles argue that structural gene evolution may not be at the basis of organismal evolution. Thus there would be two types of molecular evolution: that involving structural genes which proceeds inexorably in a predominantly time-dependent fashion and that involving regulatory systems which parallels organismal evolution. These conclusions stem from the observations that the rates of loss of hybridization potential and development of chromosomal and organismal differentiation are highly correlated in frogs, mammals, and birds, although the rate in mammals is markedly more rapid than in either birds or frogs. Thus the average time required for two lineages to lose their potential for hybridization is about 3 m.y. for mammals and about 20 m.y. for birds and frogs. Similarly, the average length of time required for a difference in chromosome number to develop is about 4 m.y. in mammals and 80 m.y. in frogs. It is also high in birds, but a precise figure cannot be given because bird chromosomes are very small, numerous, and difficult to count. Similarly, organismal evolution has been accelerated in mammals relative to birds and frogs, and this is clearly recognized in the taxonomies where placental mammals are divided into 16–20 orders while the frogs are confined to one. Birds, of course, are also divided into many orders, but no one could seriously argue that the degree of anatomical and physiological differentiation in birds is comparable to that seen in mammals; in other words, it is probable that bird taxonomic levels are inflated, with a bird order broadly equivalent to a mammalian family.

The loss of hybridization potential is a symptom and difficult to study directly, but chromosomal evolution may be an integral part of organismal evolution, and, with the advent of banding techniques, should be much more amenable to study. Our hypothesis would then predict that lineages which have experienced very rapid adaptive evolution should show numerous chromosome morphology changes, and periods of quiescence in adaptive evolution would be associated with stable chromosome morphologies. Thus we would expect to see, for example, many more changes separating our karyotype from the ancestral hominine one than would be the case for those of chimpanzees and gorillas. Data have been presented by Vogel *et al.* (this volume) indicating that this is indeed the case.

8. Formal Taxonomies

It should now be abundantly evident that rates of molecular evolution are entirely independent of rates of morphological evolution. Thus the fact that two taxa may be very similar at the molecular level does not yet directly impinge on the matter of the taxonomic level at which they should be separated. Granting this, we still feel that the general emphasis on recognizing rapid adaptive change by raising the taxonomic ranking of that group relative to its more conservative sister groups at best distorts and at worst misleads. The distortion stems from the fact that any such system is then geared to the exception rather than the rule; the misleading stems from the noncongruence of the taxonomy and the cladistics.

No one seriously doubts, for example, that the human lineage has experienced the most profound adaptive shift in mammalian history. That conclusion has to be weighed, however, against the fact that this is but one lineage among many in the Anthropoidea and thus judged in terms of the distortion of anthropoid taxonomy caused by giving man his "proper" rank. Once the human line becomes the Hominidae, its sister group becomes Pongidae, and we are at once misled. There is now no possibility of a taxon which includes only man, chimpanzee, and gorilla—a trio clearly monophyletic relative to the orangutan and gibbons (as the molecules, among other lines of evidence, tell us). The lack of such a taxon then tends to facilitate the ignoring of the important terrestrial adaptation basic to later human evolution which took place along that common African ape lineage. The often expressed sentiment that anagenesis must be balanced against cladogenesis in the construction of a taxonomy is fine in the abstract, but it appears that every time such a choice is actually made cladogenesis loses. What it comes down to with man, as with most other cases of this type, such as the pinnipeds relative to the other carnivores, is whether the emphasis should be on "recognizing" what he is (a clearly divergent form) or on figuring out how he got to be that way. How a species got to be what it is today is the ultimate evolutionary question, and systems of classifying our present state of knowledge ought not to impede our progress toward answering that question. If we could but agree on a basic rule that all monophyletic lineages along which something *adaptively significant* can be seen to have occurred should be named, we might be a long way toward effecting a realistic, and probably still necessary, balance between the demands of cladogenesis and those of anagenesis. There seems to be little point in making taxonomies purely cladistic since the formal naming of a lineage along which nothing significant can be seen as having happened contributes little to our understanding.

The distortion is perhaps a less serious matter, but one cannot help wondering if the separation of the apes and Old World monkeys at the superfamily level and of the New World monkeys from them at the infraordinal level—required by giving the hominids family status—could be justified on morphological grounds. We doubt that it could be, and suggest that an interesting and informative exercise would be to carry out a classification of all primates but man, and to then place man in an appropriate position within that framework.

9. Epilogue

All of this started with Nuttall, the tradition was carried on by Boyden and his co-workers, and finally the efforts of Goodman—along with the development of protein sequencing techniques—brought us into the modern era of molecular systematics. We have now developed, some 15 years into that era, a very good appreciation of the strengths and limitations of the various molecular approaches to systematics. We have attempted in this chapter to provide such an assessment for the mammals generally, and the primates specifically.

If full use is made of the array of molecular techniques available—and if appropriate precautions are taken—one should be able to place, with a cladistic and temporal precision equal to any demands that might be put upon it, almost any extant lineage. The documentation of the validity of this claim has only just begun, of course, and convincing the interested scientific community of it remains the most important task of the molecular systematist. As this is accomplished, we may expect a burgeoning interest in, and use of, immunological, DNA hybridization, and electrophoretic techniques to solve problems of evolutionary relationship. Those problems which remain refractory to these more readily applied techniques, such as many higher taxon relationships among the mammals and the positions of the birds and turtles among the reptiles, could then be attacked using amino acid sequencing of the appropriate proteins.

The beginning has been fitful and slow—and not unmarked by unnecessary acrimony—but the momentum is now there. We look forward to a time in the near future when any systematic discussion of an extant group will routinely open with the molecular framework and then proceed to structure upon that framework the development of the evolutionary understanding of the data at issue. The very large body of molecular information already available for the primates suggests that this group might be admirably suited for an exemplification of this proposed model for future efforts in systematics.

Acknowledgments

The sources of most of the sera used in these studies have been given in previous publications. The various new lemuriform sera were provided by Y. Rumpler of the Université de Madagascar, *C: albigena* by the Delta Regional Primate Center, *C. aterrimus* and *P. paniscus* by K. Benirschke of the San Diego Zoo, *G. g. berengei* by U. Rahm, the various *Papio* by J. Moor-Jankowski of LEMSIP, and *Mandrillus* and *C. ascanius* by W. Mottram of the San Francisco Zoo. The work was supported by NSF grants to V. M. Sarich and A. C. Wilson and a predoctoral fellowship to J. E. Cronin. We thank A. C. Wilson and the Biochemistry Department, University of California, Berkeley, for making the laboratory facilities available.

10. Notes Added in Proof

Since the submission of this manuscript, we have carried out further studies bearing on important issues in primate systematics. More detailed immunological comparisons plus a consideration of the globin sequence data now lead us to place the divergence of the Tupaiidae about midway along the lineage (Figure 3) linking the node from which the Chiroptera stem and that from which the *Tarsius,* lemuriform–lorisiform, anthropoid, and *Cynocephalus* lineages radiate. We believe that this is the first instance among the primates where sequence data truly contribute to the more precise placement of a lineage.

Recently Dr. Adrian Friday of Cambridge made available to us a sample of *Ptilocercus* serum with which we have obtained the following albumin plus transferrin distances from *Tupaia: Urogale*, 90; *Ptilocercus*, 131; with any non-tupaiid at least 260 units away. Thus the *Ptilocercus* and *Tupaia–Urogale* lineages diverged some 35–40 m.y. ago.

Dr. Yves Rumpler has continued to provide Malagasy material, with the latest samples being *Indri* and *Hapalemur simus.* We find that *Indri* is at least as closely related to *Propithecus* and *Avahi* as they are to one another and may indeed be somewhat closer to the former. *Hapalemur simus* and *griseus* are clearly part of the same monophyletic clade with the *simus–griseus* distance being half that separating *Hapalemur* from the *Lemur catta* and *macaco–fulvus–mongoz* lineages (Figure 4).

Finally, extensive comparisons of the plasma proteins, esterases, and intracellular enzymes in *Thomomys* have provided a more precise estimate of the differences in the rates of electrophoretic differentiation for our two sets of loci (pp. 161–162). We find that the rapidly evolving group of proteins is accumulating D values at about 12 times the rate for the slowly evolving group.

11. References

Benveniste, R. E., and Todaro, G. J., 1974, Evolution of type C viral genes. 1. Nucleic acid from baboon type C virus as a measure of divergence among primate species, *Proc. Natl. Acad. Sci. USA* **71**:4513.

Boyer, S. H., Noyes, A. N., Timmons, C. F., and Young, R. A., 1972, Primate hemoglobins: Polymorphisms and evolutionary patterns, *J. Hum. Evol.* **1**:515.

Champion, A. B., Soderberg, K. L., Wilson, A. C., and Ambler, R. P., 1975, Immunological comparisons of azurins of known amino acid sequence, *J. Mol. Evol.* **5**:291.

Cronin, J. E., and Sarich, V. M., 1975, Molecular systematics of the New World monkeys, *J. Hum. Evol.* **4**:357–375.

Davidson, E. H., and Britten, R. J., 1973, Organization, transcription, and regulation in the animal genome, *Quart. Rev. Biol.* **48**:565.

Dobzhansky, T., 1970, *Genetics of the Evolutionary Process,* Columbia University Press, New York.

Fitch, W. M., and Margoliash, E., 1967, Construction of phylogenetic trees, *Science* **155**:279.

Ford, E. B., 1965, *Genetic Polymorphism,* MIT Press, Cambridge, Mass.

Goodman, M., 1960*a*, On the emergence of intraspecific differences in the protein antigens of human beings, *Am. Nat.* **94**:153.

Goodman, M., 1960*b*, The species specificity of proteins as observed in the Wilson comparative analysis plates, *Am. Nat.* **94**:184.

Goodman, M., 1961, The role of immunological differences in the phyletic development of human behavior, *Hum. Biol.* **33**:131.

Goodman, M., 1962, Evolution of the immunologic species specificity of human serum proteins, *Hum. Biol.* **34**:104.

Goodman, M., 1963, Serological analysis of the systematics of recent hominoids, *Hum. Biol.* **35**:377.

Goodman, M., 1967, Deciphering primate phylogeny from macromolecular specificities, *Am. J. Phys. Anthropol.* **26**:255.

Goodman, M., 1975, Protein sequence and immunological specificity; their role in phylogenetic studies of primates, in *Phylogeny of the Primates* (W. P. Luckett and F. S. Szalay, eds.), pp. 219–248, Plenum Press, New York.

Gorman, G. C., Wilson, A. C., and Nakanishi, M., 1971, A biochemical approach towards the study of reptilian phylogeny: Evolution of albumin and lactic dehydrogenase, *Syst. Zool.* **20**:167.

Greenfield, L. O., 1974, Taxonomic reassessment of two *Ramapithecus* specimens, *Folia Primatol.* **22**:97.

Hoffstetter, R., 1974, Phylogeny and geographical deployment of the primates, *J. Hum. Evol.* **3**:327.

Hoyer, B. H., van de Velde, N. W., Goodman, M., and Roberts, R. B., 1972, Examination of hominoid evolution by DNA sequence homology, *J. Hum. Evol.* **1**:645.

King, M.-C., and Wilson, A. C., 1975, Evolution at two levels in humans and chimpanzees, *Science* **188**:107.

Johnson, W. E., and Selander, R. K., 1971, Protein variation and systematics in kangaroo rats (genus *Dipodomys*), *Syst. Zool.* **20**:377.

Kohne, D. E., Chiscon, J. A., and Hoyer, B. H., 1972, Evolution of primate DNA sequences, *J. Hum. Evol.* **1**:627.

Lillegraven, J. A., 1974, Biogeographical considerations of the marsupial–placental dichotomy, *Annu. Rev. Ecol. Syst.* **5**:215.

Lovejoy, O., Burnstein, H., and Heiple, K. H., 1972, Primate phylogeny and immunological distance, *Science* **176**:803.

McHenry, H. M., 1975, Fossils and the mosaic nature of human evolution, *Science* **190**:425.

Maxson, L. R., and Wilson, A. C., 1975, Albumin evolution and organismal evolution in tree frogs (Hylidae), *Syst. Zool.* **24**:1.

Maxson, L. R., Sarich, V. M., and Wilson, A. C., 1975, Continental drift and the use of albumin as an evolutionary clock, *Nature (London)* **225**:397.

Ohno, S., 1973, Ancient linkage groups and frozen accidents, *Nature (London)* **244**:259.

Prager, E. M., and Wilson, A. C., 1974, Slow evolutionary loss of the potential for interspecific hybridization in birds: A manifestation of slow regulatory evolution, *Proc. Natl. Acad. Sci. USA* **72:**200.

Prager, E. M., Brush, A. H., Nolan, R. A., Nakanishi, M., and Wilson, A. C., 1974, Slow evolution of transferrin and albumin in birds according to micro-complement fixation analysis, *J. Mol. Evol.* **3:**243.

Read, D. W., 1975, Primate phylogeny, neutral mutations, and "molecular clocks," *Syst. Zool.* **24:**209.

Read, D. W., and Lestrel, P. E., 1970, Hominid phylogeny and immunology: A critical appraisal, *Science* **168:**578.

Reichlin, M., 1974, Quantitative immunological studies on single amino acid substitutions in human hemoglobin: Demonstration of specific antibodies to multiple sites, *Immunochemistry* **11:**21.

Sarich, V. M., 1967, A quantitative immunochemical study of the evolution of primate albumins, Ph.D. dissertation, University of California, Berkeley.

Sarich, V. M., 1968, Human origins: An immunological view, in: *Perspectives on Human Evolution* (S. L. Washburn and P. C. Jay, eds.), pp. 94–121, Holt, Rinehart and Winston, New York.

Sarich, V. M., 1969, Pinniped origins and the rate of evolution of carnivore albumins, *Syst. Zool.* **18:**286.

Sarich, V. M., 1970, Primate systematics with special reference to Old World monkeys: A protein perspective, in: *Old World Monkeys* (J. R. Napier and P. H. Napier, eds.), pp. 175–226, Academic Press, New York.

Sarich, V. M., 1973, Just how old is the hominid line? *Yearb. Phys. Anthropol.* **17:**98.

Simons, E. L., 1969, The origin and radiation of the primates, *Ann. N.Y. Acad. Sci.* **167:**319.

Soulé, M., 1973, The epistasis cycle: A theory of marginal populations, *Annu. Rev. Ecol. Syst.* **4:**165–187.

Tashian, R. E., Tanis, R. J., Ferrell, R. E., Stroup, S. K., and Goodman, M., 1972, Differential rates of evolution in the carbonic anhydrase isozymes of catarrhine primates, *J. Hum. Evol.* **1:**545.

Wallace, B., and Kass, T. L., 1974, On the structure of gene control regions, *Genetics* **77:**541.

Wallace, D. G., King, M. C., and Wilson, A. C., 1972, Albumin differences among ranid frogs: Taxonomic and phylogenetic implications, *Syst. Zool.* **22:**1.

Weigle, W. O., 1961, Immunological properties of the crossreactions between anti BSA and heterologous albumins, *J. Immunol.* **87:**599.

Wilson, A. C., 1975, Evolutionary importance of gene regulation, in: *Stadler Symposium 7,* pp. 117, University of Missouri, Columbia, Mo.

Wilson, A. C., Maxson, L. R., and Sarich, V. M., 1974a, Two types of molecular evolution: Evidence from studies of interspecific hybridization, *Proc. Natl. Acad. Sci. USA* **71:**2843.

Wilson, A. C., Sarich, V. M., and Maxson, L. R., 1974b, The importance of gene rearrangement in evolution: Evidence from studies on rates of chromosomal, protein, and anatomical evolution, *Proc. Natl. Acad. Sci. USA* **71:**3028.

Wooding, G. L., and Doolittle, R. F., 1972, Primate fibrinopeptides: Evolutionary significance, *J. Hum. Evol.* **1:**555.

Immunodiffusion Evidence on the Phylogeny of the Primates

9

HOWARD T. DENE, MORRIS GOODMAN, and WILLIAM PRYCHODKO

1. Introduction

All modern classifications of the Primates are based to one degree or another on concepts concerning the phylogeny of the order. The widely used approach of Simpson (1945) in emphasizing grades of evolutionary development divides the Primates into suborders Prosimii and Anthropoidea. Prosimii consists of small-brained primates arranged serially into the infraorders Lemuriformes (Malagasy lemurs), Lorisiformes (lorises), and Tarsiiformes (tarsiers). Tree shrews can also be included at the base of Prosimii either as the first taxon in Lemuriformes or as the separate infraorder Tupaiiformes. In turn, Anthropoidea consists of large-brained primates arranged into the superfamilies Ceboidea (New World monkeys), Cercopithecoidea (Old World monkeys), and Hominoidea (the manlike apes and man). In this scheme, Anthropoidea is the younger of the two suborders, and any fossil primates considered to be ancestral to Anthropoidea are placed in Prosimii if they show small brains and other primitive features. To the extent that the anthropoid grade was reached independently in different lineages, Simpson's Prosimii and Anthropoidea are both polyphyletic assemblages.

HOWARD T. DENE and WILLIAM PRYCHODKO • Department of Biology, Wayne State University, Detroit, Michigan 48201.   MORRIS GOODMAN • Department of Anatomy, Wayne State University School of Medicine, Detroit, Michigan 48201.

Alternative schemes of primate classification place more emphasis on the phyletic branches (clades) within the order. They attempt to depict monophyletic groupings in which the members of a group are presumed to share a more recent common ancestry with one another than with any member of a sister group. The classification used by Hill (1953) illustrates this approach. It divides the Primates into two branches, Strepsirhini and Haplorhini. Strepsirhini has two sister branches, the suborder Lemuroidea (Simpson's Lemuriformes minus tree shrews) and Lorisioidea (Lorisiformes). Haplorhini also has as sister groups two suborders, Tarsioidea and Pithecoidea (equivalent to Anthropoidea). The latter, in turn, divides into Platyrrhini (Ceboidea) and Catarrhini (Cercopithecoidea and Hominoidea). The classification of Romer (1966) is also based on the clade approach. It has four suborders for the extant primates, Lemuroidea (equivalent to Hill's Strepsirhini), Tarsioidea, Platyrrhini, and Catarrhini. However, it does not depict degrees of relationship among these four taxa, presumably because Romer felt the evidence did not warrant doing so. In this connection, most of the evidence concerning grades and clades in the order Primates has come from comparative anatomical investigations. As just indicated, the findings in these investigations have led to conflicting or inconclusive views on the phylogenetic branching pattern of the order. The view which we present here comes from a different kind of investigation, one using the immunological properties of proteins. First we describe our findings from analyzing a large body of immunodiffusion data. Then we summarize our taxonomic conclusions in a classification which is meant to be simply a written description of the genealogical tree of Primates.

The rationale for immunological investigations is that they can estimate in an easy and reasonably reliable way the phylogenetic distances separating species. In these investigations, rabbits or other animals known to be good at producing antibodies are injected with proteins from another species. The foreign proteins act as immunizing antigens. Configurations of amino acids at their surfaces which are different from configurations on the native proteins of the animal being immunized elicit antibodies. The serum obtained from the animal producing these antibodies is called an antiserum. Tests can be carried out with it to compare the proteins used as immunizing antigens (the proteins from the *donor* or *homologous* species) to proteins from other species (called *heterologous* species). This is done by observing the cross-reactions or amounts of antibody-antigen precipitation yielded by the heterologous species in relation to the homologous reaction. Inasmuch as the surfaces or antigenic properties of proteins are progressively altered as amino acid substitutions accumulate over evolutionary time, the cross-reactions in immunological tests decrease with increasing evolutionary separation between homologous and heterologous species. These tests, therefore, provide a way to measure phylogenetic distances.

The main findings from our investigations of primate relationships have been presented in previous articles (Goodman and Moore, 1971; Baba *et al.*, 1975; Dene *et al.*, 1976). In the present chapter, these findings are supported and extended on the basis of many hundreds of additional immunodiffusion comparisons.

2. Collection and Analysis of the Immunodiffusion Data

A detailed account of the immunodiffusion technique used in our work has already been published (Goodman and Moore, 1971); therefore, only a brief description of the main steps will be given here. Typically, antiserum is produced against the whole serum or plasma of homologous species. This antiserum is then used in immunodiffusion plates to compare the antigens (serum proteins) of homologous and heterologous species. An immunodiffusion plate has three wells in a trefoil arrangement separated from one another by a field of agar. A solution of antigens from one species is placed in the left upper well. A comparable solution from a second species is placed in the right upper well. The antiserum is placed in the lower well equidistant from the two antigenic mixtures. The antigens and antibodies are allowed to diffuse through the agar field over a period of 3 days. When antibodies specific to particular antigenic sites meet these sites on an antigen, a reaction occurs resulting in a precipitin band. Since different proteins tend to diffuse in agar at different rates, the precipitin reactions of the individual proteins in an antigenic mixture can be seen separately. This is the reason we can use to good advantage antisera produced against the whole serum of a species rather than just against purified proteins.

Reactions of identity occur when the two antigenic mixtures compared in an immunodiffusion plate are identical with respect to their antibody-combining sites. All precipitin bands formed by the antigens diffusing from the left well will fuse with the corresponding bands formed by the antigens diffusing from the right well. If the two mixtures differ in their spectra of antigenic sites, then precipitin lines from opposite wells will form spurs when they intersect rather than merge together. A spur represents reactions of the antiserum with antigenic sites present on the antigen from one of the wells but not on the corresponding antigen from the other well. Such spurs increase in size as the differences in reactive antigenic sites become greater between the two samples. The scores obtained on the sizes of the spurs in a series of comparison are processed by the computer algorithm (IMDFN) of Moore and Goodman (1968) to determine the phylogenetic distances of the various species in the series from the homologous species.

2.1. Strategy for Calculating Phylogenetic Distances

Spurs are scored from 0 to 5. A "5" spur reaches all the way to the edge of the opposite well. It is expected that at least some "5" spurs would increase in size if there were sufficient agar for their continued growth. Therefore, in order to determine the theoretical spur size of the homologous species against a very distant species, a comparison network must be set up in which the homologous species is first compared to a number of suspected closely related heterologous species. The most divergent of these species is then used in further comparisons to identify those species which are even more divergent. This procedure, in which the most distant species are approached in a stepwise manner, measures the order and amount of divergence of the different heterologous species from the homologous species more accurately than direct comparisons of the homologous species to the very distant ones. Thus the best spur size measurements of antigenic divergence are given by comparison networks containing numerous species at varying phylogenetic distances from the homologous species. However, certain homologous species lack moderately distant relatives. Thus their comparison networks do not have proper distributions of heterologous species. The maximum distance values obtained from spur size data on these deficient networks tend to be much less than the values from adequate comparison networks. For example, *Erythrocebus patas* is separated from some of its extant relatives by moderate distances, whereas *Tarsius syrichta* belongs to a genus whose closest extant relatives are already very distant. The unadjusted immunological distance between these two species is 7.05 using antisera to *E. patas* (Table 1) but just 4.52 using antisera to *T. syrichta* (Table 2). Lack of reciprocity between tables for different antisera may also result from differences in the potency of the antisera or in the numbers of antibodies to "species-specific" antigenic determinants. To solve this problem where some

Table 1. Rabbit Anti-*Erythrocebus* Serum Antisera (5.31 Lines)[a]

OTU	Unadjusted divergence value	Adjusted divergence value	OTU	Unadjusted divergence value	Adjusted divergence value
Cercopithecus	0.000	0.000	*Perodicticus*	7.302	9.193
Papio	0.000	0.000	*Galago*	7.330	9.228
Cercocebus	0.000	0.000	*Loris*	7.491	9.431
Theropithecus	0.340	0.428	*Lemur*	7.581	9.544
Macaca	0.503	0.633	*Nycticebus*	7.644	9.624
Colobus	1.301	1.298	*Tupaia*	8.230	10.362
Presbytis	1.420	1.788	*Bos*	8.282	10.427
Homo	2.391	3.010	*Citellus*	8.340	10.540
Pongo	2.767	3.484	*Dasypus*	8.721	10.980
Hylobates	2.937	3.698	*Potos*	8.957	11.277
Pan	3.314	4.193	*Nasilio*	9.025	11.362
Saimiri	4.355	5.483	*Tenrec*	9.172	11.548
Aotes	4.638	5.839	*Suncus*	9.172	11.548
Chiropotes	4.644	5.847	*Eutamias*	9.172	11.548
Callicebus	5.162	6.489	*Loxodonta*	9.172	11.548
Ateles	5.200	6.547	*Atelerix*	9.234	11.626
Cacajao	5.209	6.558	*Erinaceus*	9.370	11.797
Oedipomidas	5.256	6.617	*Hemiechinus*	9.407	11.842
Lagothrix	5.492	6.914	*Tachyglossus*	9.485	11.942
Alouatta	5.492	6.914	*Rattus*	9.590	12.074
Cebus	5.633	7.092	*Didelphis*	9.798	12.336
Tarsius	7.047	8.872			

[a]Adjustment factor = 1.25894.

Table 2. Rabbit Antiserum to *Tarsius syrichta* Serum (8.00 Lines)[a]

OTU	Unadjusted divergence value	Adjusted divergence value	OTU	Unadjusted divergence value	Adjusted divergence value
Hylobates	3.371	6.265	*Aotes*	5.318	9.883
Pan	4.479	8.250	*Microcebus*	5.389	10.015
Pongo	4.439	8.250	*Saimiri*	5.398	10.032
Gorilla	4.439	8.250	*Lagothrix*	5.405	10.045
Theropithecus	4.489	8.343	*Galago*	5.476	10.177
Papio	4.520	8.400	*Lemur*	5.514	10.247
Erythrocebus	4.520	8.400	*Perodicticus*	5.514	10.247
Presbytis	4.536	8.430	*Loris*	5.571	10.353
Macaca	4.603	8.554	*Nycticebus*	5.774	10.731
Cercocebus	4.679	8.696	*Tupaia*	5.780	10.742
Cercopithecus	4.708	8.750	*Cynocephalus*	6.280	11.671
Homo	4.833	8.982	*Dasypus*	6.351	11.803
Colobus	4.945	9.190	*Citellus*	6.351	11.803
Cebus	5.151	9.573	*Bos*	6.601	12.268
Cacajao	5.179	9.625	*Bradypus*	6.976	12.965
Propithecus	5.264	9.783	*Petrodromus*	6.976	12.965
Ateles	5.286	9.824	*Atelerix*	7.164	13.314

[a]Adjustment factor = 1.86.

distance tables are better able to capture the full extent of divergence than other tables, Moore has developed a reciprocity adjustment procedure. Each table is multiplied by an appropriate factor with the aim of minimizing among the tables the differences in the values of reciprocal comparisons. The procedure is described in Baba *et al.* (1974), and also Dene *et al.* (1976), and its mathematical validity has now been proven (Norton and Moore, personal communication).

2.2 Arranging the Primates into Groups

For the present chapter, we used the spur size results from over 6200 immunodiffusion plate comparisons developed with 49 rabbit antisera to 38 species, of which 29 were primates, 6 were tree shrews, 2 were elephant shrews, and 1 was the flying lemur. These results were converted by the procedure described above into 38 phylogenetic distance tables, one for each homologous species. A divergence tree was then constructed from the distance values by the unweighted pair-group method (UWPGM) of Sokal and Michener (1958). On the assumption that, on the average, increasing antigenic distances between species indicate increasing periods of time since divergence, such an UWPGM tree approximates a cladogram (Moore, 1971). The primate portions of the tree are shown in Figs. 1 and 2 and the distance values in the comparison series of the antisera to the 29 primate species are shown as matrices in Tables 3 and 4.

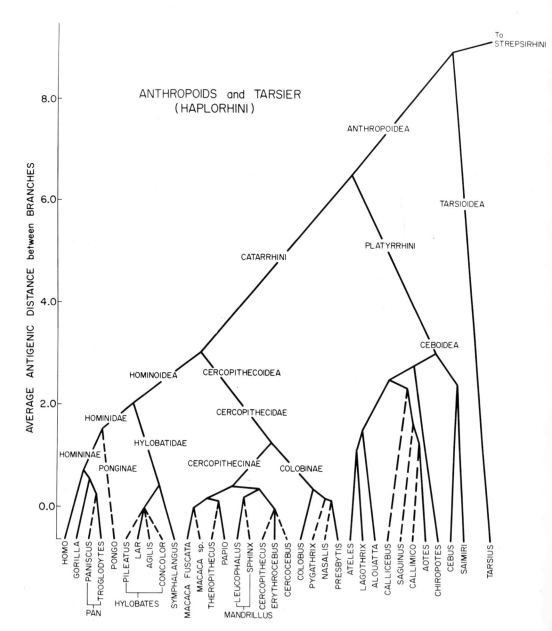

Fig. 1. Heavy lines descend to taxa used as homologous species; dashed lines descend to taxa used only as heterologous species. Haplorhini and Strepsirhini (Fig. 2) unite at an antigenic distance level of 10.31.

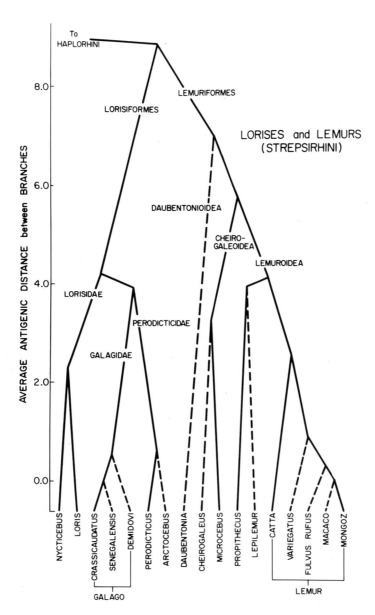

Fig. 2. Heavy lines descend to taxa used as homologous species; dashed lines descend to taxa used only as heterologous species. Strepsirhini and Haplorhini (Fig. 1) unite at an antigenic divergence level of 10.31.

Table 3. Adjusted Dissimilarity

	Homo	*Gorilla*	*Pan*	*Hylobates*	*Symphalangus*	*Macaca.*	*Erythrocebus*	*Mandrillus*	*Papio*
Homo	0	7	5	19	18	42	30	26	33
Gorilla	8	0	4	20	21	36		21	
Pan troglodytes	9	8	0	23	20	39	42	25	35
Pan paniscus				3					
Pongo	20	15	10	19	20	39	35	28	46
Hylobates	24	18	20	0	7	31	37	30	39
Symphalangus	25		19	2	0			24	
Macaca	34	26	32	32	25	0	6	4	3
Cercocebus	36	27	26	32	23	6	0	5	1
Erythrocebus	38	26	26	32	23	1		2	4
Cercopithecus	40	24	23	27	25	9	0	4	4
Mandrillus leucophaeus	33				25	7		0	
Mandrillus sphinx								2	
Theropithecus	44	26	29	32	23	2	4	4	1
Papio	45	24	35		25	0	0	3	0
Colobus	44	24	30	33	29	16	13	9	12
Nasalis				35					
Pygathrix					32				7
Presbytis	38	27	29	26	32	11	18	9	18
Lagothrix	66	66			92	54	69	57	
Chiropotes	68	59	65		89	56	58	64	53
Aotes	74	61		52	88	60	58		57
Cebus	71	64	67		99	58	71	60	
Callicebus	73	63			98	57	65	64	
Callimico									
Alouatta	74				100	61	69	64	
Ateles	79	66			90	52	65	60	
Saimiri	76	62	71	59	102	64	55	66	58
Saguinus	65	63			95	60	66	57	
Tarsius	91	89				84	89		
Galago	107	99	110	103	190	93	92		99
Loris	107	100	110	104		115	94		107
Nycticebus	108	94	112	103	151	103	96		100
Perodicticus	110	98	107	111		110	92		
Arctocebus	137								
Propithecus	105	94				101			
Lemur	108	94	110	98		99	95		99
Microcebus	110	99				103			
Urogale	121								
Tupaia	124	109	117	101		111	104		100
Cynocephalus	115	108				116			
Carnivora	131	112	140	118			113		
Edentata	143	111	133	112			110		
Chiroptera	138								
Proboscidea	141			115			115		
Rodentia	143	105	113	126			114		89
Ungulata	147	104	130	115			104		89
Erinaceidae	148	122	140	130			118		
Tenrecidae				139			115		
Soricidae				141			115		
Rhynchocyon	120								
Petrodromus	121	113							
Nasilio				137			114		
Elephantulus									
Metatheria	188			139			123		
Prototheria	223						119		

[a]Antisera are identified in the column headings of this table by genus. The actual species to which antisera were made are as follows: *Homo sapiens, Gorilla gorilla, Pan troglodytes, Hylobates lar, Symphalangus syndactylus, Macaca fuscata, Erythrocebus patas, Mandrillus leucophalus, Papio doguera, Colobus polykomos, Presbytis entellus, Lagothrix lagothricha, Chiropotes satanas, Aotes trivergatus, Alouatta* sp., *Cebus albifrons, Callicebus* sp., *Ateles geoffroy, Saimiri sciurius, Saguinus filligeri, Tarsius syrichta.*

Matrix for Anthropoid Antisera[a,b]

Colobus	*Presbytis*	*Lagothrix*	*Chiropotes*	*Aotes*	*Alouatta*	*Cebus*	*Callicebus*	*Ateles*	*Saimiri*	*Saguinus*	*Tarsius*
25	33	54	63	56	55	75	68	60	59	56	90
26	33										82
28	36	53						76			82
24	35			56					59		82
27	33	59	61	56	57	69	64	66	53	54	79
27											
13	19	61	59	59	55	72	61	73	56	56	86
10	17										87
9	17										84
11	14							71			87
13											
10	17										83
10	18										84
0	5									57	92
	2										
5	2										
0	0	65	67	59	61	73	73	61	62		84
		0	30	25	21	28	27	18	34	26	100
		25	0	27	30	23	30	27	33	28	
47	57	23	28	0	30	27	28	27	33	22	99
60		23	30	25	32	0	36	37	30	35	96
		22	30	14	27	37	0	35	38	27	
				13					28		
		7	29	15	0	29	24	19	29	19	
42		5	27	20	16	32	28	0	37	26	98
55	60	27	27	21	29	19	34	36	0	32	100
51		23	27	11	33	23	26	33	31		
									98		0
106	101		110						108		102
90	101		110					102	109		104
89		99	99	111	76	113	104	107	113	90	107
107				114				111	108		102
											98
87	101	102	106		74	107	102	92		82	102
											100
129		125	113	114	83	117	119	116	118	91	107
											117
144	129	186	128		86	131	129			95	
	126										124
	124										118
146	124	174	127		80	108	121			97	123
152	129	215	135		88	135	131	132		99	133
		139									
		131									
											130
160		196	140		83	130	131	136		102	
			141		86	143				95	

[b]Rows in this table usually represent genera. All species within a genus were grouped together for computer analysis when used in comparisons with antisera to other genera. In two cases (*Pan, Mandrillus*), we have identified rows according to species. The first-listed species in both cases is for a species to which an antiserum was made. In all columns except that in which it is the homologous species, values represent all members of the genus grouped together. Only in the column in which it is the homologous species are values shown for other member species of that genus because, in this case, the species within the genus were not grouped together in the computer analysis.

Table 4. Adjusted Dissimilarity Matrix for Strepsirhine Antisera[a,b]

	Lemur mongoz	Lemur catta	Propithecus	Microcebus	Galago	Nycticebus	Loris	Perodicticus
Lemur mongoz	0	30	41	71	96	97	83	87
L. macaco	0	25						
L. fulvus	3	33						
L. catta	6	0						
L. variegatus	9	31						
Lepilemur	33	51	40	56	86		78	87
Propithecus	32	51	0	53	96		74	87
Cheirogaleus	46	61	57	33				73
Microcebus	55	66	55	0	95		83	81
Daubentonia	70		67	75				
Galago crassicaudatus	80	96	80	92	0	35	39	40
G. genegalesis					0			
G. demidovi					5			
Nycticebus	81	97	93	92	37	0	28	42
Loris	82	91	96	93	52	18	0	46
Arctocebus	89		90	97	37	41		6
Perodicticus	90	101	90	97	37	43	53	0
Tarsius	91		96	94	110	110	112	112
Hylobates							104	
Gorilla							101	
Homo	100	101	97	91	108	102	97	108
Pan							101	
Macaca	92	96	98	89	102	108	105	109
Papio							103	
Erythrocebus							106	
Presbytis	98	96	98	89	105	100	107	111
Colobus	102		103	98	105	102	108	113
Aotes	103		100	104	103	103	109	101
Ateles	105		105	98	104	99	104	117
Lagothrix	105		107	96	106	100	108	117
Cebus	109		102	107	108	102	106	111
Saimiri	111	107	107	104	108	103	106	111
Urogale	116		106	96				118
Tupaia	116	112	111	111	120	112	122	129
Cynocephalus	113		104	106	122	112	122	137
Chiroptera							119	
Carnivora	124	101	107	112	119		120	126
Rodentia	122	112	108	104			125	120
Tenrecidae					123			
Ungulata	125	128	108	112	136		131	132
Erinaceidae	134	120	117	114	134		134	134
Soricidae		128						
Elephantulus	130	128			137			139
Edentata							141	
Metatheria	151	139	125	114	142		143	152
Prototheria	162	141	128	114	140		143	152

[a] Antisera are identified in the column headings of this table by genus with the exception of antisera to *L. mongoz* and *L. catta*. The actual species to which the other antisera were made are as follows: *Propithecus verreauxi, Microcebus murinus, Galago crassicaudatus, Nycticebus coucang, Loris tardigradus, Periodicticus potto.*

[b] Rows in this table usually represent genera. All species within a genus were grouped together for computer analysis when used in comparisons with antisera to heterologous genera. In two cases (*Lemur, Galago*), we have identified rows according to species. The first-listed species in each of both cases contains grouped values for all members of the genus when compared with antisera to other genera. Rows for other members of the genus have values entered only in columns for the members' antisera to members of that genus.

Table 5. Distances of Various Nonprimate Groups from Primates

	Distance value
Tupaioidea	11.30
Dermoptera	11.55
Rodentia	11.96
Proboscidea	12.27
Ungulata	12.28
Carnivora	12.47
Edentata	12.88
Tenrecidae	12.98
Erinaceidae	13.30
Soricidae	13.63
Chiroptera	13.77
Marsupialia	14.26
Macroscelidea	14.51
Monotremata	16.23

The tree shrew and elephant shrew regions of the tree are shown in Figures 3 and 4, and the average antigenic distances of the various nonprimate groups from primates depicted in this tree are shown in Table 5. The antigenic distances between lineages for various primate and nonprimate groups at comparable taxonomic levels are compared in Tables 6–9.

We also analyzed spur size results from about 3500 plate comparisons developed with 36 chicken antisera to 18 species which included 15 primates, a tree shrew, a hedgehog, and an elephant shrew. The distance tables from these

Table 6. Levels of Divergence within Genera

	Before adjustment	After adjustment
Pan (troglodytes vs. *paniscus)*	0.27	0.28
Hylobates[a]	0.00	0.00
Macaca[b]	0.00	0.00
Mandrillus (leucophalus vs. *sphinx)*	0.14	0.20
Galago (crassicaudatus, senegalensis vs. *demidovi)*	0.26	0.51
Lemur[c]	1.34	2.66
Tupaia[d]	1.08	2.97
Elephantulus (myurus vs. *intufi)*	1.09	1.95

[a] *H. lar* vs. *H. pileatus, H. concolor, H. agilis.*
[b] All heterologous species of macaque were grouped together for analysis because of incomplete networks of comparisons within the genus. *M. fuscata* vs. *M. mulatta, fascicularis, nemestrina, speciosa, maurus, radiata, cyclopus.*
[c] *L. catta* vs. *L. mongoz, macaco, fulvus, variegatus.*
[d] *T. palawensis* vs. *T. chinensis, balengeri, glis, longipes, montana, tana, minor.*

Table 7. Levels of Divergence within Subfamilies

	Before adjustment	After adjustment
Homininae (*Homo* vs. *Pan, Gorilla*)	0.68	0.72
Hylobatinae (*Hylobates* vs. *Symphalangus*)	0.20	0.45
Cercopithecinae[a]	0.33	0.43
Colobinae[b]	0.28	0.39
Atelinae (*Ateles* vs. *Lagothrix*)	0.50	1.16
Perodicticinae (*Periodicticus* vs. *Arctocebus*)	0.31	0.60
Tupaiinae (*Tupaia* vs. *Urogale*)	1.39	3.97
Elephantulus vs. *Nasilio*	1.60	3.28

[a]*Macaca, Papio, Theropithecus* vs. *Mandrillus, Cercopithecus, Erythrocebus, Cercocebus.*
[b]*Colobus* vs. *Presbytis, Nasalis, Pygathrix.*

Table 8. Levels of Divergence within Families

	Before adjustment	After adjustment
Hominidae (Homininae vs. *Pongo*)	1.33	1.54
Cercopithecidae (Cercopithecinae vs. Colobinae)	0.93	1.29
Lorisidae (*Loris* vs. *Nycticebus*)	1.42	2.30
Cheirogaleidae (*Microcebus* vs. *Cheirogaleus*)	1.23	3.27
Macroscelididae (*Elephantulus, Nasilio* vs. *Petrodromus*)	2.40	4.99
Sciuridae (ground squirrels vs. tree and flying squirrels)	4.2[a]	

[a]Taken from Hight *et al.* (1974).

Table 9. Levels of Divergence within Superfamilies

	Before adjustment	After adjustment
Hominoidea (Hominidae vs. Hylobatidae)	1.53	2.05
Ceboidea[a]	1.54	3.05
Lorisoidea (Galagidae, Perodicticidae vs. Lorisidae)	2.56	4.22
Lemuroidea (Indriidae, Lepilemuridae vs. Lemuridae)	2.24	4.15
Macroscelideoidea (*Elephantulus, Nasilio, Petrodromus* vs. *Rhynchocyon*)	5.40	10.97

[a]*Cebus, Saimiri* vs. *Ateles, Lagothrix, Alouatta, Aotes, Callimico, Callicebus, Chiropotes.*

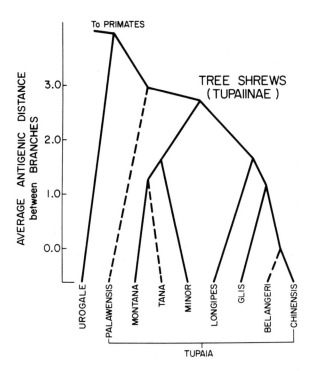

Fig. 3. Heavy lines descend to taxa used as homologous species; dashed lines descend to taxa used only as heterologous species. Tupaioidea unites with Primates at an antigenic divergence level of 11.30.

results were converted into the UWPGM cladogram shown in Fig. 5. The arrangement of primate genera into presumed cladistic or monophyletic groups in the classification shown in Table 10 is based on both rabbit and chicken antisera results. In describing these groups, however, we shall emphasize the rabbit results because the network of comparisons using rabbit antisera was more complete and the antisera generally showed stronger cross-reactions than was true in comparisons with chicken antisera.

3. The Phylogenetic Relationship of the Living Primates

We believe that our analysis of the immunodiffusion data provides evidence primarily on the clade relationships of the primates, not on their grade relationships. We cannot always rule out the possibility, however, that certain lineages might be grouped together by our data by virtue of retention of primitive antigenic features from very ancient common ancestors rather than by inheritance of derived antigenic features from more recent shared ancestry.

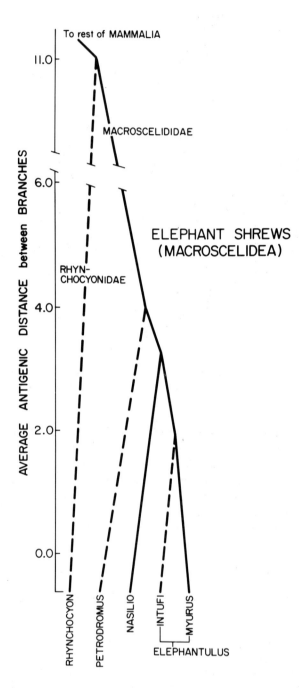

Fig. 4. Heavy lines descend to taxa used as homologous species; dashed lines descend to taxa used only as heterologous species. Macroscelidea unite with other therian mammals at an antigenic divergence level of 14.51.

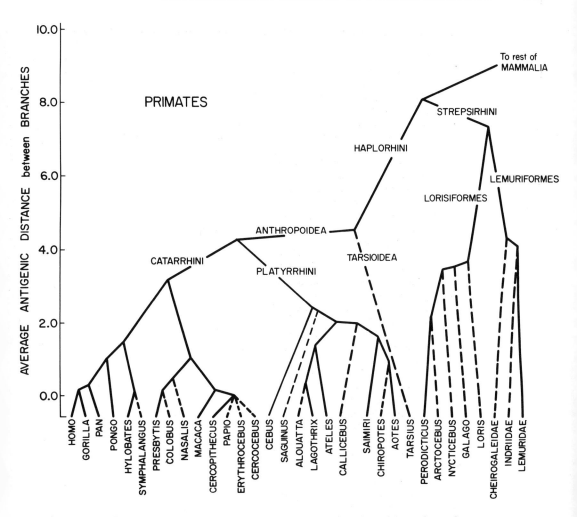

Fig. 5. Heavy lines descend to taxa used as homologous species; dashed lines descend to taxa used only as heterologous species.

Thus, for example, our immunodiffusion UWPGM trees might have grouped together in one branch all the lineages found in Simpson's Prosimii if primitive antigenic features had actually predominated in these lineages. The fact that such a grouping did not occur suggests that most of the groupings observed in our data are indeed due to derived shared antigenic features more than to any other factor.

3.1. Major Branches of Primates

Both rabbit and chicken antisera results depict a clade arrangement for the major primate branches which agrees with the groupings in Hill's classification

Table 10. A Classification of Primates Based on Immunodiffusion Evidence[a]

Order Primates	10.31
Semiorder Strepsirhini (Geoffroy, 1812)	8.87
Suborder Lemuriformes (Gregory, 1915)	7.04
Superfamily Lemuroidea (Mivart, 1864)	4.15
Family Lemuridae (Gray, 1821)	
Genus *Lemur* (Linnaeus, 1758) common lemur	2.66
Family Lepilemuridae (Rumpler and Rakotosamimanana, 1972) (new rank)	
Genus *Lepilemur* (Geoffroy, 1851) sportive lemur	
Family Indriidae (Burnett, 1828)	
Genus *Propithecus* (Bennett, 1832) sifaka	
Superfamily Cheirogaleoidea	
Family Cherogaleidae (Gregory, 1915)	3.27
Genus *Microcebus* (Geoffroy, 1812) mouse lemur	
Cheirogaleus (Geoffroy, 1828) dwarf lemur	
Superfamily Daubentonioidea (Gill, 1872)	
Family Daubentoniidae (Gray, 1870)	
Genus *Daubentonia* (Geoffroy, 1795) aye-aye	
Suborder Lorisiformes (Gregory, 1915)	
Superfamily Lorisoidea (Tate Regan, 1930)	4.23
Family Galagidae (Hill, 1953)	
Subfamily Galaginae (Gray, 1825)	
Genus *Galago* (Geoffroy, 1796) bushbaby	0.51
Family Lorisidae (Gregory, 1915)	2.30
Genus *Nycticebus* (Geoffroy, 1812) slow loris	
Loris (Geoffroy, 1796) slender loris	
Family Perodicticidae	
Subfamily Perodicticinae	0.60
Genus *Perodicticus* (Bennett, 1831) potto	
Arctocebus (Gray, 1863) angwantibo	
Semiorder Haplorhini (Pocock, 1918)	8.90
Suborder Tarsioidea (Elliot Smith, 1907)	
Family Tarsiidae (Gill, 1872)	
Genus *Tarsius* (Storr, 1780) tarsier	
Suborder Anthropoidea (Mivart, 1864)	6.52
Infraorder Platyrrhini (Geoffroy, 1812) New World monkeys	
Superfamily Ceboidea (Simpson, 1931)	3.05
Genus *Ateles* (Geoffroy, 1806) spider monkey	
Logothrix (Geoffroy, 1812) woolly monkey	
Alouatta (Lacepede, 1799) howler monkey	
Saguinus (Hoffmansegg, 1807) tamarin	
Aotes (Humbolt, 1811) night monkey	
Callicebus (Thomas, 1903) titi	
Chiropotes (Lesson, 1840) saki	
Cebus (Erxleben, 1777) capuchin	
Saimiri (Voigt, 1831) squirrel monkey	
Infraorder Catarrhini	3.06
Superfamily Cercopithecoidea (Gray, 1821)	
Family Cercopithecidae (Gray, 1821)	1.29
Subfamily Colobinae (Elliot, 1913)	0.39
Genus *Colobus* (Illiger, 1811) guerezas	

Table 10. *(continued)*

Presbytis (Eschschlotz, 1821) langur	
Pygathrix (Geoffroy, 1812) douc langur	
Nasalis (Geoffroy, 1812) proboscis monkey	
Subfamily Cercopithecinae (Blanford, 1888)	0.43
Genus *Macaca* (Lacepede, 1799) macaque	0.00
Papio (Erxleben, 1777) baboon	
Theropithecus (Geoffroy, 1843) gelada	
Mandrillus (Ritgen, 1824) drill	
Cercocebus (Geoffroy, 1812) mangabey	
Erythrocebus (Trovessart, 1897) patas	
Cercopithecus (Linnaeus, 1758) guenon	
Superfamily Hominoidea (Simpson, 1931)	2.05
Family Hylobatidae (Gill, 1872)	
Subfamily Hylobatinae (Gill, 1872) gibbon	0.45
Genus *Hylobates* (Illiger, 1811) gibbons	0.00
Symphalangus (Gloger, 1841) siamang	
Family Hominidae (Gray, 1825)	1.54
Subfamily Ponginae (Allen, 1925)	
Genus *Pongo* (Lacepede, 1799) orangutan	
Subfamily Homininae	0.72
Genus *Gorilla* (Geoffroy, 1852) gorilla	
Pan (Oken, 1816) chimpanzee	
Homo (Linnaeus, 1758) man	

[a]Values shown in the column to the right represent the average antigenic distance from Fig. 1 and 2 between members of the two most divergent sister groups making up the taxa to which the values are assigned.

but not in Simpson's. Tree shrews must be excluded from Lemuriformes. They can also be excluded from the Primates as a whole in that the four suborders Lemuriformes, Lorisiformes, Tarsiiformes, and Anthropoidea diverge less from one another than from tree shrews and any other eutherians. Lemuriformes and Lorisiformes group together to form one major branch of Primates, the Strepsirhini, and Tarsiiformes and Anthropoidea group together to form the other major branch, Haplorhini. Anthropoidea here is equivalent to Hill's Pithecoidea and includes all members of Simpson's Anthropoidea. Moreover, if Anthropoidea is to be a strict monophyletic taxon, as we believe it should be, then any fossil primates in Simpson's Prosimii judged to be cladistically closer to extant anthropoids than to either tarsiers or extant strepsirhines would have to be shifted into Anthropoidea.

The rabbit cladogram separates haplorhines and strepsirhines (Figs. 1 and 2) by an antigenic distance of 10.31. In turn, tupaiids (tree shrews) and Dermoptera (flying lemur) diverge from these primates by distances of 11.30 and 11.55, respectively, but are nevertheless closer to Primates than to other mammals (Table 5). The fact that elephant shrews diverge very markedly from both tree shrews and Primates argues strongly against older views which grouped tree shrews and elephant shrews in Menophyla as a division of the primitive eutherian order Insectivora. Microcomplement fixation tests with rabbit antisera to albumins and transferrin place tupaiids and flying lemur especially close to Primates, so close in fact that their distance from Anthropoidea is no greater than that of either tarsier or strepsirhines (Sarich and Cronin, this volume). Their greater distance in our test results from examination of a broad range of serum proteins rather than just albumin and transferrin. While our data are not inconsistent with the view that tree shrews are an early branch of Primates, they are just as readily reconciled with schemes in which tree shrews represent the closest nonprimate order to Primates.

Primates, Tupaioidea, and Dermoptera might be grouped together in the superorder Archonta as was suggested by Gregory (1910) and more recently by Butler (1956) and McKenna (1975). However, this superorder, as we would recognize it, excludes elephant shrews (included by Butler) and Chiroptera (included by McKenna). Some amino acid sequence data (Romero-Herrera *et al.*, this volume) are in disagreement with these conclusions, however. In these studies, definite similarities between tree shrew and hedgehog myoglobin sequences have been seen. This observation is particularly relevant to the issue at stake, because the amino acid sequence of tree shrew myoglobin is phenetically closer in fact to catarrhine primate myoglobin sequences than to any other mammalian myoglobin (i.e., shows fewer amino acid differences and smaller minimum mutation distances from the catarrhines), in part because of the retention of primitive features of mammalian myoglobin by both tree shrews and catarrhines. In turn, this suggests that the high degree of antigenic similarity of tree shrews to anthropoids with respect to albumin and transferrin might simply reflect a patristic relationship, i.e., primitive albumin and transferrin specificities, rather than recent common ancestry. No sequence data are available on Dermoptera at present. At the very least, immunodiffusion data are opposed to the inclusion of tree shrews and flying lemur in a taxonomic group which includes Strepsirhini but not Haplorhini.

Our results do not place elephant shrews (Macroscelidea) close to any member of Archonta. In fact, their distance from all other eutherians is even slightly greater than the Marsupialia-Eutheria distance. No particular significance, however, is attributed to this observation. Marsupial samples react very poorly with eutherian antisera and therefore placement of this group is very sensitive to the nature of the comparison networks employed. As a result, marsupial samples were not part of all comparison series and happened not to have been used with either elephant shrew antiserum. It is best then to state only that Macroscelidea and Marsupialia are the most divergent from the Primates of the therians tested and not attempt any more specific conclusions.

3.2. Relationships of Strepsirhine Primates

The relationships of strepsirhine primates are discussed in detail in a recent article (Dene *et al.*, 1976). Both chicken and rabbit antisera divide

Strepsirhini into two subgroups, Lorisiformes and Lemuriformes. In turn, the lorisiforms subdivide into three groups separated by sizable antigenic distances. It seems best to us to treat these groups as separate families, Lorisidae for Asian lorises (*Nycticebus* and *Loris*), Galagidae for African bushbabies (*Galago*), and Perodicticidae for African pottos (*Perodicticus* and *Arctocebus*). In the rabbit cladogram (Fig. 2), African pottos and African bushbabies show slightly less divergence from each other than from Asian lorises. However, in the individual immunodiffusion comparisons no consistent trend emerged; while the spurs of certain precipitin bands placed lorises and bushbabies closer together, spurs of other bands placed lorises and pottos together, and yet other spurs bushbabies and pottos together. Our findings raise the possibility that the striking similarity between Asian lorises and African pottos in their slothlike locomotor adaptations may be due as much to convergent evolution as to shared common ancestry.

Our data depict the Malagasy Lemuriformes as a monophyletic taxon. *Daubentonia,* so aberrant morphologically with its rodentlike incisors, is also antigenically quite different from other lemuriforms. Its lemuriform placement in the rabbit cladogram, while probably accurate, is based on quite inadequate comparisons because we had only a trace of plasma from this species to use in the immunodiffusion tests. The more adequately tested lemuriforms divided into two major groups which we treat as superfamilies, Cheirogaleoidea for mouse and dwarf lemurs (*Microcebus* and *Cheirogaleus*) and Lemuroidea for medium- and large-sized lemurs (*Lemur, Lepilemur,* and *Propithecus*). Contrary to recent suggestions by many morphologists that Cheirogaleoidea be returned to Lorisiformes where it was placed by earlier taxonomists (e.g., Flower, 1892), our cladogram strongly supports its inclusion in Lemuriformes.

3.3. Relationships of Haplorhine Primates

The evidence for the common origin of Tarsiiformes and Anthropoidea, i.e., for the validity of Haplorhini in a clade classification, is quite clear in our opinion. In the antigenic distance table for rabbit antisera to *Tarsius* (Table 2), there are 19 anthropoid genera and 16 of these are closer to *Tarsius* than are any strepsirhines and nonprimates. Moreover, among the seven strepsirhine genera tested in this comparison series, five are more distant than any anthropoid. (Photographs of a few of the immunodiffusion comparisons from which this table was calculated are shown in Fig. 6.) With antisera to anthropoid species, *Tarsius* always reacts better than strepshirhine species (Table 3); with antisera to strepsirhine species (Table 4), *Tarsius* on the average reacts to about the degree typical of anthropoids. The rabbit cladogram (Figs. 1 and 2) indicates that *Tarsius* is about as distant from Anthropoidea (8.90) as the two strepsirhine groups are from one another (8.87). On the chicken cladogram (Fig. 5), *Tarsius* is considerably closer to Anthropoidea but appears in very few distance tables. The evidence from hemoglobin amino acid sequences also

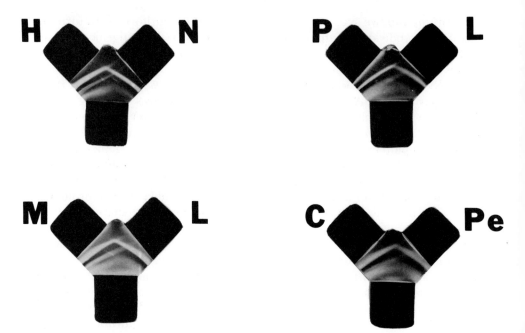

Fig. 6. Photographs of trefoil Ouchterlony plate comparisons using anti-*Tarsius* antisera. Test antigens were from the following primates: *Cebus albifrons* (C), *Homo sapiens* (H), *Loris tardigradus* (L), *Macaca fuscata* (M), *Nycticebus coucang* (N), *Pan troglodytes* (P), *Perodicticus potto* (Pe).

supports the haplorhine placement of *Tarsius* (Beard and Goodman, this volume).

Both rabbit and chicken cladograms divide the Anthropoidea as in Hill's classification into Platyrrhini and Catarrhini. A preliminary report of the relationships among platyrrhines from immunodiffusion data has been presented (Baba and Goodman, 1975) and a detailed account is in preparation. Although levels of antigenic divergence within Platyrrhini as a whole are similar to those within Catarrhini (3.05 compared to 3.06 in the rabbit cladogram), the ceboid genera among themselves show levels of divergence roughly equivalent to that found between catarrhine families. One traditionally recognized ceboid subfamily is confirmed as a monophyletic group by immunodiffusion data; Atelinae for spider monkey (*Ateles*) and woolly monkey (*Lagothrix*). Howler monkey (*Alouatta*) appears to be closely related to Atelinae, but otherwise the various genera are sharply separated from one another. On morphological grounds, Ceboidea divides into two families, Callithricidae for marmosets and Cebidae for the more simian-appearing group. Our rabbit cladogram, however, depicts certain cebids (e.g., *Aotes* and *Callicebus*) as closer to marmosets

(*Saguinus*) than to other cebids. This suggests that although the larger taxon Ceboidea is clearly monophyletic, the smaller one Cebidae as currently constructed is polyphyletic.

Rabbit and chicken antisera results both divide the Catarrhini into the sister groups Cercopithecoidea and Hominoidea. They also agree with all currently used primate classifications in separating Cercopithecoidea into Colobinae for the leaf-eating Old World monkeys and Cercopithecinae for the more omnivorous and terrestrial ones. The subbranches within these Old World monkey subfamilies shown in Figs. 1 and 5, however, are not considered significant as these regions of the tree are represented by only a few homologous species and the antisera produced to them detected only slight divergencies at the subfamily level.

In the hominoid region of the rabbit and chicken cladograms, the most ancestral separation is between gibbons and all other Hominoidea. The nongibbon hominoids then separate into two sister groups, one for orangutan (*Pongo*) and the other for *Gorilla*, chimpanzee (*Pan*), and *Homo*. The small antigenic distances among the three genera in this African ape-human group are comparable to those observed within a subfamily of Old World monkeys and significantly less than that which separates the group from *Pongo*. Classifications of Primates which weight the adaptive significance of particular anatomical traits invariably place *Pongo*, *Gorilla*, and *Pan* together in the subfamily Ponginae of the family Pongidae while assigning *Homo* as the only extant primate to a separate family Hominidae. The genealogy of the Hominoidea can be better described, in our opinion, by eliminating Pongidae and redefining Hominidae. This places the extant gibbons (Hylobatinae) in the family Hylobatidae, *Pongo* in Hominidae as the only living genus of Ponginae, and *Gorilla*, *Pan*, and *Homo* together in Homininae as the other subfamily of Hominidae.

4. Slow Antigenic Evolution in Catarrhines

On the basis of the genealogical arrangement of primates deduced from the immunodiffusion data, the relative amounts of antigenic evolution can be compared in a meaningful way in different groups at comparable taxonomic ranks. This is done for rabbit antisera results in Tables 6–9 and also in Table 10, in which we use a strictly clade approach to classify the Primates. The levels of divergence within groups were calculated from the unadjusted distance values in the various comparison series and also from the values after adjustment for better reciprocity. Both sets of results are given in Tables 6–9. In general, more adequate networks of comparisons progressing gradually from close to intermediate to distant species had been carried out with antisera to catarrhine species than with antisera to noncatarrhines as the latter often lacked the proper range of intermediately distant species. Consequently, in the unadjusted distance tables using antisera to catarrhine species noncatarrhines often showed a greater distance from catarrhines than in the tables for the reciprocal comparisons using antisera to the noncatarrhines. The adjustment procedure in achieving better overall reciprocity between tables might tend in some tables to overrate the distance of the homologous to its more closely related heterologous species. Because of this possible artifact, we consider the unadjusted values more accurate in assessing levels of divergence at the lowest taxonomic ranks such as within genera, but the adjusted values more accurate at higher taxonomic ranks such as within families. Furthermore, since the adjustment procedure tends to accentuate divergence levels at the lower taxonomic ranks in noncatarrhine groups as compared to catarrhine groups, the unadjusted values more stringently test the hypothesis of a slowdown in rates of antigenic evolution in catarrhines.

The results summarized in Table 6 indicate that levels of antigenic divergence between the species within genera are considerably less for catarrhines than for other mammalian groups. In some cases, *Hylobates* in particular, no antigenic differences were detected between member species. This is seldom the case in noncatarrhine genera. In Catarrhini, levels of divergence comparable to those seen within such noncatarrhine genera as *Lemur, Tupaia,* and *Elephantulus* can be found only between the most anciently separated branches within families (compare Tables 6 and 8). The lower degree of antigenic divergence for catarrhines than for noncatarrhine groups also holds on comparing the distant branches within superfamilies (Table 9).

Either the evolutionary origin of the Catarrhini was more recent than would be expected for a primate infraorder or antigenic evolution slowed in catarrhine lineages. Firm fossil evidence bearing on *Lemur,* tree shrew, or elephant shrew divergence is lacking. However, evidence for catarrhine groups is better. If assignment of certain species of *Dryopithecus* to either the *Pan* or the *Gorilla* ancestry is valid (i.e., Simons, 1972) and if *Ramapithecus* is cladistically closer to *Homo* than to *Pan* or *Gorilla,* then at least 14 million years has elapsed since the different genera of Homininae separated. If one assumes constant rates of molecular evolution, species separations within *Lemur, Tupaia,* or *Elephantulus* according to the observed antigenic divergences would be 1.5–2 times more ancient than the separations within Homininae. Morphological data for *Tupaia* presented by Lyon (1913) do not indicate a very ancient separation between different species in this genus. In fact, many species distinctions within *Tupaia* have recently been rejected by Napier and Napier (1967). It may also be noted that the fossil record stretches back no farther than the Oligocene on macroscelidoid elephant shrews (Patterson, 1965), also on sciuroids (Black, 1963), and no farther than the Miocene on lorisoids (Walker, 1974). This suggests that these groups are no more ancient than Hominoidea in that its fossil record also stretches back into the Oligocene (Simons, this volume). We feel that the best explanation for our findings on levels of antigenic divergence which still takes into account the extensive work done in morphology and paleontology is that antigenic evolution of serum proteins has slowed considerably in catarrhines as compared to nonanthropoid primates or to nonprimate mammals.

A strong note of caution is necessary, however, because of the incompleteness of the mid-Tertiary fossil record on early catarrhines and because mistakes are frequently made in determining the phylogenetic affinities of fossil specimens. Moreover, Sarich and Cronin (this volume) support with substantial (although not conclusive) evidence the more recent catarrhine branch times which they propose from calculations based on the clock model of molecular evolution. In their immunological tests, lemuriforms and lorisiforms diverge on the average as much in their albumins and transferrins from carnivores, bats, and rodents as do hominoids and other anthropoids. They argue cogently that the logical way to handle such data indicative of equidivergences is to have the branch times within the Anthropoidea directly proportional to the degrees of divergence between the branches, i.e., to use the clock model's assumption of uniform evolutionary rate.

Methodological and biological considerations, however, tend to weaken their argument. First, if the conclusion from immunodiffusion tests based on a range of serum proteins that anthropoids are cladistically closer to lemuriform and lorisiform primates than to tree shrews proves to be correct, then a cladogram reflecting this conclusion ought not to reveal in equidivergence of anthropoids and strepsirhines (lemuriforms and lorisiforms) from their common ancestor in the type of calculations used by Sarich and Cronin, but instead ought to show less divergence of anthropoids if tree shrews are taken as the outside reference group. Second, and more important, this method of calculation is based on the "additive" hypothesis of evolution, and as discussed by Moore (this volume) the values calculated could be incorrect or misleading because certain of the assumptions underlying this additive hypothesis do not appear to reflect what actually happens in the evolutionary process. Finally, an argument can be developed (Goodman, this volume) that in the early stages of an adaptive radiation molecular evolution as well as morphological evolution might progress at a faster rate than later in descent of the lineages. Specifically, positive natural selection might have speeded up rates of molecular evolution in early Tertiary Primates, especially on the line ancestral to the Catarrhini, and then stabilizing or negative selection could have drastically slowed these evolutionary rates in later anthropoids.

5. Conclusion

The classification of Primates presented in Table 10 summarizes the immunodiffusion evidence on the genealogy of the Primates and on the amounts of antigenic divergence between branches.

ACKNOWLEDGMENTS

In addition to those scientists acknowledged in previous articles, the authors would like to express their appreciation to Dr. Sarwonono Prawiro-hardjo, Director, The Bureau for International Relations (L.I.P.I.), Djakarta, Mrs. Sjamsjah Achmad (L.I.P.I.), and H. M. Kamil Oesman and Mr. L. Darsono, Fa. Primex, Djakarta, for their help in obtaining blood samples from Indonesia, to M. Baba and L. L. Darga for their analyses of New and Old World monkey materials represented, to Dr. G. W. Moore for designing the computer programs used in this study, and to Mr. Walter Farris for his frequent help. This research was supported by NSF grant GB 36157.

6. References

Baba, M., and Goodman, M., 1975, Phylogenetic relationships of the Ceboidea viewed immunologically, *Am. J. Phys. Anthropol.* **42:**288.

Baba, M., Goodman, M., Dene, H., and Moore, G. W., 1975, Origins of the Ceboidea viewed from an immunological perspective, *Hum. Evol.* **4:**89.

Black, C. C., 1963, A review of the North American Tertiary Sciuridae, *Bull. Mus. Comp. Zool.* **130:**109.

Butler, P. M., 1956, The skull of *Ictops* and the classification of the Insectivora, *Proc. Zool. Soc. London* **126:**453.

Dene, H., Goodman, M., Prychodko, W., and Moore, G. W., 1976, Immunodiffusion systematics of the primates. III. The Strepsirhini, *Folia Primatol.* **25:**35.

Flower, W. H. 1892, Lemur, *Encyclopaedia Britannica*, 9th ed. **14:**440.

Goodman, M., and Moore, G. W., 1971, Immunodiffusion systematics of the primates. I. The Catarrhini, *Syst. Zool.* **20:**19.

Gregory, W. K., 1910, The orders of mammals, *Bull. Am. Mus. Nat. Hist.* **27:**1.

Hight, M. E., Goodman, M., and Prychodko, W., 1974, Immunological studies of the Sciuridae, *Syst. Zool.* **23:**12.

Hill, W. C. O., 1953, *Primates—Comparative Anatomy & Taxonomy: I—Strepsirhini*, University Press, Edinburgh.

Lyon, M. W., 1913, Tree shrews: An account of the mammalian family Tupaiidae, *Proc. U.S. Natl. Mus.* **45:**1.

McKenna, M. C., 1975, The phylogenetic classification of the Mammalia, in: *Phylogeny of the Primates: An Interdisciplinary Approach* (W. P. Luckett and F. S. Szalay, eds.), pp. 21–46, Plenum Press, New York.

Moore, G. W., 1971, A mathematical model for the construction of cladograms, North Carolina University Institute of Statistics, mimeo ser. No. 731.

Moore, G. W., and Goodman, M., 1968, A set theoretical approach to immunotaxonomy: Analysis of species comparisons in modified Ouchterlony plates, *Bull. Math. Biophys.* **30:**279.

Napier, J. R., and Napier, P. H., 1967, *A Handbook of Living Primates*, Academic Press, London.

Patterson, B., 1965, The fossil elephant shrews (family Macroscelididae), *Bull. Mus. Comp. Zool.* **133:**295.

Romer, A. S., 1966, *Vertebrate Paleontology*, 3rd ed., University of Chicago Press, Chicago.

Simons, E. L., 1972 *Primate Evolution: An Introduction to Man's Place in Nature*, Macmillan, New York.

Simpson, G. G., 1945, The principles of classification and a classification of mammals, *Bull. Am. Mus. Nat. Hist.* **85:**1.

Sokal, R. R., and Michener, C. D., 1958, A statistical method for evaluating systematic relationships, *Kan. Univ. Sci. Bull.* **38:**1409.

Walker, A., 1974, A review of Miocene Lorisidae of East Africa, in: *Prosimian Biology* (R. D. Martin, G. A. Doyle, and A. C. Walker, eds.), pp. 435–448, Duckworth, London.

Evolutionary Rates in Proteins: Neutral Mutations and the Molecular Clock

10

WALTER M. FITCH and CHARLES H. LANGLEY

1. Introduction

There is an interesting relationship between neutral mutations and the molecular clock. The theory of neutral mutations requires that the rate of fixation of a mutation be equal to the neutral mutation rate. Thus the fixation of neutral mutations should be clocklike, with each "tick" of the clock representing another fixation. Naturally the clock will not be metronomic but, like a radioactive clock, stochastic, with fixation events in the unit time interval showing a Poisson distribution. Although a test of the clock hypothesis is a test of the neutral hypothesis, the existence of a clock does not depend on the correctness of the neutral hypothesis (see also Sarich and Cronin, this volume). This has resulted in past confusion. In a similar fashion, the covarion (concomitantly variable codons) concept (Fitch and Markowitz, 1970) is also independent of the correctness of the neutral hypothesis. We shall consider in turn a statistical model to test evolutionary rates, the results of that test, a comparison of our estimated rates with those from other sources, and some problems in testing evolutionary clocks.

WALTER M. FITCH • Department of Physiological Chemistry, University of Wisconsin Medical School, Madison, Wisconsin 53706. CHARLES H. LANGLEY • National Institute of Environmental Health Services, Research Triangle Park, North Carolina 27709.

2. The Neutral Theory

Kimura (1968) originally proposed the theory of neutral mutations to account for a discrepancy between the maximum rate at which evolution was expected to be able to occur according to Haldane's (1957) calculation of the cost of natural selection and the rate at which mutations were apparently being fixed. Since the observed rate was estimated at 2 orders of magnitude greater than the estimated maximum rate of Haldane, some adjustments were required. The number of unproved assumptions underlying both estimates is considerable, and probably every one of them has been challenged in attempts to resolve the conflict. It is indicative of Kimura's insight that, like a supreme court trying to do justice with as little disturbance as possible to the structure of the law, he found a way to reconcile the two calculations while preserving both. In particular, he noted that Haldane's calculations applied only to those advantageous mutations fixed by selection acting on that advantage. It was necessary only to assume that 99% of the fixations proved neither beneficial nor detrimental to the organism at the time they were fixed. A selection rate of 1% of the total rate nicely accounts for 2 orders of magnitude. But what was intended to provide for the common defense of both estimates did not promote the domestic tranquility of a biological community which firmly believed that, if there were a difference, there had to be a reason for it.

The preceding argument provides the motivation for the original formalization of the theory, but it should be emphasized that the existence of neutral mutations is a relevant question independent of any "cost" considerations, and those considerations are not much used currently to support the existence of neutral mutations. This particular view of the question does, however, lend itself to a readily understandable test of the neutral mutation theory. If 99% of all mutations fixed during evolution of proteins are neutral, then the evolution

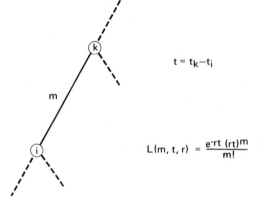

Fig. 1. Likelihood of observing x nucleotide substitutions. Node k is ancestral to its descendent node i, the elapsed time in that interval is t, and λ is the rate of amino acid changing nucleotide substitutions per unit time. The likelihood of observing x substitutions, given t and λ, is simply the Poisson expression on the right.

$$t = t_k - t_i$$

$$L(m, t, r) = \frac{e^{-rt}(rt)^m}{m!}$$

of these proteins should be proceeding at a uniform rate equal to the neutral mutation rate. Indeed, an apparent uniformity of evolutionary rate was one of the major pieces of evidence provided by Kimura in support of his hypothesis. That evidence was not, however, really strong, and a more critical test would be of value in two ways: (1) the more difficult the test, the more plausible the theory if it passes the test; and (2), if passed, the test would provide an estimate of the neutral mutation rate for those mutations that change the encoded amino acid. We will show that there is significant nonuniformity of rates of nucleotide substitutions during the evolution of several mammalian structural proteins. As a consequence, the molecular clock is not as accurate as would be expected. We will also show, however, that averaged over sufficient time and proteins the estimates of divergence times, measured in nucleotide substitutions, correspond reasonably to those estimated from the paleontological record.

3. The Neutral Model

The expected number of amino acid changing nucleotide substitutions in a given interval of time is simply λt, where t is the length of the interval ($= t_k - t_i$, where t_k and t_i are the times at the beginning and end, respectively, of the ith evolutionary interval) and λ is the substitution rate. If x substitutions are in fact observed, the likelihood of observing that number is $L(x) = (\lambda t)^x \exp(-\lambda t)/x!$ (see Fig. 1). This computation can be repeated for every interval i and every protein m. The preceding equation then becomes

$$L(x_{m,i}, m, i) = [\lambda_m(t_k - t_i)]^{x_{m,i}} \exp[-\lambda_m(t_k - t_i)]/(x_{m,i})!$$

the only difference being additional subscripts to particularize the protein and interval being examined. The likelihood, L, of all observations is simply the product of all of the likelihoods for all of the observations, i.e.,

$$L = \prod_{m,i} L(x_{m,i}, m, i)$$

Since the null hypothesis being tested is that nucleotide substitutions are uniform over time, it is clear that we may measure time in nucleotide substitutions. Moreover, since the interval is arbitrary, we may define its unit as the number of years that, on an average, are required to fix one nucleotide substitution in the collection of proteins being examined. As a consequence, $\Sigma_m \hat{\lambda}_m = 1$. A computer is used to obtain those values of λ_m and t_i, designated $\hat{\lambda}_m$ and $\hat{t}_i$, that maximize L. This likelihood value is

$$L_{max} = \prod_{m,i} \{[\hat{\lambda}_m(\hat{t}_k - \hat{t}_i)]^{x_{m,i}} \exp[-\hat{\lambda}_m(\hat{t}_k - \hat{t}_i)]/(x_{m,i})!\}$$

Because it is the product of many probabilities, L will be very small. The crucial point is its relationship to the maximum value that L could assume if the observations were in fact the expected values. This value

$$\hat{L} = \prod_{m,i} (x_{m,i})^{x_{m,i}} \exp(-x_{m,i})/(x_{m,i})!$$

is a measure of the probability of the goodness of fit of the observations under the uniform-rate hypothesis and is obtained by treating the value of $-2 \ln(L_{max}/\hat{L})$ as a χ^2 estimate with degrees of freedom equal to the number of observations less the number of intervals plus rates that were estimated. Finally, it is possible to apportion χ^2 into two components, one of which (among legs over proteins) indicates the extent to which the total rate of substitution for all proteins combined is the same among the various intervals and the other of which (among proteins within legs) indicates the extent to which the relative rates of substitution among the proteins are the same for every interval. A more complete exposition of the method is given in Langley and Fitch (1973, 1974).

4. Counting Nucleotide Substitutions

The nucleotide substitutions are counted by the method of Fitch (1971) and Fitch and Farris (1974) (see also Moore, this volume). In the present case, we assume that the phylogeny or ancestral relationship is, with respect to the order of divergence, as shown in Fig. 2. The amino acid sequences of the proteins of those taxa are then translated into their messenger RNA sequence through the knowledge of the genetic code. The nucleotide substitutions are counted by the procedure shown in Fig. 3, where initially one has the topology but nucleotides are present only at the tips. One works upward from the tips, placing in a given node those nucleotides in common to both immediately descendent nodes, i.e., the intersection of the descendent nucleotide sets. If there are none in common (the intersection is empty), then one assigns to their immediate ancestor all nucleotides in either descendant, i.e., the union of the two sets. The minimum number of nucleotide substitutions required to account for the known sequence is the number of times a union was required. In the case shown, only one union was required (G/U node). The procedure is shown for but a single nucleotide position and accordingly must be repeated for each nucleotide in the message or 3 times the number of different amino acid positions.

There are three additional points about the procedure. If we wish to know not only how many substitutions are required but also in which intervals they occurred, we need to perform some additional operations. In the case of Fig. 3, the upper ambiguous position must be reduced to guanine, while the two descendant ambiguous positions must be reduced to uracil, thus revealing that the required substitution was a G $\rightarrow$ U change in the descent from the penultimate node. There is insufficient time to explain the algorithm for

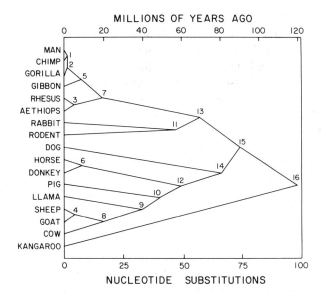

Fig. 2. Assumed phylogeny of the species for which sequences were examined. The order of branching was assumed. Thoe nodes depicting speciation are placed on the abscissa according to the maximum-likelihood solution for the number of nucleotides substituted in all seven proteins. The upper scale does not give the true times of divergence but rather the estimated times of divergence if the marsupial-placental divergence were placed at 120×10^6 years ago and time were directly proportional to the nucleotide substitution scale. Numbers are by increasing abscissal value. From Fitch and Langley (1976).

performing this task, but it is not difficult and the details can be found in the previous reference (Fitch, 1971).

The second point is that serine, leucine, and arginine codons occasionally lead to complications that need to be recognized because of their potential for ambiguity in more than one nucleotide position. This is handled in Fitch and Farris (1974). An alternative procedure can be found in Moore *et al.* (1973).

The third point is that not all the historical substitutions are necessarily

Fig. 3. Parsimony operation. A single nucleotide position for a gene is shown for each of five taxa that are presumed to be divergently related according to the branching network provided. This could arise if this were the first nucleotide position of three codons for valine (G), one for leucine (C/U), and one for serine (A/U). The method, explained briefly in the text and more extensively elsewhere (Fitch, 1971), reveals that only one nucleotide substitution (G → U) is required in this position. To account for the

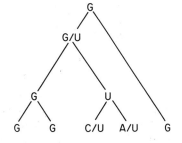

serine, a U → C substitution in the second position of the codon would also be required, and the third position of the codon would have to be a purine if the amino acids in the descendants were to be accounted for in only two substitutions.

accounted for. If a given position is alanine in all taxa examined, then no nucleotide substitutions are required. But the alanine codons are GCX, where X may be any one of the four bases. Clearly, substitutions in the third position would not be observable in these data. Moreover, consider the case where the taxa represented previously were mammals and invertebrates and we now examine three more taxa, a fish, a frog, and a finch, with the result that all prove to have glycine in this position. It is clear that parsimony requires of the new data that there was an alanine-to-glycine replacement in the descent from the invertebrate to the vertebrate ancestor and a glycine-to-alanine replacement from the amniote to the mammalian ancestor. This requires that the alanine, previously viewed as unchanged between the invertebrate and mammalian ancestors, now be considered as having been substituted but to have subsequently back-mutated. Thus we can also miss nucleotide substitutions because not all the pertinent taxa were represented in the sample. Indeed, it is possible that the relevant taxa may now be extinct, in which case they can never be examined. Since the number of such missed cases in any given interval will increase as the length of that interval increases, there is a bias that must be corrected. The extent of that bias is seen by examining the nonlinearity of substitutions observed and expected under the uniform rate hypothesis (Fig. 4) and correcting the expected substitutions accordingly. This procedure was detailed in Langley and Fitch (1974).

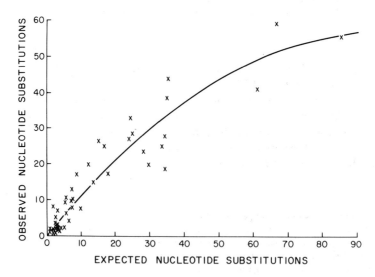

Fig. 4. Underestimate of observed nucleotide substitutions for the longer intervals. The observed number of amino acid changing nucleotide substitutions is plotted against the maximum-likelihood number of expected substitutions for the same data. Their relation is clearly nonlinear. The line shown is the best-fitting quadratic through the origin and is $y = 1.168x - 0.006x^2$, where y is the observed and x is the expected number of substitutions. Higher-order terms do not significantly improve the fit. The quadratic expression was used to correct the expected data for nonlinearity. It is intended only for that purpose and is clearly not of theoretical significance since y would become negative at very large values of x. From Fitch and Langley (1976).

Table 1. Species and Amino Acid Sequences Used in This Study[a]

		Cyt$c$	FibA	FibB	Hbα	Hbβ	InsC	Mb
1	Man	+	+	+	+	+	+	+
2	Chimpanzee	+	+	+	+	+		+
3	Gorilla		+	+	+	+		
4	Gibbon		+	+		+		+
5	Rhesus	+	+		+	+		M. irus
6	Vervet		+	+	+	+	+	
7	Rabbit	+	+	+	+	+		
8	Rodent	mouse	rat	rat	mouse	mouse	rat	
9	Dog	+	+	+	+	+		badger
10	Horse	+	+	+	+	+	+	+
11	Donkey	+	+	+	+			
12	Pig	+	+	+	+	+	+	
13	Llama	guanaco	+	+	+	+		
14	Sheep	+	+	+	+	+	+	+
15	Goat		+	+	+	+		
16	Cow	+	+	+	+	+	+	+
17	Kangaroo	+	+	+	+	+		+
18	Chicken	+			+	+	duck	
19	Other	frog	lizard		carp	frog		

[a]The plus sign indicates a taxon sequence used in this study. Sequences may be found in Dayhoff (1972) except as follows: cytochrome c, mouse (E. Margoliash, unpublished) and guanaco (Niece, Margoliash, and Fitch, unpublished); α and β chains of vervet hemoglobin (Matsuda *et al.*, 1973); proinsulins (Tager and Steiner, 1972) except duck (Markusen and Sundby, 1973); and myoglobins (Romero-Herrera *et al.*, 1973) except badger (Tetaert *et al.*, 1974).

5. Data

Seven proteins were examined from 18 taxa (plus four others used only as an outside marker to aid in establishing the nature of the bird-mammal divergence). These are presented in Table 1. Blank spaces indicate taxa for which no sequence was available. This gave a total of 687 nucleotide substitutions distributed as follows: 38, 35, 45, 198, 255, 43, and 73 among the proteins cytochrome c, fibrinopeptides A and B, α and β chains of hemoglobin, insulin C-peptide, and myoglobin, respectively. These were found among 104, 13, 11, 129, 146, 22, and 153 codons, respectively, for a total of 578 codons. The fibrinopeptides, α chain of hemoglobin, and insulin C-peptide all have less than the full complement of codons because the procedure is strictly valid only if all the amino acid positions in a protein for which λ is estimated are present in the sequence. It was therefore necessary to eliminate from consideration those positions that were represented by gaps in most proteins or those taxa whose proteins were incomplete or whose presence would introduce many gaps when homologously aligned to the other sequences. This required the removal, for example, of positions 6, 7, and 8 of bovine fibrinopeptide A because of their absence in taxa other than artiodactyls. We also omitted positions 128–139 of α chain of hemoglobin because they have not been sequenced for the llama, but it seemed preferable to include the llama, especially since these positions are, for the most part, not particularly variable among the mammals. On the other hand, rhesus fibrinopeptide B, for which the five N-terminal amino acids are missing, and the dog insulin C-peptide, missing positions 4–11, were not included because these positions are quite informative in other taxa and it seemed preferable to omit the sequence rather than those positions.

The dates of divergence were provided by L. Van Valen (personal communication), except for those among the primates (nodes 1, 2, 3, and 5) following the hominoid-cercopithecoid divergence, which were provided by M. Goodman (personal communication).

6. Test of Fit to Data

Table 2 shows the relative rates of change among the several proteins. It should be remembered that these rates apply strictly only to the collection of positions present in the data. They are reflective of the total protein for cytochrome c, β chain of hemoglobin, and myoglobin, but, of the other proteins, only to the extent that one is willing to assume that the rates of change in the positions examined are representative of those not examined. If one is willing (or finds it necessary) to make this assumption, then the rates of the latter group need to be increased in proportion to the fraction of the sequence not examined and then all rates reduced proportionately so that their sum will equal 1. This is reflected in the row labeled "adjusted" in Table 2. These rates are not greatly different from the previous relative rates found by Langley and

Table 2. Relative Evolutionary Rates of Amino Acid Changing Nucleotide Substitutions[a]

	Cyt$c$	FibA	FibB	Hbα	Hbβ	InsC	Mb
Per gene ($\hat{\lambda}$)	0.049	0.056	0.073	0.249	0.321	0.081	0.171
Per gene (adjusted)	0.042	0.072	0.121	0.238	0.280	0.099	0.148
Per gene (previous)	0.044	0.086	nd	0.216	0.286	nd	nd
Per codon	0.023	0.212	0.326	0.095	0.108	0.181	0.055
Per codon per 10^9 yr	0.38	3.55	5.40	1.58	1.80	3.01	0.91

[a]The proteins are Cyt$c$, cytochrome c; FibA and FibB, fibrinopeptides A and B; Hbα and Hbβ, α and β chains of hemoglobin; InsC, insulin C-peptide; Mb, myoglobin. Rates are relative such that the row sums add up to 1.00. The top row is the result of the maximum-likelihood fit. The next row (adjusted) is an adjustment to include the unexamined positions in four of the proteins and should be viewed with the caution expressed in the text. The adjustment is done solely to facilitate comparison with the data in the third row ("previous"), which are from previous articles (Langley and Fitch, 1973, 1974) that examined only four of these proteins and which account for only 63.2% of the present data. The entry "nd," means that the rate was not determined. Thus the adjusted relative rates, which did sum to 1.00, were multiplied by 0.632 to get the comparisons to the second row shown here. The next to the bottom row shows the relative rates per codon. The bottom row shows absolute rates rather than relative rates and therefore does not add up to 1. It assumes the placental-marsupial divergence to have occurred 120 m.y. ago. Values are obtained by multiplying 98.17 times $\hat{\lambda}$ and dividing that by 0.12. From Fitch and Langley (1976).

Table 3. Relative Distribution of Nucleotide Substitution among the Three Codon Positions during the Evolution of Seven Proteins

Nucleotide position		Cyt*c*	FibA	FibB	Hbα	Hbβ	InsC	Mb	Total
1	Expected	18.04	16.65	21.30	93.79	120.63	20.36	34.56	325.34
	Found	18.00	19.00	21.00	95.69	121.65	15.00	35.00	
2	Expected	15.80	14.58	18.65	82.11	105.61	17.82	30.26	284.84
	Found	17.11	13.00	21.00	82.08	101.65	25.00	25.00	
3	Expected	4.27	3.94	5.04	22.20	28.55	4.82	8.18	77.00
	Found	3.00	3.17	3.00	20.33	31.50	3.00	13.00	
	Total	38.11	35.17	45.00	198.10	254.79	43.00	73.00	687.18

$$\chi_{12}^2 = 11.67, \ p \sim 0.5$$

Fitch (1974) after their multiplication by 0.632 to reflect the fact that these proteins account for only 63.2% of the substitutions in these data. Our unwillingness to advocate accepting the above assumption of the sample's representativeness of all positions stems from the idea that substitutions are more often likely to be acceptable in positions that are not even present in other taxa than they are in the average position that is present in all taxa. Positions that have been inserted or deleted in a few taxa constitute the bulk of the missing data. Moreover, cursory examination of those positions suggests these positions may indeed be more variable.

Of the total number of amino acid changing nucleotide substitutions observed, 0.473, 0.415, and 0.112 of them were in the first, second, and third codon positions, respectively. One can ask whether this distribution varies significantly among the seven proteins examined. Table 3 shows the comparison between what would be expected if this were true and what was found. It is clear that the relative rate of amino acid changing nucleotide substitutions among the three codon positions does not vary significantly among these seven proteins, although 25% of the χ^2 value is attributable to the insulin C-peptide's second nucleotide position.

The results of the maximum likelihood fit are shown in Table 4. Looking at the corrected results, one sees that there is significant nonrandomness in both the relative and overall rates and that the null hypothesis—that amino acid changing nucleotide substitutions are accumulating at uniform rates—is rejectable, even after correcting for the nonlinearity bias. That correction was necessary since χ^2 dropped 17.3 with the removal of only 2 degrees of freedom, showing that nonlinearity was a significant bias in the uncorrected data. This means that the molecular clock is not as regular as a radioactive clock and that there are substantial selective effects involved in the accumulation of amino acid replacements.

Table 4. Tests of Hypotheses[a]

	Uncorrected			Corrected		
	χ^2	df	p	σ^2	df	p
Among proteins within legs (relative rates)	166.3	123	6×10^{-3}	166.3	123	6×10^{-3}
Among legs over proteins (total rates)	99.7	33	8×10^{-8}	82.4	31	4×10^{-6}
Total	266.0	156	6×10^{-7}	248.7	154	6×10^{-6}

[a] Initial maximum-likelihood fit gave the results shown as uncorrected. Allowance for the nonlinearity shown in Fig. 4, as discussed in the text, produced the corrected results. The p value is the probability that results this far removed from expectation would arise by chance if the null hypothesis were true, i.e., if amino acid changing nucleotide substitutions are accumulating uniformly over time. From Fitch and Langley (1976).

The result with the lower p (4×10^{-6}, corrected) shows that the total rate at which all seven proteins together evolve varies significantly in different lines of descent from a common ancestor (among legs over proteins). The other result ($p = 6 \times 10^{-3}$, corrected) shows that the rate at which one protein evolves relative to another varies significantly in different intervals in the same line of descent (among proteins within legs).

7. Selective Implications of Nonuniformity of Rates

The hypothesis being tested assumes a uniform rate over geological time as Kimura (1968) originally assumed. It would be possible to argue that the uniformity of rate ought to be in terms of generations rather than years. Alternatively, Vogel et al. (this volume) have suggested it ought to be in terms of germ-line cell divisions. To answer this contention precisely requires a knowledge of ancestral generation times or germ-line divisions. There is a hint of support for these ideas in that the primate side of the phylogeny shows fewer substitutions than the nonprimate side. However, it is not necessary to take great pains to exclude a failure to compute time in generations or germ-line divisions as the explanation for the uniform rate hypothesis being rejected. The test of relative rates (among proteins within legs) depends only on a comparison of data for which the times are identical regardless of whether measured in years, generations, or germ-line divisions. These times are identical because the comparison is among the proteins in the same interval of descent. Thus the nonuniformity of rates is rejectable apart from any consideration of the way time should be measured. The nonuniformity of evolutionary rate was first given statistical support by Ohta and Kimura (1971). Their result was statistically significant despite a small sample size. This conclusion was later given more substantial support by Langley and Fitch (1973, 1974) and is greatly expanded here. Others have also noted nonuniformity of rates but with evidence less well substantiated statistically. Frequently those analyses depend on an estimate of divergence times, and what might otherwise be significant differences could disappear if the times of divergence are in error in the right way. Noteworthy are articles on hemoglobins from Goodman's laboratory (1974, Moore *et al.*, 1975, and this volume), where several effects are commented on including a nonuniformity of rates.

Those articles also note a significant inequality (not dependent on divergence times) in the rate at which different codons evolve, a result in conformance with the earlier concepts of Fitch and Markowitz (1970) on cytochrome *c* and of Fitch (1972*a,b*) on α and β chains of hemoglobin. More importantly, Goodman and his collaborators noted that many of the changes occurring in positions where they were advantageous functionally (e.g., cooperativity and diphosphate binding) occurred at elevated rates in the historical intervals during which those functions apparently came into being. That evidence, like the data of this chapter, strongly supports a selective mechanism as being involved in a substantial portion of the amino acid replacements.

8. Primate Slowdown

Still another aspect of those articles (Goodman *et al.*, 1974, Moore *et al.*, 1975, and Goodman, this volume) is an apparent slowing down of the rate of substitution among the primates. Tashian *et al.* (1972, and this volume) see a similar slowing down of carbonic anhydrase evolution among the primates. Although the rate estimates do depend on divergence time estimates, it does not seem possible that their error can amount to the factor of 10 needed to make the rates equal. Our data lend support to the suggestion of slower primate protein evolution. We have previously suggested that this may be because most of the substitutions examined derive from α and β chains of hemoglobin so that we may be confirming the analysis of their hemoglobin data rather than the generality of their observation. We have, however, examined the rates of change for the nonhemoglobin sequences among the primate lines and find that all five show excellent agreement with the estimate of the expected amounts of change in the overall analysis (Table 5), indicating that they too are all showing reduced rates among the primates. Note that the total number of observed replacements deviates less than 2% from expectation. Since eight out of eight proteins (these seven plus carbonic anhydrase) show a slowing down among primates, the phenomenon would appear to be general.

The times of divergence in substitutions, shown in Fig. 2, can be plotted against estimated times of divergence, with the result shown in Fig. 5. It can be seen, as just mentioned, that the primates, falling significantly below the line, show a lower evolutionary rate unless we would be willing to make some kind of adjustment. Otherwise, the data fit quite well to a straight line. We infer that, despite significant variability in the rate at which each individual protein replaces its amino acids over moderate evolutionary time intervals, averaged over many proteins the estimates are reasonable; that the clock is erratic from moment to moment but satisfactory if enough ticks are counted, in the same sense that two mean value estimates can be equally good provided that the sample size is appropriately increased for the sample with the larger variance. How much should the sample size (number of substitutions counted) be increased? We note that the number of degrees of freedom (Table 4) is slightly greater than half the value of χ^2. This permits us to estimate the variance of our process as roughly (but for our purposes conservatively) twice that of a normal Poisson process. For the latter, the variance equals the number of observed events so that a standard error (equal to the square root of the variance) of 10 percent may be achieved by counting 100 disintegrations in a radioactive sample. But if the variance is double that of a Poisson process, a sample of size 200 has a variance of 400 and a standard error of 20, which is again a 10% error. Thus an increase in the variance requires a comparable increase in sample size.

The utility of this conclusion is only to put a lower bound on the estimated error since there is an implicit assumption that the errors are randomly distributed among the data. The primate results give us some reason to doubt the validity of that assumption. It is a great virtue of this test that we can reject the uniformity of evolutionary rates without knowledge of divergence times, having found $\hat{t}$'s that best fit the data. To the extent that the true times of divergence are not linearly related to our t's, the variance in the rate of the amino acid changing nucleotide substitutions is greater than estimated.

Table 5.   Primate Rates of Evolution in Nonhemoglobin Proteins[a]

		Cytc		FibA		FibB		InsC		Mb		Σ	
E (all)	Leg(s)	E	F	E	F	E	F	E	F	E	F	E	F
0.9352	MAN	0.0455	0	0.0527	0	0.0679	0			0.160	0	0.3261	0
0.9352	CHI	0.0455	0	0.0527	0	0.0679	0			0.160	1	0.3261	1
1.1580	GOR			0.0653	0	0.0841	0					0.1494	0
6.8916	GIB			0.3887	0.67	0.5003	0			1.176	0.5	2.0650	1.267
3.9043	RHE			0.2202	0							0.2202	0
3.9043	AET			0.2202	0							0.2202	0
0.2228	1			0.0126	0	0.0162	0					0.0288	0
5.7336	2			0.3234	1.33	0.4163	2					0.7397	3.333
9.1874	5			0.5182	0	0.6670	0			1.567	5	2.7520	5
12.1747	3			0.6866	0.33							0.6866	0.333
16.0790	MAN, 1,2,5							1.3020	1			1.3020	1
16.0790	RHE, 3	0.7814	0							2.743	1.5	3.525	1.500
16.0790	AET, 3					1.1673	2	1.3020	0			2.4693	2
15.1438	1,2,5	0.7360	1									0.7360	1
5.9564	1,2									1.016	0.5	1.016	0.500
	Σ	1.608	1	2.539	2.33	2.987	4	2.604	1	6.822	8.5	16.560	16.833

[a] E (all) is the expected total number of substitutions in all seven proteins from the maximum-likelihood solution and, multiplied by the λ's of Table 2, gives the expected number of substitutions (E) for individual proteins in the interval(s) shown under leg(s). F is the number observed. A leg name or number is the interval immediately above the node of that name or number in Fig. 2. Multiple legs arise because of missing sequences. For example, the absence of a cytochrome c for *aethiops* means that any substitutions observed for the *rhesus* leg could have occurred at any time since node 7. Accordingly, the expectation must be based on the sum for the *rhesus* and leg 5 intervals.

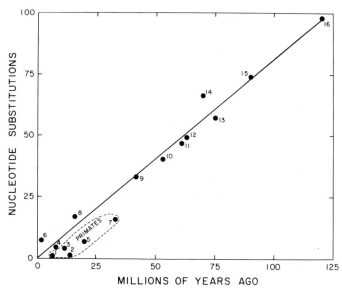

Fig. 5. Linear relation between time elapsed and nucleotide substitutuions. Numbers on points identify the nodes of Fig. 2 that have been plotted according to the number of nucleotide substitutions expected from the maximum-likelihood solution. The divergence times were provided by Goodman (points 1, 2, 3, and 5) and by Van Valen (all other points) without knowledge of our results. The line was simply drawn through the origin and point 16. From Fitch and Langley (1975).

Note that the controversy regarding the primates is quite the same whether one analyzes immunoglobulins as Wilson and Sarich did (1969), analyzes hemoglobins as Goodman's group did (1974), or examines Fig. 5. The problem is how we can get the primate points to fall on the line. The biochemists, such as Goodman, having abundant faith in the work of the anthropologists, assume that the primate points need to be raised vertically and the extent of the needed correction shows how much variability resides in rates of evolution. The anthropologists, such as Sarich, having abundant faith in the work of the biochemists, assume that the primate points need to be moved horizontally and the extent of the needed leftward correction shows how much variability resides in the dating of phyletic divergences. In the spirit of compromise, Tobias (1975), possibly as a paleontologist having abundant faith in no one, has suggested that both sides are wrong and that, upon correction of both parameters, we shall find that we have moved them diagonally.

9. Comparison to Other Rates

In combination with other results, we are in a position to make several comparisons of various kinds of nucleotide substitutions. If we accept the line drawn in Fig. 5, its slope is the rate of amino acid changing nucleotide substitutions in these seven proteins during mammalian evolution. The maximum-likelihood solution gives the t value of node 16 (Fig. 2) as 98.17 substitutions distributed over 1734 nucleotide positions and, accepting 120 million

Table 6. Rates of Silent Nucleotide Substitution[a]

		α	β	H	Total
A	Codons (third position)	11	19	18	48
B	Number of A unchanged	5	12	12	29
C	Substitutions/site [$= -\ln (B/A)$]	0.788	0.460	0.405	0.504
D	Million years of evolution	122	122	120	121
E	Rate$_1$ ($= C/D \times 10^{-3}$)	6.46	3.77	3.38	4.17
F	Rate$_2$ ($= E/0.716$)	9.02	5.27	4.72	5.82

[a]The column headings stand for α and β chains of hemoglobins and histone F2a$_1$. Messenger RNAs were partially sequenced for the human hemoglobins (Forget *et al.*, 1974, and personal communication), the rabbit hemoglobins (Salser *et al.*, 1975, and personal communication), and two species of sea urchin, *Lytechinus pictus* and *Strongylocentrotus purpuratus* (M. Grunstein, personal communication). Row A gives the number of codons whose third-position nucleotide is known in both species. Row B gives the number of third-position nucleotides that are identical. It does not necessarily follow that they are unchanged, but the mathematical treatment operates as if they were unchanged. Row C is the average number of changes per site (B/A) corrected for sites containing multiple changes using the Poisson distribution, where $B/A = e^{-r}$ and r is the corrected value wanted. Row D is twice the number of years (in millions) since the common ancestor of the two species. Row E is the rate in silent nucleotide substitutions per third nucleotide position per 10^9 yr. Row F is a lower bound of the mutation rate in mutations per nucleotide per 10^9 yr assuming that all third-codon position silent mutations are neutral and 0.716 of all third-position mutations are silent. From Fitch (1976).

years ago as the time of the placental-marsupial divergence, gives an evolutionary rate of 0.47×10^{-9} amino acid changing nucleotide substitutions per year per nucleotide position.

We can also estimate the rate of nucleotide substitutions that have no effect on the amino acid encoded (Table 6). The rate, 4.2 silent nucleotide substitutions per year per third nucleotide position, is about 10 times that of those that change the amino acid. It might be noted from this comparison that any attempt to estimate the total number of fixations separating two differing but homologous amino acid sequences on the basis of those sequences alone and on the assumption that fixations are equiprobable in all three positions is certain to be an underestimate.

If one is willing, as a first approximation, to assume that third-position silent mutations are selectively neutral, then the silent substitution rate is the neutral mutation rate. Moreover, the 61 codons can mutate $61 \times 3 = 183$ ways in the third position, of which seven are to terminators and may be excluded as being almost invariably lethal or malefic. Of the remaining 176, 126 do not change the amino acid encoded; i.e., 0.716 of all possible, nonterminating, third-nucleotide changes are silent. (A better estimate might be obtained if we knew the frequency of codon utilization, but it is probably not of great moment here.) Thus the total mutation rate in the third position can be obtained by dividing the silent substitution rate by 0.716, which gives 5.8×10^{-9} mutations per year per third nucleotide position. If the mutation process does not distinguish among the three coding positions, then this is the rate for all nucleotide positions in the structural genes of proteins.

It should be recognized that this rate estimate is a lower bound since the assumption that silent substitutions are neutral may well be false, and the greater the degree to which it is indeed false the greater will be the underestimate of the above mutation rate. The conversion of this number to the more familiar one of mutations per locus per generation is fraught with even greater peril than taken to this point. Apart from not knowing the average size of a structural gene, or the frequency with which these mutations are phenotypically observable, or, worse, the number of generations separating humans and rabbits, there is the perhaps even more tenuous assumption that the mutation of structural genes is either predominant or typical of the process underlying most phenotypic changes.

Knowing the rate for those substitutions that both do and do not change the amino acid, we can determine a total rate of substitution as shown in Table 7. To get the rates on a per nucleotide basis, the amino acid changing rates per codon of Table 2 are divided by 3. Since the silent substitution rates of Table 6 apply to only one-third of the positions, they too are divided by 3. The hemoglobins are given separately because they represent the only genes for which the data are available in both categories. Fibrinopeptide B is given separately because it is the most rapidly changing. Its rate is considerably less

Table 7. Evolutionary Rates of Nucleotide Substitution[a]

Type	Rate	Remarks and assumptions
Amino acid changing		Marsupial-placental split 120 m.y. ago
α, β chains of hemoglobin	0.6	
Fibrinopeptide B	1.8	
Overall	0.5	Including cytochrome c, insulin C-peptide, and myoglobin
Silent substitutions		Corrected for multiple substitutions in any one site
α, β chains of hemoglobin	1.6	From human and rabbit messenger RNA; split 61 m.y. ago
Overall	1.4	Including also two sea urchin histone messenger RNAs
Total		
α, β chains of hemoglobin	2.2	
Overall	1.9	
Single-copy DNA hybridization	1.5	1.5% base change/1°C ΔT_m; rhesus-man split 34 m.y. ago
28 S ribosomal DNA hybridization	0.8	1.5% base change/1°C ΔT_m; baboon-man split 34 m.y. ago

[a]Rates in nucleotide substitutions per nucleotide position per 10^9 yr. The amino acid changing rates are the rates per codon of Table 2 divided by 3. Silent substitution rates are from Table 6 and were divided by 3 since they apply to only the third nucleotide position, whereas we wish the rate for all positions on average. This ignores those few silent substitutions that do not occur in the third position.

than the third-position silent rate and implies that most amino acid alternatives are not evolutionarily acceptable, even here. The total is then found simply by adding the two since they are mutually exclusive.

It is interesting to compare these results to those from DNA hybridization, of which there are two: one on single-copy (unique-sequence) DNA from Kohne (1970) and another on 28 S ribosomal DNA from Jones and Purdom (1975). For reasons of conservatism to be considered later, their raw data have been recomputed on the basis of the assumptions shown in Table 7. It is clear that the α and β chains of hemoglobin are evolving faster than single-copy DNA (2.2 vs. 1.5) and that fibrinopeptide B is evolving twice as fast ($1.8 + 1.4 = 3.2$ vs. 1.5). This is in distinction to suggestions that the fibrinopeptide rate is equal to the single-copy DNA rate of substitution and that these approach the neutral rate. While a factor of 2 or less would not normally be significant in such estimates as these, we believe that the discrepancy is even greater because the protein rates are probably underestimates and the DNA rates probably overestimates, for reasons that will now be discussed. It would be interesting to know how much of the single-copy DNA codes for proteins.

There are at least six sources of uncertainty that should keep us from being overly confident of the preceding conclusions: (1) The calculations depend on the uncertain accuracy of the paleontological datings. The estimate of 34 m.y. ago for the Old World monkey–anthropoid ape split is near the lower bound of a range that goes as high as 50. Any value larger than that used will lower the substitution rate for the DNAs in Table 7. The date of 120 m.y. ago for the placental-marsupial split would have seemed large some years ago and lowering it would increase the amino acid changing substitution rate. It does not seem so large today, however. The 61 m.y. ago for the man-rabbit split was used by Salser, and, although not unusual, is the only one that is not conservative in this context. A figure of 70 would have placed it right on the line of Fig. 5. (2) Leucine and arginine are capable of silent substitutions in the first nucleotide position. These were ignored. Allowance for these would increase, albeit slightly, the silent and total protein gene substitution rate. (3) The correction for missing data changed the expected estimate rather than the observed data, and thus the true amino acid changing substitution rate should be somewhat (perhaps 25%?) greater than given. (4) The relationship between base mispairing and the lowering of the melting point is not very accurately known (see Fig. 3 of Kohne, 1970). Thus Kohne uses 1.5% base changes per 1°C lowering of the hybrid melting point, while Jones and Purdom (1975; Jones, the volume) use 1% per 1.5°C, a difference by a factor of 2.25. All alternatives to the present assumption will lower the DNA substitution rates of Table 7. (5) For any pair of Old World monkey and anthropoid ape, the rates of base-pair substitution in the two lines from their common ancestor differ from their mean by 14% (from Fig. 14, Kohne, 1970). (6) The rate of change of single-copy DNA among the primates may not apply to other lineages such as the rabbit.

Note that the first four sources of uncertainty are such as to underestimate the total nucleotide substitution rate in protein genes (except possibly the time of the rabbit-man divergence) and to overestimate the nucleotide substitution rate in the DNA samples. The overall result is that, contrary to previous assumptions, protein genes are very probably evolving faster than single-copy or 28 S ribosomal DNA. The difference lies in the failure of previous estimates to include the silent substitutions in their estimates of protein gene rates. An interesting conclusion must be that the constraints on DNA change outside of protein structural genes are clearly considerable. They are evolving at less than a quarter of the rate that silent substitutions occur.

It is also possible to estimate a lower bound to the fraction of amino acid alternatives that are evolutionarily acceptable where alternatives must be obtainable from the wild-type gene by a single nucleotide substitution and the sample set for determining acceptability consists of all proteins differing from the wild type by a single amino acid. There are $61 \times 9 = 549$ ways that the various codons can mutate. Of these, 23 are to terminators and 134 are silent, leaving 392 ways of substituting one amino acid for another. Of these, 37 are alternative ways of getting the same amino acid replacement so that there are, on the average, $(392 - 37)/61 = 5.82$ alternative amino acids per codon. Now there must be, by definition, at least one acceptable alternative for each covarion present. For these sequences, excluding myoglobin and fibrinopeptide B, for which the data are not available, the covarions make up 0.31 of the sequences (131 covarions out of 441 codons, Fitch, 1973). Thus for every 5.82 alternative amino acids at least 0.31 or >5% of them are evolutionarily acceptable. This seems contradictory to the prevailing view that the vast preponderance of mutations are deleterious, but it should be remembered that the 5% figure we are here considering applies only to nucleotide substitutions in genes for proteins, whereas it appears, apropos of the prevailing view, that a large fraction of naturally occurring mutations are frameshifts, and, additionally, it may well be that phenotypically important mutants are frequently outside protein structural genes.

An independent estimate of the fraction of amino acid alternatives that are evolutionarily acceptable is available in the rate data already presented. Of the 526 nonterminating nucleotide substitutions possible over the 61 codons for amino acids, the 392 that are not silent constitute 0.745 of them. Thus if 5.8×10^{-9} mutations per year per nucleotide position is the mutation rate then there are 5.8×0.745 or 4.3×10^{-9} amino acid changing mutations per year per nucleotide position. The overall rate of fixation is, however, 0.5×10^{-9} amino acid changing nucleotide substitutions per year per nucleotide position. The latter value is 11.6% of the total mutations and suggests that an even larger fraction of alternative amino acids are evolutionarily acceptable. The two estimates may be reconciled by assuming either that the mutation rate is twice that estimated or that the average number of acceptable alternative amino acids is two per covarion. Of course, estimation errors could also account for the discrepancy. If one were to assume an increased mutation rate, it would also be necessary to assume, subject again to uncertainties in the estimates, that there was selection against silent substitutions. The other explanations, two acceptable alternatives per covarion or estimation errors between 5 and 12, would leave the silent substitutions as effectively neutral.

10. Evaluating Clocks

The following material is basically a résumé of a report published else-where (Fitch, 1976), in which more complete details may be found. A molecular clock requires that there be a measurable event occurring with regularity. This has three components: an event, its measurability, and its regularity, and each will be considered in turn.

The events are to date of only two varieties, amino acid and nucleotide changes inferable from measurements on two or more taxa.

Measurability involves a procedure and its attendant problems. The proce-dures are direct methods such as sequencing either proteins or nucleic acids or, alternatively, indirect methods that determine sequence-dependent properties. Examples of the latter include immunological distance for amino acid differ-ences (Prager and Wilson, 1971; Sarich and Cronin, this volume) and nucleic acid hybridization (Kohne, 1970; Jones, this volume). Attendant problems include the following. *Reliability* is the degree to which the fundamental mea-surement is as stated. This subsumes reproducibility and includes sequencing errors and, for indirect methods, the degree of linearity between the (possibly transformed) data and the number of changes in the sequence as well as the variance in those data. *Sensitivity* is the extent to which sequence changes are picked up and is related to the slope of the linearity mentioned in the previous sentence. *Reciprocity* is the extent to which the measurement of the distance between two species is the same in both directions. It is necessarily identical for the direct methods but not for the indirect ones. Immunological distances in particular have this problem. The standard deviation of the reciprocal pairs of albumin distances is 15% (Maxson and Wilson, 1975) and of lysozyme distances is 20% (Prager and Wilson, 1971). An extreme case is represented by the human-baboon lysozymes, which, depending on which species of lysozyme is used as the antigen, have an immunological distance of 66 or 127 (Wilson and Prager, 1974). Were it not for pooling antisera, the problem would be even greater.

The third part of the clock requirement, regularity, is what is to be tested and depends on some evolutionary hypothesis within which to examine the matrix containing one's distance measurements relating the pairs of taxa exam-ined. The phylogenetic relationships among the taxa may be assumed on the basis of other biological and/or paleontological information or derived from the data themselves. In either case, the method of Fitch and Margoliash (1967) or some variant of it (see, for example, Sarich and Cronin, this volume) is generally used to assign units of change to the intervals of that tree, although other numerical taxonomic techniques have been described (Sneath and Sokal, 1973). The parsimony procedures already discussed are yet another method, but the present discussion is intended to apply primarily to methods appropri-ate to distance matrices.

It is quite true that procedures of the Fitch and Margoliash type (which Sarich uses) do not assume equal amounts of change in two lines of descent

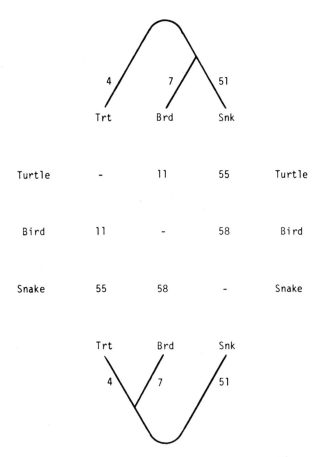

	Trt	Brd	Snk	
Turtle	-	11	55	Turtle
Bird	11	-	58	Bird
Snake	55	58	-	Snake

Fig. 6. Evolutionary rates as a function of the phylogeny. The data are from Jukes and Holmquist (1972), are for cytochrome *c,* are expressed as "relative evolutionary hits" (REH), and are the estimated numbers of nucleotide changes per 100 codons occurring since the common ancestor of the pair. The upper phylogeny conforms to current biological opinion, and the numbers assigned to the legs of the tree by the Fitch and Margoliash (1967) procedure sum to the data in the matrix. Note that 7 and 51 are grossly unequal amounts of change for equal time intervals. The lower tree is the result obtained if the data are allowed to determine the structure of the tree. Note that 4 and 7 are not significantly different amounts of change for equal time intervals.

from a common ancestor. The upper part of Fig. 6, using data on cytochrome *c* from Jukes and Holmquist (1972), admirably illustrates the point on a phylogeny adhering to current biological opinion. Critics who fault Sarich for using a procedure that assumes equality of rates are therefore incorrect so far as the preceding result illustrates their complaint.

There is, however, a more subtle way in which the criticism may become true. If the data themselves are used to construct the phylogeny (as is usually done by Sarich), one employs the time-honored dictum that the most similar taxa are the most closely related. Applying this to the data of Fig. 6, one discovers that the most similar pair is turtle-bird, with the result that the phylogeny becomes instead that shown at the bottom of Fig. 6. The rates of evolution are still unequal, but the difference is no longer statistically significant. Since neither external sources nor the data themselves are always historically correct, the choice of the phylogeny must ultimately depend on the questions asked, and, if they are significance tests, that tree form should be used that biases the result in the direction opposite to the desired conclusion so that the conclusion cannot be attributed to the bias.

It is common to treat the original data so as to increase their value for various purposes. For example, protein sequences are most commonly compared on the basis of the number of their amino acid differences or that number per 100 codons. As discussed fully elsewhere with simple examples (Fitch, 1976), there are three additional, successively higher levels of information available in the raw data. The first is the minimum coding distance, which uses the genetic code to infer the minimum number of base changes to account for the amino acid differences between two proteins. The next higher level notes that further differences are detectable simply by requiring that the same codon for taxon A be used in all its comparisons to other taxa in determining the minimum number of base differences between every gene pair. The highest level is that of the parsimony procedure discussed earlier. Each of the four methods gives a minimum estimate of the amount of change and each in turn gives a larger minimum. If the true value is known to be at least 6 and also at least 4, it seems strange that one must still argue today for the superiority of using 6 rather than 4 as an estimate of that value.

All of the preceding estimates deal only with changes necessary to explain the data. It is clear that additional changes will have occurred that are not seen in the data, especially some parallel and back mutations and multiple changes at a single site. There are four methods for estimating these: (1) the Poisson correction (see, for example, Margoliash and Fitch, 1968; Dickerson, 1971), (2) the PAM correction (Dayhoff, 1972), (3) the REH correction (Holmquist, 1972), and (4) the augmentation correction (Goodman *et al.*, 1974). The first procedure can be applied to any data that are discrete and has even been used on immunological data on the assumption that for every 1% by which the sequences differ the immunological distance increases by 5 (Wilson and Prager, 1974). It is important to recognize that this correction procedure adds to the observed amount of change an additional amount equal to that which has been hidden, ostensibly to get the true total. For statistical purposes, however, comparisons should be between the observations and the expected total reduced by the hidden change. Fitch (1976) has shown how the Poisson (and presumably the Dayhoff and Holmquist) procedure has an inherent property that tends to exaggerate inequalities in rates to such a pronounced extent that frequently one of the two legs will suffer a decrease in the absolute amount of change even while all pairwise distances are being increased.

Finally, the various tests of molecular clocks are of two varieties, calibration dependent and calibration free. The calibration-dependent tests require an estimate of divergence times provided by paleontologists. Almost all examples to date (see, for example, Dickerson, 1971 or Fig. 5 of this chapter) suffer multiple defects for statistical purposes. Apart from those many already listed whose effects, even when known, are unquantifiable, most of the data are not independent. The amount of change estimated in the descent of the cow from its common ancestor with the perissodactyls is not independent of the estimate for the amount of change in the descent of the cow from its common ancestor with the pig. And, of course, if the null hypothesis is that rates of change are uniform over time, rejection of the null hypothesis may be due to the failure of any of the underlying assumptions, including the "truth" of the paleontological dates, as well as to the nonexistence of true regularity of the molecular changes. Conversely, failure to reject the null hypothesis cannot prove that the clock exists, and, given the problems already alluded to, one cannot even give good confidence limits to the presumed clock's accuracy.

Calibration-free tests hardly avoid all these problems, but three of the most formidable—dependency, accuracy of paleontological dating, and correction of the expected rather than of the observed—disappear in the method used here. We therefore feel quite confident that the molecular clock does not have the accuracy expected of a random Poisson process.

11. Summary

The rates of amino acid changing nucleotide substitutions are significantly nonuniform in the same protein in different lines of descent and among proteins in the same lines of descent. This removes a major source of support for the neutral mutation theory as originally presented. But while it rules out that the vast majority of such substitutions are neutral, it by no means excludes the possibility that a sizable fraction of them may be neutral; and it would appear to us that the time has come to try to put upper and lower bounds to the fraction that might be neutral. The rate of fixation of silent mutations in the third position of the codon (4.2×10^{-9} silent substitutions per year per third nucleotide position) is about 3 times that for single-copy DNA and 7 times that for amino acid changing nucleotide substitutions in the hemoglobin genes. The silent substitution rate, corrected for the fraction of third-position mutations that are not silent, is a lower bound to the general mutation rate and is 5.8×10^{-9} mutations per year per nucleotide position. If not all silent mutations are neutral, the mutation rate is greater.

ACKNOWLEDGMENTS

The valuable assistance of Dr. R. L. Niece is gratefully noted. This research was supported by National Science Foundation grant GB-32274 and National Institutes of Health grant AM 15282.

12. Note Added in Proof

The data for the histone H4 (F2al) referred to in Table 6 has now been published [Grunstein, M., Schedl, P., Kedes, L., 1976, Sequence Analysis and Evolution of Sea Urchin (*Lytechinus pictus* and *Strongylocentrotus purpuratus*) Histone H4 Messenger RNA, *J. Mol. Biol.* **104**:351]. There are now 27 (rather than 18) third codon positions known for both sequences of which 18 (rather than 12) are unchanged. Therefore column H of Table 6 would be unchanged in rows C–F. The final column is only slightly affected with Rate$_2$ now being 5.63 rather than 5.82.

13. References

Dayhoff, M. O., 1972, *Atlas of Protein Sequences,* National Biomedical Research Foundation, Washington, D.C.

Dickerson, R. E., 1971, The structure of cytochrome *c* and the rates of molecular evolution, *J. Mol. Evol.* **1**:26.

Fitch, W. M., 1971, Toward defining the course of evolution: Minimum change for a specific tree topology, *Syst. Zool.* **20**:406.

Fitch, W. M., 1972*a*, Evolutionary variability in hemoglobins, *Haematol. Bluttransfus.* **10**:199.

Fitch, W. M., 1972*b*, Does the fixation of neutral mutations form a significant part of observed evolution in proteins? *Brookhaven Symp. Biol.* **23**:186.

Fitch, W. M., 1973, Aspects of molecular evolution, *Annu. Rev. Genet.* **7**:343.

Fitch, W. M., 1976, An evaluation of molecular evolutionary clocks, in: *Molecular Study of Biological Evolution* (F. J. Ayala, ed.) Sinauer Assoc., Sunderland, Mass.

Fitch, W. M., and Farris, J. S., 1974, Evolutionary trees with minimum nucleotide replacements from amino acid sequences, *J. Mol. Evol.* **3**:263.

Fitch, W. M., and Langley, C. H., 1976, Protein evolution and the molecular clock, *Fed. Proc.* **35**:2092.

Fitch, W. M., and Margoliash, E., 1967, The construction of phylogenetic trees—A generally applicable method utilizing estimates of the mutation distance obtained from cytochrome *c* sequences, *Science* **155**:279.

Fitch, W. M., and Markowitz, E., 1970, An improved method for determining codon variability in a gene and its application to the rate of fixations of mutations in evolution, *Biochem. Genet.* **4**:579.

Forget, B. G., Marotta, C. A., Weisman, S. M., Verma, I. M., McCaffrey, R. P., and Baltimore, D., 1974, Nucleotide sequences of human globin messenger RNA, *Ann. N.Y. Acad. Sci.* **241**:290.

Goodman, M., Moore, G. W., Barnabas, J., and Matsuda, G., 1974, The phylogeny of human globin genes investigated by the maximum parsimony method, *J. Mol. Evol.* **3**:1.

Haldane, J. B. S., 1957, The cost of natural selection, *Genetics* **55**:511.

Holmquist, R., 1972, Empirical support for a stochastic model of evolution, *J. Mol. Evol.* **1**:211.

Jones, K. W., and Purdom, J. F., 1975, The evolution of defined classes of human and primate DNA, in: *Proceedings of the Society for the Study of Evolutionary Biology* (A. J. Boyce, ed.), pp. 39–51, Taylor and Francis, London.

Jukes, T. H., and Holmquist, R., 1972, Evolutionary clock: Nonconstancy of rate in different species, *Science* **177**:530.

Kimura, M., 1968, Evolutionary rate at the molecular level, *Nature (London)* **217**:624.

Kohne, D. E. 1970, Evolution of higher-organism DNA, *Quart. Rev. Biophys.* **33**:327.

Langley, C. H., and Fitch, W. M., 1973, The constancy of evolution: A statistical analysis of the α and β hemoglobins, cytochrome *c* and fibrinopeptide A, in: *Genetic Structure of Populations* (N. E. Morton, ed.), pp. 246–262, University Press of Hawaii, Honolulu.

Langley, C. H., and Fitch, W. M., 1974, An examination of the constancy of the rate of molecular evolution, *J. Mol. Evol.* **3**:161.

Margoliash, E., and Fitch, W. M., 1968, Evolutionary variability of cytochrome *c* primary structures, *Proc. N.Y. Acad. Sci.* **151**:359.

Markusen, J., and Sundby, F., 1973, Isolation and amino acid sequence of the C-peptide of duck proinsulin, *Eur. J. Biochem.* **34**:401.

Matsuda, G., Maita, J., Watanabe, B., Araya, A., Morokuma, K., Goodman, M., and Prychodko, W., 1973, The amino acid sequences of the α and β polypeptide chains of adult hemoglobin of the savannah monkey *(Cercopithecus aethiops)*, *Physiol. Chem.* **354**:1153.

Maxson, L. R., and Wilson, A. C., 1975, Relationship between albumin evolution and organismal evolution in tree frogs (Hylidae), *Syst. Zool.* **24**:1.

Moore, G. W., Barnabas, J., and Goodman, M., 1973, A method for constructing maximum parsimony ancestral amino acid sequences on a given network, *J. Theor. Biol.* **38**:459.

Moore, G. W., Barnabas, J., and Goodman, M., 1975, Darwinian evolution in the genealogy of haemoglobin, *Nature (London)* **253**:603.

Ohta, T., and Kimura, M., 1971, On the constancy of the evolutionary rate of cistrons, *J. Mol. Evol.* **1**:18.

Prager, E. M., and Wilson, A. C., 1971, The dependence of immunological cross-reactivity upon sequence resemblance among lysozymes, *J. Mol. Biol.* **246**:5978.

Romero-Herrera, A. E., Lehmann, H., Joysey, K. A., and Friday, A. E., 1973, Molecular evolution of myoglobin and the fossil record: A phylogenetic synthesis, *Nature (London)* **246**:389.

Salser, W., Bowen, S., Browne, D., El Adli, F., Federoff, N., Fry, K., Heindell, H., Paddock, G., Poon, R., Wallace, B., and Whitcome, P., 1975, Investigation of the organization of mammalian chromosomes at the DNA sequence level, *Fed. Proc.* **35**:23.

Sneath and Sokal, 1973, *Numerical Taxonomy*, W. H. Freeman and Co., San Francisco, 573 pp.

Tager, H. S., and Steiner, D. F., 1972, Primary structures of the proinsulin connecting peptides of the rat and the horse, *J. Biol. Chem.* **247**:7936.

Tashian, R. E., Tanis, R. J., Ferrell, R. E., and Stroup, S. K., 1972, Differential rates of evolution in the carbonic anhydrase isozymes of catarrhine primates, *J. Hum. Evol.* **1**:545.

Tetaert, D., Han, K.-K., Plancot, M.-T., Dautrevaux, M., Ducastaing, S., Hombrados, I., and Neuzil, E., 1974, The primary sequence of badger myoglobin, *Biochim. Biophys. Acta* **351**:317.

Tobias, P. V., 1975, Long or short hominid phylogenies? Palaeontological and molecular evidences, in: *The Role of Natural Selection in Human Evolution* (F. M. Salzano, ed.) 34 pp., Wenner-Gren Foundation for Anthropological Research, New York.

Wilson, A. C., and Prager, E. M., 1974, Antigenic comparison of lysozymes, in: *Lysozyma* (E. F. Osserman, R. E. Canfield, and S. Beychok, eds.), pp. 127–141, Academic Press, New York.

Wilson, A. C., and Sarich, V. M., 1969, A molecular time scale for human evolution, *Proc. Natl. Acad. Sci. USA* **63**:1088.

Primate Evolution Inferred from Amino Acid Sequence Data

IV

Evolution of the Primary Structures of Primate and Other Vertebrate Hemoglobins

11

GENJI MATSUDA

1. Introduction

Since the first living organisms appeared on the earth, the genetic information contained in genes in the form of DNA has been successfully transmitted from generation to generation, up to the present time. It is thought that these genes evolve through the process of mutation and natural selection, resulting in the diversity of organisms in the world. Assuming that the innumerable organisms now existing on the earth evolved from one source, what was the mechanism that gave rise to this diversity?

In 1859, Darwin published *The Origin of Species,* in which he discusses organic evolution and presents the theory of natural selection. The studies of organic evolution at that time were primarily concerned with comparing the forms of various organisms, and the information which this comparative morphology yielded is still very important.

At the present time, as comparative studies on the primary structures of various homologous proteins accumulate, the subject of molecular evolution is becoming increasingly important. Since DNA is extremely stable, it permits genetic traits to be passed from generation to generation, with minimal altera-

GENJI MATSUDA • Department of Biochemistry, Nagasaki University School of Medicine, Nagasaki 852, Japan.

tion. On the other hand, and this may seem paradoxical, the DNA in a gene is always subject to changes which are reflected in the evolution of organisms. How have these genes mutated during organic evolution? The most effective means to date of understanding the progress of these gene mutations is through the comparison of the primary structures of homologous proteins obtained from various kinds of organisms.

Early studies in comparative morphology acknowledged the diversity in the organic world and tried to find a common denominator in this diversity. Later, comparative studies on the metabolisms of the various kinds of organisms revealed that basic chemical reactions existing in these organisms were the same. In comparing the primary structures of homologous proteins, evolutionists are aware of the diversity of these proteins, and are searching for a unifying aspect.

2. Studies on the Primary Structures of the Homologous Proteins of Hemoglobin

Hemoglobin is a hemeprotein, existing in most animals and some plants. Furthermore, almost all of the vertebrate hemoglobins are contained in erythrocytes, so that their isolation and purification are comparatively easy. There have been many studies on the primary structure, tertiary structure, quaternary structure, and inherited variants of hemoglobin, and, consequently, it has come to be viewed as a very useful protein for the study of molecular evolution. Erythrocruorin, a respiratory protein found in the blood of certain invertebrates, is sometimes distinguished from hemoglobin, but not very clearly, so it will be incorporated here in my discussion of hemoglobin.

Hemoglobins typically are globular proteins with molecular weights of approximately 65,000 and consist of two α chains and two non-α chains containing one heme each. We now know that the heme parts in all types of hemoglobins are identical and that the primary structures of the various globin chains differ among species. Braunitzer et al. (1961) determined the primary structures of the α and β polypeptide chains of human hemoglobin, and since that time the primary structures of various kinds of other hemoglobins have been determined, attracting considerable attention in the area of molecular evolution. Table 1 lists those species in which the primary structures of hemoglobins have been determined. Myoglobin is contained in muscle and usually exists as a monomer with a molecular weight of approximately 17,000, one-fourth that of higher vertebrate hemoglobin. Both hemoglobin and myoglobin are hemeproteins and combine reversibly with the oxygen molecule. Detailed studies of the tertiary structures of hemoglobin (Perutz et al., 1960) and myoglobin (Kendrew et al., 1961) indicate quite conclusively that myoglobin is homologous to the various globin chains of hemoglobin.

At present, the primary structures of approximately 60 different α and β globins have been determined from the hemoglobins of the species listed in

Table 1. Goodman *et al.* (1975) arranged the amino acid residues in the primary structures of these homologous proteins in sequence from 1 to 175, using a system of homology which recognizes not only the α and β hemoglobin chains obtained from Mammalia, Aves, Amphibia, and Teleostei but also hemoglobins and myoglobins obtained from Agnatha, Arthropoda, Mollusca, and Annelida. Thus it can be postulated that these homologous proteins have evolved from a common primitive protein. According to Schopf (1967), Protozoa, presumably the lowest form of animal life, parted from Bacteriophyta and Cyanophyta, and lowest forms of unicellular life, about 1000×10^6 years ago at the latest. Since hemoglobin exists not only in the animal kingdom but also as leghemoglobin in the vegetable kingdom, it is assumed that the gene responsible for the production of primitive hemoglobin had already appeared before animals separated from plants.

3. Comparison of the Primary Structures of Primate Hemoglobins

The study of primate evolution has long been of great interest because of its relevance to the human species. Primates may be classified, broadly speaking, into the Prosimii (prosimians) and Anthropoidea (monkeys, great apes, and

Table 1. Species Whose Hemoglobin Primary Structures Have Been Determined

Animalia
    Vertebrata
        Mammalia
            Primates: human *(Homo sapiens)*, hanuman langur *(Presbytis entellus)*, rhesus macaque *(Macaca mulatta)*, Japanese macaque *(Macaca fuscata)*, savannah monkey *(Cercopithecus aethipos)*, spider monkey *(Ateles geoffroyi)*, capuchin monkey *(Cebus apella)*, slow loris *(Nycticebus coucang)*
            Lagomorpha: rabbit *(Orycytolagus caniculus)*
            Rodentia: mouse *(Mus musculus)*
            Carnivora: dog *(Canis familiaris)*
            Perissodactyla: horse *(Equus cabalus)*
            Artiodactyla: ox *(Bos taurus)*
            Marsupialia: kangaroo *(Macropus giganteus)* potoroo *(Potorous tridactylus)*
            Monotremata: echidna *(Tachyglossus aculeatus)*
          Aves: chicken *(Gallus gallus)*
          Amphibia: frog (Rana esculeuta)
          Osteichthyes: carp *(Cyprinus carpia)*, Catostomus clarkii
          Agnatha: *Petromyzon morinus, Lampetra Fluviatilis*
        Arthropoda: *Chironomus thumi*
        Mollusca: *Aplysia limacina*
        Annelida: blood worm *(Glycera dibranchiata)*
    Vegetabilia
        Tracheophyta: soybean *(Glycine soja)*

man), and the Anthropoidea may be divided into the Platyrrhini (New World monkeys), Catarrhini containing Cercopithecoidea (Old World monkeys), and Hominoidea (great apes and man). There may be some disagreement on the phylogeny, but hypothetically it can be stated that primates diverged from the other mammals near the end of the Cretaceous Epoch, about $70-80 \times 10^6$ years ago; following that, about 60×10^6 years ago, the Prosimii parted from the Anthropoidea. From the Anthropoidea, the divergence of Platyrrhini took place in the latter period of the Eocene Epoch, about 40×10^6 years ago, and also in the early period of the Oligocene Epoch. It is thought that Cercopithecoidea and Hominoidea parted about 35×10^6 years ago.

Among the primates, the primary structures of hemoglobin have been determined for the following species: human (Braunitzer *et al.*, 1961; Konigsberg *et al.*, 1961) belonging to Hominoidea; *Presbytis entellus* (Matsuda *et al.*, 1973d), *Macaca mulatta* (Matsuda *et al.*, 1968a), *Macaca fuscata* (Matsuda *et al.*, 1973b), and *Cercopithecus aethiops* (Matsuda *et al.*, 1973e) belonging to Cercopi-

Table 2. Comparison of the Primary Structures of α Chains of Primate Hemoglobins

```
                          1         2         3         4         5         6         7        8
                 12345678901234567890123456789012345678901234567890123456789012345678901234567890
Human            V LSPADKTNVKAAWG  KVGAHAGEYGAEALERMFLSFPTTKTYFPHF DLSH   GSAQVKGHGKKVA
Langur           V LSPADKTNVKAAWG  KVGGHGGEYGAEALERMFLSFPTTKTYFPHF DLSH   GSAQVKGHGKKVA
Rhesus           V LSPADKSNVKAAWG  KVGGHAGEYGAEALERMFLSFPTTKTYFPHF DLSH   GSAQVKGHGKKVA
Japanese         V LSPADKSNVKAAWG  KVGGHAGEYGAEALERMFLSFPTTKTYFPHF DLSH   GSAQVKGHGKKVA
Savannah         V LSPADKSNVKAAWG  KVGGHAGEYGAEALERMFLSFPTTKTYFPHF DLSH   GSAQVKGHGKKVA
Spider           V LSPADKSNVKAAWG  KVGGHAGDYGAEALERMFLSFPTTKTYFPHF DLSH   GSAQVKGHGKKVA
Capuchin         V LSPADKTNVKTAWG  KVGGHAGDYGAEALERMFLSFPTTKTYFPHF DLSH   GSAQVKGHGKKVA
Slow loris       V LSPADKTNVKAAWE  KVGSHAGDYGAEALERMFLSFPTTKTYFPHF DLSH   GSAQVKAHGKKVA
Primate ancestor V LSPADKTNVKAAWG  KVGGHAGEYGAEALERMFLSFPTTKTYFPHF DLSH   GSAQVKAHGKKVA

                   1         1         1         1         1         1         1         1
                   0         1         2         3         4         5         6         7
                 9
         12345678901234567890123456789012345678901234567890123456789012345678901234567890123456789012345
DALTNAVAHVDDMPN A    LSALSDLH    AHKLRVDPVNFKLLSHCLLVTLAAHLPAEFTPAVHASLDKFLASVSTVLTSKYR
DALTNAVAHVDDMPH A    LSALSDLH    AHKLRVDPVNFKLLSHCLLVTLAAHLPAEFTPAVHASLDKFLASVSTVLTSKYR
DALTLAVGHVDDMPN A    LSALSDLH    AHKLRVDPVNFKLLSHCLLVTLAAHLPAEFTPAVHASLDKFLASVSTVLTSKYR
DALTLAVGHVDDMPN A    LSALSDLH    AHKLRVDPVNFKLLSHCLLVTLAAHLPAEFTPAVHASLDKFLASVSTVLTSKYR
DALTLAVGHVDDMPH A    LSALSDLH    AHKLRVDPVNFKLLSHCLLVTLAAHLPAEFTPAVHASLDKFLASVSTVLTSKYR
DALTNAVAHVDDMPN A    LSALSDLH    AHKLRVDPVNFKLLSHCLLVTLAAHHPADFTPAVHASLDKFLASVSTVLTSKYR
DALSNAVAHVDDMPN A    LSALSDLH    AHKLRVDPVNFKLLSHCLLVTLAAHHPADFTPAVHASLDKFLASVSTVLTSKYR
DALTNAVSHVDDMPS A    LSALSDLH    AHKLRVDPVNFKLLSHCLLVTLACHHPADFTPAVHASLDKFLASVSTVLTSKYR
DALTNAVAHVDDMPS A    LSALSDLH    AHKLRVDPVNFKLLSHCLLVTLAAHHPADFTPAVHASLDKFLASVSTVLTSKYR
```

Table 3. Comparison of the Primary Structures of β Chains of Primate Hemoglobins

	1	2	3	4	5	6	7	8
	1234567890	1234567890	1234567890	1234567890	1234567890	1234567890	1234567890	1234567890

Human	VHLTPEEKSAVTALWG	KV	NVDEVGGEALGRLLYVYPWTQRFFESFGDLSTPDAVMGNPKVKAHGKKVL
Langur	VHLTPEEKAAVTALWG	KV	NVDEVGGEALGRLLVVYPWTQRFFESFGDLSSPDAVMGNPKVKAHGKKVL
Rhesus	VHLTPEEKNAVTTLWG	KV	NVDEVGGEALGRLLLVYPWTQRFFESFGDLSSPDAVMGNPKVKAHGKKVL
Japanese	VHLTPEEKNAVTTLWG	KV	NVDEVGGEALGRLLVVYPWTQRFFESFGDLSSPDAVMGNPKVKAHGKKVL
Savannah	VHLTPEEKTAVTTLWG	KV	NVDEVGGEALGRLLVVYPWTQRFFESFGDLSSPDAVMGNPKVKAHGKKVL
Spider	VHLTGEEKAAVTALWG	KV	NVDEVGGEALGRLLVVYPWTQRFFESFGDLSTPDAVMSNPKVKAHGKKVL
Capuchin	VHLTAEEKSAVTTLWG	KV	NVDEVGGEALGRLLVVYPWTQRFFDSFGDLSTPDAVMNNPKVKAHGKKVL
Slow loris	VHLTGEEKSAVTALWG	KV	NVDDVGGEALGRLLVVYPWTQRFFESFGDLSSPSAVMGNPKVKAHGKKVL
Primate ancestor	VHLTAEEKSAVTALWG	KV	NVDEVGGEALGRLLYVYPWTQRFFESFGDLSSPDAVMSNPKVKAHGKKVL

	1	1	1	1	1	1	1	1
	9	0	1	2	3	4	5	6	7
	1234567890	1234567890	1234567890	1234567890	1234567890	1234567890	1234567890	123456789012345

GAFSDGLAHLDNLKG	T	FATLSELH	CDKLHVDPENFRLLGNVLVCVLAHHFGKEFTPPYQAAYQKVVAGVANALAHKYH
GAFSDGLAHLDNLKG	T	FAQLSELH	CDKLHVDPENFRLLGNVLVCVLAHHFGKEFTPQVQAAYQKVVAGVANALAHKYH
GAFSDGLNHLDNLKG	T	FAQLSELH	CDKLHVDPENFKLLGNVLVCVLAHHFGKEFTPQVQAAYQKVVAGVANALAHKYH
GAFSDGLNHLDNLKG	T	FAQLSELH	CDKLHVDPENFKLLGNVLVCVLAHHFGKEFTPQVQAAYQKVVAGVANALAHKYH
GAFSDGLAHLDNLKG	T	FAQLSELH	CDKLHVDPENFKLLGNVLVCVLAHHFGKEFTPQVQAAYQKVVAGVANALAHKYH
GAFSDGLAHLDNLKG	T	FAQLSELH	CDKLHVDPENFRLLGNVLVCVLAHHFGKEFTPQLQAAYQKVVAGVANALAHKYH
GAFSDGLTHLDNLKG	T	FAQLSELH	CDKLHVDPENFRLLGNVLVCVLAHHFGKEFTPQVQAAYQKVVAGVATALAHKYH
SAFSDGLNHLDNLKG	T	FAKLSELH	CDKLHVDPENFRLLGNVLVYVLAHHFGKDFTPQVQSAYQKVVAGVANALAHKYH
AAFSDGLNHLDNLKG	T	FAKLSELH	CDKLHVDPENFRLLGNVLVIVLAHHFGKEFTPQVQAAYQKVVAGVANALAHKYH

thecoidea; *Ateles geoffroyi* (Matsuda *et al.*, 1973*g;* Boyer *et al.*, 1971) and *Cebus apella* (Matsuda *et al.*, 1973*f*) belonging Platyrrhini; and *Nycticebus coucang* (Matsuda *et al.*, 1973*c*) belonging to Prosimii. The primary structures of their α chains and β chains are shown in Tables 2 and 3. The amino acid residues in the primary structures of these hemoglobins are arranged from 1 to 175 by homology according to Goodman *et al.* (1975) in single-letter codes. The assumed primary structures of α and β chains of ancestral hemoglobin of primate are also shown in Tables 2 and 3.

In the primate globin chains that have been sequenced (Tables 2 and 3), 14 amino acid exchanges were found in the 141 amino acid residues of the α chain, and 19 amino acid exchanges were found in the 146 amino acid residues of the β chain. These exchanging residues are indicated by a dot (·) beneath each in the tables. Perutz *et al.* (1960, 1968; Perutz, 1969) and Kendrew *et al.* (1961) studied the tertiary structures of hemoglobin and myoglobin in detail and made clear which amino acid residues are important in maintaining the structure and function of hemoglobin.

Upon investigating those exchanging residues in the α and β chains of each of the primate hemoglobins, it was found that none of the residues involved heme contacts, $\alpha_1\beta_2$ interchain contacts, the Bohr effect, or the binding of 2,3-diphosphoglycerate. Almost all of these exchanging residues are located on the outside of the hemoglobin molecule, but the exchanging residue 139(G18) of the α chain (Table 2) and residues 45(B15), 135(G14), 148(H3), and 151(H6) of the β chain (Table 3) are involved in $\alpha_1\beta_1$ contact positions.

The matrices of the numbers of amino acid exchanges and the minimum mutation distances between each pair of these amino acid sequences in the α and β chains of these primate hemoglobins are compared in Tables 4 and 5. When the α and β chains are compared, the β chains show more amino acid exchanges and higher minimum mutation distances than the α chains in all of the primate hemoglobins. In the case of an α chain, there is an increase in the number of amino acid exchanges in the primary structure as the two animals diverge from one another, and among those animals which have almost the same divergence points the number of amino acid exchanges of their hemoglobins is also almost the same. In the case of the β chain, the regularity decreases in comparison with the α chain. For example, as the species diverge, the number of amino acid exchanges does not necessarily increase; among the Platyrrhini, a greater degree of similarity is not necessarily shown.

Table 4. Amino Acid Exchanges and Minimum Mutation Distances in α Chains of Primate Hemoglobins

	Human	Langur	Rhesus	Japanese	Savannah	Spider	Capuchin	Slow loris	Primate ancestor
Human	0	3/3	5/4	5/4	6/5	5/5	6/6	10/9	5/5
Langur	3/3	0	6/5	6/5	5/4	6/6	7/7	12/10	6/5
Rhesus	5/4	6/5	0	0	1/1	6/5	9/8	13/11	8/7
Japanese	5/4	6/5	0	0	1/1	6/5	9/8	13/11	8/7
Savannah	6/5	5/4	1/1	1/1	0	7/6	10/9	14/11	9/7
Spider	5/5	6/6	6/5	6/5	7/6	0	3/3	8/7	4/4
Capuchin	6/6	7/7	9/8	9/8	10/9	3/3	0	9/8	5/5
Slow loris	10/9	12/10	13/11	13/11	14/11	8/7	9/8	0	6/5
Primate ancestor	5/5	6/5	8/7	8/7	9/7	4/4	5/5	6/5	0

Table 5. Amino Acid Exchanges and Minimum Mutation Distances in β Chains of Primate Hemoglobins

	Human	Langur	Rhesus	Japanese	Savannah	Spider	Capuchin	Slow loris	Primate ancestor
Human	0	5/4	10/8	9/7	7/6	8/6	10/8	15/11	10/8
Langur	5/4	0	7/5	6/4	3/3	5/4	9/8	14/10	9/7
Rhesus	10/8	7/5	0	1/1	4/3	12/9	10/9	15/12	10/9
Japanese	9/7	6/4	1/1	0	3/2	11/8	9/8	14/11	9/8
Savannah	7/6	3/3	4/3	3/2	0	8/7	9/8	16/12	11/9
Spider	8/6	5/4	12/9	11/8	8/7	0	8/8	15/12	10/8
Capuchin	10/8	9/8	10/9	9/8	9/8	8/8	0	17/14	10/9
Slow loris	15/11	14/10	15/12	14/11	16/12	15/12	17/14	0	9/8
Primate ancestor	10/8	9/7	10/9	9/8	11/9	10/8	10/9	9/8	0

4. Rate of Molecular Evolution of Homologous Proteins of Hemoglobin

In the morphological studies of living organisms, the change of body length and numerous other factors have been used to measure the rate of evolution. It is now felt that ascertaining the rate of evolution at the molecular level may be considerably more accurate than guessing at it through morphological studies. Zuckerkandl and Pauling (1962) calculated that 14.5×10^6 years was needed for the substitution of one amino acid of hemoglobin for another. This was the first report to show that the rate of evolution could be calculated by comparing the amino acid sequence of homologous proteins. Later, Zuckerkandl and Pauling (1965) calculated that 7×10^6 years was needed for the substitution of one amino acid for another during the evolution of mammalian hemoglobins.

Kimura (1968, 1969) proposed the neutral theory of gene mutation in the molecular evolution of proteins. He calculated the rate of change (k_{aa}) for each amino acid position in the molecular evolution of a protein and found a value of 1×10^{-9} per year for hemoglobin. The term "1 Pauling" was assigned to this unit and it has become a standard measure in molecular evolution. Dickerson (1971) defined the period needed for a 1% change in a protein within 1 million years as a "UEP(m.y.) unit." In the case of hemoglobin, the period needed for a 1% change is 5.8 m.y. These calculations were done on the assumption that the rates of molecular evolution of each protein are almost uniform.

Goodman *et al.* (1971, 1974, 1975) Barnabas *et al.* (1972) and Moore *et al.* (1973*a,b*) have constructed genealogical trees using the maximum parsimony method and have calculated the rate of molecular evolution of hemoglobins in a different way. The mean value obtained by Goodman *et al.* was 31 NR%. The unit "NR%" indicates the rate of change of nucleotides per 100 codons during 1 $\times$ 10^8 years. More importantly, their studies show that the rate of molecular evolution of hemoglobins is different with each evolutionary period.

For example, a very high value, 108.9 NR% for 100×10^6 years, is calculated for the period when the vertebrate ancestor evolved to the teleost-tetrapod α ancestor, but it decreases to 19.9 NR% (about one-fifth of the former) for the next period of 400×10^6 years, when the teleost-tetrapod α ancestor evolved to the present α chain group. Very high values are also found for the period of the vertebrate ancestor to the tetrapod β ancestor (65.7 NR%, 160×10^6 years) and for the period of the tetrapod β ancestor to the amniote β ancestor (73.6 NR% for 40×10^6 years). In the succeeding period, however, from amniote β ancestor to the present β chain group of mammals, it decreases to 16.3 NR% (about one-fifth of the former) for 300×10^6 years.

Table 6. Comparison of Rates of Molecular Evolution of Hemoglobins after Eutherian Ancestor[a]

Species	α chain		β chain	
	MMD	NR%	MMD	NR%
Primates				
Human	17	13.4	22	16.7
Langur	18	14.2	21	16.0
Rhesus	20	15.8	28	21.3
Japanese	20	15.8	27	20.5
Savannah	21	16.6	24	18.3
Spider	16	12.6	26	19.8
Capuchin	17	13.4	26	19.8
Slow loris	20	15.8	23	17.5
	average 14.7		average 18.7	
Other mammals				
Rabbit	47	37.0	23	17.5
Mouse	22	17.3	52	39.6
Dog	40	31.5	15	11.4
Horse	21	16.6	47	35.8
Ox	21	16.6	40	30.4
	average 23.8		average 26.9	

[a]Abbreviations: MMD, minimum mutation distance; NR%, nucleotide replacements per 100 codons per 1×10^8 years. The values of MMD were calculated from the genealogical tree of Goodman *et al.* (1975).

Table 7. Comparison of Rates of Molecular Evolution of Hemoglobins after Primate Ancestor[a]

Species	α chain		β chain	
	MMD	NR%	MMD	NR%
Human	5	5.9	9	10.3
Langur	6	7.1	8	9.1
Rhesus	8	9.5	15	17.1
Japanese	8	9.5	14	16.0
Savannah	9	10.6	11	12.6
Spider	4	4.7	13	14.8
Capuchin	5	5.9	13	14.8
Slow loris	6	7.1	10	11.4
	average 7.5		average 13.3	

[a]Abbreviations: MMD, minimum mutation distance; NR%, nucleotide replacements per 100 codons per 1×10^8 years. The values of MMD were calculated from Tables 2 and 3.

Analogously, the rapid and slow periods in the rates of molecular evolution of homologous globin chains of hemoglobin were distinguished. It is assumed that the period of rapid evolution corresponds to the period when the structure of the hemoglobin molecule was being formed, and the slow period corresponds to the time when its structure had become stabilized. Generally speaking, it is assumed that when very rapid evolution has taken place at an early stage, a protein will be in the process of assuming a new function. When the function has stabilized, it suggests that the rate of the evolution of the protein is decreasing.

Matsuda *et al.* (1968*b*), after studying the primary structures of the α and β chains in the hemoglobins of rhesus and Japanese macaques, supposed that the rates of molecular evolution of primate hemoglobins were becoming slower than those of the other mammalian species. After the primary structures of the α and β chains of five more primate hemoglobins had been determined, this impression was strengthened.

The rates (NR%) of molecular evolution of hemoglobins in primates and the other mammals after the eutherian ancestor are compared in Table 6. The eutherian ancestor is supposed to have emerged 90×10^6 years ago. Generally speaking, the rates of molecular evolution of primate hemoglobins are becoming slower (Table 6). Moreover, comparing α and β chains, the β chains show generally more rapid evolution in both primate and the other mammalian hemoglobins, although there are exceptions such as the hemoglobins of rabbit and dog.

Table 7 shows the rates (NR%) of molecular evolution of hemoglobins after the primate ancestor, which is supposed to have emerged 60×10^6 years ago. As indicated, the average evolutionary rate of primate α and β chains in their descent from the primate ancestor is much slower than in their descent from the eutherian ancestor. These calculations indicate that the rate of molecular evolution of primate hemoglobins is slowing. It is shown also in Table 7 that the β chains have evolved more rapidly than the α chains in all of the primate hemoglobins.

5. Mechanism of Molecular Evolution of the Hemoglobins

Darwin compared the forms of living organisms and advanced the theory of natural selection as the primary factor in evolution. Can this hypothesis of natural selection be applied at the molecular level? Kimura (1968) and King and Jukes (1969) presented the neutral theory of evolution, which states that molecular evolution occurs primarily through the random fixation of selectively neutral mutations rather than natural selection. If we admit to this hypothesis, the question then arises as to how the structures of various proteins which seem to be so well fitted for their functions in various kinds of proteins occurred.

Since 1971, Goodman *et al.* (1971, 1974, 1975) have said that natural selection would affect the evolution of the homologous globin chains of hemoglobin. Furthermore, as previously mentioned, they feel that the rates of molecular evolution of the α and β chains of hemoglobin were not necessarily uniform. That is, the rates in the period of 100×10^6 years from the vertebrate ancestor to the teleost-tetrapod α ancestor and the 200×10^6 years from the vertebrate ancestor to the amniote β ancestor were very rapid. After that, the rates slowed down considerably. The rapid period can be said to correspond to the period of "macroevolution" and the slow period to "microevolution."

5.1. Macroevolution of Homologous Globin Chains of Hemoglobin

The myoglobin existing in the muscles of vertebrates has a very strong affinity for oxygen, its oxygen equilibrium curve is hyperbolic, and it does not show a Bohr effect. These characteristics are advantageous in the storage of oxygen in muscular myoglobin. Primitive hemoglobins probably had an oxygen-storing physiological function similar to that of myoglobin. Leghemoglobin in soybeans and myxine hemoglobin have the same oxygen-binding characteristics as myoglobin.

As life gradually evolved from unicelluar organisms to multicellular organisms, the specialization of tissues and organs was also developing. Gradually, respiratory organs evolved which functioned in initiating the transfer of external oxygen to various tissues of the body by way of body fluids. The oxygen equilibrium curve of almost all of the vertebrate hemoglobins is sigmoidal. This means that these hemoglobins are highly efficient in transporting oxygen throughout the organism because in the tissues of the respiratory organs, with their higher partial oxygen pressures, hemoglobin easily combines with oxygen, and in the tissues of the other organs, with lower partial oxygen pressures, the oxygen is easily released from the hemoglobin. Moreover, these hemoglobins do show a Bohr effect. That is, as carbonic acid in the blood of tissues increases, oxygen is more easily released from hemoglobin. When blood returns to the respiratory organs and the carbonic acid in it decreases, the oxygenation of hemoglobin is accelerated. With the development of the respiratory system, it is assumed that the globins evolved from their role as oxygen storers to their new function as oxygen carriers.

As previously described, the different hemoglobin chains such as myoglobin, α and β chains seem to have evolved most rapidly during the periods of 100×10^6 years from the vertebrate ancestor to the teleost-tetrapod α ancestor and 200×10^6 years from the vertebrate ancestor to the amniote β ancestor. It is assumed that during these periods the globin genes underwent positive natural selection, evolving from the primitive oxygen-storing role to the present one with its function of oxygen transport. Naturally, during these periods, mutations in the globin genes as well as duplications frequently occurred.

The oxygen equilibrium curve of vertebrate myoglobin is hyperbolic, while that of almost all normal hemoglobins from all vertebrates except cyclostomes is sigmoidal. The generation of this sigmoidal curve is attributed to the heme-heme interaction among four hemes contained in two α chains and two β chains in these hemoglobins. Cyclostome hemoglobin has the ability to form a homopolymer, but its oxygen equilibrium curve is hyperbolic. In the case of HbH, an abnormal human hemoglobin, a homotetramer of the β chain only is formed; the same thing happens with cyclostome hemoglobin. That is, hemoglobin does not generate the sigmoidal oxygen equilibium curve until it forms a heterotetramer with two kinds of polypeptide chains, an α chain and a non-α chain. It is apparently at this stage that the characteristics of the primary structures of the α chain and β chain involved in the contacts between the different kinds of polypeptide chains and between the same kinds of polypeptide chains evolved. The stage was then set for the physiological function of hemoglobin to be established. The same thing can be said of the amino acid residues involved in the Bohr effect; these residues were selected for because of their important role in establishing the physiological function of hemoglobin. This era, then, was one of "macroevolution" of the homologous proteins of hemoglobin.

5.2. Microevolution of the Homologous Proteins of Hemoglobin

The rate of evolution of the hemoglobin molecule decreased remarkably for 400×10^6 years, the period of the teleost-tetrapod α ancestor to the present α chain group of vertebrates, and for 300×10^6 years, the period of the amniote β ancestor to the present β chain group of mammals. This is the period after the structural characteristics of hemoglobin were established, and it can be viewed as one of microevolution for the homologous globin chains of hemoglobin. No genuinely clear distinction can be made between the period of macroevolution and that of microevolution because the former transformed to the latter very gradually.

It is assumed that once each protein has evolved a particular function it has a stable structure with a definite limitation. This condition of structural limitation can be called "the structural-free degree" of the protein. Any mutation which exceeds the capacity of this structural limitation would be selected against and ultimately vanish from the organism; however, some minor mutations within the structural limitation might be constructive and undergo positive selection.

Can natural selection act on these minor mutations? We presume that living organisms have always been evolving, with their internal physiology changing by degrees. For example, in the human erythrocyte there is a large amount of 2,3-diphosphoglycerate, which acts on hemoglobin as an allosteric effector, combining with $\beta1$(Val), $\beta2$(His), $\beta82$(Lys), and $\beta143$(His) of the β chains of human adult deoxyhemoglobin and helping to release oxygen from the hemoglobin (Perutz, 1970). In the erythrocyte of birds, inositol-1,3,4,5,6-pentaphosphate instead of 2,3-diphosphoglycerate may act as an allosteric effector (Johnson and Tate, 1969). Only the primary structure of chicken hemoglobin has been determined in birds (Matsuda et al., 1971, 1973a), and it was assumed that inositol pentaphosphate combines with $\beta1$(Val), $\beta2$(His), $\beta82$(Lys), $\beta135$(Arg), $\beta139$(His), or $\beta143$(Arg) of the β chains of chicken hemoglobin (Matsuda et al., 1973a; Arnone and Perutz, 1974). It is known that 2,3-diphosphoglycerate does not exist in the erythrocyte of sika, a kind of artiodactyl, and does not affect the affinity for oxygen of sika hemoglobin. This may be because $\beta1$(Val) is deleted and $\beta2$(His) is substituted for Met in sika hemoglobin (Maeda and Enoki, 1974).

This example assumes that the primary structure of hemoglobin has undergone natural selection and subsequent molecular evolution by coadapting closely to the internal physiology of living organisms. In this manner, proteins, once having evolved certain structures, seem to become coadapted to both complicated internal and external circumstances. Minor mutations within the limits of their structural-free degree would then be able to undergo natural selection. This can be thought of as a kind of stabilizing selection.

The same thing can be said of the phenomena governing the rates of evolution of primate hemoglobins, which appear to be gradually slowing down (Tables 6 and 7). As Tables 2 and 3 show, the mutated residues in the α chain and the β chain from primate hemoglobins are not at the positions concerned with the transport of oxygen. But it is known that the positions of amino acid residues 139(G18) in the α chain and 45(B15), 135(G14), 148(H3), and 151(H6) in the β chain, which mutated in primate hemoglobins, participate in the $\alpha_1\beta_1$ contact. These residues may not have strong effects on the physiological function of hemoglobin, but they are useful for stabilizing the spatial structure of hemoglobin.

For example, abnormal human hemoglobins where mutations have occurred in amino acid residues participating in the $\alpha_1\beta_1$ contact have the ability to combine with oxygen in the normal fashion, but their spatial structures are not stable and are easily denatured by heating. The mutated residues of primate hemoglobins are different from the mutated residues of these abnormal human hemoglobins, but they are also the residues participating in the $\alpha_1\beta_1$ contact. In primates, hemoglobin molecules seem to be evolving in the direction of more balanced stabilities, coadapting to the inside and outside circumstances of the animal.

Table 3 shows four kinds of amino acid residues, Ser, Ala, Asn, and Thr, appearing at position 17(A6) in the β chain. These four kinds of residues on the average are fixed throughout mammalian hemoglobins. On the other hand, in the case of 148(H3) in the β chain, only Pro and Gln have appeared in primate hemoglobins, while six different amino acid residues are found at this position in the hemoglobins of other mammals. That is, the degrees of limitation of certain amino acid residues which appear at position 17(A6) in the β chain are similar in mammalian hemoglobins, but in the case of 148(H3) in the β chain the degree of limitation became clearly stronger in primate hemoglobins (Maita *et al.*, 1973). In other words, it is felt that in the structural-free degree of primate hemoglobins not only the number of mutated positions but also the degree of variability of each mutated position has decreased.

Accordingly, primate hemoglobins have undergone minor mutations, coadapted to inside and outside circumstances, decreased their structural-free degree, and consequently undergone microevolution with a kind of stabilizing selection.

6. References

Arnone, A., and Perutz, M. F., 1974, Structure of inositol hexaphosphate–human deoxyhaemoglobin complex, *Nature (London)* **249**:34.

Barnabas, J., Goodman, M., and Moore, G. W., 1972, Descent of mammalian alpha globin chain sequences investigated by the maximum parsimony method, *J. Mol. Biol.* **69**:249.

Boyer, S. H., Crosby, E. F., Noyes, A. N., Fuller, G. F., Leslie, S. E., Donaldson, L. J., Vrablik, G. R., Schaefer, E. W., Jr., and Thurmon, T. F., 1971, Primate hemoglobin: Some sequences and some proposals concerning the character of evolution and mutation, *Biochem. Genet.* **5**:405.

Braunitzer, G., Gehring-Müller, R., Hilschman, N., Hilse, K., Hobom, G., Rudloff, V., and Wittman-Liebold, B., 1961, Die Konstitution des normalen adulten human Hämoglobins, *Hoppe-Seyler's Z. Physiol. Chem.* **325**:283.

Dickerson, R. E., 1971, The structure of cytochrome *c* and the rate of molecular evolution, *J. Mol. Evol.* **1**:26.

Goodman, M., Barnabas, J., Matsuda, G., and Moore, G. W., 1971, Molecular evolution in the descent of man, *Nature (London)* **233**:604.

Goodman, M., Moore, G. W., Barnabas, J., and Matsuda, G., 1974, The phylogeny of human globin genes investigated by the maximum parsimony method, *J. Mol. Evol.* **3**:1.

Goodman, M., Moore, G. W., and Matsuda, G., 1975, Darwinian evolution in the genealogy of haemoglobin, *Nature (London)* **253**:603.

Johnson, L. F., and Tate, M. E., 1969, Structure of phytic acid, *Can. J. Chem.* **47**:63.

Kendrew, J. C., Watson, H. C., Strandberg, B. D., Dickerson, R. E., Phillips, D. C., and Shore, V. C., 1961, A partial determination by X-ray method, and its correlation with chemical data, *Nature (London)* **190:**666.

Kimura, M., 1968, Evolutionary rate at the molecular level, *Nature (London)* **217:**624.

Kimura, M, 1969, The rate of molecular evolution considered from the standpoint of population genetics, *Proc. Natl. Acad. Sci. USA* **63:**1181.

King, T. L., and Jukes, T. H., 1969, Non-Darwinian evolution, *Science* **164:**788.

Konigsberg, W., Guidotti, G., and Hill, R. J., 1961, The amino acid sequence of the α chain of human hemoglobin, *J. Biol. Chem.* **236:**pc55.

Maeda, N., and Enoki, Y., 1974, The amino acid sequence of sika hemoglobin, *Seikagaku* **46:**706.

Maita, T., Nakashima, Y., and Matsuda, G., 1973, The primary structures of the α and β polypeptide chains of the hanuman langur hemoglobin, *Proc. XIVth Protein Struct. Conf. Jpn.*, p. 65.

Matsuda, G., Maita, T., Takei, H., Ota, H., Yamaguchi, M., Miyauchi, T., and Migita, M., 1968*a*, The primary structure of adult hemoglobin from *Macaca mulatta* monkey, *J. Biochem.* **64:**279.

Matsuda, G., Maita, T., and Ota, H., 1968*b*, The primary structures of primate hemoglobins and their phylogenetic implications, *Proc. VIIth Int. Congr. Anthropol. Ethnol. Sci. (Tokyo)*, Vol. 1, p. 376.

Matsuda, G., Takei, H., Wu, K. C., and Shiozawa, T., 1971, The primary structure of the α polypeptide chain of AII component of adult chicken hemoglobin, *Int. J. Peptide Protein Res.* **3:**173.

Matsuda, G., Maita, T., Mizuno, K., and Ota, H., 1973*a*, Amino acid sequence of a β chain of AII component of adult chicken haemoglobin, *Nature New Biol.* **244:**244.

Matsuda, G., Maita, T., Ota, H., Araya, A., Nakashima, Y., Ishii, U., and Nakashima, M., 1973*b*, The primary structures of α and β chains of adult hemoglobin of the Japanese monkey *(Macaca fuscata fuscata)*, *Int. J. Peptide Protein Res.* **5:**405.

Matsuda, G., Maita, T., Watanabe, B., Ota, H., Araya, A., Goodman, M., and Prychodko, W., 1973*c*, The primary structures of the α and β polypeptide chains of adult hemoglobin of the slow loris *(Nycticebus coucang)*, *Int. J. Peptide Protein Res.* **5:**419.

Matsuda, G., Maita, T., Nakashima, Y., Barnabas, J., Ranjekar, P. K., and Gandhi, N. S., 1973*d*, The primary structures of the α and β polypeptide chains of adult hemoglobin of the hanuman langur *(Presbytis entellus)*, *Int. J. Peptide Protein Res.* **5:**423.

Matsuda, G., Maita, T., Watanabe, B., Araya, A., Morokuma, K., Goodman, M., and Prychodko, W., 1973*e*, The amino acid sequence of the α and β polypeptide chains of adult hemoglobin of the savannah monkey *(Cercopithecus aethiops)*, *Hoppe-Seyler's Z. Physiol. Chem.* **354:**1153.

Matsuda, G., Maita, T., Watanabe, B., Araya, A., Morokuma, K., Ota, Y., Goodman, M., Barnabas, J., and Prychodko, W., 1973*f*, The amino acid sequences of the α and β polypeptide chains of adult hemoglobin of the capuchin monkey *(Cebus apella)*, *Hoppe-Seyler's Z. Physiol. Chem.* **354:**1513.

Matsuda, G., Maita, T., Suzuyama, Y., Setoguchi, M., Ota, Y., Araya, A., Goodman, M., Barnabas, J., and Prychodko, W., 1973*g*, Studies on the primary structures of α and β polypeptide chains of adult hemoglobin of the spider monkey *(Ateles geoffroyi)*, *Hoppe-Seyler's Z. Physiol. Chem.* **354:**1517.

Moore, G. W., Goodman, M., and Barnabas, J., 1973*a*, An iterative solution from the standpoint of the additive hypothesis to the dendrogram problem posed by molecular data sets, *J. Theor. Biol.* **38:**423.

Moore, G. W., Barnabas, J., and Goodman, M., 1973*b*, An approach to constructing maximum parsimony ancestral amino acid sequences on a given network, *J. Theor. Biol.* **38:**459.

Perutz, M. F., 1969, The Croonian Lecture, 1968: The haemoglobin molecule, *Proc. R. Soc. London Ser. B* **173:**113.

Perutz, M. F., 1970, Stereochemistry of cooperative effects in haemoglobin, *Nature (London)* **228:**726.

Perutz, M. F., Rossman, M. G., Cullis, A. F., Muirhead, H., Will, G., and North, A. C. T., 1960, Structure of haemoglobin: A three-dimensional Fourier synthesis at 5.5 Å resolution obtained by X-ray analysis, *Nature (London)* **185:**416.

Perutz, M. F., Muirhead, H., Cox, J. M., and Goaman, L. C. G., 1968, Three-dimensional Fourier synthesis of horse oxyhaemoglobin at 2.8 Å resolution. II. The atomic model, *Nature (London)* **219:**131.

Schopf, J. W., 1967, Antiquity and evolution of Precambrian life, *McGraw-Hill Yearbook of Science and Technology,* p. 47.

Zuckerkandl, E., and Pauling, L., 1962, Molecular disease, evolution, and genetic heterogeneity, in: *Horizons in Biochemistry* (M. Kasha and B. Pullman, eds.), pp. 189–225, Academic Press, New York.

Zuckerkandl, E., and Pauling, L., 1965, Evolutionary divergence and convergence in proteins, in: *Evolving Genes and Proteins* (V. Bryson and H. J. Vogel, eds.), pp. 97–166, Academic Press, New York.

The Hemoglobins of *Tarsius bancanus*

12

JAN M. BEARD and MORRIS GOODMAN

1. Introduction

The position of *Tarsius* in the evolution of the order Primates has long been controversial. Gadow (1898) put it in a separate suborder, but Pocock (1918) allied it with the monkeys, apes, and man in the group Haplorhini, assigning the lemurs, lorisoids, and tree shrews to the Strepsirhini. This latter classification was later accepted by Hill (1955). On the other hand, Simpson (1945) in his influential classification of mammals divided the Primates into two suborders, Anthropoidea and Prosimii, placing *Tarsius* in the latter together with lemurs, lorisoids, and tree shrews, although as a separate infraorder.

At first glance, *Tarsius* looks quite like the lesser bushbaby, *Galago senegalensis*, with its large eyes and elongated hind limb, especially the foot. It is clear, however, that these resemblances indicate no close phylogenetic affinities but are due to the fact that both forms are highly adapted for leaping in the trees under nocturnal or crepuscular conditions, a propensity which is also developed to varying degrees in certain lemurs.

The main point at issue is whether *Tarsius* and the anthropoids had a common ancestor in the early Tertiary and owe some of their resemblances to this. We can only mention here some salient points that have been held to favor this view.

The term "haplorhine" refers to the structure of the end of the snout. In strepsirhines and many other mammals, the nostrils lie in an area of moist,

JAN M. BEARD • Department of Anthropology, University College London, London WCIE 6BT, England. MORRIS GOODMAN • Department of Anatomy, Wayne State University School of Medicine, Detroit, Michigan 48201.

hairless skin (rhinarium) which is prolonged down the midline of the upper lip. In *Tarsius* and other haplorhines, however, the rhinarium is absent and the nostrils are surrounded by hairy skin which extends over the more mobile upper lip.

The eyes of *Tarsius* are exceptionally large, no doubt in relation to nocturnal vision; but despite the fact that it has an all-rod retina there is a small, localized depression that resembles the fovea of anthropoids. The occipital area of the brain devoted to vision is relatively large and the visual cortex has clear lamination. The very large orbits are partially separated from the temporal fossa by a lamina formed by the same bony elements that compose the more complete partition in anthropoids. The size of the orbits also results in a compression of the nasal region and a brain case that is broad in relation to its length; the resemblance to the anthropoid head with short face and rounded cranium may therefore be spurious.

There is a marked contrast between the Haplorhini and the Strepsirhini in that the former have a discoidal, hemochorial placenta, while the latter have the diffuse, epitheliochorial type. Moreover, in the early stages of placentation *Tarsius* has a short allantoic stalk from the mesoderm of which the chorion is precociously vascularized as in anthropoids. There are some differences in the implantation and early formation of fetal membrane in *Tarsius,* but the similarity to higher primates is on the whole impressive.

Tarsius is often cited as a good example of a "living fossil," i.e., a contemporary form that closely resembles some remote fossil ancestor. The implication is that evolutionary change stopped or has been very much slowed down. We know of no paleontological evidence about the ancestry of *Tarsius* until we come to certain middle Eocene forms (50 million years ago). There have been many divergent views about the interpretation of various early Tertiary primates. However, in a recent account Simons (1972) places four European genera (*Necrolemur, Nannopithex, Microchoerus,* and *Pseudoloris*) in the family Tarsiidae. The skull of *Necrolemur* is particularly well known from a dozen very complete specimens and is said to resemble that of *Tarsius* in many respects, while the palatal form and dentition of *Pseudoloris* are extremely similar to those of the living tarsier. We know very little about the postcranial skeletons of these forms, but it is known that the calcaneum of *Nannopithex* was elongated, suggesting adaption for leaping as in *Tarsius.*

So far, the results of biochemical methods of classification, i.e., immunodiffusion studies and comparisons of DNA, have tended to group *Tarsius* closer to anthropoids than the strepsirhines (reviewed by Goodman, 1974). It may be that data on the primary structure of proteins can make an important contribution to the solution of the *Tarsius* problem. We have chosen to work on the hemoglobin α and β chains because these globins can be obtained in adequate amount from a small animal. Ideally one might hope to find substitutions that, for instance, occur in *Tarsius* and anthropoids but not in prosimians. This implies that we need to have characterized these chains adequately in both anthropoids and prosimians by sequencing a fair proportion of the living forms, so chosen as to represent taxonomic diversity of the suborders without bias.

2. Primate Sequences and Analytic Procedures

The data on hemoglobin chains in anthropoids are moderately good, a number of sequences having been determined by Matsuda and co-workers. Both α and β chain sequences are known for *Macaca mulatta* (Matsuda *et al.*, 1968), *Macaca fuscata* (Matsuda *et al.*, 1973a), *Presbytis entellus* (Matsuda *et al.*, 1973c), *Cercopithecus aethiops* (Matsuda *et al.*, 1973d), *Cebus apella* (Matsuda *et al.*, 1973e), and *Ateles geoffroyi* (Matsuda *et al.*, 1973f). The partial data on *Ateles* β chain confirmed the results of Boyer *et al.* (1971). The sequences of human α chain (Braunitzer *et al.*, 1961; Konigsberg *et al.*, 1961) and β chain (Braunitzer *et al.*, 1961) are also known. The position is less satisfactory with regard to prosimians: sequence data on four genera have been published *(Lemur, Propithecus, Galago, Nycticebus)*, but the work on three has not been published in detail and is open to some doubt, leaving only *Nycticebus* (Matsuda *et al.*, 1973b) as a reliable example.

According to Kimura's (1969) views, most of the evolutionary divergence in proteins is due to random processes and proceeds at a more or less constant rate. This view predicts that the proteins of *Tarsius* should not be exempt from change despite tarsier's anatomical resemblances to Eocene forms. Moreover, even if natural selection produces nonuniformities in rates of protein evolution, as appears to be the case (e.g., Goodman *et al.*, 1975; Fitch and Langley, this volume; Matsuda, this volume; Hewett-Emmett *et al.*, this volume), the evolutionary factors selecting for subtle changes in a protein would not necessarily select for gross changes in anatomical features. Greater conservatism of tarsier compared to anthropoids at the anatomical level need not have to mean greater conservatism of hemoglobin sequences. Given adequate data, it is possible to reconstruct the most probable sequence of a particular protein chain in the ancestor of anthropoids or in primates as a whole (Goodman *et al.*, 1974, 1975; Dayhoff, 1972). We can then see to what extent the sequence of *Tarsius* resembles this putative ancestral sequence. We stress the word *putative* because such reconstructions are based on assumptions that may not always be true and even so some amino acid sites are apt to yield several equally likely solutions.

We can determine further whether the tarsier sequence shows more residues in common with the anthropoid sequence or with the prosimian *(Nycticebus)* sequence. Moreover, we can directly investigate the phylogenetic position of *Tarsius* by including it in the sequence data sets and then search by the maximum parsimony algorithms (Moore *et al.*, 1973; Goodman *et al.*, 1974;

Moore, this volume) for the evolutionary trees which describe the descent of the sequences. Both approaches have been followed, and a summary has been included in a preliminary report (Beard *et al.*, 1976).

3. Primary Structure of Tarsier Hemoglobin Chains

In the last few years, two samples of blood were acquired from *Tarsius bancanus.* One animal was captured in Sarawak by J. Rollinson and kept in the Laboratory of Physical Anthropology at University College London for about 6 months. A small amount of blood was obtained immediately after the animal died. The second sample was sent by C. Niemitz. This animal had also been captured in Sarawak.

As reported previously (Barnicot and Hewett-Emmett, 1974, hemoglobin samples from each animal were studied by starch gel electrophoresis at pH 8.6 (tris-EDTA-borate system). The mobilities of the major and minor components corresponded to those of human HbA and HbA₂. However, the relative concentration of the minor component in *Tarsius* was considerably higher than that of HbA₂; the minor component represented 18% of the total in both samples.

The two hemoglobins could be readily separated by ion-exchange chromatography on DEAE-Sephadex (Huisman and Dozy, 1965), and the major component was used for sequencing studies. The α- and β-globin chains were separated using chromatography on CM-cellulose in 8 M urea (Clegg *et al.*, 1968) and the chains were aminoethylated.

The methods used for sequence determination will not be described in detail here but involved tryptic digestion, fractionation of the peptides by paper electrophoresis and chromatography, and sequence determination by the dansyl-Edman degradation. Where necessary, tryptic peptides were redigested with thermolysin or chymotrypsin. Amides were assigned using electrophoretic mobility. The aim was to determine the sequences as completely as possible with the material available without having to resort to amino acid analysis and alignment of large sections by homology with other primate chains.

The sequence of the α chain differed from that of the human α chain in 11 positions (Table 1). The β chain of the tarsier showed 15 differences from the human chain (Table 2). Presently, only the amino acid composition of residues 129–136 in the α chain is known. There was one extra valine residue in the tarsier, and this has been assigned to residue 129 rather than residue 136, as a change from leucine has been observed previously at this position (dog; cited in Dayhoff, 1972). The sequence data on the β chain are incomplete with respect to the region covering residues 67–75. The sequence suggested in Table 2 is based on amino acid compositions of two thermolytic peptides.

The sequence differences among primates in α and β chains are summarized in Tables 1 and 2, respectively. Only those residues which are variable are shown, together with the two sets of hypothetical ancestral primate and anthropoid sequences, the one set based on prior studies of globin phylogeny (Good-

Table 1. Variable Residues in Primate α-Globin Chains[a]

Residue No.	8	12	15	19	21	23	53	57	67	68	71	73	78	111	113	116	129
Primate ancestor[b]	Thr	Ala	Gly	Gly	Ala	Asp	Ala	Ala	Thr	Asn	Ala	Val	Ser	Ser	His	Asp	Leu
Anthropoid ancestor[b]	—	—	—	—	—	—	—	Gly	—	—	—	—	Asn	Ala	—	—	—
Nycticebus	—	—	Glu	Ser	—	—	—	—	—	—	Ser	—	—	Cys	—	—	—
Tarsius	—	—	Asp	—	—	—	Ser	Gly	—	Thr	Gly	Ile	Asn	Cys	—	—	Val
Ateles	Ser	—	—	—	—	—	—	Gly	—	—	—	—	Asn	Ala	—	—	—
Cebus	—	Thr	—	—	—	—	—	Gly	Ser	—	—	—	Asn	Ala	—	—	—
Cercopithecus	Ser	—	—	—	—	Glu	—	Gly	—	Leu	Gly	—	His	Ala	Leu	Glu	—
M. mulatta	Ser	—	—	—	—	Glu	—	Gly	—	Leu	Gly	—	Asn	Ala	Leu	Glu	—
M. fuscata	Ser	—	—	—	—	Glu	—	Gly	—	Leu	Gly	—	Asn	Ala	Leu	Glu	—
Presbytis	—	—	—	—	Gly	Glu	—	Gly	—	—	—	—	His	Ala	Leu	Glu	—
Homo	—	—	—	Ala	—	Glu	—	Gly	—	—	—	—	Asn	Ala	Leu	Glu	—
Ancestors from tree[c]																	
Primate	Thr	Ala	Gly	Gly	Ala	Asp	Ala	Ala	Thr	Asn	Ala	Val	Ser	Ser	His	Asp	Leu
Haplorhine	Thr	Ala	Gly	Gly	Ala	Asp	Ala	Gly	Thr	Asn	Ala	Val	Asn	Ser	His	Asp	Leu
Anthropoid	Thr	Ala	Gly	Gly	Ala	Asp	Ala	Gly	Thr	Asn	Ala	Val	Asn	Ala	His	Asp	Leu

[a]Residues which are variable among the listed primates and which differ from the primate ancestor are shown.
[b]From the parsimony construction of Goodman *et al.* (1975); these constructions also include the following alternative parsimony solutions at positions 23 (primate Glu, anthropoid Glu) and 111 (Ala, Ala). However, with these alternatives inclusion of the tarsier sequence would add more mutations to the tree than with the solution employed here.
[c]These ancestral sequences are taken from the tree illustrated in Fig. 1. Whenever alternative maximum parsimony solutions existed at a residue position, the A-solution (as described in Goodman *et al.*, 1974) was utilized in the tree. A mutation which may appear in a lineage leading either to few or many contemporary species will choose the lineage with the fewest species in the A-solution, since this decreases the number of times the mutation is counted among lineages between the most ancestral point of the tree and each contemporary species. The A-solution counteracts more than any other solution the bias toward a greater underestimation of change on the lineages with few nodal points.

man *et al.*, 1975) and the other found in this study by including the tarsier sequences in the reconstruction procedure.

In the α chain, the sequence of tarsier is unique among known primates at five positions. Only one of these, position 129 (H12), is involved in the function of hemoglobin (Perutz, 1969; Goodman *et al.*, 1975). The presence of valine instead of leucine would not be expected to interfere with the heme contact. Phenylalanine is present in the dog sequence at this position.

In the β chain, there are eight positions at which tarsier is unique among the primates. One of these, position 75 (E19) is internal and helps stabilize the tertiary structure. Methionine is present at this position in sheep (Dayhoff, 1972). All the other changes occur at nonfunctional sites and most of the changes are conservative, e.g., Glu → Asp. Tarsier is unusual in having an aspartic acid residue at residue 19 in the β chain instead of asparagine.

4. Minor Component of Tarsier Hemoglobin

Before discussing the possible evolutionary position of *Tarsius* as suggested by the sequence data, we wish to report our findings to date on the minor component. Minor components have been shown to exist in man, apes, and New World monkeys but not in Old World monkeys, except for *Macaca*

Table 2. Variable Residues in

Residue No.	5	6	9	13	19	21	22	33	43	50	52
Primate ancestor[b]	Ala	Glu	Ser	Ala	Asn	Asp	Glu	Val	Glu	Ser	Asp
Anthropoid ancestor[b]	—	—	—	—	—	—	—	—	—	Thr	—
Nycticebus	Gly	—	—	—	—	—	Asp	—	—	—	Ser
Tarsius	—	Asp	Ala	—	Asp	Glu	Asp	—	Asp	Thr	Ala
Ateles	Gly	—	Ala	—	—	—	—	—	—	Thr	—
Cebus	—	—	—	Thr	—	—	—	—	Asp	Thr	—
Cercopithecus	Pro	—	Thr	Thr	—	—	—	—	—	—	—
M. mulatta	Pro	—	Asn	Thr	—	—	—	Leu	—	—	—
M. fuscata	Pro	—	Asn	Thr	—	—	—	—	—	—	—
Presbytis	Pro	—	Ala	—	—	—	—	—	—	—	—
Homo	Pro	—	—	—	—	—	—	—	—	Thr	—
Ancestors from tree[c]											
Primate	Ala	Glu	Ala	Ala	Asn	Asp	Glu	Val	Glu	Ser	Asp
Haplorhine	Ala	Glu	Ala	Ala	Asn	Asp	Glu	Val	Glu	Thr	Asp
Anthropoid	Ala	Glu	Ala	Ala	Asn	Asp	Glu	Val	Glu	Thr	Asp

[a]Residues which are variable among the listed primates and which differ from the primate ancestor are shown.

[b]From the parsimony construction of Goodman *et al.* (1975); these constructions also include the following alternative parsimony solutions at positions 9 (primate Ala, anthropoid Ala), 50 (Ser, Ser), 52 (Asn, Asp), 69 (Ala, Gly), 87 (Lys, Gln), and 112 (Ile, Cys and Val, Cys). However, with any of

fascicularis. The variant chain in this case has been shown to be an α-like chain and not a β or δ chain homologue. The orthology of the δ chain in man and the apes with the "δ" chain in New World monkeys is equivocal.

Following the observation of a minor hemoglobin in tarsier which has A₂-like mobility, Goodman (1973) suggested that a β-protodelta duplication may have occurred in a tarsioid line ancestral to both *Tarsius* and Anthropoidea.

The δ chains in New World monkeys, apes, and man all have arginine and asparagine in positions 116 and 117, respectively, instead of histidine and histidine, which occur in these positions of the corresponding β chains. However, glycine is present at position 5 in both the β and δ chains of the New World monkeys (except for *Cebus,* which has alanine in the β chain), whereas proline is present at position 5 of β and δ chains of apes and man (Boyer *et al.,* 1971, 1972). These observations and the numbers of differences between β and δ chains of the same and different species suggest that multiple gene duplications occurred during approximately the same evolutionary epoch (Boyer *et al.,* 1971).

Starch gel electrophoresis, in 6 M urea, of the globins of *Tarsius* showed that the minor component had an altered β chain. Efforts to separate the chains by CM-cellulose chromatography were not successful, so the whole globin was digested with trypsin and the digests were fingerprinted. The peptides were eluted and hydrolyzed for amino acid analysis. Table 3 gives the results of

Primate β-Globin Chains[a]

56	58	69	73	75	76	87	104	112	121	125	126	128	139
Gly	Pro	Gly	Asp	Leu	Asn	Lys	Arg	Phe	Glu	Gln	Val	Ala	Asn
—	—	—	—	—	Thr	—	—	Cys	—	—	—	—	—
—	—	Ser	—	—	—	—	—	Val	Asp	—	—	Ser	—
—	Ala	Asn	Glu	Met	Ala	—	—	Cys	—	—	—	—	Thr
Ser	—	—	—	—	Ala	Gln	—	Cys	—	—	Leu	—	—
Asn	—	—	—	—	Thr	Gln	—	Cys	—	—	—	—	Thr
—	—	—	—	—	Ala	Gln	Lys	Cys	—	—	—	—	—
—	—	—	—	—	—	Gln	Lys	Cys	—	—	—	—	—
—	—	—	—	—	—	Gln	Lys	Cys	—	—	—	—	—
—	—	—	—	—	Ala	Gln	—	Cys	—	—	—	—	—
—	—	—	—	—	Ala	Thr	—	Cys	—	Pro	—	—	—
Gly	Pro	Ser	Asp	Leu	Asn	Lys	Arg	Ile	Glu	Gln	Val	Ala	Asn
Gly	Pro	Ser	Asp	Leu	Ala	Lys	Arg	Cys	Glu	Gln	Val	Ala	Asn
Gly	Pro	Gly	Asp	Leu	Ala	Lys	Arg	Cys	Glu	Gln	Val	Ala	Asn

these alternatives inclusion of the tarsier sequence would add more mutations to the tree than the solution employed, except at β9 and β87 where the alternatives could be used equally well as the anthropoid ancestors.

[c] These ancestral sequences are taken from the tree illustrated in Fig. 1. This tree utilized only A-solution residues.

analysis of the tryptic peptides containing the sites important in deciphering the origin of the δ chains of primates (positions 5, 116, and 117). The analyses of both the major and minor component peptides are shown. The compositions of the corresponding peptides were identical. Two histidines are present at positions 116 and 117, and position 5 is alanine in both cases. There is, therefore, no direct evidence that the β-like chain of the minor hemoglobin of *Tarsius* is orthologous with either man-ape or New World monkey δ chain.

One of the differences in the minor component contributing to the change in mobility at pH 8.6 of the whole hemoglobin is in the tryptic peptide βTp3 (positions 18–30). This peptide had an altered electrophoretic mobility on fingerprinting, and the amino acid analysis indicated that asparagine is present at position 19 instead of aspartic acid. The other charge change has not been characterized.

The other important point which should be noted is that the "minor" component of tarsier is present in relatively high concentration—18% of the total. In other species, the concentration of minor components is between 0.5 and 6% (Boyer *et al.*, 1971). It would be of interest to examine the hemolysates of other tarsier blood samples and look at the frequency of occurrence of the minor component in all species of tarsier. It is also possible that the proportions of the hemoglobins may vary. Two major hemoglobins have been observed in several other prosimians (reviewed by Sullivan, 1971; Barnicot and Hewett-Emmett, 1974).

Table 3. Amino Acid Compositions[a] of Selected Tryptic Peptides from Tarsier β Chain (Major Component) and "β" Chain (Minor Component)

	βTpl	"β"Tpl	βTpl2B	"β"Tpl2B
Asp	1.1	1.0		
Thr	1.1	1.0		
Glu	1.0	1.1		
Gly			1.1	1.2
Ala	1.0	0.9	1.0	1.0
Val	0.8	0.9	1.0	0.8
Leu	1.0	1.1	1.0	1.2
Phe			1.0	1.0
Lys	1.0	1.1	1.0	0.8
His	0.9	1.0	1.9	2.0
Residues in sequence	1–8	1–8	113–120	113–120

[a]Amino acid analyses were carried out after hydrolysis of the peptides with 6 N HCl for 24 hr at 110°C. Compositions are expressed as moles per mole of peptide.

5. Evidence on Phylogenetic Position of Tarsius

5.1. Using Ancestral Sequences Constructed without Tarsier

The original aim of the work on tarsier hemoglobin was to acquire sequence information which could possibly be used as evidence for the position of tarsier in primate evolution. If one assumes that the ancestral sequence is known correctly, it may be possible, by examining the known primate sequences, to find changes which are common to tarsier and prosimians only, or which have been acquired by tarsier and the anthropoids after the divergence of the prosimians. If tarsier diverged from the anthropoid branch very soon after the prosimians, there may not have been time for substitutions to occur which would be inherited by both lineages. Moreover, as another possibility, the most ancient splitting in the Primates might have separated tarsier from all other extant primates.

The putative primate and anthropoid ancestral sequences constructed from data lacking tarsier are shown in the top rows of Tables 1 and 2. These data refer only to globin chains in which a majority of residues had been placed by actual sequencing procedures. The data on several Old World monkey hemoglobins compiled by Hewett-Emmett et al. (this volume) were not used, because a majority of their residues were placed by inferred homology to known sequences.

The only prosimian sequence which influenced the ancestral reconstructions was that of *Nycticebus*. Sequences have been reported for *Lemur fulvus*,

Propithecus verrauxi, and *Galago crassicaudatus* and attributed to Nishizaki and Hill (Sober, 1968; Sullivan 1971). However, the evidence for these sequences may not be adequate. Perutz (personal communication) doubts the validity of the *Lemur* sequence because it conflicts with invariant or semi-invariant features of mammalian hemoglobins and would introduce unlikely variations in tertiary structure. Certainly the *Lemur* and *Propithecus* β chains seem to be considerably different from other primate β chains. They apparently have fixed many more mutations since divergence than other primates.

In the α chains (Table 1), position 15 provides some evidence for a common ancestor of *Tarsius* and *Nycticebus.* If such an ancestor existed, only two mutations are required to produce the known changes in the sequences. If there is no common ancestor, three mutations are needed. There is one residue which only *Tarsius* and *Nycticebus* have in common among the primates, cysteine at position 111 instead of the ancestral serine. On the other hand, there are two positions at which *Nycticebus* retains the ancestral sequence while tarsier and most of the anthropoids share a common change (57—Ala to Gly; 78—Ser to Asn). Both are single base changes.

If one considers the β chains in Table 2, there is one position (22) at which tarsier and *Nycticebus* share a common change from the ancestral sequence. Also, at positions 52 and 69 fewer mutations need have occurred if they shared a common ancestor after diverging from the anthropoids. However, at position 112 a cysteine is present in tarsier and the anthropoids instead of the ancestral phenylalanine, while *Nycticebus* has a valine residue at this position. Also, at positions 50 and 76 *Nycticebus* has retained the ancestral sequence, while tarsier and several anthropoids have acquired the same change.

Thus, with respect to both α and β chain data, the hypothetical ancestral residues constructed independently of tarsier fail to indicate whether *Tarsius* is closer phylogenetically to Anthropoidea or to Prosimii. It is likely that the tarsier sequence contains information needed for a closer approximation to the ancestral sequence. If so, tarsier needs to be included in the sequence data sets for a more adequate test of its phylogenetic position.

5.2. Maximum Parsimony Analysis of Data Sets Including Tarsier

The most extensive analysis has been carried out with a combined α- and β-globin alignment. The sequences in it represent a "hybrid" amphibian (newt α plus frog β), a chicken, a monotreme, a marsupial, and 16 eutherian mammals, of which nine are the primates in Tables 1 and 2. The criterion for selecting these tetrapods was that in each case a majority of the residues in at least one of the two hemoglobin chain types had been placed by actual sequencing procedures. Various computer programs, the principles of which are described in Moore (this volume), have been used to search for the most parsimonious evolutionary tree. Over 360 dendrograms have been examined, testing a wide range of phylogenetic possibilities. So far, no tree with a length less than 664 mutations or nucleotide replacements (NR) has been found. There are nine trees with this most parsimonious NR length. Of these alternatives, tarsier is always closer to Anthropoidea than to *Nycticebus,* or, conversely *Nycticebus* is always more anciently separated from the anthropoids than is tarsier. This supports the inclusion of *Tarsius* in the Haplorhini subdivision of Primates. The ancestral primate, haplorhine, and anthropoid sequences presented in Tables 1 and 2 (bottom rows) are based on the maximum parsimony tree shown in Fig. 1, a tree which conforms closely to traditional evidence on tetrapod phylogeny.

The variations in branching order among the nine parsimony trees are

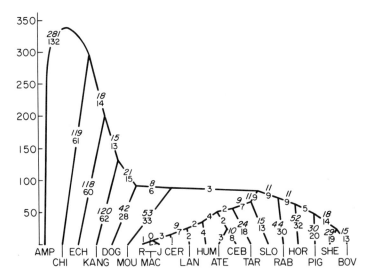

Fig. 1. Parsimony tree requiring 664 nucleotide replacements for 20 taxa on using a combined α- and β-globin alignment. Link lengths are the numbers of nucleotide replacements between adjacent ancestral and descendant sequences; italicized numbers are link lengths corrected for super-imposed replacements by the augmentation algorithm of Moore and Goodman (Goodman *et al.,* 1974). The ordinate scale, in millions of years, is inferred from fossil evidence on ancestral splitting times of the taxa represented by the sequences. The taxa are as follows: AMP (amphibian) newt α (R. T. Jones, personal communication) and frog β (Chauvet and Acher, 1972); CHI (chicken) α (Matsuda *et al.,* 1971) and β (Matsuda *et al.,* 1973*g*); ECH (echidna) α (Whittaker *et al.,* 1973) and β (Whittaker *et al.,* 1972); KANG (kangaroo) α (Air and Thompson, 1969); DOG (dog) β (Jones *et al.,* 1971); MOU (mouse) α (Popp, 1967) and β (Popp, 1973); R-MAC (rhesus macaque) α and β (Matsuda *et al.,* (1968); J-MAC (Japanese macaque) α and β (Matsuda *et al.,* 1973*a*); CER (*Cercopithecus*) α and β (Matsuda *et al.,* 1973*d*); LAN (langur) α and β (Matsuda *et al.,* 1973*c*); HUM (human) α and β (Braunitzer *et al.,* 1961); ATE (*Ateles*) α and β (Matsuda *et al.,* 1973*f*); CEB (*Cebus*) α and β (Matsuda *et al.,* 1973*e*); TAR (*Tarsius*) α and β (this chapter) SLO (slow loris or *Nycticebus*) α and β (Matsuda *et al.,* 1973*b*); RAB (rabbit) α (Von Ehrenstein, 1966) and β (Best *et al.,* 1969); HOR (horse) α (Matsuda *et al.,* 1963) and β (Smith, 1968; Dayhoff, 1972); PIG (pig) α (Yamaguchi *et al.,* 1965; Dayhoff, 1972) and β (Braunitzer and Kohler, 1966; Dayhoff, 1972); SHE (sheep) α (Beale, 1967; Dayhoff, 1972) and β (Boyer *et al.,* 1967; Dayhoff, 1972); BOV (ox) α and β (Schroeder *et al.,* 1967*a,b;* Dayhoff, 1972).

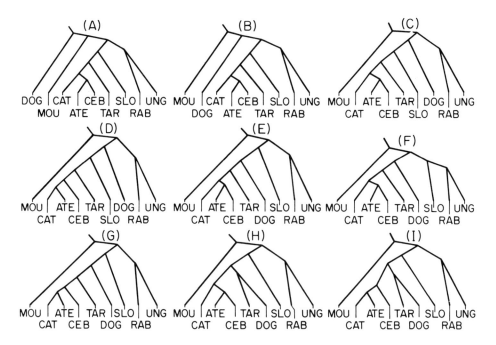

Fig. 2. Variations in branching order among the nine alternative parsimony dendrograms with nucleotide replacement lengths of 664. As the topological variations in these dendrograms were all confined to the branching order among placental mammals, only this region in the alternative dendrograms is shown. The alternative diagrammed in (A) represents the tree used for Fig. 1. The taxa depicted are DOG (dog), MOU (mouse), CAT (catarrhine primates), ATE *(Ateles)*, CEB *(Cebus)*, TAR *(Tarsius)*, SLO (slow loris or *Nycticebus*), RAB (rabbit), and UNG (ungulates).

diagrammed in Fig. 2. It may be noted that all trees preserve the general features of tetrapod phylogeny. The avian (chicken) branch separates from the mammalian branch and the latter splits into monotreme (echidna) and therian divisions. In turn, the therians split into Metatheria (kangaroo) and Eutheria. Moreover, within the Eutheria the branching arrangement among catarrhines, and also among rabbit and ungulates, is constant in all nine trees. The principal variation in these trees is in the position of dog. Note that it can change position with mouse and then join the primates. When it does the latter (Fig. 2C–I), either joining with the common ancestor of primates or just of haplorhines or with tarsier itself, the branching order within Anthropoidea becomes ambiguous in that the cebid *Ateles* can now separate from *Cebus* and join the catarrhines without increasing the NR length of the parsimony reconstruction (Fig. 2D,G,I). Moreover, when dog joins the common haplorhine ancestor it becomes as parsimonious for slow loris to join the rabbit-ungulate branch as to remain with the dog-haplorhine branch. When species genealogies are constructed from combined sequence alignments of hemoglobin α and β chains plus either just myoglobins or myoglobins, fibrinopeptides, and cytochromes c, the most ancestral splitting in the Eutheria in all parsimony alternatives always separates dog from a common ancestor of primates and ungulates (Goodman, this volume). Thus the branching arrangement shown in Figs. 1 and 2A, which we consider the preferred arrangement for dog, primates, and ungulates, can be supported by the overall parsimony evidence from protein sequences.

Another argument, again based on parsimony, can be offered for preferring the branching topology in Figs. 1 and 2A to the other eight alternatives. Namely, there is sufficient evidence from other sources, such as immunological evidence, to rule out a species genealogy which removes either tarsier or slow loris or both from the primates. To so remove these primates would require that their hemoglobin sequences be paralogously rather than orthologously related to other primate hemoglobin sequences. In turn, the assumption of paralogous relationship requires prior events of gene duplication to produce the nonallelic loci. If the mutational cost of such events is considered, then trees which join dog to the haplorhine common ancestor or to tarsier itself are not as parsimonious as trees which do not, all else being equal.

If the positions of tarsier and slow loris in the tree shown in Fig. 1 are reversed, six more mutations are added to the path length of the tree. On the other hand, only two mutations are added if tarsier and slow loris joined each other (as prosimians) before the divergence of Anthropoidea from prosimian primates. While this prosimian position of tarsier compared to its haplorhine position subtracts four mutations from the NR length of the tree (at residues positions $\alpha15$, $\alpha111$, $\beta22$, and $\beta52$), it also adds six mutations (at $\alpha57$, $\alpha78$, $\beta50$, and $\beta112$ and two mutations at $\beta76$). It would seem, therefore, that the hemoglobin parsimony evidence for placing tarsier in the Haplorhini resides solely in the β sequences. This, however, is not the case. When the 20 α sequences of the taxa in Fig. 1 were analyzed separately by the maximum parsimony method, tarsier and Anthropoidea in the most parsimonious topologies always possessed a common ancestor after divergence of *Nycticebus*. The NR length of these topologies is 321 compared to 324 for the topology in Fig. 1. Of 218 dendrograms examined, nine had the 321 NR length. In all nine, *Nycticebus* joined the haplorhine (tarsier-Anthropoidea) branch; in seven of the nine, dog and primates shared a common ancestry after their divergence from other eutherians, in the other two, dog joined the rabbit-ungulate branch.

When the 20 β sequences were analyzed separately, tarsier again was closer to Anthropoidea than to *Nycticebus* in the most parsimonious topologies. The NR length of these topologies is 338 compared to 340 for the topology in Fig. 1. Of 240 dendrograms examined, seven had the 338 NR length. In one of these (identical to the topology in Fig. 2H), a dog-tarsier branch joined the Anthropoidea; the other six also had tarsier, Anthropoidea, and dog share a common ancestry after their divergence from other eutherians, and then had dog separate from the common ancestor of haplorhines (tarsier and Anthropoidea).

Thus from the aggregate of results obtained by analyzing α and β sequences separately and together it seems reasonable to conclude that *Tarsius* and Anthropoidea are related as haplorhines. However, this conclusion is tenuous. Clearly, sequences from more prosimian species than just *Nycticebus* need to be included in the analyses to adequately test the conclusion. For that matter, inclusion of a wider range of species from carnivores, lagomorphs, and other nonprimate mammals would also help provide a more decisive answer. Indeed, three more species, mangabey, opossum, and platypus, have recently been added to the data set of combined α- and β-globin chains, and the most parsimonious trees (only two have been found) confirm the haplorhine placement of *Tarsius* depicted in Fig. 1 (see Appendix).

With regard to our present estimate of putative ancestral sequences, it is worth noting that the haplorhine ancestral sequence is closer to the ancestral anthropoid than to the ancestral primate sequence (bottom rows of Tables 1 and 2). Mutations on the link joining the primate and haplorhine ancestors occur at $\alpha57$ (Ala $\rightarrow$ Gly), $\alpha78$ (Ser $\rightarrow$ Asn), $\beta50$ (Ser $\rightarrow$ Thr), $\beta76$ (Asn $\rightarrow$ Ala), and $\beta112$ (Ile $\rightarrow$ Cys), whereas mutations occur only at $\alpha111$ (Ser $\rightarrow$ Ala) and $\beta69$ (Ser $\rightarrow$ Gly) on the link joining the haplorhine and anthropoid ancestors. It is also evident that tarsier diverges more from the haplorhine ancestor than do most of the Anthropoidea. This clearly supports our impression that, at the molecular level, tarsier is no "living fossil."

6. Appendix

The collection of sequences in the combined α- and β-globin alignment was enlarged to represent three additional vertebrate species or taxa: mangabey *(Cercocebus atys)*, opossum *(Didelphus marsupialis)*, and platypus *(Ornithorhynchus anatinus)*. We thoroughly retested the nine topologies which had previously been the most parsimonious for the original 20 vertebrate taxa (Fig. 2) in an examination of 164 different network topologies for the enlarged collection of 23 taxa. Only two alternative maximum parsimony topologies, each with a length of 771 nucleotide replacements, were found. Mangabey must either join the macaque branch (as in Fig. 3) or switch positions with *Cercopithecus*, i.e., join a macaque-*Cercopithecus* branch. Our reason for preferring the alternative shown in Fig. 3 is discussed by Goodman (this volume). The two marsupials, opossum and kangaroo, must join together as a monophyletic branch; also, the two monotremes, platypus and echidna, must so join together. Aside from these three additions, the topology in Fig. 3 with respect to the original 20 taxa is the same as in Fig. 1. All other topologies tested for them were less parsimonious. In particular, one or two nucleotide replacements were added to the length of the genealogy when topologies which had previously been alternative maximum parsimony ones (Fig. 2B–I) were examined.

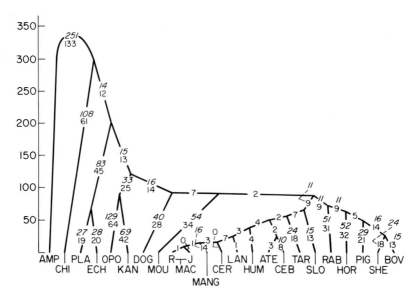

Fig. 3. Parsimony tree requiring 771 nucleotide replacements for 23 taxa on using a combined α- and β-globin alignment. Link lengths are the numbers of nucleotide replacements between adjacent ancestral and descendant sequences; italicized numbers are link lengths corrected for superimposed replacements by the augmentation algorithm of Moore and Goodman. The ordinate scale, in millions of years, is inferred from fossil evidence on ancestral splitting times of the taxa represented by the sequences. The taxa are the 20 shown in Fig. 1 and in addition PLA (platypus) α (Whittaker and Thompson, 1974) and β (Whittaker and Thompson, 1975); OPO (opossum) α (Stenzel, 1974) and β (R. T. Jones, personal communication); and MANG (mangabey) α and β (Hewett-Emmett *et al.,* this volume).

ACKNOWLEDGMENTS

We wish to acknowledge the unfailing support of the late Professor N. A. Barnicot, who introduced one of us (J. M. B.) to the *Tarsius* problem and discussed the morphological and paleontological evidence. Dr. David Hewett-Emmett carried out some of the initial observations on tarsier hemoglobin and helped in the early stages of the sequence determination. We also thank Mr. E. Crutcher and Mrs. D. M. Fisher for technical assistance. Dr. J. Bridgen of the MRC Laboratory of Molecular Biology, Cambridge, provided a sequenator analysis of residues 1–20 of the β chain of tarsier which agreed with the foregoing results. The Medical Research Council (U.K.) and the Systematic Biology Section of the National Science Foundation (U.S.A.) gave financial support. The upkeep of the tarsiers captured by Dr. June Rollinson was financed by MRC (UK) support to Dr. R. D. Martin.

7. References

Air, G. M., and Thompson, E. O. P., 1969, Studies on marsupial proteins. II. Amino acid sequence of the β-chain of haemoglobin from the grey kangaroo, *Macropus giganteus, Aust. J. Biol. Sci.* **22:**1437.

Barnicot, N. A., and Hewett-Emmett, D., 1974, Electrophoretic studies on prosimian blood proteins, in: *Prosimian Biology* (R. D. Martin, G. A. Doyle, and A. C. Walker, eds.), pp. 891–902, Duckworth, London.

Beale, D., 1967, A partial amino acid sequence of sheep hemoglobin A, *Biochem. J.* **103:**129.

Beard, J. M., and Thompson, E. O. P., 1971, Studies on marsupial proteins. V. Amino acid sequence of the α-chain of haemoglobin from the grey kangaroo, *Macropus giganteus, Aust. J. Biol. Sci.* **24:**765.

Beard, J. M., Barnicot, N. A., and Hewett-Emmett, D., 1976, α and β chains of the major haemoglobin and a note on the minor component of *Tarsius, Nature (London)* **259:**338.

Best, J. S., Flamm, U., and Braunitzer, G., 1969, Haemoglobins. XVII. The primary structure of the β-chains of rabbit haemoglobin, *Hoppe-Seyler's Z. Physiol. Chem.* **350:**563.

Boyer, S. H., Hathaway, P., Pascasio, F., Bordley, J., Orton, C., and Naughton, M. A., 1967, Differences in the amino acid sequences of tryptic peptides from three sheep hemoglobin beta chains, *J. Biol. Chem.* **242:**2211.

Boyer, S. H., Crosby, E. F., Noyes, A. N., Fuller, G. F., Leslie, S. E., Donaldson, L. J., Vrablik, G. R., Schaefer, E. W., and Thurmon, T. F., 1971. Primate hemoglobins: Some sequences and some proposals concerning the character of evolution and mutation, *Biochem. Genet.* **5:**405.

Boyer, S. H., Noyes, A. N., Timmons, C. F., and Young, R. A., 1972, Primate hemoglobins and evolutionary patterns, *J. Hum. Evol.* **1:**515.

Braunitzer, G., and Kohler, H., 1966, Zur Phylogenie des Hämoglobinmoleküla: Untersuchungen am Hämoglobin des Schweines, *Hoppe-Seyler's Z. Physiol. Chem.* **343:**290.

Braunitzer, G., Gehring-Müller, R., Hilschman, N., Hilse, K., Hobom, G., Rudloff, V., and Wittman-Liebold, B., 1961, Die Konstitution des normalen adulten Human Hämoglobins, *Hoppe-Seyler's Z. Physiol. Chem.* **325:**283.

Chauvet, J. P., and Acher, R., 1972, Phylogenie of hemoglobins: β chain of frog *(Rana esculenta)* hemoglobin, *Biochemistry* **11:**916.

Clegg, J. B., Naughton, M., and Weatherall, D. J., 1968, Separation of the α- and β-chains of human hemoglobin, *Nature (London)* **219:**69.

Dayhoff, M. O., 1972, *Atlas of Protein Sequence and Structure,* Vol. 5, National Biomedical Research Foundation, Washington, D.C.

Gadow, H., 1898, *A Classification of Vertebrata, Recent and Extinct,* A. and C. Black, London.

Goodman, M., 1973, The chronicle of primate phylogeny contained in proteins, *Symp. Zool. Soc. London* **33:**339.

Goodman, M., 1974, Biochemical evidence on hominid phylogeny, *Annu. Rev. Anthropol.* **3:**203.

Goodman, M., Moore, G. W., Barnabas, J., and Matsuda, G., 1974, The phylogeny of human globin genes investigated by the maximum parsimony method, *J. Mol. Evol.* **3:**1.

Goodman, M., Moore, G. W., and Matsuda, G., 1975, Darwinian evolution in the genealogy of haemoglobin, *Nature (London)* **253:**603.

Hill, W. C. O., 1955, *Primates Comparative Anatomy and Taxonomy,* Vol. II: *Haplorhini: Tarsioidea,* University Press, Edinburgh.

Huisman, T. H. J., and Dozy, A. M., 1965, Studies on the heterogeneity of hemoglobin. IX. The use of tris(hydroxymethyl)aminomethane-HCl buffers in the anion-exchange chromatography of hemoglobins, *J. Chromatogr.* **19:**160.

Jones, R. T., Brimhall, B., and Duerst, M., 1971, Amino acid sequence of the α and β chains of dog hemoglobin, *Fed. Proc. Fed. Am. Soc. Exp. Biol.* **30(2):** abst. 1207.

Kimura, M., 1969, The rate of molecular evolution considered from the standpoint of population genetics, *Proc. Natl. Acad. Sci. USA* **63**:1181.

Konigsberg, W., Guidotti, G., and Hill, R. J., 1961, The amino acid sequence of the α chain of human hemoglobin, *J. Biol. Chem.* **236**:pc55.

Matsuda, G., Gehring-Mueller, R., and Braunitzer, G., 1963, The total sequence of the alpha chain of the slower component of horse hemoglobin in Germany, *Biochemistry* **2(338)**:669.

Matsuda, G., Manita, T., Takei, H., Ota, H., Yamaguchi, M., Miyauchi, T., and Migita, M., 1968, The primary structure of adult hemoglobin from *Macaca mulatta* monkey, *J. Biochem. (Tokyo)* **64**:279.

Matsuda, G., Takei, H., Wu, K. C., and Shiozawa, T., 1971, The primary structure of the α polypeptide chain of AII component of adult chicken hemoglobin, *Int. J. Peptide Protein Res.* **3**:173.

Matsuda, G., Maita, T., Ota, H., Araya, A., Nakashima, Y., Ishii, V., and Nakashima, M., 1973*a*, The primary structures of α and β chains of adult hemoglobin of the Japanese monkey *(Macaca fuscata fuscata)*, *Int. J. Peptide Protein Res.* **5**:405.

Matsuda, G., Maita, T., Watanabe, B., Ota, H., Araya, A., Goodman, M., and Prychodko, W., 1973*b*, The primary structures of the α and β polypeptide chains of adult hemoglobin of the slow loris *(Nycticebus coucang)*, *Int. J. Peptide Protein Res.* **5**:419.

Matsuda, G., Maita, T., Nakashima, Y., Barnabas, J., Ranjekar, P. K., and Gandhi, N. S., 1973*c*, The primary structures of the α and β polypeptide chains of adult hemoglobin of the hanuman langur *(Presbytis entellus)*, *Int. J. Peptide Protein Res.* **5**:423.

Matsuda, G., Maita, T., Watanabe, B., Araya, A., Morokuma, K., Goodman, M., and Prychodko, W., 1973*d*, The amino acid sequence of the α and β polypeptide chains of adult hemoglobin of the savannah monkey *(Cercopithecus aethiops)*, *Hoppe-Seyler's Z. Physiol. Chem.* **354**:1153.

Matsuda, G., Maita, T., Watanabe, B., Araya, A., Morokuma, K., Ota, Y., Goodman, M., Barnabas, J., and Prychodko, W., 1973*e*, The amino acid sequences of the α and β polypeptide chains of adult hemoglobin of the capuchin monkey *(Cebus apella)*, *Hoppe-Seyler's Z. Physiol. Chem.* **354**:1513.

Matsuda, G., Maita, T., Suzuyama, Y., Setoguchi, M., Ota, Y., Araya, A., Goodman, M. Barnabas, J., and Prychodko, W., 1973*f*, Studies on the primary structures of α and β polypeptide chains of adult hemoglobin of the spider monkey *(Ateles geoffroyi)*, *Hoppe-Seyler's Z. Physiol. Chem.* **354**:1517.

Matsuda, G., Maita, T., Mizuno, K., and Ota, H., 1973*g*, Amino acid sequence of a β chain of AII component of adult chicken haemoglobin, *Nature (London) New Biol.* **244**:244.

Moore, G. W., Barnabas, J., and Goodman, M., 1973, A method for constructing maximum parsimony ancestral amino acid sequences on a given network, *J. Theor. Biol.* **38**:459.

Perutz, M. F., 1969, The haemoglobin molecule, *Proc. R. Soc. London Ser. B* **173**:113.

Pocock, R. I., 1918, On the external characters of lemurs and *Tarsius*, *Proc. Zool. Soc. London*, p. 19.

Popp, R. A., 1967, Hemoglobins of mice: Sequence and possible ambiguity at one position of the alpha chain, *J. Mol. Biol.* **27**:9.

Popp, R. A., 1973, Sequence of amino acids in the β chain of single hemoglobins from C57BL, SWR, and NB mice, *Biochim. Biophys. Acta* **303**:52.

Schroeder, W. A., Shelton, J. R., Shelton, J. B., Robberson, B., and Babin, D. R., 1967*a*, Amino acid sequence of the α-chain of bovine fetal hemoglobin. *Arch. Biochem. Biophys.* **120**:1.

Schroeder, W. A., Shelton, J. R., Shelton, J. B., Robberson, B., and Babin, D. R., 1967*b*, A comparison of amino acid sequences in the β-chains of adult bovine hemoglobins A and B, *Arch. Biochem. Biophys.* **120**:124.

Simons, E. L., 1972, *Primate Evolution: An Introduction to Man's Place in Nature*, Macmillan, New York.

Simpson, G. G., 1945, The principles of classification and a classification of mammals, *Bull. Am. Mus. Nat. Hist.* **85**:1.

Smith, D. B., 1968, Amino acid sequences of some tryptic peptides from the β-chain of horse hemoglobin, *Can. J. Biochem.* **46**:825.

Sober, H., 1968, *Handbook of Biochemistry*, Chemical Rubber Company, Chicago.

Stenzel, P., 1974, Opossum Hb chain sequence and neutral mutation theory, *Nature (London)* **252:**62.

Sullivan, B., 1971, Comparison of the hemoglobins in non-human primates and their importance in the study of human hemoglobins, in: *Comparative Genetics in Monkeys, Apes, and Man* (A. B. Chiarelli, ed.), pp. 213–256, Academic Press, London.

Von Ehrenstein, G., 1966, Translational variations in the amino acid sequence of the α-chain of rabbit hemoglobin, *Cold Spring Harbor Symp. Quant. Biol.* **31:**705.

Whittaker, R. G., and Thompson, E. O. P., 1974, Studies on monotreme proteins. V. Amino acid sequence of the α-chain of haemoglobin from the platypus, *Ornithorhynchus anatinus, Aust. J. Biol. Sci.* **27:**591.

Whittaker, R. G., and Thompson, E. O. P., 1975, Studies on monotreme proteins. VI. Amino acid sequence of the β-chain of haemoglobin from the platypus, *Ornithorhynchus anatinus, Aust. J. Biol. Sci.* **28:**353.

Whittaker, R. G., Fisher, W. K., and Thompson, E. O. P., 1972, Studies on monotreme proteins. I. Amino acid sequence of the β-chain of haemoglobin from the echidna, *Tachyglossus aculeatus aculeatus, Aust. J. Biol. Sci.* **25:**989.

Whittaker, R. G., Fisher, W. K., and Thompson, E. O. P., 1973, Studies on monotreme proteins. II. Amino acid sequence of the α-chain in haemoglobin from the echidna, *Tachyglossus aculeatus aculeatus, Aust. J. Biol. Sci.* **26:**277.

Yamaguchi, Y., Horie, H., Matsuo, A., Sasakawa, S., and Satake, K., 1965, On the chemical structure of porcine globin α, *J. Biochem (Tokyo)* **58:**186.

Old World Monkey Hemoglobins: Deciphering Phylogeny from Complex Patterns of Molecular Evolution

13

DAVID HEWETT-EMMETT,
CHRISTOPHER N. COOK, and N. A. BARNICOT

1. Introduction

Some of the pertinent questions being addressed by those carrying out research in molecular anthropology can be summarized as follows:

1. Are most of the DNA mutations that become fixed selectively neutral (or almost so)?
2. What effect do gene duplications have on the rate of mutation acceptance?
3. Can cladistic relationships of use to taxonomists be elucidated, despite any limitations arising from (1) and (2)?
4. If so, is the minority view that proteins can be used as "molecular clocks" tenable?

DAVID HEWETT-EMMETT, CHRISTOPHER N. COOK, and N. A. BARNICOT • Department of Anthropology, University College London, London WC1E 6BT, England. Present address of David Hewett-Emmett is Department of Biochemistry, University of Bristol, Bristol, England. Professor Barnicot died May 14, 1975.

With many experts on these very fields writing chapters in this volume, the aim of this chapter is to provide some new data on a single protein (hemoglobin) among a closely related group of primate species (Old World monkeys). Within this narrow focus, a surprising degree of complexity can be discerned with implications for each of the four questions posed above.

2. Old World Monkey Hemoglobins

Long before the controversy concerning the occurrence of equal rates of mutation in all lineages began to develop, both Hill *et al.* (1963) and Barnicot *et al.* (1965) observed that the tryptic fingerprints (peptide maps) of globins from *Macaca* species differed more from those of *Papio* species than from that of human HbA ($\alpha_2\beta_2$). Since *Papio* and *Macaca* are closely related genera from the tribe Papionini (Kuhn, 1967; Jolly, 1966, 1970), this was a surprise. The finding was pursued by Barnicot and Wade (1970), who demonstrated that the tryptic fingerprint of an arboreal mangabey *(Cercocebus albigena)* globin was almost identical to that of *Papio*. Since fingerprinting detects mainly charge changes (but also gross changes in hydrophobicity and additional tryptic cleavage points), it was apparent that some Old World monkeys had fixed a large number of substitutions involving charged amino acids (Asp, Glu, His, Arg, Lys).

In 1970, a study was initiated of the hemoglobins of the monkeys from the tribe Papionini. This tribe is comprised of four mainly African (in faunal terms Ethiopian) genera: mangabeys *(Cercocebus)*, baboons *(Papio)**, mandrills and drills *(Mandrillus)**, gelada baboon *(Theropithecus)*, and a fifth genus *(Macaca)*, species of which range from North Africa throughout much of Southeast Asia. These monkeys all possess a 42-chromosome karyotype (reviewed in Chiarelli, 1971). At the outset of the study, the complete amino acid sequences of the rhesus monkey *(Macaca mulatta)* α and β chains of hemoglobin were known (Matsuda *et al.*, 1968) and the inferred α and β chain sequences of the crab-eating macaque *(Macaca fascicularis)* had been determined (Wade *et al.*, 1970). During the course of the work, some details of the *Papio anubis* α chain became known (Sullivan 1971a; Sullivan and Nute, personal communication). Later, Matsuda and his colleagues published the α and β chain sequences from the Japanese macaque *(Macaca fuscata)* (Matsuda *et al.*, 1973a) and from two nonpapionine Old World monkeys, the vervet or green monkey *(Cercopithecus aethiops)* (Matsuda *et al.*, 1973b) and the hanuman langur *(Presbytis entellus)* (Matsuda *et al.*, 1973c).

The occurrence of hemoglobin polymorphism among Old World monkeys

---

*Dr. C. J. Jolly has advised us that a definitive ruling may soon be handed down on the nomenclature of these genera. *Papio* may in the future revert to its original use for the mandrills and drills, with *Chaeropithecus* adopted for the baboons.

is unevenly distributed (see reviews by Sullivan, 1971a; Nute, 1974). Thus four macaque species (*M. fascicularis, M. nemestrina, M. speciosa,* and *M. niger*) show complex polymorphisms, whereas other species are monomorphic by starch gel electrophoresis, including *M. mulatta,* which has been sampled in large numbers. However, Basch (1972) has reported that isoelectric focusing reveals polymorphism in *M. mulatta* (see review by Nute, 1974). *Papio* species, also sampled widely, are lacking in hemoglobin polymorphisms, as is *Theropithecus gelada,* judged by the 26 individuals screened by Wade (1969). All of the 15 *Mandrillus* hemolysates examined displayed two components (Hb^{Fast} and Hb^{Slow}) in approximately equal concentrations (Buettner-Janusch *et al.*, 1970; Barnicot, 1969, and unpublished results). *Cercocebus* species exhibit complex polymorphisms (Jacob and Tappen, 1957, 1958; Barnicot and Hewett-Emmett, 1972; Hewett-Emmett *et al.*, 1973), and this is described in detail below. Among other Old World monkey species, polymorphism has not been noted. Of course, the screening of samples by electrophoresis (usually at pH 8.6) does not generally detect polymorphism resulting from amino acid substitutions which do not alter the charge of the molecule.

3. Mangabey Hemoglobins

Jacob and Tappen (1957, 1958) examined 22 mangabey hemolysates and noted polymorphism in the four *Cercocebus galeritus* examined, but not the 18 *Cercocebus albigena.* In London, we examined 28 hemolysates drawn from the five recognized mangabey species, and the results of electrophoretic comparisons are illustrated in Fig. 1. Barnicot and Hewett-Emmett (1972) showed that the heterogeneity in the terrestrial mangabey hemoglobins (*C. atys* and *C. galeritus*) was due to α chain polymorphism. The small sample size made

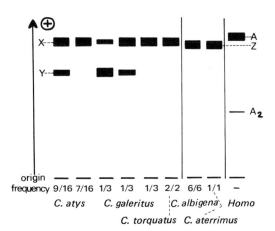

Fig. 1. Diagram (to scale) representing hemoglobin phenotypes from *Cercocebus* hemolysates subjected to starch gel electrophoresis (pH 8.6). The frequency of each phenotype is indicated.

predictions about the allelic or nonallelic nature of the polymorphism impossible. It was noted however that the α and β chain mobilities of the arboreal mangabey, *C. albigena,* differed from those of the terrestrial mangabeys, which (for the X hemoglobin) were closely similar to those of the human chains. Based on their electrophoretic mobilities, the mangabey hemoglobins were named X and Y (terrestrial species) and Z (arboreal species). Thus *C. atys* exhibited the phenotypes X$^+$Y (X stronger than Y) and X; the three *C. galeritus* sampled were X$^+$Y, X, XY$^+$; the third terrestrial species, *C. torquatus,* was X. Both arboreal mangabey species (*C. albigena* and *C. aterrimus*) were Z.

Fingerprinting studies of the purified α^X chain from a *C. atys* X individual (referred to as CFU 11 in our records) demonstrated the presence of two α chains. Amino acid analysis of the tryptic peptides demonstrated that these two chains, inseparable by electrophoretic or chromatographic techniques prior to trypsin digestion, differed because of a pair of linked changes in sequence. The important sequence changes were at α15 (located in tryptic peptide 3 = αT3) and α19 (located in αT4). Trypsin cleaved at the Lys (α16) located between these sites. The two chains were named α^{XA} and α^{XB}

	15 (αT3)	19 (αT4)
CFU 11 α^{XA}	Gly	Glu$^\ominus$
CFU 11 α^{XB}	Asp$^\ominus$	Gly

Thus αT4^{XA} is more acidic than αT4^{XB} and αT3^{XB} is more acidic than αT3^{XA}; these charge changes balance each other in the intact α chains. Qualitatively, the α^{XA} peptides are approximately twice as strong as the α^{XB} peptides.

Table 1. α **Chains of the Terrestrial Mangabeys: Qualitative Relative Concentrations of α^{XA} and α^{XB} Judged by Single-Dimension Electrophoresis Experiments at pH 6.4**[a]

	Phenotype	Constituent α chain ratio in X	% X
C. atys			
CFU 7	X$^+$Y	AB$^+$	—
CFU 11	X	A$^+$B	100
CFU 72	X$^+$Y	AB	66[b]
CFU 113	X	B	100
CFU 114	X	AB$^+$	100
CFU 170	X$^+$Y	AB	60[b]
CFU 171	X	B	100
CFU 173	X	A$^+$B	100
C. galeritus			
17048	XY$^+$	AB$^+$	42[b]
PAI 22 (= PAI 2)	X	A$^+$B	100
C. torquatus			
PAI 21 (= PAI 3)	X	B	100

[a]The % X in certain animals containing both Hb X and Hb Y is also shown.
[b]Starch block band elution ratio (monitored at 540 nm).

Suggested haplotypes
(i) B → approximately 2 chain copies   i.e., B^2
(ii) $A\,B$ → approximately 3 chain copies (2:1)   i.e., A^2B^1
(iii) $C\,B\,A$ → approximately 7 chain copies (4:2:1)   i.e., $C^4B^2A^1$

Expected genotypes	Predicted %C	Predicted A:B strength	Mangabeys with these properties
A^2B^1/A^2B^1	0	A^+B	2 C. atys 1 C. galeritus
A^2B^1/B^2	0	AB^+	1 C. atys
B^2/B^2	0	B	2 C. atys 1 C. torquatus
$C^4B^2A^1/A^2B^1$	40	AB	2 C. atys (40% & 34%C)*
$C^4B^2A^1/B^2$	44	AB^+	1 C. atys
$C^4B^2A^1/C^4B^2A^1$	57	AB^+	1 C. galeritus (58%C)*

Fig. 2. A model explaining the inheritance of the hemoglobin phenotypes (and relative concentrations of hemoglobin components) in terrestrial mangabeys. *Starch block Hb band elution ratio (monitored at 540 nm).

A *C. atys* X^+Y individual (CFU 72) was then examined. Again a tryptic fingerprint of the purified α^X chain showed $\alpha T3^{XA}$, $\alpha T3^{XB}$, $\alpha T4^{XA}$, and $\alpha T4^{XB}$, but this time the ratio of $\alpha^{XA}:\alpha^{XB}$ was approximately equal. Therefore, this animal (phenotype X^+Y) had three α chains. Since the X:Y ratio was 2:1 (by quantitative methods), the ratio $\alpha^{XA}:\alpha^{XB}:\alpha^{YC}$ was approximately 1:1:1.

This complex situation was further studied by examination (by single-dimension high-voltage paper electrophoresis) of the α^X chain tryptic peptides from a total of eight *C. atys,* two *C. galeritus,* and a single *C. torquatus.* This allowed approximate quantitative estimation of the $\alpha^{XA}:\alpha^{XB}$ ratio. From these data (Table 1), a working hypothesis was constructed for the mode of inheritance of α chain genes in the terrestrial mangabeys (Fig. 2). The model is self-consistent (i.e., all "expected" and no "unexpected" phenotypes have been observed), but the small number of individuals sampled makes the exclusion of other more complex models impossible. However, this model does suggest suitable breeding experiments which, for instance, would determine whether the A, B, and C genes are indeed closely linked on the same chromosome. On this model, three haplotypes are postulated: B, AB, and CBA. Thus two gene duplications are implicated. It should be noted (see Fig. 2) that in the haplotypes AB and CBA the different gene loci must produce different amounts of globin. This contrasts with the situation in *Mandrillus,* where the α duplicates produce equal amounts of globin, but is similar to the situation in *Macaca fascicularis* (Barnicot et al., 1966, 1970) and other macaques (see Nute, 1974).

The amino acid sequences of the α^{XA} and α^{XB} chains of *C. atys* have been almost completely determined (Cook and Barnicot, in preparation) from work on CFU 11 (phenotype X, where $\alpha^{XA} > \alpha^{XB}$) and CFU 113 (phenotype X, only

α^{XB}). Some of the tryptic peptide compositions have been determined for α^{YC} from CFU 72 (phenotype X$^+$Y, where $\alpha^{XA} \frown \alpha^{XB} \frown \alpha^{YC}$) (Hewett-Emmett, 1973). The differences between the three α chains are located at a minimum of five sites:

	5	9	15	19	78
α^{XA}	Asp	His	Gly	Glu	His
α^{XB}	Asp	His	Asp	Gly	His
α^{YC}	Ala	Asn	Asp	Gly	Gln(?)
Catarrhine ancestor	Ala	Asn	Gly	Gly	Asn

It would seem that α^{YC} is the most "primitive" of the three α chains and α^{XA} is the most "evolved" [assuming that α15(Gly) is a reversion in α^{XA}]. The possibility that α^{XB} resulted from a crossover between site 9 of α^{XA} and site 15 of α^{YC} is discussed below.

Examination of the arboreal mangabey hemoglobins resulted in no hint of such heterogeneity. Some of the tryptic peptides of a *C. albigena* α chain were analyzed (Hewett-Emmett, 1973). At residues α 5, 9, 15, and 19, the inferred sequence was identical to that of the α^{XA} chain of *C. atys* (the most "evolved" of the three chains). At other sites, however, substitutions were noted that were unique (among the mangabeys) to the arboreal mangabeys [α56(Asx) and α57(Lys)]. Surprisingly, however, the inferred sequence of the *Theropithecus gelada* α chain (see below) has α57(Lys) and that of *Papio anubis* (Sullivan and Nute, personal communication) has both α56(Asx) and α57(Lys). These substitutions are unique among the primates. The phylogenetic implications of the "sequence affinity" of the arboreal mangabey with the gelada baboon and baboon are discussed below.

4. Catarrhine α Chain Sequences

In addition to the studies on *Cercocebus* hemoglobin α chains, sequences of the α chains from *Theropithecus gelada* and *Colobus badius* were constructed from tryptic peptide compositions. Some ambiguous changes and the insoluble "core peptides" were analyzed by further proteolytic digestion with thermolysin or chymotrypsin (Hewett-Emmett, 1973). Some tryptic peptides were also analyzed from the α^{Fast} and α^{Slow} chains from *Mandrillus sphinx*. These data, together with those on *Cercocebus* and published data on other catarrhine primates, are included in Table 2.

From this compilation of sequences, putative sequences, and partial sequences, dendrograms were constructed. These represent mixed gene and species phylogenies. The original phylogenies (Hewett-Emmett, 1973) have

Table 2. Catarrhine α Chain Sequences: Variable Sites[a]

	5	8	9	12	15	19	21	23	47	53	54	56	57	68	71	73	78	82	113
Homo	Ala	Thr	Asn	Ala	Gly	ALA	Ala	Glu	Asp	Ala	Gln	Lys	Gly	Asn	Ala	Val	Asn	Ala	Leu
Macaca mulatta	Ala	SER	Asn	Ala	Gly	Gly	Ala	Glu	Asp	Ala	Gln	Lys	Gly	LEU	GLY	Val	Asn	Ala	Leu
C. atys α^{XA}	ASP	LYS	HIS	Ala	Gly	GLU	Ala	Glu	ASN	ASP[b]	Gln[b]	Lys	Gly	LEU	GLY	Val	HIS	LYS	Leu
C. atys α^{XB}	ASP	LYS	HIS	Ala	ASP	Gly	Ala	Glu	ASN	ASP[b]	Gln[b]	Lys	Gly	LEU	GLY	Val	HIS	LYS	Leu
C. atys α^{YC}	Ala	LYS	Asn	Ala	ASP	Gly	Ala	Glu	ASN[c]	ASP[c]	Gln[c]	Lys	Gly	(LEU)	(GLY)	(Val)	(GLN)	LYS	—
Mandrillus α^{Fast}	ASP	LYS	HIS	Ala	ASP	Gly	Ala	Glu	—	—	—	—	—	—	—	—	(GLN)	(LYS)	—
Mandrillus α^{Slow}	Ala	LYS	Asn	Ala	ASP	Gly	Ala	Glu	ASN[c]	ASP[c]	Gln[c]	Lys	Gly	—	—	—	(GLN)	(LYS)	—
Papio	ASP	LYS	HIS	Ala	Gly	GLU	Ala	Glu	Asp[d]	ASP[d]	Gln[d]	ASN[d]	LYS	LEU	GLY	Val	GLX	LYS	—
Theropithecus[e]	ASP	LYS	HIS	ASP	Gly	GLU	Ala	GLN	Asp	ASP	Gln	Lys	LYS	LEU	GLY	Val	GLN	LYS	Leu
C. albigena	ASP	LYS	HIS	Ala	Gly	GLU	Ala	Glu	Asp[d]	ASP[d]	Gln[d]	ASN[d]	LYS	—	—	—	—	LYS	—
Cercopithecus[d]	Ala	SER	Asn	Ala ·	Gly	Gly	Ala	Glu	Asp	Ala	Gln	Lys	Gly	LEU	GLY	Val	HIS	Ala	Leu
Colobus	Ala	Thr	Asn	THR	Gly	Gly	GLY	Glu	Asp	Ala	Gln	Lys	Gly	LEU	Ala	ALA	SER	Ala	HIS
Presbytis	Ala	Thr	Asn	Ala	Gly	Gly	GLY	Glu	Asp	Ala	Gln	Lys	Gly	Asn	Ala	Val	HIS	Ala	Leu
Proposed catarrhine ancestor	Ala	Thr	Asn	Ala	Gly	Gly	Ala	Glu	Asp	Ala	Gln	Lys	Gly	Asn	Ala	Val	Asn	Ala	Leu

[a]References are in the text. Residues underlined have been sequenced or placed with certainty because they are either the NH$_2$- or COOH-terminus of a peptide. Residues in capitals (e.g., ALA) are different from those of the proposed catarrhine ancestor. Asx, Glx represent residues that could be either Asp/Asn or Glu/Gln. Residues in parentheses are extremely tentative. A dash indicates that there is no information on these sites.
[b]Alternatively α53(ASN) and α54(GLU).
[c]One acid residue at α47(Asx) α53(Asx) α54(Glx) α56(Asx).
[d]Two acid residues at α47(Asx) α53(Asx) α54(Glx) α56(Asx).
[e]Automated sequencing (in collaboration with D. A. Waltz) identified residues 1–36 and 33–54 (Cyanogen bromide derived fragment). This confirmed assignments at residues 5, 8, 9, 12, 15, 19, 21, 23, 47, 53 and 54.

been amended to include the published data on the hanuman langur *(Presbytis entellus)* and the vervet *(Cercopithecus aethiops)* (see Matsuda, this volume). α8 and α78 present difficulties since they are such hypervariable sites and the new information on *Presbytis* and *Cercopithecus* has altered the most parsimonious distribution of mutations at these sites (Figs. 3 and 4). These placements remain tentative, particularly as α78 is located in a large tryptic peptide (αT9) which is difficult to purify. In the case of *Theropithecus, Colobus, C. atys* (α^{YC} chain), and *Mandrillus,* the assignments need further analysis. α68 is another problem. The assignment of Leu at this site in *Colobus* was made before the full sequencing data on another leaf-eating monkey, *Presbytis entellus,* became known. Leu (α68) and Gly (α71) had been suggested as possible Old World monkey markers (Sullivan, 1971a), and therefore the extra Leu in *Colobus* (located between α67 and α80) was placed at α68. Full sequencing will be necessary on this region of the *Colobus* α chain before any cladistic interpretation can be based on it. However, neither leaf-eating monkey has α71(Gly), which must now be regarded as a marker for the Cercopithecinae. α21(Gly) can be regarded as a marker for the Colobinae. One explanation of the α68 (Asn or Leu) problem is that the site exhibited allelic polymorphism in an ancestral cercopithecoid at the time the Colobinae became separate and that different leaf-eating monkey species fixed either Asn (ancestral allele) or Leu (mutant allele). This possibility is discussed below in relation to two similar problem sites (β13 and β104). It would be extremely surprising if all ancestral species had nonheterogeneous proteins at the time new species diverged, convenient though it would be for our attempts to reconstruct what happened!

The two alternative phylogenies are similar and equally parsimonious but (1) invokes two successive gene duplications (Fig. 3) whereas (2) invokes a single gene duplication followed by a crossover event located between α9 and α15 (Fig. 4). The trees were constructed by hand following an intuitive parsimonious approach. To be more objective, the original trees (Hewett-Emmett, 1973) were subsequently used as inputs for computer-aided branch-swapping methods, using the approaches of Goodman's group (Barnabas *et al.*, 1972) and Dayhoff (1972) (program based on her approach but written by A. A. Young, University of Durham, England). Other primate and nonprimate sequences were included at this stage. Significantly, the arrangement of the papionine α chains was not altered (i.e., could not be made more parsimonious). More complete details can be found elsewhere (Hewett-Emmett, 1973) and trees. including some of these inferred sequences are presented by Goodman (this volume). It should be noted that α8(Lys) and α82(Lys) are unequivocal markers for the African papionines.

The main finding of taxonomic interest was the suggestion that *Cercocebus albigena* may be more closely related to *Papio* and *Theropithecus* than to the terrestrial mangabeys. Hewett-Emmett (1973) stated: "speculation about the

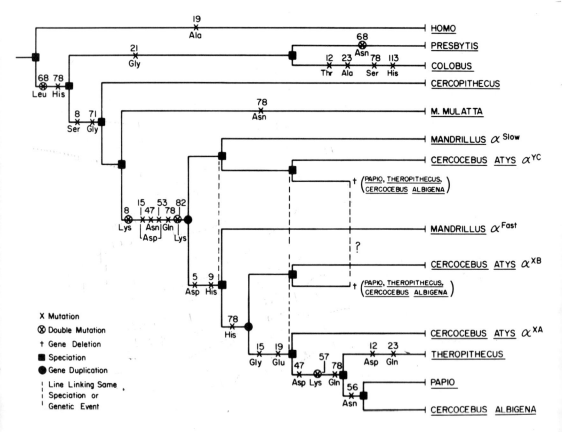

Fig. 3. Serial gene duplication hypothesis for catarrhine hemoglobin α chain evolution. Sequences taken from Table 2.

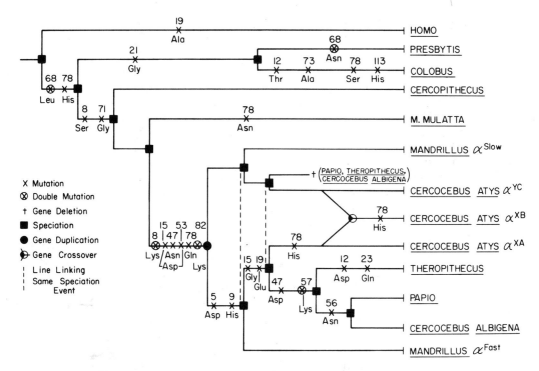

Fig. 4. Gene crossover hypothesis for catarrhine hemoglobin α chain evolution. Sequences taken from Table 2.

taxonomic affinities of *C. albigena* should perhaps await further studies (including the β-chain sequences), but, if the α-chain sequence similarity of this species with *Papio* should provoke a closer examination of the mangabeys with reference to *Papio* and *Mandrillus,* then a definitive taxonomic judgement may eventually result." Significantly, marked behavioral anatomical and ecological differentiation between the arboreal and terrestrial mangabeys has been noted (Rollinson, 1975, and personal communication).

Perhaps while contemplating alterations to the current view of papionine classification we should recall Robert Frost's poem*:

> The rose is a rose,
> And was always a rose.
> But the theory now goes
> That the apple's a rose,
> And the pear is, and so's
> The plum, I suppose.
> The dear only knows
> What will next prove a rose.

*From "The Rose Family," in *The Poetry of Robert Frost* (1969), edited by Edward Connery Lathem. By permission of Holt, Rinehart and Winston.

5. Catarrhine β Chain Sequences

A more limited set of cercopithecoid β chain sequences and inferred sequences is available (Table 3). These include new data on *Theropithecus* and *Colobus* (Hewett-Emmett, 1973), based largely on peptide compositions, and more rigorous data on *Cercocebus atys* (Cook and Barnicot, unpublished). Again, dendrograms have been constructed by hand (Figs. 5 and 6). Unlike the α chain situation, there are no substitutions which define subdivision of the Papionini. However, there are two substitutions which appear to define the Papionini [β9(Asn) and β76(Asn)]. No information is available on the *Cercocebus albigena* β chain, but there are possibilities for resolving its affinity either with *Papio* and *Theropithecus* or with the terrestrial *Cercocebus* species. At β 9, 43, 50, and 52 there are substitutions which could decide the issue. It should be noted that βT5 (residues β41–59) is less acidic in *Theropithecus* than in *Macaca* or *Homo* due to the change at β52 (Asp → Ala). In the fingerprints of *Papio* and *Cercocebus (albigena)* (Barnicot and Wade, 1970), both these species have a less acidic βT5 than *Macaca (fascicularis)*. Whether this is due to the same change at β52 remains to be elucidated. It is clear that the charge changes observed in fingerprints of *Papio* and *Cercocebus albigena* were almost exclusively α chain changes.

The two trees differ by two mutations. When *Colobus* and *Presbytis* share an ancestor (Fig. 6), the distribution of mutations is less parsimonious. This is due to the distribution of mutations at β13 and β104. Of all the cercopithecoids, only *Presbytis* has Arg at β104 and Ala at β13, which are both the most probable catarrhine ancestral residues. The possibility has been discussed briefly above (in connection with α68) that either *Presbytis* and *Colobus* do not represent a true grouping or, possibly, the α and β chains in a cercopithecoid ancestor were polymorphic at α68, β13, and β104, and whereas *Presbytis* fixed the ancestral

Table 3. Catarrhine β Chain Sequences: Variable Sites[a]

	6	9[b]	13	33	43	50[b]	51[c]	52	76	87	104	125
Homo	Glu	SER	Ala	Val	Glu	Thr	Pro	Asp	Ala	THR	Arg	PRO
Macaca mulatta	Glu	ASN	THR	LEU	Glu	SER	Pro	Asp	ASN	Gln	LYS	Gln
Macaca fuscata	Glu	ASN	THR	Val	Glu	SER	Pro	Asp	ASN	Gln	LYS	Gln
Macaca fascicularis[d]	Glu	ASN	THR	Val	Glu	SER	Pro	Asp	ASN	Gln	LYS	Gln
Theropithecus[e]	Glu	ASN	THR	Val	ASP	SER	Pro	ALA	ASN	Gln	LYS	Gln
Cercocebus atys[f]	Glu	VAL	THR	Val	Glu	ASN	Pro	Asp	ASN	Gln	LYS	Gln
Cercopithecus	Glu	THR	THR	Val	Glu	SER	Pro	Asp	Ala	Gln	LYS	Gln
Colobus[d]	ASP	Ala	ASN	Val	ASP	Thr	ALA	Asp	Ala	Gln	LYS	Gln
Presbytis	Glu	Ala	Ala	Val	Glu	SER	Pro	Asp	Ala	Gln	Arg	Gln
Catarrhine ancestor	Glu	Ala	Ala	Val	Glu	Thr	Pro	Asp	Ala	Gln	Arg	Gln

[a]References are in the text. Residues underlined have been fully sequenced or placed with certainty because they are the NH₂- or COOH-terminus of a peptide. Residues in block letters (e.g., ASP) represent changes from the catarrhine ancestral sequence.
[b]The catarrhine ancestor has been arbitrarily assumed at these sites, as alternatives are possible.
[c]Alternatively site 58.
[d]Changes based mainly on peptide compositions and alignment with fully sequenced β chains.
[e]Automated sequencing (in collaboration with D. A. Walz) identified residues 1–35 and residues 56–76 of a cyanogen bromide derived fragment. This confirmed assignments at 6, 9, 13, 33, and 76. Remainder based on tryptic and thermolytic peptide compositions.
[f]111 out of the 146 residues sequenced (Cook and Barnicot, unpublished).

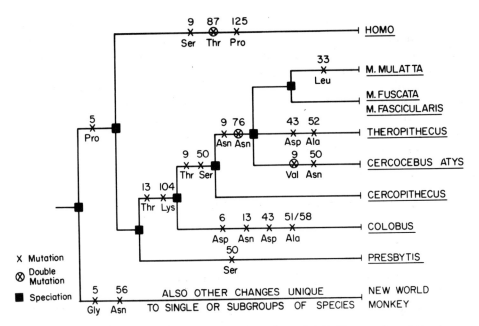

Fig. 5. Catarrhine hemoglobin β chain evolution (alternative 1). Most parsimonious arrangement of the Colobinae. Sequences taken from Table 3.

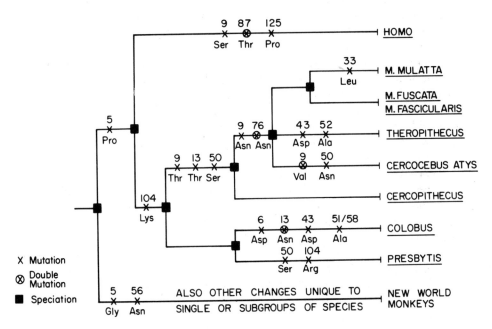

Fig. 6. Catarrhine hemoglobin β chain evolution (alternative 2). Arrangement of Colobinae to correlate with immunological and other biological data. Sequences taken from Table 3.

alleles *Colobus* fixed the mutant alleles. Immunological work (Sarich, 1970; Goodman and Moore, 1971) has shown that the leaf-eating monkeys (Colobinae) shared an ancestor subsequent to the divergence of the other cercopithecoid lineage (Cercopithecinae).

The substitutions in the β chain trees are all neutral with respect to charge, except that noted at $\beta52$ in *Theropithecus* (Asp → Ala). This, together with the more even distribution of substitutions throughout the tree(s), makes hemoglobin β chain a more "conventional" representation of what the neutral mutation theorists might predict (see Section 6.2).

6. Functional Aspects of the Catarrhine Hemoglobin Sequences

6.1. New Salt Bridges

When the substitutions found in the papionine hemoglobins were examined with respect to the three-dimensional models of human and horse hemoglobins constructed from X-ray crystallographic data in the MRC Laboratory of Molecular Biology, Cambridge, by M. F. Perutz and colleagues, a number of surprising interactions were noted.

$\alpha5$(Asp) and $\alpha9$(His) can form a salt bridge at physiological pH. The proximity of the β-COOH group of $\alpha5$ would in all probability raise the pK of the imidazole nitrogen of $\alpha9$(His). There is a human variant, Hb J Toronto α_2(5 Ala → Asp) β_2 (Crookston *et al.*, 1965), corresponding to this change at $\alpha5$. In this heterozygous individual (20% Hb J Toronto), very slight anisocytosis with a 2% incidence of reticulocytes was noted. Could it be that in the primates the substitution at $\alpha9$ (Asn → His) occurred first, thus "allowing" the substitution at $\alpha5$ (Ala → Asp) to occur by neutralizing its possible harmful effects? $\alpha9$(His) would be only partially charged, but once the substitutions at $\alpha5$ occurred, the pK would be raised and a salt bridge formed. This salt bridge occurs in *Cercocebus atys* (α^{XA} and α^{XB} but not α^{YC}), *Cercocebus albigena*, *Mandrillus* (α^{Fast} but not α^{Slow}), *Theropithecus*, and *Papio*.

$\alpha53$(Asp) and $\alpha57$(Lys) can also form a salt bridge. This occurs in *Cercocebus albigena*, *Papio*, and *Theropithecus*, but not in the terrestrial mangabeys or *Mandrillus*. However, in these latter species the $\alpha53$ change (Ala → Asp) fixed during papionine evolution is compensated for, not by formation of a salt bridge with $\alpha57$ (Gly → Lys), but by the loss of an acid group at $\alpha47$ (Asp → Asn). In the former group of species, $\alpha47$ is Asp* as in man, rhesus monkey, and other anthropoids. No human variants corresponding exactly to the changes at $\alpha53$ and $\alpha57$ are known, but Hb L Persian Gulf α_2(57Gly → Asn) β_2 (Rahbar *et al.*, 1969) and Hb J Norfolk α_2(57 Gly → Asp)β_2 (Baglioni, 1962) are harmless in the heterozygous state.

---

*It has not been proved that this situation does not represent $\alpha47$(Asn) $\alpha54$(Glu) instead of $\alpha47$(Asp) $\alpha54$(Gln). Recently, however, sequenator analysis of α-chain peptides from *Theropithecus* has confirmed that $\alpha47$ is Asp and $\alpha54$ is Gln.

A third salt bridge has been noted that is exclusive to *Theropithecus*. $\alpha12$(Asp), a substitution found only in *Theropithecus*, forms a salt bridge with $\alpha16$(Lys). A human variant, Hb J Paris α_2(12 Ala $\rightarrow$ Asp)β_2, has been identified and the patient had some clinical symptoms; however, a relative, also heterozygous for this variant, displayed a normal hematological profile (Rosa *et al.*, 1966; Rosa, personal communication, 1973).

It seems that within α-helical regions, acid residue N can form a salt bridge to basic residue $N + 4$. In *Theropithecuus,* three such intrachain salt bridges not found in the human α chain occur.

6.2. Proportion of Charge-Altering Substitutions

It is a remarkable fact that all 17 hemoglobin α chain substitutions detected in the African papionines have resulted in a change of charge (histidine is regarded as basic and of equivalent charge to arginine and lysine). Admittedly, $\alpha78$ is a hypervariable site and the distribution of substitutions illustrated in Figs. 3 and 4 may not be correct, in which case an Asn-Gln substitution may have occurred. If, therefore, this site is ignored, a total of 14 substitutions (17 DNA base mutations) occurred. On the remaining internal and terminal branches of the catarrhine tree, only nine substitutions (11 DNA base mutations) have been fixed; once again, $\alpha78$ is removed from the analysis. All but one of these [$\alpha113$ (Leu $\rightarrow$ His) in *Colobus*] are, in terms of charge, neutral substitutions. In the catarrhine β chains (Fig. 5), only one charge-altering substitution [$\beta52$ (Asp $\rightarrow$ Ala) in *Theropithecus*] is noted among a total 19 substitutions (22 DNA base mutations).

The structure of the genetic code is such that, of detectable single-base substitutions (e.g., Leu, coded by the triplet CUA, mutated to Leu, coded by UUA, is not detected), only 37.7% involve charge changes (His equivalent to Arg, Lys). Of detectable two-base substitutions, essentially the same proportion (38%) involve charge changes. As noted above, only one of 19 β chain substitutions in catarrhine primates involves a charge change. This is significantly less than 38% ($\chi^2 = 8.14, p < 0.005$, 1 df) and may be due to restrictions imposed by negative selection.

To test whether the conservatism of the catarrhine β chains and the apparent atypical nature of the African papionine α chains were similar to globin chain evolution as a whole or whether globin evolution mirrored estimates based on the genetic code, an examination was made of the eutherian α- and β-globin phylogenies constructed by Goodman *et al.* (1974). These trees were selected because the reconstructed ancestral nucleotide sequences were given. Only external sites (those not listed as either functional or internal in Goodman *et al.*, 1975) were considered, because functional and internal sites are presumably the subject of severely restrictive substitutions. Of the 73 substitutions noted at the 32 external α-globin sites which varied, 25 involved a charge change (34.2%). This does not differ significantly from the proportion predicted (38%) from examination of the genetic code ($\chi^2 = 0.52, p > 0.40$, 1 df). Of the 93 substitutions noted at the 37 variable external β-globin sites, 27 involved a charge change (29.0%). This does not differ significantly from the proportion predicted from examination of the genetic code ($\chi^2 = 2.93, 0.10 > p > 0.05$, 1 df).

It would seem that the catarrhine β chains have been subjected to greater negative selection than the eutherian β chains as a whole. The proportion of charge changes is significantly lower than that expected from examination of either the genetic code ($\chi^2 = 8.14, p < 0.005$, 1 df) or the eutherian β chains as a whole ($\chi^2 = 4.34-6.09, 0.05 > p > 0.01$, 1 df). This correlates well with the long-established view of Goodman's group that at the molecular level there has been a marked deceleration in the fixation of substitutions in the higher primates. In addition to the conservatism in terms of quantity, there is now evidence of a significant reduction in the proportion of charge-altering substitutions from around 29% to 5%.

This makes the finding that all of the α-globin substitutions detected in the African papionines are charge changes even more extraordinary. Needless to say, the deviation from the proportion expected (34.2% for eutherian α chains as a whole, or 38% from the genetic code) is highly significant ($\chi^2 = 7.78, p < 0.01$, 1 df). It should of course be noted that most of these charge-altering substitutions balance each other; in some cases, intrachain salt bridges result from these balanced or "paired changes" (see Section 6.1).

6.3. Structure–Function Relationships

Sullivan (1971a,b) began an ambitious attempt to correlate the oxygen-binding properties of primate hemoglobins with their structure. He and his colleagues have contributed a chapter to this volume on their work on *Papio*.

Sullivan (1971b) reported that among the catarrhines *Theropithecus* was unique in having "a hemoglobin characterized by a sharply decreased acid Bohr effect." He argued that "this has caused a reduction of the oxygen affinity in the physiological range in spite of a decrease in the base-level oxygen affinity." The acid Bohr effect describes the release of protons as oxygen is unloaded below pH 6.0, but according to Perutz (1970) "may have no physiological significance." A mechanistic approach to the alkaline (normal) and acid Bohr effects explains them by invoking charged groups whose pK's change as a result of the conformational change occurring during oxygenation of the hemoglobin tetramer. Whereas most of the charged groups responsible for the alkaline Bohr effect have been identified, it has proved impossible to find such groups to explain the acid Bohr effect. Perutz believes that the acid Bohr effect may represent another conformational charge occurring below pH 6.0.

Beetlestone and his co-workers (Bailey *et al.*, 1970a,b) have noted a correlation among the "characteristic pH" (pH_{ch}) of a hemoglobin, the magnitude of the acid Bohr effect, and the relative members of positively and negatively charged amino acids. The pH_{ch} is experimentally determined and it is the pH at which methemoglobin maximally binds azide ions. It seems to represent the pH at which a conformational change takes place. The hemoglobin of *Theropithecus* is the most basic found in the catarrhine primates (apart from *Mandrillus* Hb^{Slow} and *Cercocebus atys* HbY, neither of which occurs independent of other less basic components). By use of the correlations published by Beetlestone's group, this basic nature of *Theropithecus* hemoglobin should be reflected in a lowered ΔH^+ (magnitude of the acid Bohr effect) via the correlation with pH$_{ch}$. This is what Sullivan (1972) observed, although he cautioned that the data needed checking.

Why the interest in *Theropithecus?* Simply that it lives at high altitude (2000–5000 m) in Ethiopia (Starck and Frick, 1958) and it is known that humans who move to comparable altitudes from sea level adjust by producing increased amounts of 2,3-diphosphoglycerate (2,3-DPG), which lowers the oxygen affinity (Benesch and Benesch, 1969). Is the gelada baboon adapted to its environment via 2,3-DPG or through its hemoglobin structure? Unfortunately, Sullivan's data were obtained on dialyzed hemolysates, before the function of 2,3-DPG became known and so it is impossible to answer the question at present. The African papionines should, in view of the sequence heterogeneity and the unusual number and nature of the α chain changes, provide fruitful material for a reinvestigation of oxygen-binding studies, taking into account 2,3-DPG and working with purified hemoglobin components. For instance, Sullivan (1971*b*) noted some discontinuity in his oxygen-binding curves for *Mandrillus* and this may have been due to the presence of the two hemoglobin components of different overall charge.

7. Conclusions

It would be convenient if positive natural selection had limited impact at the molecular level and we simply catalogued the relentless accumulation of neutral mutations. In the papionine hemoglobin α chains, at least, the opportunity for relatively spectacular bursts of evolution seems to have occurred. A minimum of two gene duplications have punctuated (or perhaps promoted) the accumulation of numerous charge-altering (but ultimately almost charge-balanced) substitutions. The strict neutral mutation theorists (and pragmatists) could perhaps find two alternative explanations:

1. The situation represents the extreme of a Poisson-type distribution of substitution rates (see Sarich, 1970). This argument can lose currency with overuse, and in the case described in this chapter, the substitutions that do occur are also atypical (100% represent charge changes).
2. The situation is explained because a further (but earlier) gene duplication must have taken place shortly after or before the divergence of the macaques (if it were before, it might explain the α chain duplications

noted in some *Macaca* species). The duplicated locus, thus freed from some of the constraints of negative natural selection, evolved to code an α chain that selectively replaced the original and presumably more "conservative" locus. This is essentially Ohno's (1972) view of evolution by gene duplication. Once there are two similar closely linked loci on a chromosome, the possibility of further duplications (and deletions) resulting from nonmatching gene loci alignments is improved (see Smithies, 1964)—the Lepore and anti-Lepore hemoglobins are witness to this.

However, the fact remains that a minimum of 14 substitutions (17 DNA base mutations) are located on the branches leading to *Theropithecus* subsequent to the divergence of the macaques (Fig. 4). If a tentative date of 10 million years ago is put on this divergence, this represents a minimum mutation rate of 4.0×10^{-9} mutations per year per nucleotide position (expressed in the units adopted by Fitch and Langley, this volume). Fitch and Langley have calculated that the silent mutation rate is 4.2×10^{-9} silent nucleotide substitutions per year per third nucleotide position. Since only 0.716 of all possible nonterminating third nucleotide positions are silent, the total mutation rate can be calculated as 5.8×10^{-9} mutations per year per third nucleotide position. They estimated that the hemoglobin α chain has evolved at 1.6×10^{-9} amino acid changing mutations per year per codon—representing a rate of 0.53×10^{-9} amino acid changing mutations per year per nucleotide position. This is 11 times slower than the silent mutation rate. By contrast, the rate for the *Theropithecus* lineage is only 1.5 times slower. This could be due to the almost complete relaxation of negative selection, but if this were the case one would expect only 34–38% of the substitutions to be charge altering, whereas 100% are. This seems therefore to be a worthy candidate for consideration as an example of positive or Darwinian natural selection. Of course, the untestable theory that improbably founder effects could have influenced these observations must not be left unmentioned.

From a taxonomic viewpoint, there are indications that *Cercocebus* might not be a monophyletic genus, but this needs careful investigation with other proteins. Sarich and Cronin (this volume) do find similar indications in work on the immunological properties of serum albumin and transferrin. Again, there is some evidence that *Colobus,* an African leaf-eater, may be more closely related to the Cercopithecinae than to *Presbytis* (an Asiatic leaf-eater). However, in this case there is immunological evidence arguing that the Colobinae are indeed more closely related to each other than to any cercopithecine genus. The tribe Papionini is clearly found to be monophyletic based on β chain evidence, thus supporting the classification of this tribe of species within the Cercopithecinae.

From a functional viewpoint, a reinvestigation of the African papionines is warranted with respect to oxygen-binding properties of the hemoglobins. Purified components should be investigated and also artificially produced hybrids. The possibility that *Theropithecus* has adapted to high altitude via its hemoglobin structure rather than by elevation of the levels of 2,3-DPG (as in man) should not be discounted.

ACKNOWLEDGMENTS

We appreciate discussions with Drs. J. M. Beard, M. Goodman, C. J. Jolly, R. D. Martin, M. A. Rosemeyer, and P. T. Wade Cohen. In particular, we are indebted to Dr. M. F. Perutz for reviewing the possibility of intrachain salt bridges in the Papionini hemoglobin α chains. Drs. B. Sullivan and P. E. Nute kindly provided us with their data on the *Papio* α chain sequence. Dr. J. Rollinson and Dr. A. Gautier encouraged us to pursue our findings on the difference between *Cercocebus* species.

Dr. D. A. Walz kindly provided spare time on a Beckman sequenator during the stay of one of us (D.H.-E.) at Department of Physiology, Wayne State University School of Medicine, Detroit. This enabled the confirmation of 16 substitutions previously postulated for *Theropithecus* α and β chains.

One of us (C. N. C.) was supported by a Science Research Council (London) award to N. A. B. One of us (D. H.-E.) was supported by University College London during the course of this research, some of which formed part of his Ph.D. thesis.

The following are thanked for providing blood samples used in this work:

Cercocebus spp.: the Director, National Center for Primate Biology, Davis, California *(C. atys);* the Director, Delta Regional Primate Research Center, Covington, Louisiana *(C. galeritus, C. albigena);* Drs. J. P. and A. Gautier, La Station Bioligique, Paimpont, nr. Rennes, France *(C. galeritus chrysogaster, C. torquatus, C. albigena);* the Director, Le Jardin des Plantes, Paris *(C. aterrimus).*

Colobus badius: Dr. T. Struhsaker, Uganda.

Mandrillus sphinx: the Director, Yerkes Regional Primate Center, Emory University, Atlanta, Georgia (animal was "Hokie"). Sample processed and globin components separated by Dr. P. T. Wade Cohen.

Theropithecus gelada: Dr. H. Kummer, then at Delta Regional Primate Center, Covington, Louisiana. Sample processed by Dr. P. T. Wade Cohen.

8. References

Baglioni, C., 1962, A chemical study of hemoglobin Norfolk, *J. Biol. Chem.* **237**:69.

Bailey, J. E., Beetlestone, J. G., Irvine, D. H., and Ogunmola, G. A., 1970a, Reactivity differences between haemoglobins. Part XVI. A correlation between the amino-acid composition of methaemoglobins and their reactivity towards azide ions, *J. Chem. Soc. A* **1970**:749.

Bailey, J. E., Beetlestone, J. G., and Irvine, D. H., 1970b, Reactivity differences between haemoglobins. Part XVII. The variability of the Bohr effect between species and the effect of 2,3-diphosphoglyceric acid on the Bohr effect, *J. Chem. Soc. A* **1970**:756.

Barnabas, J., Goodman, M., and Moore, G. W., 1972, Descent of mammalian α-globin chain sequences investigated by the maximum parsimony, *J. Mol. Biol.* **69**:249.

Barnicot, N. A., 1969, Some biochemical and serological aspects of primate evolution, *Sci. Prog. (London)* **57**:459.

Barnicot, N. A., and Hewett-Emmett, D., 1972, Red cell and serum proteins of *Cercocebus, Presbytis, Colobus* and certain other species, *Folia Primatol.* **17**:442.

Barnicot, N. A., and Wade, P. T., 1970, Protein structure and the systematics of Old World monkeys in: *Old World Monkeys* (J. R. Napier and P. H. Napier, eds.), pp. 227–260, Academic Press, New York.

Barnicot, N. A., Jolly, C. J., Huehns, E. R., and Dance, N., 1965, Red cell and serum protein variants in baboons, in: *The Baboon in Medical Research*, Vol. 1 (H. Vagtborg, ed.), pp. 323–338, University of Texas Press, Austin.

Barnicot, N. A., Huehns, E. R., and Jolly, C. J., 1966, Biochemical studies on haemoglobin variants of the irus macaque, *Proc. R. Soc. London Ser. B* **165**:224.

Barnicot, N. A., Wade, P. T., and Cohen, P., 1970, Evidence for a second haemoglobin α-locus duplication in *Macaca irus, Nature (London)* **228**:379.

Basch, R. S., 1972, Haemoglobin polymorphism in the rhesus macaque, *Nature (London) New Biol.* **238**:238.

Benesch, R., and Benesch, R. E., 1969, Intracellular organic phosphates as regulators of oxygen release by haemoglobin, *Nature (London)* **221**:618.

Buettner-Janusch, V., Buettner-Janusch, J., and Mason, G. A., 1970, Multiple haemoglobins of mandrills, *Papio sphinx, Int. J. Biochem.* **1**:322.

Chiarelli, A. B., 1971, Comparative cytogenetics in primates and its relevance for human cytogenetics, in: *Comparative Genetics in Monkeys, Apes and Man* (A. B. Chiarelli, ed.), pp. 273–308, Academic Press, London.

Crookston, J. H., Beale, D., Irvine, D., and Lehmann, H., 1965, A new haemoglobin, J Toronto (α5 alanine-aspartic acid), *Nature (London)* **208**:1059.

Dayhoff, M. O., 1972, *Atlas of Protein Sequence and Structure,* Vol. 5, pp. D367 and D370–371, National Biomedical Research Foundation, Silver Springs, Md.

Goodman, M., and Moore, G. W., 1971, Immunodiffusion systematics of the primates. I. The Catarrhini, *Syst. Zool.* **20**:19.

Goodman, M., Moore, G. W., Barnabas, J., and Matsuda, G., 1974, The phylogeny of human globin genes investigated by the maximum parsimony method, *J. Mol. Evol.* **3**:1.

Goodman, M., Moore, G. W., and Matsuda, G., 1975, Darwinian evolution in the genealogy of haemoglobin, *Nature (London)* **253**:603.

Hewett-Emmett, D., 1973, Ph.D. thesis, University of London.

Hewett-Emmett, D., Cook, C. N., and Barnicot, N. A., 1973, Haemoglobin genetics of Old World monkeys, *Heredity* **31**:429 (abst.).

Hill, R. L., Buettner-Janusch, J., and Buettner-Janusch, V., 1963, Evolution of hemoglobin in primates, *Proc. Natl. Acad. Sci. USA* **50**:885.

Jacob, G. F., and Tappen, N. C., 1957, Abnormal haemoglobins in monkeys, *Nature (London)* **180**:241.

Jacob, G. F., and Tappen, N. C., 1958, Haemoglobins in monkeys, *Nature (London)* **181**:197.

Jolly, C. J., 1966, Introduction to the Cercopithecoidea, with notes on their use as laboratory animals, *Symp. Zool. Soc. London* **17**:427.

Jolly, C. J., 1970, The large African monkeys as an adaptive array, in: *Old World Monkeys* (J. R. Napier and P. H. Napier, eds.), pp. 139–174, Academic Press, London.

Kuhn, H. J., 1967, Zur Systematik der Cercopithicidae, in: *Neue Ergebnisse der Primatologie* (Progress in Primatology) (D. Starck, D. Schneider, and H. J. Kuhn, eds.), pp. 25–46, Gustav Fischer, Stuttgart.

Matsuda, G., Maita, T., Takei, H., Ota, H., Yamaguchi, M., Miyanchi, T., and Migita, M., 1968, The primary structure of adult haemoglobin from *Macaca mulatta* monkey, *J. Biochem. (Tokyo)* **64**:279.

Matsuda, G., Maita, T., Ota, H., Araya, A., Nakashima, Y., Ishii, V., and Nakashima, M., 1973a, The primary structures of the α and β chains of adult hemoglobin of the Japanese monkey *(Macaca fuscata fuscata), Int. J. Peptide Res.* **5**:405.

Matsuda, G., Maita, T., Watanabe, B., Araya, A., Morokuma, K., Goodman, M., and Prychodko, W., 1973b, The amino acid sequence of the α and β chains of adult hemoglobin of the savannah monkey *(Cercopithecus aethiops), Hoppe-Seyler's Z. Physiol. Chem.* **354**:1153.

Matsuda, G., Maita, T., Nakashima, Y., Barnabas, J., Ranjekar, P. K., and Gandhi, N. S., 1973c, The primary structures of the α and β polypeptide chains of adult hemoglobin of the hanuman langur *(Presbytis entellus), Int. J. Peptide Protein Res.* **5:**423.

Nute, P. E., 1974, Multiple hemoglobin α-chain loci in monkeys, apes, and man, *Ann. N.Y. Acad. Sci.* **241:**39.

Ohno, S., 1972, An argument for the genetic simplicity of man and other mammals, *J. Hum. Evol.* **1:**651.

Perutz, M. F., 1970, Stereochemistry of cooperative effects in haemoglobin, *Nature (London)* **228:**726.

Rahbar, S., Kinderlerer, J. L., and Lehmann, H., 1969, Haemoglobin L Persian Gulf: α57(E6)glycine-arginine, *Acta Haematol.* **42:**169.

Rollinson, J., 1975, Ph.D. thesis, University of London.

Rosa, J., Maleknia, N., Vergoz, D., and Dunet, R., 1966, Une nouvelle hémoglobine anormale: l'Hémoglobine J α Paris. 12 Ala → Asp, *Nouv. Rev. Fr. Hematol.* **6:**423.

Sarich, V. M., 1970, Primate systematics with special reference to Old World monkeys: A protein perspective, in: *Old World Monkeys* (J. R. Napier and P. H. Napier, eds.), pp. 175–226, Academic Press, London.

Smithies, O., 1964, Chromosomal rearrangements and protein structure, *Cold Spring Harbor Symp. Quant. Biol.* **29:**309.

Starck, D., and Frick, H., 1958, Beobachtungen an aethiopischen Primaten, *Zool. Jahrb.* **86:**41.

Sullivan, B., 1971a, Comparison of the hemoglobins in non-human primates and their importance in the study of human hemoglobins, in: *Comparative Genetics in Monkeys, Apes and Man* (B. Chiarelli, ed.), pp. 213–256, Academic Press, London.

Sullivan, B., 1971b, Structure, function and evolution of primate hemoglobins. II. A survey of the oxygen-binding properties, *Comp. Biochem. Physiol.* **40B:**359.

Sullivan, B., 1972, Variation in protein structure and function: Primate hemoglobins, *J. Mol. Evol.* **1:**295.

Wade, P. T., 1969, Ph.D. thesis, University of London.

Wade, P. T., Barnicot, N. A., and Huehns, E. R., 1970, *Biochim. Biophys. Acta* **221:**450.

Structure and Function of Baboon Hemoglobins

14

BOLLING SULLIVAN, JOSEPH BONAVENTURA,
CELIA BONAVENTURA, and PETER E. NUTE

1. Introduction

This chapter details some preliminary findings concerning the structural and functional properties of hemoglobin from the olive baboon, *Papio cynocephalus* (= *anubis*). Although the amino acid sequence characterizations are not complete, nevertheless they do locate many of the presumed interchanges and even at this stage may possibly contribute to our understanding of human and primate hemoglobins.

Initially, our intent was to look at the structural properties of the α chain of *Papio cynocephalus* hemoglobin in order to obtain data on the phylogenetic position of this and related species. Our data soon indicated that there were an unusual number of substitutions, and so we expanded our studies to include the β and γ chains. Such unusual structural variation might be indicative of a substantial change in the oxygen-binding properties. Accordingly, we have made equilibrium and kinetic measurements of oxygen binding to stripped hemoglobin in the presence of the organic phosphate, inositol hexaphosphate (IHP). Hemoglobin is stripped of accompanying organic and inorganic ions because these ions are not removed by dialysis and because many are known to

BOLLING SULLIVAN, JOSEPH BONAVENTURA, and CELIA BONAVENTURA • Department of Biochemistry, Duke University Medical Center, and Duke University Marine Laboratory, Beaufort, North Carolina 28516.   PETER E. NUTE • Regional Primate Research Center and Department of Anthropology, University of Washington, Seattle, Washington 98195.

bind to hemoglobin and to change its oxygen-binding properties. The organic ion of greatest importance in human erythrocytes is 2,3-diphosphoglycerate (2,3-DPG). In many ways, IHP mimics the effects of 2,3-DPG and is a more effective modulator of hemoglobin function. For this reason, we chose to study both stripped hemoglobin by itself and stripped hemoglobin in the presence of known amounts of IHP.

The species used in this study, *Papio cynocephalus*, has a single major hemoglobin. Blood samples were obtained from the Regional Primate Research Center, University of Washington in Seattle. Isolation of the hemoglobin, separation of the chains, aminoethylation, cyanogen bromide fragmentation, tryptic and thermolysin digestion, and isolation of the peptides by ion-exchange chromatography were done as described previously (Nute and Sullivan, 1971). The amino acid substitutions reported in this chapter are based on the compositions of small peptides and their presumed homology to the known sequences of human α-, β-, and γ-polypeptide chains. The functional properties were characterized by methods described in Bonaventura *et al.* (1974).

2. Structural Characterization of the α Chain

Several fragmentation procedures were used to obtain the peptides from the α chain. Tryptic digestions of both the whole chain and some of the CNBr fragments were carried out. We were able to locate all peptides except peptide αT12b. Table 1 summarizes the amino acid compositions of the 14 peptides which were isolated to reasonable purity. They are compared to the homologous peptides from human globin. Peptide αT1 contains an additional amino acid presumably resulting from the substitution of Lys for Thr at the amino-terminal end of αT2. Alarine at position 5 is substituted by Asx, probably aspartic acid since Ala → Asn requires two codon changes. One very disconcerting observation was the isolation of a peptide, from a tryptic digestion of a CNBr fragment of the α chain, which matches the composition of human αT1. This may be indicative of allelic variation. However, the shortened αT2 was isolated in good yield and the two substitutions in αT1 are also found in *Theropithecus* (Hewett-Emmett, 1973). Peptide αT2 contains His in place of Asn. Peptide αT3 is normal but αT4 contains an additional Glx with the loss of an Ala residue. Isolation of the thermolysin peptides produced from αT4 (Table 2) locates the substitution at position 19. Since Ala → Gln is a two-step mutation, the substitution is most likely glutamic acid as it is in *Theropithecus*. Peptide αT5 is normal, but αT6 contains additional Asx residues. This peptide was slightly contaminated as evidenced by the high values for Ala (should be 0.0) and Leu (should be 1.0). Isolation of the thermolysin peptides from αT6 (Table 2) positioned the following substitutions: Ala → Asx at 53, Lys → Asx at 56, and Gly → Lys at 57. Position 53 is probably Asp since Ala → Asn requires two base changes. Lysine at 56 could mutate to Asp or Asn in a single step, so no choice

Table 1. Composition of Tryptic Peptides from α Chain of *P. cynocephalus*[a]

Peptide No.:	1	2	3	4	5	6	7
Residues:	1–8	9–11	12–16	17–31	32–40	41–57	58–60
Lys	1.96(1)	1.11(1)	1.00(1)		1.08(1)	1.32(1)	1.10(1)
His		0.95		1.01(1)		1.83(2)	0.85(1)
Arg				0.92(1)			
AECys							
Asp	1.96(1)	0.00(1)				2.63(1)	
Thr		0.00(1)			1.70(2)	1.17(1)	
Ser	1.01(1)				1.02(1)	2.48(2)	
Glu				4.02(3)		1.21(1)	
Pro	1.08(1)				1.02(1)	1.11(1)	
Gly			1.00(1)	2.67(3)		1.31(1)	1.00(2)
Ala	0.00(1)		1.83(2)	2.94(4)		0.95(1)	
Val	0.99(1)	0.91(1)		0.99(1)		1.36(1)	
Met					0.87(1)		
Ile							
Leu	1.07(1)			1.14(1)	1.08(1)	1.93(1)	
Tyr				0.91(1)		0.75(1)	
Phe					1.72(2)	1.51(2)	
Trp			+(1)				

Peptide No.:	8	9a	9b	10	11	12	13	14
Residues:	61	62–82	83–90	91–92	93–99	100–127	128–139	140–141
Lys	1.00	0.94	1.17(1)		1.15(1)		1.18(1)	
His		0.74(1)	1.87(2)					
Arg				1.05(1)				1.05(1)
AECys								
Asp		2.50(5)	1.23(1)		1.24(2)			
Thr		0.97(1)					1.88(2)	
Ser		1.09(1)	1.26(1)				3.05(3)	
Glu		1.10						
Pro		1.23(1)			0.97(1)			
Gly		1.29						
Ala		3.94(6)	1.07(1)		1.09		1.31(1)	
Val		2.54(3)			1.26(2)		2.03(2)	
Met		0.78(1)						
Ile								
Leu		2.70(2)	2.07(2)	0.87(1)	1.24		2.00(2)	
Tyr								0.88(1)
Phe					0.75(1)		0.95(1)	
Trp								

*Values in parentheses refer to the human α chain. Values denoting differences between the α chains of man and *P. cynocephalus* are underlined.

Table 2. Composition of Peptides Produced from Thermolysin Digestion of Isolated *P. cynocephalus* α Chain Tryptic Peptides

Tryptic peptide No.	Thermolysin peptides
4	(Val, Gly, Glx, His)(Ala, Gly, Glx, Tyr, Gly)(Ala, Glx, Ala)(Leu, Glx, Arg)
6	(Thr, Tyr, Phe, Pro, His)(Phe, Asx)(Leu, Ser, His, Gly, Ser, Asx, Glx)
	(Val,Asx,Lys)
9a	(Val, Ala, Asx)(Ala)(Leu,Thr)(Leu,Ala,Val,Gly,His)
	(Ala,Val,Gly,His)
	(Val,Gly,His)
	(Val,Asx,Asx,Met,Pro,Glx,Ala)(Leu,Ser,Lys)
9b	(Leu, Ser, Asx)(Leu, His)(Ala, His, Lys)[a]

[a]Peptide not isolated.

can be made. Interestingly, Gly → Lys at position 57 requires two base substitutions. Peptide αT7 is shortened, but otherwise identical to its human homologue. Human peptide αT8 is free Lys. Free Lys was found and it could derive from residues 61 and/or 8. Peptide αT9 has a substitution at position 82 inserting Lys, hence the isolation of αT9a and αT9b. Peptide αT9a lacks two residues of Asx and Ala and contains additional residues of Lys, Glx, Gly, and Leu. Isolation of the thermolysin peptides from αT9a (Table 2) positions three of the four substitutions: Asn → Leu at position 68, Ala → Gly at position 71, and Ala → Lys at position 82 (a two-step mutation). The Asx → Glx substitution could occur at position 74, 75, or 78 and has been located at position 78 in *Theropithecus* (Hewett-Emmett, 1973). No substitutions were found in peptides αT9b and αT10. Peptide αT11 contains additional residues of Ala and Leu and lacks single residues of Asx and Val. Conservative substitutions would be Val → Leu at position 93 or 96 and Asp → Ala at position 94. Peptide αT12a (100–104) was isolated impure, but appeared to be the same as the human peptide. Peptide αT12b was not found. Peptides αT12 and αT14 had amino acid compositions identical to those of the homologous human peptides.

3. Structural Characterization of the β Chain

All of the β chain tryptic peptides were isolated, although CNBr fragments had to be utilized to obtain some of the core peptides. They are compared to the homologous human globin peptides. Peptide βT1 (Table 3) contains a single interchange, Glu → Asx. This substitution occurs at either position 6 or 7 and is probably Glu → Asp. Peptide βT2 contains Asx and no serine. The substitution is placed at position 9 by thermolysin digestion products (Table 4). Only one base change is needed to go to Asn, but two are required to go to Asp. Peptides βT3 and βT4 had the same amino acid compositions as their human homologues. Peptide βT5 lacks Thr and Glu but has additional residues of Ser and Ala. Isolation of the thermolysin peptides derived from βT5 shows that

Table 3. Composition of Tryptic Peptides from β Chain of P. cynocephalus[a]

Peptide No.:	1	2	3[b]	4[b]	5	6	7	8
Residues:	1–8	9–17	18–30	31–40	41–59	60–61	62–65	66
Lys	1.41(1)	0.96(1)			1.04(1)	1.00(1)	1.04(1)	1.00(1)
His	0.83(1)						0.95(1)	
Arg			0.90(1)	1.00(1)				
AECys								
Asp	0.86	1.16	2.05(2)		2.83(3)			
Thr	0.83(1)	1.02(1)		1.00(1)	0.00(1)			
Ser		0.20(1)			2.87(2)			
Glu	1.30(2)		1.83(2)	0.90(1)	0.42(1)			
Pro	1.03(1)			1.16(1)	1.87(2)			
Gly		0.96(1)	2.82(3)		2.24(2)		0.91(1)	
Ala		1.88(2)	0.97(1)		1.94(1)		1.02(1)	
Val	0.86(1)	1.32(1)	2.67(3)	1.94(2)	1.13(1)	1.00(1)		
Met					1.02(1)			
Ile								
Leu	1.22(1)	1.34(1)	1.08(1)	2.30(2)	1.08(1)			
Tyr				0.92(1)				
Phe					2.69(3)			
Trp[c]		0.82(1)		0.96(1)				

Peptide No.:	9	10	11	12a	12b	13	14[b]	15
Residues:	67–82	83–95	96–104	105–112	113–120	121–132	133–144	145–146
Lys	0.94(1)	1.08(1)	1.10(0)		1.17(1)	0.94(1)	1.00(1)	
His	1.09(1)	0.94(1)	0.98(1)		1.72(2)			0.90(1)
Arg			0.00(1)				0.99(1)	
AECys		0.65(1)		0.60(1)				
Asp	3.89(3)	1.17(1)	1.91(2)	1.00(1)			1.02(1)	
Thr		0.88(2)				0.99(1)		
Ser	0.84(1)	0.85(1)						
Glu		2.00(1)	1.12(1)			3.83(3)		
Pro						1.04(2)		
Gly	1.88(2)	0.84(1)		1.00(1)	1.19(1)		0.95(1)	
Ala	1.08(2)	0.97(1)			1.00(1)	2.11(2)	3.95(4)	
Val	1.11(1)		1.00(1)	2.05(2)	0.98(1)	1.08(1)	2.82(3)	
Met								
Ile								
Leu	3.92(4)	1.93(2)	1.00(1)	2.79(3)	1.00(1)		0.95(1)	
Tyr							0.90(1)	1.06(1)
Phe	0.95(1)	0.93(1)	1.00(1)		0.86(1)	0.97(1)		
Trp[c]								

[a]Values in parentheses refer to the human β chain. Values denoting differences between the β chains of man and P. cynocephalus are underlined.
[b]72-hr hydrolysate.
[c]Thioglycolic acid included in hydrolysis.

Table 4. Composition of Peptides Produced from Thermolysin Digestion of Isolated *P. cynocephalus* β Chain Tryptic Peptides

Tryptic peptide No.	Thermolysin peptides
2	(Asx, Ala)(Val, Thr, Ala)(Leu, Trp, Gly, Lys)
3	(Val, Asx)(Val, Asx, Glx, Val, Gly, Gly, Glx, Ala)(Leu, Gly, Arg)
5	(Phe, Phe, Asx, Ser)(Phe, Gly, Asx)(Leu, Ser, Ser, Pro, Ala, Ala)(Val, Met, Gly, Asx, Pro, Lys)
8,9	(Val, Leu, Gly, Ala)(Phe, Ser, Asx, Gly)(Leu, Asx, His)(Leu, Asx, Asx)(Leu, Lys)[a]
10	(Gly, Thr)(Phe, Ala, Glx)(Leu, Ser, Glx)(Leu, His, Cys, Asx, Lys)
11	(Leu, His)(Val, Asx, Pro, Glx, Asx)(Phe, Lys)
12a	(Leu, Leu, Gly, Asx)(Val, Leu)[a](Val, Cys)
13	(Glx, Phe, Thr, Pro, Glx)(Val, Glx, Ala)(Ala, Tyr, Glx, Lys)
14	(Val, Val, Ala, Gly)(Val, Ala, Asx, Ala)(Leu, Ala, His, Lys)

[a]Peptide not isolated.

there are actually three interchanges (Table 4). At position 43 the interchange is Glu → Asx, probably Asp for Asn would require two base changes. At position 50, Ala or Ser replaces Thr and at position 52 Ser or Ala replaces Asp. Substitution of Ser at 50 and Ala at 52 is most likely as this arrangement involves fewer base changes and has also been observed in *Theropithecus* (Hewett-Emmett, 1973). Peptides βT6, 7, and 8 show no substitutions. Peptide βT9 has an additional Asx and one less Ala. Thermolysin digestion of βT8,9 positions this substitution at 76, and electrophoretic analysis indicates that the actual interchange is Ala → Asn, a two-step mutation which is common to other Old World monkey β chains. Peptide βT10 contains one less Thr and one additional Glx. Thermolysin digestion of βT10 places the substitution at position 87. Although two mutational changes are required for the substitution Thr → Gln, it is the amide which is present in almost all primate β chains. Peptide βT11 contains Lys instead of Arg, but otherwise its composition is the same as human βT11. Not surprisingly, the thermolysin digestion products fix this substitution at position 104. The only other substitution found is located in βT13, where Pro → Glx is indicated. Thermolysin digestion of βT13 positions the substitution at 124 or 125. Position β125 is Gln in most primate species. Electrophoresis indicates that Gln is the substitution, and the thermolysin cleavage between residues 125 and 126 is consistent with positioning the substitution at 125.

4. Structural Characterization of the γ Chain

The CNBr fragments from the γ chain were prepared and tryptic digestions of all but the *C*-terminal fragment were done. Peptides γT1, 2, 8, 11, 12, and 13 were never located. Peptides γT15 and γT16 were characterized as an intact CNBr fragment. Peptides γT3, 4, 5b (CNBr), 6, and 7 had compositions identical to those of their human counterparts (Table 5). Peptide γT4 had a low value for Val, presumably because of the Val-Val bond which hydrolyzes slowly. No prolonged hydrolysis was done. Peptide γT5a (CNBr) contained one less Ser and Phe and one additional Glx and Gly. No thermolysin digestion was done to more precisely locate these interchanges. Peptide γT9 shows an Ile → Glx substitution presumably at position 75. However, Ile → Glx is a two-step mutation, the Leu value is low, and so other substitutions may be involved. In γT10, His is absent and an additional residue of Asx appears. This substitution probably occurs at 77 and either Asn or Asp could result from a single base change. The *C*-terminal CNBr fragment contains no Thr and one additional Ala. This substitution likely occurs at 135.

5. Functional Characterization of Papio cynocephalus Hemolysates

The oxygen-binding properties of stripped hemolysates and stripped hemolysates plus inositol hexaphosphate (IHP) are summarized in Fig. 1. *Papio* hemolysates, like those of other Old World monkeys, are characterized by a

Table 5.   Composition of Tryptic or CNBr Peptides from the γ Chain of *P. cynocephalus*[a]

| Peptide No.: | 3 | 4 | 5a | 5b | 6 | 7 | 9 | 10 | CNBr-3 |
Residues:	18–30	31–40	41–55	56–59	60–61	62–65	67–76	77–82	134–146
Lys				1.02(1)	1.02(1)	1.00(1)	1.08(1)	1.06(1)	
His						0.89(1)		0.00(1)	0.95(1)
Arg	1.00(1)	1.08(1)							0.99(1)
AECys									
Asp	1.97(2)		1.75(2)	0.98(1)			1.24(1)	2.76(2)	
Thr	0.98(1)	1.11(1)					0.80(1)		0.00(1)
Ser			3.00(4)				1.06(1)		2.74(3)
Glu	2.00(2)	1.37(1)	1.27				0.88		
Pro		0.87(1)		0.94(1)					
Gly	2.96(3)		2.13(1)	1.01(1)		0.98(1)	1.30(1)		1.20(1)
Ala	1.00(1)		1.80(2)			1.05(1)	0.86(1)		2.95(2)
Val	2.19(2)	0.89(2)			0.90(1)		0.74(1)		2.27(2)
Met			+ (1)						
Ile			0.83(1)				0.00(1)		
Leu	1.06(1)	1.92(2)	1.33(1)				1.44(2)	2.02(2)	1.20(1)
Tyr		0.97(1)							0.86(1)
Phe			1.87(3)						
Trp		+(1)							

[a]Values in parentheses refer to the human γ chain. Values denoting differences between the γ chains of man and *P. cynocephalus* are underlined.

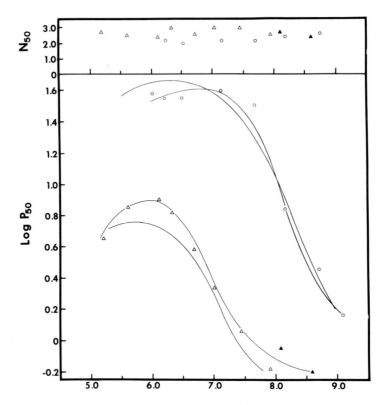

Fig. 1. Oxygen affinity as a function of pH for stripped hemolysates from *Papio cynocephalus* in the presence (circles) and absence (triangles) of a 100-fold excess of IHP over hemoglobin tetramers. The experiments were done at 20°C in 0.05 M bis-tris (open symbols) or 0.05 M tris (closed symbols) buffer. N_{50} is a measure of the cooperativity at 50% saturation. The curves without data points represent human hemoglobin and are taken from Bonaventura *et al.* (1974).

decreased oxygen affinity compared to human hemoglobin (Sullivan, 1971). *Papio cynocephalus* also shows an increased acid Bohr effect. The point of minimum oxygen affinity is shifted from pH 5.75 for human hemolysates to pH 6.0 for *Papio* hemolysates. This shift is seen more clearly in the experiments done in the presence of a 100-fold excess of IHP. The decrease in oxygen affinity caused by IHP is considerably less than in comparable experiments with human hemoglobin (Fig. 1). Cooperativity as measured from Hill plots at 50% saturation is 2.5–3.0 for stripped hemoglobin but somewhat lower for hemolysates with added IHP (2.0–2.5).

6. Kinetics

The dissociation of oxygen from baboon hemoglobin was studied by measuring the rate of the spectral change which occurs when an air-equili-

Table 6. Effects of 2,3-Diphosphoglycerate (DPG) on the Process of Oxygen Dissociation from Stripped Baboon Hemoglobin in 0.05 M Bis Tris (pH 7) or 2% Borate (pH 9) Buffers at 20°C.[a]

Protein	Cofactors	7 (sec⁻¹)	9 (sec⁻¹)
Baboon Hb	None	30	16
Baboon Hb	2,3-DPG	54	16
Human HbA	None	22	10

[a]The rates given represent the final (dominant) phase of the reaction. Data on human HbA are given for comparison.

brated hemoglobin solution is rapidly mixed with an equal volume of buffer containing sodium hydrosulfite (dithionite). The time course of deoxygenation is autocatalytic, increasing in rate as progressive oxygen molecules are released. This increase, which is more pronounced than in the case of human hemoglobin, is a reflection of cooperative interactions between the subunits of baboon hemoglobin. The dominant rates (corresponding to about 70% of the reaction) which characterize the oxygen dissociation process are dependent on pH and cofactors, as shown in Table 6. The rates of oxygen dissociation from baboon hemoglobin are slightly greater than for human hemoglobin and thus contribute to its lower oxygen affinity.

The process of ligand combination was studied by flash photolysis of the CO derivative of baboon hemoglobin. The second-order velocity constant for CO combination at pH 7.0 is 2.3×10^5 M^{-1} sec^{-1}, somewhat lower than the corresponding rate constant for human hemoglobin ($2.9-3.2 \times 10^5$ M^{-1} sec^{-1}). This indicates that the "on" constant for baboon hemoglobin also contributes slightly to its lower oxygen affinity.

The kinetics of ligand combination with baboon hemoglobin differ most strikingly from those of human hemoglobin in the dependence of the flash photolysis kinetics on protein concentration. In flash photolysis studies, the percentage of quickly reacting material may be used as an indication of the extent of dissociation. Plots of the percentage of the quickly reacting material vs. protein concentration for baboon hemoglobin and human hemoglobin suggest that baboon hemoglobin is substantially more dissociated. At a protein concentration of 3 μM (heme), a solution of the CO derivative of human hemoglobin in 0.05 M bis-tris at pH 7.0 will contain approximately 24% quickly reacting material (dimers) whereas 52% is found for baboon hemoglobin.

7. Discussion

When these structural data were obtained, the only amino acid sequences of cercopithecoid hemoglobins were those determined for *Macaca mulatta*

Table 7. Numbers of Amino Acid Sequence and Base Codon Changes between Cercopithecoid and Human Hemoglobins

	Human α chain to cercopithecoid α chain		Human β chain to cercopithecoid β chain	
	Replacements	Mutations	Replacements	Mutations
Macaca mulatta	4	5	8	10
Cercocebus atys	10	12	7	10
Papio	13	16	9	11
Theropithecus	12	15	9	11
Cercopithecus	5	6	6	7
Colobus	7	8	8	11
Presbytis	3	3	4	6

(Matsuda *et al.*, 1970). It became immediately obvious to us that *Papio* α chain was divergent. Subsequent studies by Barnicot and co-workers have shown that the α chains of some other cercopithecoids (e.g., *Cercocebus, Theropithecus,* and *Colobus*) are also divergent. These findings are summarized in Table 7. For most mammalian species, substitutions in the β chain are fixed at a rate 1.5–2.0 times as fast as in the α chain. For *Papio, Theropithecus,* and *Cercocebus,* it is clearly the α chain which is evolving most rapidly. This is most interesting because the necessity of pairing with γ and β chains is thought to restrict substitutions in the α chain. Another interesting feature of the α chain substitutions is that 8 of the 11 result in a charge change on the molecule. The net charge change is quite small, which may reflect restrictions placed on the net charge. The β chain substitutions are more conservative; only one charge change is indicated (Table 8).

The increased rate of substitution has provided useful phylogenetic markers (for a discussion see Hewett-Emmett *et al.*, this volume). Additionally, it provides strong evidence that the rates of amino acid substitution in homologous proteins are not always equal. When compared to human α chains, those from *Papio, Theropithecus,* and *Cercocebus* are evolving 2–5 times as fast as those of *Presbytis, Macaca, Colobus,* and *Cercopithecus.* Such unequal rates of substitution have been inferred from amino acid compositional data on fish proteins (Rama Rao *et al.*, 1976), but are contrary to most molecular data (Langley and Fitch, 1974).

The striking structural changes in baboon hemoglobin are accompanied by smaller but significant functional changes. Most obvious are the lowered oxygen affinity, the decreased effects of IHP, and the increased tendency to dimerize. Dimerization may be favored by substitution at position α94 (probably) of Ala for Asp. Residue 94 is involved in the contact plane $\alpha_1\beta_2$, and loss of a negative charge could conceivably lead to increased dimerization. Although the substitutions of Lys for Gly at α57 and Lys for Ala at α82 do not involve residues thought to be important for hemoglobin function, they are adjacent to important residues and conceivably could be affecting the oxygen-binding properties.

Table 8. Amino Acid Substitutions in the α- and β-Globin Chains of
Papio cynocephalus

Chain	Sequence position	Human chain	*Papio* chain	Most probable substitution
α	5	Ala	Asx	Asp
	8	Thr	Lys	
	9	Asn	His	
	19	Ala	Glx	Glu
	53	Ala	Asx	Asp
	56	Lys	Asx	?
	57	Gly	Lys	
	68	Asn	Leu	
	71	Ala	Gly	
	74,75,78	Asp,Asn	Glx	78 Gln
	82	Ala	Lys	
	93,96	Val	Leu	
	94,97	Asp,Asn	Ala	94 Ala
β	6,7	Glu	Asx	Asp
	9	Ser	Asx	Asn
	43	Glu	Asx	Asp
	50	Thr	Ser	
	52	Asp	Ala	
	76	Ala	Asx	Asn
	87	Thr	Glx	Gln
	104	Arg	Lys	
	125	Pro	Glx	Gln

Little is known concerning other respiratory factors which might influence selection of hemoglobins with different functional properties. It is known that the erythrocyte life span in *Papio* is about half that measured for man (Rowe and Davis, 1972). This could be attributable to an instability of hemoglobin structure, with the result that *Papio* must expend considerably more energy than man to maintain its level of red blood cells. Hemoglobin concentration may be 20% lower in *Papio* than in other cercopithecoid species (Tartour and Idris, 1973), although other data do not report such differences (Balasch *et al.,* 1975). The concentration of 2,3-diphosphoglycerate is similar in human and *Papio* erythrocytes (Bunn *et al.,* 1974) and the decreased oxygen affinity observed for baboon compared to human hemolysate preparations is also found in studies on whole blood (Paton *et al.,* 1971).

The structural studies of the *Papio* γ chain are very incomplete, yet they do indicate that numerous substitutions have occurred in this chain, also. Comparative studies of cercopithecoid globin chains clearly are important for our understanding of primate phylogeny and the evolution of structure-function relationships in hemoglobins.

ACKNOWLEDGMENTS

This work was supported by grants from the National Institutes of Health (GM-15253, 2R01-HE-06400 and RR-00166).

8. References

Balasch, J., Musquera, S., Palacios, L., Jimenez, M., and Palomeque, J., 1975, Hematological values, serum proteins and hemoglobins of *Mandrillus, Comp. Biochem. Physiol.* **51A:**335–340.

Bonaventura, C., Sullivan, B., and Bonaventura, J., 1974, Effects of pH and anions on functional properties of hemoglobin from *Lemur fulvus fulvus, J. Biol. Chem.* **249:**3768–3775.

Bunn, H. F., Seal, J. S., and Scott, A. F., 1974, The role of 2,3-diphosphoglycerate in mediating hemoglobin function of mammalian red cells, *Ann. N.Y. Acad. Sci.* **241:**498–512.

Hewett-Emmett, D., 1973, Structural studies of Old World monkey haemoglobins in relation to phylogeny, Ph.D. thesis, University College London.

Langley, C. H., and Fitch, W. M., 1974, An examination of the constancy of the rate of molecular evolution, *J. Mol. Evol.* **3:**161–177.

Matsuda, G., Maita, T., Ota, H., and Takei, H., 1970, Biochemical studies on hemoglobins and myoglobins. IV. The primary structure of the α and β polypeptide chains of adult hemoglobin of the rhesus monkey *(Macaca mulatta), Int. J. Protein Res.* **2:**99–108.

Nute, P. E., and Sullivan, B., 1971, Primate hemoglobins: Their structure, function and evolution. I. Amino acid compositions of the tryptic peptides from the beta chain of *Cebus albifrons, Comp. Biochem. Physiol.* **39B:**797–814.

Paton, J. B., Peterson, E., Fisher, D. E., and Behrman, R. E., 1971, Oxygen dissociation curves of fetal and adult baboons, *Respir. Physiol.* **12:**283–290.

Rama Rao, K. V., Mahajan, C. L., Pennell, L., Sullivan, B., Bonaventura, J., and Bonaventura, C., 1976, The structure and evolution of parvalbumins. II. Differential rates of evolution among sciaenid isoparvalbumins, submitted for publication.

Rowe, A. W., and Davis, J. H., 1972, Erythrocyte survival in chimpanzees, gibbons and baboons, *J. Med. Primatol.* **1:**86–89.

Sullivan, B., 1971, Structure, function and evolution of primate hemoglobins. II. A survey of the oxygen-binding properties, *Comp. Biochem. Physiol.* **40B:** 359–380.

Tartour, G., and Idris, O. F., 1973, Haematological studies in monkeys and baboons, *Sunda J. Vet. Sci. Anim. Husb.* **14:**24–32.

Evolution of Myoglobin Amino Acid Sequences in Primates and Other Vertebrates

15

A. E. ROMERO-HERRERA, H. LEHMANN,
K. A. JOYSEY, and A. E. FRIDAY

1. Introduction

We have previously (Romero-Herrera *et al.*, 1973) sought to combine evidence from the myoglobin sequences of 18 mammals with a phylogenetic pattern based on zoological evidence. By minimizing the number of mutations, we constructed a cladogram showing the possible molecular evolution of myoglobin and generated a hypothetical myoglobin chain for a mammalian common ancestor. We proposed that the term "average" rate is more appropriate than "constant" rate, in the context of molecular evolution. Our phylogenetic pattern was based on the fossil record, and when the available evidence did not resolve the sequence of successive dichotomies the branching points were superimposed and shown as a trichotomy. The evidence from myoglobin resolved these trichotomies in a pattern that would be acceptable to comparative anatomists.

In a more recent publication (Lehmann *et al.*, 1974), we used the myoglo-

A. E. ROMERO-HERRERA and H. LEHMANN • University Department of Clinical Biochemistry, Addenbrooke's Hospital, Cambridge, England. K. A. JOYSEY and A. E. FRIDAY • University Museum of Zoology, Cambridge, England.

bin sequence of the tree shrew *(Tupaia)* to investigate its phylogenetic position. Evidence from this animal led to adjustments in our reconstruction of myoglobin phylogeny; however, the tree shrew could be accommodated equally well as a primate or a nonprimate because the alternative interpretations led to an identical set of changes from the postulated ancestral chain.

The complete myoglobin sequences of 29 mammals and two birds are now available and have been used in the present study. We have constructed a number of cladograms, broadly based on the parsimony criterion, but constrained by zoological considerations. It was hoped that a comparison of alternative disputed phylogenies might provide some indication of their relative merits. The combination of parsimony methods with zoological information has been found to be profitable in this type of study.

As a development of the present work, we have drawn on comparative anatomy and the fossil record for estimates of possible dates of divergence and on this basis we have extended our study of rates of molecular evolution. A number of procedures are now available which allow some correction for undetectable mutational events (Holmquist, 1972a,b; Holmquist et al., 1972; Jukes and Holmquist, 1972). Because the three-dimensional structure of the myoglobin molecule is known, it has been possible to allocate a functional role to many of its residues (Kendrew et al., 1960; Perutz et al., 1965; Watson, 1969). This has permitted us to speculate on the functional significance of differences at homologous sites and we have also paid particular attention to those parts of the molecule which appear to have remained invariant. Our recent completion of the amino acid sequence of rabbit myoglobin has prompted some investigation of regions involved in antigenicity of the whole molecule. In recent years there have been extensive studies on the antigenic determinants of sperm whale myoglobin (Atassi, 1973), and these have provided a point of reference for comparison between species. The results of these further studies are being prepared for publication elsewhere.

In the present chapter, we have taken into account all the myoglobin sequence information currently available, but the results presented below are largely concerned with primate phylogeny.

2. The Phylogenetic Pattern

In our previous report (Lehmann et al., 1974), we indicated lineages leading to well-defined groups, while we left other groups combined in a single line labeled "other eutherians." As a result of this method of presentation, the differentiation of Carnivora from "other eutherians" left a line of other eutherians including the tupaiids. Such a presentation was possible because there are no undisputed fossil tupaiids. This gave the impression that tupaiids and the Carnivora had a common ancestry within the "other eutherian" line, and although this cannot be entirely discounted it is unlikely. On these grounds, the line labeled "other eutherians" should be thought of as a bundle of lines of separate descent united only by the method of presentation. Hence the mutation shown as common to both tree shrew and harbor seal might well be a result of parallel evolution in lines which had already become differentiated.

In an earlier publication (Romero-Herrera *et al.*, 1973), the phylogenetic chart gave emphasis to points of divergence which could be substantiated and dated from fossil evidence. These points also dictated the pattern of branching. In the present work, the phylogenetic pattern has been established prior to the consideration of probable dates of divergence. In Fig. 1 the living species for which the primary structure of myoglobin is known have been arranged to show the probable pattern of branching, but the length of each branch is not related to time. The species are clustered on the basis of comparative anatomy of both living and fossil forms, and inevitably the resulting pattern reflects conventional systematic classification. The pentachotomy represents the high degree of uncertainty of relationships at this level. The lineage leading to the tree shrew deserves special comment. It is placed outside the Primates because we did not wish to preempt the investigation by linking this animal either with the Primates or with the hedgehog.

3. The Cladogram

With the object of achieving the most parsimonious and acceptable solution of the pentachotomy, and taking due account of the genetic code, several cladograms were constructed by minimizing the number of base changes needed to account for the myoglobin sequences of the living species. Parallel and back mutations were assigned only when they could not be avoided and single base changes were favored over double changes, and these over triple changes. The availability of sequences of a number of closely related species in several parts of the cladogram allowed confident assignment of most substitutions. However, the analysis of the pentachotomy involves considerable uncertainty. The assignment of residues to the bird and mammal lines which emerge from the root of the cladogram involves further uncertainty, and we have adopted the convention of assigning all of the mutations to the bird line (Fig. 2). It is sometimes necessary to choose between two alternative solutions because a back mutation may be equivalent in cost to two mutations in parallel. These alternatives have been explored in two complementary cladograms, only one of which is presented here. For such positions, we cannot at present choose between the possible alternatives on grounds of either function or parsimony.

The cladogram shown in Fig. 2 is one of the two most parsimonious solutions, both scoring 278 hits overall. In the following description, only the Primates will be considered, under several sections.

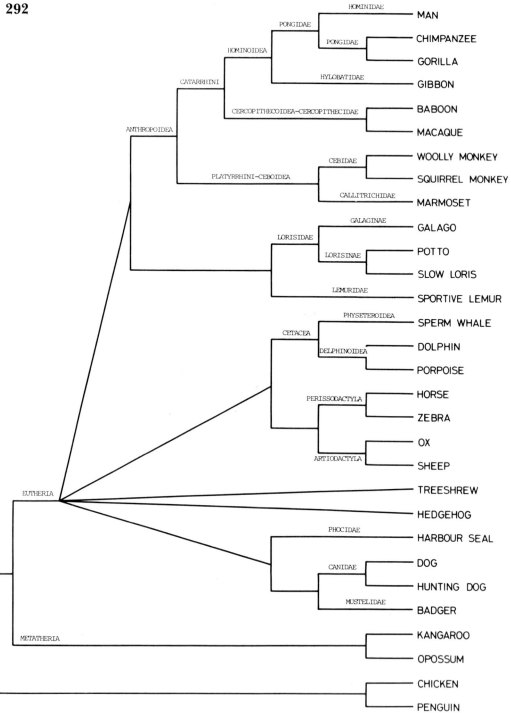

Fig. 1. A phylogenetic pattern giving the relationships of those species whose myoglobin amino acid sequence has so far been investigated. The pattern is based on evidence from comparative anatomy of living and fossil animals.

3.1. Superfamily Hominoidea

There are four amino acid residues indicating the common ancestry of the Hominoidea: Cys-110, Lys-140, Ser-144, and Asn-145. Two, Cys-110 and Asn-145, are peculiar to this superfamily, and although Ser-144 also appears in the birds, among mammals it is found only in the Hominoidea. Residue Lys-140 belongs only to man, chimpanzee, gorilla, and gibbon within the order Primates. The first dichotomy within the Hominoidea is between the branch leading to the Hylobatidae (gibbon), which from the myoglobin viewpoint did not introduce any amino acid change, and the branch leading to the Pongidae fixing residue Gly-23. The Pongidae gave rise to the branch leading to the Hominidae, and within the Pongidae there is a dichotomy leading to chimpanzee and gorilla, introducing the substitutions His-116 and Ser-22, respectively. However, the relationship among man, gorilla, and chimpanzee is debatable and deserves attention because there is no fossil evidence to elucidate whether the branches leading to chimpanzee and gorilla shared a common ancestor after the divergence of the Hominidae or whether one of them diverged at some earlier time while the other shared a longer period of common ancestry with man. The myoglobins could not resolve this problem because man and chimpanzee share residue Pro-22, while the gorilla has Ser at this position, and man and gorilla share residue Gln-116, a position occupied by His in the chimpanzee.

If one turns to the known amino acid sequences of other proteins, again no decision can be made. The fibrinopeptides A and B of man, chimpanzee, and gorilla are the same. Comparison of the δ chain of hemoglobin A$_2$ also yields no decisive information because the chimpanzee and the gorilla share residue Val-126, whereas man has Met at this position. However, this valine is also found in *Hylobates, Ateles, Saimiri,* and *Saguinus* and can be considered as the ancestral state.

One might expect that a comparison of the adult hemoglobins could help. Man and chimpanzee have the same α and β chains, and differ from the gorilla by one residue in both chains. This might be thought to indicate common ancestry for chimpanzee and man after the divergence of the gorilla lineage. On the other hand, it could also come about if a common ancestor of chimpanzee and gorilla diverged from the stem leading to man and the hemoglobin mutations became fixed in the gorilla after this common pongid stock separated further into the branches leading to the living forms. This could indicate the same phylogenetic distance from man for either of the two pongids. As is well recognized by comparative anatomists, such problems can be resolved only when the ancestral and derived states have been distinguished.

3.2. Superfamily Cercopithecoidea

Only two representatives of Cercopithecoidea, *Papio anubis* and *Macaca fascicularis,* have been studied, both belonging to the subfamily Cercopithe-

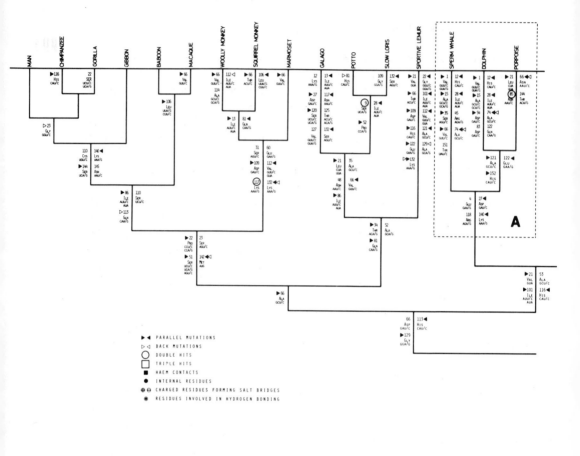

Fig. 2. A possible ancestral myoglobin chain together with a cladogram showing a reconstruction of the course of accepted amino acid substitutions based on minimum pathways to 30 living animals.

cinae. The common ancestry between baboon and macaque is indicated by the shared residue Leu-106. This residue appears elsewhere as a parallel mutation only in the squirrel monkey lineage.

The two cercopithecoids differ from each other in only one residue: position 66 is Ala in the baboon and Val in the macaque. In the context of this cladogram, Val-66 has been allocated to the macaque branch, because Ala has been chosen for the primate common ancestor at this position (see below). It is of interest to mention that in several of the computed solutions in which the programs designed by Moore and Goodman were used, the common ancestry between baboon and macaque was rejected, and these two genera emerged as

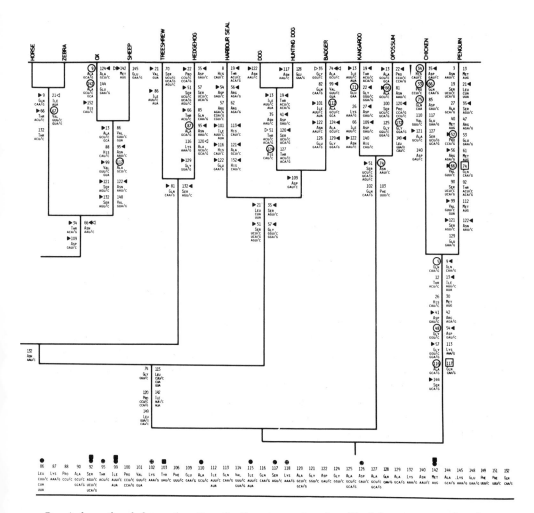

Box A shows the phylogenetic pattern for the cetaceans based on Fig. 1, but a more parsimonious pattern is obtained by linking the sperm whale and dolphin.

independent branches from the catarrhine stem. It is equally parsimonious when Val-66 is considered to be the primate common ancestral residue instead of Ala. This leads to the assignment of two independent double hits, those of Thr-66 for the sportive lemur and Thr-66 for the squirrel monkey, and one single point mutation, Ala-66, for a common ancestor of Hominoidea and baboon. On the other hand, this arrangement assigns Leu-106 as three single parallel events in the lineages of baboon, macaque, and squirrel monkey. In this alternative solution, the overall cost within the Primates is eight mutational events, the same cost as that found when Ala is regarded to be the primate common ancestral residue for position 66, and Leu-106 is introduced in the cercopithecoid common branch. This second solution (Fig. 2) seems to be slightly advantageous because it does not incorporate any double hit with an unknown intermediate.

3.3. Infraorder Catarrhini

The common ancestry between cercopithecoids and hominoids is given by the shared residues Ile-86, Ser-110, and Gln-113.

Among the Primates, residue Ile-86 is found also in the common ancestor of the Lorisidae. Residue 110 is of special interest because in all mammalian myoglobins so far studied, with the exception of those of the hominoids and the cercopithecoids, this position is occupied by Ala. To arrive from alanine to the cysteine of the Hominoidea, more than a single point mutation in one of the codons for alanine is required. However, GCU or GCC for alanine can be converted by one step into the codon UCU or UCC for serine and a further step can change the serine codon into UGU or UGC for cysteine. Thus the serine in the cercopithecoids provides evidence of the intermediate between the alanine in other mammals and the cysteine found in the hominoids. The third residue shared by the Hominoidea and Cercopithecoidea is Gln-113, which, among the Primates, belongs only to these two superfamilies. The sequence of divergence for the Ceboidea, Cercopithecoidea, and Hominoidea cannot at present be deduced from the fossil record. The myoglobin information solves the problem of the radiation of the Anthropoidea by forming first a dichotomy between Platyrrhini and Catarrhini and a subsequent divergence of the latter into Cercopithecoidea and Hominoidea, in agreement with other molecular and anatomical information.

3.4. Infraorder Platyrrhini

Residues Ser-31, Glu-60, and Lys-117 give evidence of the common ancestry of the Ceboidea because, among the myoglobins investigated, they appear only in the woolly monkey, squirrel monkey, and marmoset. As the myoglobin common ancestor for position 117 is Ser, AGU/C, then the Lys, AAA/G, assigned to the ceboid branch is the outcome of two nucleotide substitutions.

Three other substitutions also appear to have taken place in this common stem: Asp-109, Val-112, and Lys-132. It is of interest to see that these three mutations are found within the Primates as parallel changes in the sportive lemur lineage. Two alternative phylogenies, equally parsimonious, were found for the three ceboids, as a result of the interplay of the residues found at four positions. In the following discussion, residues Ala-114 in the woolly monkey and Leu-106 in the squirrel monkey have been left out because they appear only in these two genera. The first ceboid phylogeny postulates a common ancestor from which two branches diverge, one leading to the marmoset and the other being the cebid common stem. The latter subsequently forks into the woolly monkey and squirrel monkey lineages. This tree allocates Val-112 to the ceboid ancestor, changing this to Ile-112 for the woolly monkey. Two mutations are introduced in the cebid common stem, Ile-13 and Gln-81. Finally, as Ala is considered to be the primate common ancestral residue for position 66, it becomes necessary to assign in this position Val to the woolly monkey and marmoset and Thr to the squirrel monkey. This arrangement of residues costs seven single point mutations.

Another possibility is to change the ceboid phylogeny as follows: from the ceboid common ancestor one branch leads to the woolly monkey lineage and another to a common stem for the remaining cebid and callitrichid, which subsequently split into branches leading to the squirrel monkey and marmoset. In this interpretation, Val-66, instead of Ala, is taken as the primate common ancestral residue; hence to reach Thr-66 in the squirrel monkey requires a double hit. Residues Ile-13 and Gln-81 are allocated to the ceboid common branch, necessarily leading to the assignment of Val-13 and Gln-81 as back mutations in the marmoset. Lastly, the common ancestor of squirrel monkey and marmoset introduces Val-112. This alternative phylogeny also costs seven hits. However, on the basis of comparative anatomy the woolly monkey and the squirrel monkey are considered to be more closely related than either of them to the marmoset, which is indicated by the first phylogeny. The myoglobins themselves favor the first solution as more parsimonious since it does not involve a double hit.

3.5. Suborder Anthropoidea

The common ancestry of the suborder Anthropoidea is strongly supported by the common residues Pro-22, Ser-51, and Met-142, which among the Primates are found only in the nine anthropoids so far studied. A fourth residue Ser-23 has been assigned to this common stem because of its presence in the three ceboids, the two cercopithecoids, and the gibbon. This Ser was eventually substituted by Gly as a back mutation in the common branch leading to man, chimpanzee, and gorilla.

3.6. Prosimians

Although it seems likely that the Anthropoidea arose from tarsiiform prosimians, the relationship of tarsiiforms to lemuriform and lorisiform prosimians is obscure. In the present context, the term "prosimian" is used to denote the latter two groups.

After the dichotomy between the anthropoid ancestors and the prosimians, the latter incorporated residues Thr-34, Gln-81, and Ala-52. The first is found in all prosimians so far studied, the second in three of them, and the third in

Lepilemur and *Galago*. Residue Ala-52 deserves a special comment because the chosen ancestral residue for this position is Glu, whose codons GAA/G can change by a single nucleotide substitution into Ala, GCA/G, assigned to the prosimian common stock, and another single step in this codon is necessary to reach the Pro, CCA/G, found in the Lorisinae.

The common ancestry of the lorisids is indicated by residues Ala-35 and Asn-48, which, among all myoglobins studied to date, are found only in this family. Residues Leu-21, Ile-86, and Val-66 are shared by the three lorisids, the first being unique to this family among the Primates. Close relationship between potto and slow loris is indicated by the residues Ser-9 and Pro-52, found only in these two Lorisinae when all available mammalian myoglobins are taken into account, and residue Ile-28, also exclusive to them when the order Primates is considered. As can be noted in Fig. 2, residue Ser-9, UCA/G, in the lorisine stem is the result of a double hit, when the ancestral residue for this position, Leu, is translated by one of the possible codons CUA/G; this is an unavoidable situation if parsimony is considered. The alternative choice of Leu, UUA/G, as the ancestral residue, which can change into Ser, UCA/G, for the potto-loris stem by a single point mutation, would cost one hit more in the overall codon phylogeny of this position.

Among the myoglobins investigated, three of the eight residues fixed along the *Galago* lineage are found only in this genus: Lys-12, Thr-125, and Val-127. Within the primates, positions Asp-27, Asn-117, and Ser-120 are unique to the galago, whereas Ile-13 and Ser-132 are shared as parallel mutations with the two cebids and the slow loris, respectively. In the potto lineage, residue His-81 appears as the only back mutation for this position with the Gln found in the prosimian ancestor as intermediate. In the present cladogram, two mutations could be allocated to the slow loris, Gly-109 unique among all myoglobins and Ser-132. The line leading to the sportive lemur incorporates many substitutions but it does not have any unique residue. All 11 changes either are the result of back mutations or are found elsewhere as parallel substitutions. Five of these parallel events occur in the lineage from condylarthran to horse and zebra as follows: sportive lemur and the condylarthran common stem share residues Val-21, Ile-101, and His-116; Asp-109 appears also in the ungulate common ancestor and Thr-66 in the horse-zebra stem. This is the reason why "phylogenies" produced by procedures such as the UWPGM (unweighted pair-group method) and some of the Wagner trees, in which no zoological guidance is given, tend to associate these two groups.

3.7. Order Primates

Before discussing the relationships within the order Primates, it is relevant to mention that in several of our computer results (obtained when the maximum parsimony method of Moore and Goodman was used) the prosimians appeared to share a common ancestry with the condylarthran derivatives (cetaceans and ungulates), instead of with the Anthropoidea. A detailed examination of this phenomenon revealed that residues Ala-22, Gly-23, Thr-51, and Ile-142 are the same in both stems (the condylarthran derivatives and the prosimians) because they form part of the eutherian common ancestral chain, whereas Pro-22, Ser-23, Ser-51, and Met-142 had been substituted in the anthropoid common ancestor. In addition, residue Thr-34 is found in the prosimian common ancestor, as well as in that of the ungulates, becoming a fifth residue which could be accommodated in a prosimian-condylarthran stem. On the other hand, the only substitution present in the common ancestor of the Anthropoidea and the prosimians is residue Ala-66, which, as mentioned above, could be Val as an alternative. The uncertainty increases when one considers that finding the correct phylogeny for position 66 is difficult because eight different amino acids have been found at this place in the 31 myoglobins so far investigated. For the present, we have interpreted the apparent polyphyletic origin of the Primates as being a result of the common retention of ancestral residues and we have chosen to construct our cladograms on the basis of the phylogeny supported, albeit weakly, by the fossil record and comparative anatomy.

The foregoing sections illustrate, with respect to the Primates, the methods of analysis which were employed to produce the whole cladogram (Fig. 2).

Consideration of amino acid sequences of myoglobin has not demanded any change in the generally accepted pattern of primate phylogeny, but in those cases where direct evidence from the fossil record is lacking the molecular evidence is consistent with that of comparative anatomy regarding the likely sequence of successive dichotomies. The analysis of the cladogram (Fig. 2) presented here has revealed that parallel evolution is commonplace. Indeed, of the 278 mutational events reconstructed, 139 changes occur in parallel (i.e., 50%). Parallelism occurs at 38 out of the 83 positions which have accepted change (and at four of these positions the parallel event was a back mutation). The extent to which this high incidence of parallelism reflects functional constraint on the possibilities for change clearly deserves further investigation, but meanwhile it is evident that the unguided use of similarity as a basis for phylogeny is likely to lead to some false conclusions.

4. References

Atassi, M. Z., 1973, in: *Specific Receptors of Antibodies, Antigens and Cells,* pp. 118–135, Third International Convocation on Immunology, Buffalo, N.Y., Karger, Basel.
Holmquist, R., 1972a, Theoretical foundations for a quantitative approach to paleogenetics. Part I. DNA, *J. Mol. Evol.* **1:**115–133.
Holmquist, R, 1972b, Theoretical foundations for a quantitative approach to paleogenetics. Part II. Proteins, *J. Mol. Evol.* **1:**134–149.
Holmquist, R., Cantor, C., and Jukes, T. H., 1972, Improved procedures for comparing homologous sequences in molecules of proteins and nucleic acids, *J. Mol. Biol.* **64:**145–161.

Jukes, T. H., and Holmquist, R., 1972, Estimation of evolutionary changes in certain homologous polypeptide chains, *J. Mol. Biol.* **64:**163–179.

Kendrew, J. C., Dickerson, R. D., Strandberg, B. E., Hart, R. G., and Davies, D. R., 1960, Structure of myoglobin, *Nature (London)* **185:**422–427.

Lehmann, H., Romero-Herrera, A. E., Joysey, K. A., and Friday, A. E., 1974, Comparative structure of myoglobin: Primates and tree-shrew, *Ann. N.Y. Acad. Sci.* **241:**380–391.

Perutz, M. F., Kendrew, J. C., and Watson, H. C., 1965, Structure and function of haemoglobin. II. Some relations between polypeptide chain configuration and amino acid sequence, *J. Mol. Biol.* **13:**669–678.

Romero-Herrera, A. E., Lehmann, H., Joysey, K. A., and Friday, A. E., 1973, Molecular evolution of myoglobin and the fossil record: A phylogenetic synthesis, *Nature (London)* **246:**389–395.

Watson, H. C., 1969, The stereochemistry of the protein myoglobin, *Prog. Stereochem.* **4:**299–333.

Evolution of Carbonic Anhydrase in Primates and Other Mammals

16

RICHARD E. TASHIAN, MORRIS GOODMAN,
ROBERT E. FERRELL, and ROBERT J. TANIS

1. Introduction

The carbonic anhydrases are especially well suited for studies on molecular evolution for a variety of reasons: (1) In many mammals, two isozymes of carbonic anhydrase which appear to be products of two closely linked genes are present in the red cells. (2) They are able to incorporate mutational changes at relatively rapid rates. (3) They can be readily isolated in pure form from hemolysates. (4) They appear to be involved in a variety of important physiological functions where their specific catalytic role is the interconversion of CO_2 and HCO_3^- . For reviews, see Lindskog *et al.* (1971) and Tashian and Carter (1976).

The two major isozymes of mammalian carbonic anhydrase have been designated CA I and CA II. The evolutionarily homologous CA II isozymes generally show higher specific CO_2 hydrase activities than the CA I isozymes, and for this reason they have been termed "high"- and "low"-activity isozymes, respectively. As yet only one main form of carbonic anhydrase has been isolated from such nonmammalian sources as plants, bacteria, sharks, bony fish, and birds, and the carbonic anhydrases from these organisms exhibit specific CO_2 hydrase activities similar to those of the so-called high-activity CA II isozymes of

RICHARD E. TASHIAN, ROBERT E. FERRELL, and ROBERT J. TANIS • Department of Human Genetics, University of Michigan Medical School, Ann Arbor, Michigan 48104. MORRIS GOODMAN • Department of Anatomy, Wayne State University School of Medicine, Detroit, Michigan 48201.

mammals. Thus it has been suggested that since the duplication of the ancestral high-activity carbonic anhydrase gene the functional properties of this enzyme have been retained in the evolutionarily homologous high-activity CA II isozymes. Greater selective constraints seem to have been exerted on the CA II enzymes for the maintenance of their characteristic high activity, whereas greater variation in activity has been noted in the CA I isozymes (cf. Tashian *et al.*, 1972).

The primary structures have now been completed for both human isozymes, CA I and CA II (Andersson *et al.*, 1972; Henderson *et al.*, 1973; Lin and Deutsch, 1973, 1974), sheep CA II (Tanis *et al.*, 1974), and ox CA II (Sciaky *et al.*, 1974). The fact that 60% of the amino acid residues were the same when the two human isozyme sequences were aligned for maximum homology (cf. Table 1) clearly confirms their descent from a common ancestral gene. In addition, the three-dimensional structures of human CA I and CA II have now been completed at high resolution (Kannan *et al.*, 1971, 1975; Liljas *et al.*, 1972). As might have been expected, the tertiary structures of the two carbonic anhydrases were found to be very similar (Notstrand *et al.*, 1974). These structural similarities notwithstanding, there are a number of differences in kinetic and physical properties between these enzymes (cf. Lindskog *et al.*, 1971), suggesting that each isozyme is associated with a different physiological role.

In this presentation, we will elaborate on the evolutionary origins of the mammalian carbonic anhydrase isozymes and compare their respective rates of evolution in different species of primates and other mammals. In addition, we will discuss the relative constancy of these rates during the descent of each isozyme.

2. Species Examined and the Methods of Analysis

2.1. The Sequences

For comparative purposes, actual or inferred sequences were used from the following sources: human CA I (Andersson *et al.*, 1972); human CA II (Henderson *et al.*, 1973); chimpanzee *(Pan troglodytes)* CA I, orangutan *(Pongo pygmaeus)* CA I, vervet *(Cercopithecus aethiops)* CA I and CA II, rhesus macaque *(Macaca mulatta)* CA I and CA II, baboon *(Papio cynocephalus)* CA I (Tashian and Stroup, 1970; Ferrell, unpublished orangutan data; Tanis, unpublished rhesus data); rabbit *(Oryctolagus cuniculus)* CA I and CA II (Ferrell, unpublished data); sheep *(Ovis aries)* CA II (Tanis *et al.*, 1974); ox *(Bos taurus)* CA II (Sciaky *et al.*, 1974); and European elk *(Alces alces)* CA II (Carlsson *et al.*, 1973).

2.2. Construction of the Phylogenetic Tree

Table 1 shows the sequences that were used to construct the evolutionary tree of mammalian carbonic anhydrases. This tree (Fig. 1) was constructed by

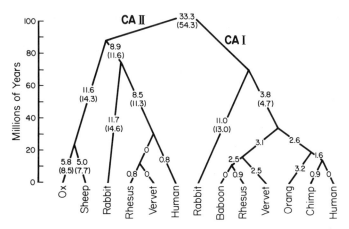

Fig. 1. Maximum parsimony phylogenetic tree of the mammalian carbonic anhydrase isozymes. The link lengths are the number of nucleotide replacements per 100 codons. The numbers in parentheses were calculated by the augmented maximum parsimony program. As can be noted from Table 1, an archetype alignment of 261 positions was required to align the CA I and CA II sequences against one another. However, because of two gaps in the CA II sequences and one gap in the CA I sequences, there were a maximum of 258 positions at which nucleotide replacements (NRs) could occur on the link separating the CA II region of the tree from the CA I region. Within the CA II region, the bovine, sheep, and human sequences were complete. Thus bovine and sheep were compared to each other and to the rabbit-primate ancestor at 259 positions. As the rabbit sequence was 12 positions short of being completed, there were 247 positions on the line to rabbit at which NRs could occur. Inasmuch as all NRs which occurred on the link between the rabbit-primate and catarrhine ancestor were at positions occupied by amino acids in both the rabbit and human sequences, this link was also considered to have had 247 positions or codons. Similarly, inspection of the catarrhine sequences revealed that, although the human CA II sequence was complete, the 1 NR on the terminal link to it occurred at a position which was occupied by a different codon in the rhesus sequence, and as amino acids had been placed at only 124 positions of rhesus CA II, the terminal link to man was also considered to have had only 124 codons. In turn, amino acids for the vervet sequence had been placed at only 119 positions, and as the 1 NR in the terminal link to rhesus CA II occurred at one of these positions this link was considered to have also had only 119 positions. In the CA I region of the tree, the human sequence was completely determined, the rhesus was done at 224 positions, and the orangutan was at 185 positions, 12 of which had not been done on the rabbit. In addition, amino acids had been placed at 118 positions of the chimpanzee, vervet, and baboon sequences, and at 100 positions of the rabbit sequence, 31 of which had not been done in chimpanzee, vervet, or baboon. Thus numbers of codon positions on the terminal links to rabbit, vervet, baboon, orangutan, and chimpanzee, at which NRs could have occurred, were, respectively, 100, 118, 118, 185, and 118. Inasmuch as the 3 NRs which occurred on the link between the hominoid and chimpanzee-human ancestors were all at positions represented by the orangutan sequence, this link was considered to contain 185 codons rather than 258 codons (the number of positions that the human CA I sequence shared with the CA II region of the tree). Some of the NRs on the link between the rabbit-primate and catarrhine CA I ancestors occurred at positions not represented by amino acids in the incomplete rabbit sequence, but all these NRs occurred at positions represented by amino acids in either the rhesus or orangutan sequence. Thus this link was considered to contain 236 codons. The link from the catarrhine to the hominoid ancestor was also considered to contain 236 codons from inspection of the positions at which the NRs occurred, whereas the link from the catarrhine to the cercopithecoid ancestor contained just the 224 codon positions occupied by amino acids in the rhesus sequence. From Tashian and Carter (1976), with permission.

Table 1. Amino Acid Sequences of Mammalian Carbonic Anhydrase Isozymes CA I and CA II[a]

```
              1          10          20          30
Human    I   A S P D W G Y D D K N G P E Q W S K L Y P I A N G N N Q S P V D I K T S E T K
Chimp    I   (A S P E W G Y D D K N G P E Q W S K L Y P I A N G N N Q S P V D I K T S E T K
Orang    I   (A S P E W G Y D D K N G P E Q W S K)L Y P I A D G(N N Q S P)V D I K T S E A K
Vervet   I   (A S P E W G Y D D K N G P E Q W S K L Y P I A N G N N Q S P V D I K T S E T K
Rhesus   I   (A S P)D W G Y D D K N G P E E W S K L Y P I A(N G N N Q S P)V D I K T S E A K
Baboon   I   (A S P D W G Y D D K N G P Q E W S K L Y P I A N G N N Q S P V D I K T S E A K
Rabbit   I               A Y D G E B G P E H W S K L Y P I A D G N K E         I K S S E V K

Human    II  - S H H W G Y G K H N G P E H W H K D F P I A K G E R Q S P V D I D T H T A K
Vervet   II  (- S H H W G Y G K H N G P E H W H K D F P I A K G E R Q S P V D I D T H T A K
Rhesus   II  (- S H H W G Y G K H N G P E H W H K D F P I A K G E R Q S P V D I D T H T A K
Sheep    II  - S H H W G Y G E H N G P E H W H K D F P I A D G E R Q S P V D I D T K A V V
Ox       II  - S H H W G Y G K H B G P Z H W H K D F P I A N G E R Q S P V N I D T K A V V
Rabbit   II  - S H H W G Y G K H N G P E H W H K D F P I A D G E R(Q S P A D I D T K)A V K

              40          50          60          70
Human    I   H D T S L K P I S V S Y N P A T A K E I I N V G H S F H V N F E D N N D R S V
Chimp    I   H D T S L K P I S V S Y N P A T A K)                               (S V
Orang    I   (H D T S L K P V S V S Y N P A T A K)E I I N V G H S F(H V)N F E D D N N R S V
Vervet   I   H D T S L K P I S V S Y N P A T A K)                               (S V
Rhesus   I   H D T S L K P I S V S Y N P A T A K E I I N V G H S F H V N F E D N D N R S V
Baboon   I   H D T S L K P I S V S Y N P A T A K)                               (S V
Rabbit   I   H D T S L F P K S V S Y N P A S     K E L I N V G H V F H V N F

Human    II  Y D P S L K P L S V S Y D Q A T S L R I L N N G H A F N V E F D D S Q D K A V
Vervet   II  Y D P S L K P L S V S Y D Q A T S L R)                             (A V
Rhesus   II  Y D P S L K P L S V S Y D Q A T S L R)                             (A V
Sheep    II  P D P A L K P L A L L Y E Q A A A S R M V N N G H S F N V E F D D S Q D K A V
Ox       II  Q D P A L K P L A L Y Y G E A T S R R M V N N G H S F N V E Y D D S Q D K A V
Rabbit   II  H D P S L F P L R L L Y E E A     S R R I I N N G H S P N V E F D D S Q D K S V
```

[a]Numbering based on human CA I sequence (Andersson *et al.*, 1972). Inferred sequences are in parentheses. Deletions are indicated by a dash(—). Orang CA I residues 67 and 68, rhesus CA I residues 64–76, rabbit CA I residues 6–18, 24–28, 40,

Table 1 (*continued*)

```
            80          90          100         110
Human  I   L K G G P F S D S Y R L F Q F H F H W G S T N E H G S E H T V D G V K Y S A

Chimp  I   L K)

Orang  I   L K G G P L S D   Y R L F Q F H F H W G

Vervet I   L K)

Rhesus I   L K G G P F S D S Y R L F Q F H F H W G S S N E Y G S E H T V D G V K Y

Baboon I   L K)

Rabbit I       G G P L S D S Y R L S Q F H F H W G S T


Human  II  L K G G P L D G T Y R L I Q F H F H W G S L D G Q G S E H T V D K K K Y A A

Vervet II  L K)

Rhesus II  L K)

Sheep  II  L K D G P L T G T Y R L V Q F H F H W G S S D D Q G S E H T V D R K K Y A A

Ox     II  L K D G P L T G T Y R L V Q F H F H W G S S B B Q G S E H T V D R K K Y A A

Rabbit II  L K E G P L E G T Y R L I Q F H F H W G S S D G E G S E H T V N K K K Y A A


            120         130         140         150
Human  I   E L H V A H W N S A K Y S S L A E A A S K A D G L A V I G V L M K V G E A N

Chimp  I                                                             (V G E A N

Orang  I                                                              K V G E A N

Vervet I                                                             (V G E A N

Rhesus I   E L H V     W N     K Y S P L A E A V S K A D G L A V I G V L M K V G E A N

Baboon I                                                             (L G E A N

Rabbit I                                                               V G E A D


Human  II  E L H L V H W N T - K Y G D F G K A V Q Q P D G L A V L G I F L K V G S A K

Vervet II                                                            (V G S A K

Rhesus II                                                            (V G S A K

Sheep  II  E L H L V H W N T - K Y G D F G T A A Q Q P D G L A V V G V F L K V G D A N

Ox     II  E L H L V H W N T - K Y G D F G T A A Q Q P D G L A V V G V F L K V G D A N

Rabbit II  E L H L V H W N T - K Y G D F G K A V Q H P D G L A V L G I F L K I G D A P
```

47–53, 58–69, 185–194 and rabbit CA II residues 28–36 were not used to construct the phylogenetic tree in Fig. 1. (A tree constructed using these additional residues differed little from that in Fig. 1.)

(*continued*)

Table 1 *(continued)*

		160		170		180		190	
Human	I	P K L Q K V L D A L Q A I K T K G K R A P F T N F D P S T L L P S S L D F W T
Chimp	I	P K L Q K V L D A L Q A I K T K G K R)
Orang	I	P K L Q K V L D A L Q A I K T K G K R A P F T N F D P S T L L P S S L D F W T
Vervet	I	P K L Q K V L D A L H A I K T K G K R)
Rhesus	I	P K L Q K V L N A L H A I K T K G K R A P F T N F D P S T L L P S S L D F W T
Baboon	I	P K L Q K V L D A L H A I K T K G K R)
Rabbit	I	P K L Q K V L D A L S A V K T K G K R A L F P N F D P S T L L P S S L D Y W T
Human	II	P G L Q K V V D V L D S I K T K G K S A D F T N F D P R G L L P E S L D Y W T
Vervet	II	P G L Q K V V D V L D S I K T K G K S)
Rhesus	II	P G L Q K V V D V L D S I K T K G K S)
Sheep	II	P A L Q K V L D V L D S I K T K G K S A D F P N F D P S S L L K R A L N F W T
Ox	II	P A L Q K V L D A L D S I K T K G K S T D F P N F D P G S L L P Q V L D Y W T
Rabbit	II	P G L Q K V V D T L S G I K T K G K T V D F P N F D P R G L L P E S L D Y W T

		200		210		220		230	
Human	I	Y P G S L T H P P L Y E S V T W I I C K E S I S V S S E Q L A Q F R S L L S N
Chimp	I	(S L L S N
Orang	I	Y P G S L T H P P L Y R S L L S N
Vervet	I	(S L L S N
Rhesus	I	Y S G S L P H P P L Y E S V T W I I S L L S N
Baboon	I	(S L L S N
Rabbit	I	Y S L L S N
Human	II	Y P G S L T T P P L L E C V T W I V L K E P I S V S S E Q V L K F R K L N F N
Vervet	II	(K L N F N
Rhesus	II	M L K F R K L N F N
Sheep	II	Y P G S L T N P P L L E S V T W V V L K E P T S V S S Q Q M L K F R S L N F N
Ox	II	Y P G S L T T P P L L E S V T W I V L K E P I S V S S E E M L K F R T L N F N
Rabbit	II	Y P G S L T T P P L L E(C)V T W I V L K E P I S V S S E Q M L K F R N L N F N

Table 1 *(continued)*

		240	250	260
Human	I	V E G D N A V P M Q H N N R P T Q P L K G R T V R A S F
Chimp	I	V E G D N A V P M Z H N N R P T Q P L K G R T V R A S F)
Orang	I	V E G D N A V P I E H N N R P T Q P(L K)G R T V R A S F
Vervet	I	V E G D N P V P I Z H N N R P T Q P L K G R T V R A S F)
Rhesus	I	V E G S N P V P I Z R N N R P T Q P L K G R T V R A S F
Baboon	I	V E G S N P V P I Z R N N R P T Q P L K G R T V R A S F)
Rabbit	I	A E G E A A V P I L H N N R P P G R T V K A S F
Human	II	G E G E P E E L M V D N W R P A Q P L K N R Q I K A S F K
Vervet	II	G E G E P E E L M V D N W R P A Q P L K N R Q I K A S F K)
Rhesus	II	(G E G E P E E L)M V(D)N(W R)P A Q P L K N R Z I K A S F K
Sheep	II	A E G E P E L L M L A N W R P A Q P L K N R Q V R V F P K
Ox	II	A E G E P E L L M L A N W R P A Q P L K N R Q V R G F P K
Rabbit	II	K E A E P E E P M V D N W R P T Q P L K G R Q V K A S F V

Key to the one-letter code

A = Ala	I = Ile	S = Ser
B = Asx	K = Lys	T = Thr
C = Cys	L = Leu	V = Val
D = Asp	M = Met	W = Trp
E = Glu	N = Asn	Y = Tyr
F = Phe	P = Pro	Z = Glx
G = Gly	Q = Gln	
H = His	R = Arg	

the maximum parsimony method (Moore *et al.*, 1973; Goodman *et al.*, 1974). The numbers on the branches are the minimum amounts of nucleotide change required to account for the descent of the different carbonic anhydrases. These amounts are represented in terms of the number of nucleotide replacements (i.e., base changes) per 100 codons. The 13 known amino acid sequences in the data set were first translated into codon sequences, including all necessary alternative synonymous codons among the 61 codons specifying the 20 amino acids. Those ancestral descendant configurations which yielded the fewest mutations over the tree were then calculated. Where alternative configurations were equally parsimonious at a particular residue position, the computer program selected the configuration which distributed the mutations to those regions of the tree with the smallest amount of sequence data. This is because a nucleotide change tends to be grossly underestimated in such relatively empty regions (Goodman *et al.*, 1974). By this type of analysis, many different branching topologies could be examined in the search for the tree of minimum nucleotide change. This tree required 269 nucleotide replacements. No other topology in those examined was more parsimonious than the tree in Fig. 1. Several topologies, however, were equally parsimonious. For instance, either vervet CA II or rhesus CA II could just as well be joined first to human CA II as to each other. Also, vervet CA I could originate either from the hominoid stem or as in Fig. 1 from the cercopithecoid stem, and orangutan CA I could originate either from the catarrhine stem or as in Fig. 1 from the hominoid stem; an orangutan-vervet branch could also be joined to a chimpanzee-human branch. However, it added mutations to the tree to first join orangutan CA I to either human or chimpanzee CA I. Thus this parsimony tree provides evidence for a closer cladistic relationship between chimpanzee and man than between chimpanzee and orangutan.

In addition, an augmented computer procedure called MNAUG (Goodman *et al.*, 1974) was used to obtain a better estimate of evolutionary change in the regions of the parsimony tree with the smallest number of ancestors.

2.3. C-Terminal Sequences

In addition to the 13 *C*-terminal sequences listed in Table 1, five additional *C*-terminal sequences were obtained from pig CA I, horse CA I and CA II, deer CA II (Ashworth *et al.*, 1971), and European elk CA II (Carlsson *et al.*, 1973). The sequences from erythrocyte carbonic anhydrases of dog, cat, mouse, pigeon, duck, turkey, goose, and chicken were determined in our laboratory (Tashian, unpublished data). In each case, the carbonic anhydrase isozyme was isolated from hemolysates of these animals by affinity chromatography using sulfonamide-coupled Sephadex columns by the method of Osborne and Tashian (1975).

3. Comparisons of Different Structural Regions

3.1. Hydrophobic Cores

As mentioned in the introduction, the tertiary structures of human CA I and CA II are strikingly similar. Thus, during the evolution of the two isozymes after their origin by gene duplication, the internal hydrophobic residues must have remained essentially unchanged. In Table 2, comparisons are made between residues assigned to the hydrophobic cores of the human isozymes and residues at the same positions in other mammalian carbonic anhydrases. As can

Table 2. Comparison of the Residues Found in the Hydrophobic Cores of Human CA I and CA II with Those at the Same Positions in Other Mammalian CA I and CA II Isozymes[a]

Species and CA isozyme		Position No.																	
		60	66	68	70	79	91	93	95	97	116	118	120	122	144	146	148	157	160
Human	I	I	F	V	F	L	F	F	F	W	A	L	V	H	I	V	M	L	V
Orang	I	I	F	V	F	L	F	F	F	W	—	—	—	—	—	—	—	L	V
Rhesus	I	L	F	V	I	L	F	F	F	W	—	L	V	—	I	V	M	L	V
Rabbit	I	I	F	V	F	L	S	F	F	W	—	—	—	—	—	—	—	L	V
Human	II	L	F	V	F	L	I	F	F	W	A	L	L	H	L	I	L	L	V
Sheep	II	V	F	V	F	L	V	F	F	W	A	L	L	H	V	V	L	L	V
Ox	II	V	F	V	Y	L	V	F	F	W	A	L	L	H	V	V	L	L	V
Rabbit	II	V	F	V	F	L	I	F	F	W	A	L	L	H	L	I	L	L	V

Species and CA isozyme		Position No.													
		161	163	164	176	179	181	184	185	186	210	212	216	218	226
Human	I	L	A	L	F	F	P	L	L	P	I	C	I	V	F
Orang	I	L	A	L	F	F	P	L	L	P	—	—	—	—	—
Rhesus	I	L	A	L	F	F	P	L	L	P	I	—	—	—	—
Rabbit	I	L	A	L	F	F	P	L	L	P	—	—	—	—	—
Human	II	V	V	L	F	F	P	L	L	P	I	L	I	V	F
Sheep	II	L	V	L	F	F	P	L	L	K	V	L	T	V	F
Ox	II	L	A	L	F	F	P	L	L	P	I	L	I	V	F
Rabbit	II	V	A	L	F	F	P	L	L	P	I	L	I	V	F

[a]See Table 1 for key to amino acid code letters.

be seen, these residues either have remained invariant or have involved only conservative changes with the exception of a lysyl residue at position 186 in sheep CA II.

3.2. Active-Site Residues

Because the three-dimensional structures of human CA I and CA II have now been determined at high resolution, it is possible to locate the positions of a number of residues located in the active sites of both enzymes. In view of the activity and binding differences between the two enzymes, it was of interest to compare some of the active-site residues from both isozymes in different mammalian species to see which have remained constant and which have changed during the evolution of the isozymes.

In Table 3, a total of 23 residues located in the active-site cavities of the homologous forms of both isozymes are compared in several mammalian species. As might be expected, the three histidyl residues forming ligands to the zinc (positions 94, 96, and 119) are the same for all enzymes. In addition, Tyr-7, Asn-61, His-64, Gln-92, Leu-141, Val-143, Gly-145, Pro-201, Pro-202, and Val-207 are common to both isozymes. Those residues which are different, and might account for the different activities and binding properties of the two isozymes, are at positions 67 (His in CA I, Asn in CA II), 69 (Asn in CA I, Glu in CA II), 121 (Ala in CA I, Val in CA II), 131 (Leu in CA I, Phe in CA II), 200 (His in CA I, Thr or Asn in CA II), and 204 (Tyr in CA I, Leu in CA II).

3.3. Aromatic Clusters

Three pronounced aromatic clusters which are probably responsible for the maintenance of the stability of the active-site cavities have been described on the basis of the three-dimensional structures of human CA I and CA II. One of these, cluster I, is located in the amino-terminal region near the active site; another, cluster II, is located in the interior of the molecule below the active site; and cluster III is in the C-terminal region. In Table 4, the residues which make up these clusters are compared from the carbonic anhydrases of six different mammals. In cluster I, the only difference occurs at position 20 where the residue is Tyr for the CA I enzymes and Phe for the CA II enzymes. In cluster II, the phenylalanyl residues at six positions have remained unchanged. In cluster III, five of the eight residues are constant.

4. Comparative Rates of Evolution

In Figure 1, an ordinate scale of millions of years is used for the period of eutherian (placental mammals) phylogeny with the times for some of the

Table 3. Comparisons of Exposed Residues Located in the Active-Site Regions of the Human CA I and CA II with Those Found at the Same Positions in Other Mammalian CA I and CA II Isozymes

Species and CA isozyme	Position No.						
	7	61	64	65	67	69	91
Human I	Tyr	Asn	His	Ser	His	Asn	Phe
Orang I	Tyr	Asn	His	Ser	His	Asn	Phe
Rhesus I	Tyr	Asn	His	Ser	His	Asx	Phe
Rabbit I	Tyr	Asn	His	Val	His	Asn	Ser
Human II	Tyr	Asn	His	Ala	Asn	Glu	Ile
Sheep II	Tyr	Asn	His	Ser	Asn	Glu	Val
Ox II	Tyr	Asn	His	Ser	Asn	Glu	Val
Rabbit II	Tyr	Asn	His	Ser	Asn	Glu	Ile

Species and CA isozyme	Position No.						
	92	94	96	119	121	131	141	143
Human I	Gln	His	His	His	Ala	Leu	Leu	Val
Orang I	Gln	His	His	—	—	—	—	—
Rhesus I	Gln	His	His	His	—	Leu	Leu	Val
Rabbit I	Gln	His	His	—	—	—	—	—
Human II	Gln	His	His	His	Val	Phe	Leu	Val
Sheep II	Gln	His	His	His	Val	Phe	Leu	Val
Ox II	Gln	His	His	His	Val	Phe	Leu	Val
Rabbit II	Gln	His	His	His	Val	Phe	Leu	Val

Species and CA isozyme	Position No.						
	145	200	201	202	204	206	207	211
Human I	Gly	His	Pro	Pro	Tyr	Ser	Val	Ile
Orang I	—	His	Pro	Pro	Tyr	—	—	—
Rhesus I	Gly	His	Pro	Pro	Tyr	Ser	Val	Ile
Rabbit I	—	—	—	—	—	—	—	—
Human II	Gly	Thr	Pro	Pro	Leu	Cys[a]	Val	Val
Sheep II	Gly	Asn	Pro	Pro	Leu	Ser	Val	Val
Ox II	Gly	Thr	Pro	Pro	Leu	Ser	Val	Val
Rabbit II	Gly	Thr	Pro	Pro	Leu	—	Val	Val

[a]Cys not exposed in human CA II.

branch points in the tree chosen from paleontological evidence on the divergence times of the eutherian lineages. Thus the eutherian ancestor from which ungulate, primate, and lagomorph lineages diverged is placed at 90 million years ago (McKenna, 1969), and the bovid ancestor at 25 m.y. ago (Gentry, 1970), the catarrhine (Old World monkeys, great apes, and man) ancestor at 35 m.y. ago (Uzzell and Pilbeam, 1971; Simons, 1967), and the human-chimpanzee ancestor at 15 m.y. ago (Pilbeam, 1968). The lagomorph-primate divergence is

Table 4. Comparison of Residues in the Three Aromatic Clusters of Human CA I and CA II with Those at the Same Positions in Other Mammalian CA I and CA II Isozymes

Species and CA isozyme	Residue positions: Cluster I				
	5	7	16	20	64
Human I	Trp	Tyr	Trp	Tyr	His
Orang I	Trp	Tyr	Trp	Tyr	His
Rhesus I	Trp	Tyr	Trp	Tyr	His
Rabbit I	Trp	Tyr	Trp	Tyr	His
Human II	Trp	Tyr	Trp	Phe	His
Sheep II	Trp	Tyr	Trp	Phe	His
Ox II	Trp	Tyr	Trp	Phe	His
Rabbit II	Trp	Tyr	Trp	Phe	His

Species and CA isozyme	Residue positions: Cluster II							
	66	70	93	95	97	176	179	226
Human I	Phe	Phe	Phe	Phe	Trp	Phe	Phe	Phe
Orang I	Phe	Phe	Phe	Phe	Trp	Phe	Phe	—
Rhesus I	Phe	Ile	Phe	Phe	Trp	Phe	Phe	—
Rabbit I	Phe	Phe	Phe	Phe	Trp	Phe	Phe	—
Human II	Phe	Phe	Phe	Phe	Trp	Phe	Phe	Phe
Sheep II	Phe	Phe	Phe	Phe	Trp	Phe	Phe	Phe
Ox II	Phe	Tyr	Phe	Phe	Trp	Phe	Phe	Phe
Rabbit II	Phe	Phe	Phe	Phe	Trp	Phe	Phe	Phe

Species and CA isozyme	Residue positions: Cluster III							
	107	114	147	191	192	194	209	260
Human I	His	Tyr	Leu	Phe	Trp	Tyr	Trp	Phe
Orang I	—	—	—	Phe	Trp	Tyr	—	Phe
Rhesus I	His	Tyr	Leu	Phe	Trp	Tyr	Trp	Phe
Rabbit I	—	—	—	Tyr	Trp	Tyr	—	Phe
Human II	His	Tyr	Phe	Tyr	Trp	Tyr	Trp	Phe
Sheep II	His	Tyr	Phe	Phe	Trp	Tyr	Trp	Pro
Ox II	His	Tyr	Phe	Tyr	Trp	Tyr	Trp	Pro
Rabbit II	His	Tyr	Phe	Tyr	Trp	Tyr	Trp	Phe

depicted in the tree as more recent than the divergence of ungulate and lagomorph-primate branches. Thus the lagomorph-primate ancestor is placed at 75 m.y. ago. Using these times and the amounts of nucleotide change shown on the branches of the tree, the rates of evolution for carbonic anhydrases I and II can be compared during the period of eutherian phylogeny (Table 5). CA I shows an average evolutionary rate during the past 75 m.y. of one base change per 100 codons every 7.9 m.y. A slightly faster average rate (one base change per 100 codons per 6.4 m.y.) is maintained in the catarrhine lineages during the past 35 m.y. However, in the human and chimpanzee lineages, the rate appears to have decreased to a base change per 100 codons about every 35 m.y. This is

Table 5. Comparative Rates (Base Changes per 100 Codons) of Descent for Carbonic Anhydrase Isozymes Since Their Divergence from Various Common Ancestors[a]

	Eutherian ancestor (90 m.y. ago) CA II		Rabbit-primate ancestor (75 m.y. ago) CA I		CA II		Catarrhine ancestor 35 m.y. ago CA I	CA II	Bovid ancestor (25 m.y. ago) CA II		Human-chimpanzee ancestor (15 m.y. ago) CA I
	O	A	O	A	O	A	O	A	O	A	A
Human	18.2	23.7	8.0	8.9	9.3	12.1	4.2	0.81	—	—	0
Chimp	—	—	8.9	9.8	—	—	5.1	—	—	—	0.85
Orang	—	—	9.6	10.5	—	—	5.8	—	—	—	—
Vervet	17.4	22.9	9.4	10.3	8.5	11.3	5.6	0	—	—	—
Rhesus	18.2	23.7	10.3	11.2	9.3	12.1	6.5	0.84	—	—	—
Baboon	—	—	9.4	10.3	—	—	5.6	—	—	—	—
Rabbit	20.6	26.2	11.0	13.0	11.7	14.6	—	—	—	—	—
Sheep	16.6	22.0	—	—	—	—	—	—	5.0	7.7	—
Ox	17.4	22.8	—	—	—	—	—	—	5.8	8.5	—
Average	18.1	23.6	9.5	10.6	9.5	12.6	5.5	0.55	5.4	8.1	0.43

Average time (m.y.) for one base change per 100 codons:

	5.0	3.8	7.9	7.1	7.7	6.0	6.4	63.6	4.6	3.1	34.9

[a]Link lengths summed from Fig. 1. Abbreviations: O, original link lengths; A, augmented link lengths; m.y., million years.

founded on the assumption that the 14-m.y.-old *Ramapithecus* specimen from Kenya (Pilbeam, 1968) is indeed an antecedent of the genus *Homo* which had already spearated in descent from the most recent common ancestor of chimpanzees and humans. If the most recent common ancestor of chimpanzees and humans still existed within the last 10 m.y., then the apparent deceleration in the rate of evolution of human CA I would not be so marked.

With respect to the CA II isozymes, an exceptionally pronounced deceleration in evolutionary rate seems to have occurred in the catarrhine primates. This isozyme shows an average rate during the past 90 m.y. of one nucleotide change per 100 codons per 5.0 m.y. A faster rate (one nucleotide change per 100 codons per 4.6 m.y.) was maintained in the bovid lineages. In the catarrhine lineages, however, the rate of evolution in the descent of the CA II isozymes appears to have decreased drastically to only one base change per 100 codons about every 64 m.y.

Using the augmented program (cf. Table 5), the average rates show an increase for the CA II isozymes during their descent from a common eutherian ancestor, and in the descent of ox and sheep from a common bovid ancestor. Since no change in the rates of CA I and CA II was found with the augmented program in the descent of these enzymes from a common catarrhine ancestor, the decrease in the evolutionary rates for the primate CA II isozymes is even more marked.

Is there anything unique about the primary structures of the CA II isozymes of the higher primates? Although the following is based on inferred sequences, it is nevertheless of interest that residues at 15 positions appear to be common to the CA II isozymes of human, vervet, and rhesus when compared to the residues at these positions from the rabbit and bovid CA II sequences.

These are Lys-24, His-36, Thr-37, Ala-38, Tyr-40, Ser-48, Val-49, Ser-50, Asp-52, Leu-57, Ser-152, Lys-154, Lys-228, Gly-233, and Ile-256. Selection for some of these residues may have been important in maintaining some aspect of the functional roles of the CA II isozymes in the higher primates.

It should be pointed out that this "constancy" in the primary structures of the CA II isozymes may not be limited to the Old World primates, since there is some evidence (based on comparative tryptic peptide patterns) that the CA II isozymes of the New World monkeys may be very similar in their primary structures to those of the Old World monkeys (Tashian *et al.*, 1968). If this is true, then the slowdown in the evolution of the CA II molecules is even more pronounced, extending back perhaps to the common ancestor of the New and Old World monkeys some 50–55 m.y. ago (see Simons, this volume).

Obviously, trends of this kind should be viewed with circumspection until more data are available; nevertheless, they pose a thought-provoking corollary to the "molecular clock" concept, which holds that homologous proteins evolve at approximately equal rates in different lines of descent (cf. Dickerson, 1971; Ohta and Kimura, 1971). A similar deceleration in primate globins has been shown by the maximum parsimony method (cf. Goodman *et al.*, 1974).

If we assume that the mutation rates for both CA I and CA II have remained essentially the same, it might indicate that selective forces were responsible for the observed differences in evolutionary rates. For example, CA II appears to have been evolving at about the same rate as CA I over the past 75 m.y.; however, in the higher primates the evolution of CA II may have been about 10 times slower than that of CA I. This would suggest that it has been physiologically important to preserve the primary structure of CA II in the higher primates. In this respect, it is of interest that among the 22 species of catarrhine primates that have been tested there appears to be greater genetic variability at the CA I locus than at the CA II locus. An examination of the number of variant alleles in Table 6 reveals a total of 20 rare and 21 polymorphic alleles at the CA I locus, and two rare and ten polymorphic alleles at the CA II locus. This disparity in the number of alleles found at the two loci may be due in part to the fact that more animals were typed for CA I than for CA II in ten of the 20 species in which both isozymes were tested. However, even if more animals had been examined in the ten species, it is doubtful that many more polymorphic alleles would have been detected. Thus the apparent deceleration in the evolution of CA II in higher primates may also be reflected in the decreased frequency of genetic variation.

Is there any evidence that any of these polymorphisms are being maintained by natural selection? Natural populations of different species of macaque monkeys show a fairly high degree of polymorphism at both the CA I and CA II loci. These data were therefore analyzed for the action of natural selection using a procedure developed by Ewens (1972). The results indicated that the frequencies of some of the alleles at both loci were being maintained by heterotic selection or directional selection (Adams and Tashian, 1976).

Table 6. Number of Electrophoretic Alleles of CA I and CA II in Various Species of Catarrhine Primates[a]

Species	Number tested	CA I		Number tested	CA II	
		Rare[b]	Polymorphic[c]		Rare[b]	Polymorphic[c]
Hominoidea						
Man (Homo sapiens)	23,000	12	0	12,561	0	1
Chimpanzee (Pan troglodytes)	295	0	0	94	0	0
Gorilla (Gorilla gorilla)	12	0	0	12	0	0
Orangutan (Pongo pygmaeus)	33	0	2	33	0	0
White-handed gibbon (Hylobates lar)	19	0	0	19	0	0
Siamang (Symphalangus syndactylus)	5	0	0	5	0	0
Cercopithecoidea						
Purple-faced langur (Presbytis senex)	12	0	3	12	0	0
Entellus monkey (Presbytis entellus)	5	0	3	5	0	0
Mandrill (Papio sphinx)	9	0	1	6	0	0
Kenya baboon (Papio cynocephalus)	847	0	1	847	1	1
Stump-tailed macaque (Macaca speciosa)	154	0	1	26	0	1
Toque macaque (Macaca sinica)	55	0	1	26	0	1
Bonnet macaque (Macaca radiata)	58	0	0	58	0	0
Pig-tailed macaque (Macaca nemestrina)	595	0	3	169	1	1
Rhesus macaque (Macaca mulatta)	449	2	0	109	0	1
Celebes macaque (Macaca maura)	7	0	2	7	0	1
Cynomolgous macaque (Macaca irus)	707	3	0	234	0	1
Japanese macaque (Macaca fuscata)	143	1	1	116	0	0
Formosan macaque (Macaca cyclopis)	92	1	0	0	—	—
Black ape (Macaca niger)	6	0	1	6	0	0
Patas monkey (Cercopithecus patas)	85	0	1	81	0	1
Vervet (Cercopithecus aethiops)	105	1	1	96	0	1
Totals	26,696	20	21	14,522	2	10

[a]For references, see Tashian and Carter (1976).
[b]Frequency < 0.02.
[c]Frequency ≥ 0.02; one less total number of alleles.

5. Duplication of the Carbonic Anhydrase Gene

We would have a better understanding of the evolutionary history of the carbonic anhydrase isozymes if we could determine the period in evolutionary time when the proposed gene duplication occurred. One approach would be to identify the oldest group of organisms in which both isozymes are known to occur. As previously mentioned, only one form of carbonic anhydrase has as yet been isolated from nonmammalian sources. If only one form is present in birds and reptiles as well as in all evolutionarily older organisms, and two forms are present in mammals, then one could speculate that the duplication took place sometime before the separation of marsupials from the ancestors of other mammals (about 120 m.y. ago) and the point of divergence of the reptile-bird line from the line leading to mammals (about 300 m.y. ago). Since it is critical to establish whether or not one high-activity form of carbonic anhydrase is characteristic for all birds, we undertook a preliminary study to examine the red cell carbonic anhydrases of pigeon, duck, turkey, goose, and chicken utilizing an affinity chromatography method which is capable of separately isolating both isozymes from mammalian hemolysates based on their different sulfonamide-binding affinities (Osborne and Tashian, 1975). Our results indicate, on the basis of differential sulfonamide-binding affinities, that pigeon, chicken, turkey, goose, and duck possess a high-activity form apparently homologous to the CA II isozymes of mammals.

In addition to these binding properties, a comparison of the C-terminal residues might also be helpful in identifying the homologies of the bird carbonic anhydrases. Table 7 lists the C-terminal CA I and CA II sequences

Table 7. Comparison of C-Terminal Sequences of Mammalian and Avian Red Cell Carbonic Anhydrases

Species	CA I isozyme	C-terminal sequence
Human	I	-Gly-Arg-Thr-Val-Arg-Ala-Ser-Phe-COOH
Chimp	I	(Gly,Arg,Thr,Val,Arg,Ala,Ser,Phe-COOH)
Orang	I	-Gly-Arg-Thr-Val-Arg-Ala-Ser-Phe-COOH
Vervet	I	(Gly,Arg,Thr,Val,Arg,Ala,Ser,Phe-COOH)
Rhesus	I	-Gly-Arg-Thr-Val-Arg-Ala-Ser-Phe-COOH
Baboon	I	(Gly,Arg,Thr,Val,Arg,Ala,Ser,Phe-COOH)
Dog	I	-Gly -Arg -Ile -Val-Lys-Ala-Ser-Phe-COOH
Rabbit	I	-Gly-Arg-Thr-Val-Lys-Ala-Ser-Phe-COOH
Horse	I	-Val-Arg-Ala-Phe-Phe-COOH
Pig	I	-Lys-Ala-Ser-Phe-COOH
Human	II	-Asn -Arg -Gln -Ile-Lys-Ala-Ser-Phe-Lys-COOH
Vervet	II	(Asn ,Arg ,Gln ,Ile,Lys,Ala,Ser,Phe,Lys-COOH)
Rhesus	II	-Asn- Arg -Gln -Ile-Lys-Ala-Ser-Phe-Lys-COOH
Sheep[a]	II	-Asn-Arg-Gln-Val-Arg-Val-Phe-Pro-Lys-COOH
Ox[a]	II	-Asn-Arg-Glu-Val-Arg-Gly-Phe-Pro-Lys-COOH
Horse	II	-Ile-Arg-Ala-Ser-Phe-Lys-COOH
Elk[a]	II	-Leu-Arg-COOH
Deer[a]	II	-Pro-Arg-COOH
Cat[a]	II	-His-Phe-Ser-Lys-COOH
Mouse	II	-His-Ser-Phe-Lys-COOH
Rabbit	II	-Gly-Arg-Gln-Val-Lys-Ala-Ser-Phe-Val-COOH
Duck[a]	?	-Ala-Ser-Phe-Ser-COOH
Pigeon[a]	?	-Val-Arg-Ala-Ser-Phe-Ser-COOH
Chicken[a]	?	-Arg-Ala-Ser-Phe-Ser-COOH
Turkey[a]	?	-Lys-Tyr-Ser-Phe-Ser-COOH
Goose[a]	?	-Arg-Ala-Ser-Phe-Ser-COOH

[a]Carbonic anhydrase present as only one form in the red cell.

from a number of mammals. The CA I forms characteristically show Ser or Phe as the penultimate residue, and Phe as the C-terminal residue. On the other hand, the CA II molecules show Ser or Phe as the residue third from the end, Phe, Pro, Leu, or Ser as the penultimate residue, and Lys or Arg as the C-terminal residue.

As shown in Table 7, only one form of carbonic anhydrase was found in the red cells of the five species of birds that were examined. Since all of the bird carbonic anhydrases showed both high CO_2 hydrase activities and high sulfonamide-binding affinities (Tashian, unpublished data), it would indicate that they are functionally more homologous to the high-activity forms of mammals. However, until more sequence data are available, and more species of birds are examined, we cannot be certain that birds characteristically possess only one form of carbonic anhydrase.

The evidence now available would suggest that the gene duplication which gave rise to the two mammalian anhydrase isozyme genes took place early in mammalian evolution, perhaps 100–150 m.y. ago. As yet we have no clear-cut evidence that both isozymes are present in a marsupial. The one species that we have looked at most carefully, the red kangaroo *(Macropus rufus)*, showed only one form (high activity) in the red cell (Tashian, unpublished data). If two isozymes of carbonic anhydrase are not found in marsupials, then the duplication may have occurred after the separation of the placental mammals. It is of interest that the two carbonic anhydrase isozyme loci appear to be closely linked in mammals (Carter, 1972*b*; DeSimone *et al.*, 1973), a fact which might indicate that the duplication was a relatively recent event.

6. Summary

Partial and complete primary sequences for the carbonic anhydrase isozymes, CA I and CA II, of man, five higher primates (chimpanzee, orangutan, vervet, rhesus macaque, and Kenya baboon), two ruminants (ox and sheep), and a lagomorph (rabbit) have been compared. A maximum parsimony statistical method was used to determine the number of nucleotide replacements that had occurred during the evolution of the CA I and CA II isozymes in these species. Over the past 75 million years, the rate of fixation of mutational change appears to have been somewhat faster in the CA II isozymes than in the CA I isozymes. However, during the past 35 million years, the average evolutionary rates of the CA II enzymes of the higher primates seem to have decreased markedly. Over the same period of time, the average evolutionary rates of the CA I isozymes show some increase in the primates, with the exception of the human and chimpanzee isozymes where the rates appear to have decelerated considerably. Preliminary studies on the C-terminal sequences of the red cell carbonic anhydrases of 17 species of mammals and five species of birds suggest that the duplication which produced the two carbonic anhydrase isozyme genes occurred early in mammalian evolution.

ACKNOWLEDGMENTS

We thank Mrs. Sharon K. Stroup and Dr. G. William Moore for their invaluable help in obtaining and analyzing the sequence data. This work was supported in part by U.S. Public Health Service research grant GM-15419.

7. References

Adams, J., and Tashian, R. E., 1976, Natural selection for carbonic anhydrase isozyme variants in macaque monkeys, *Isozyme Bull,* (in press).

Andersson, B., Nyman, P. O., and Strid, L., 1972, Amino acid sequence of human erythrocyte CA B, *Biochem. Biophys. Res. Commun.* **48:**670.

Ashworth, R. B., Brewer, J. M., and Stanford, R. C., Jr., 1972, Composition and carboxy-terminal amino acid sequences of some mammalian erythrocyte carbonic anhydrases, *Biochem. Biophys. Res. Commun.* **44:**667.

Carlsson, U., Hannestad, U., and Lindskog, S., 1973, Purification and some properties of erythrocyte carbonic anhydrase from the European moose, *Biochim. Biophys. Acta* **327:**515.

Carter, M. J., 1972a, Carbonic anhydrase: Isoenzymes, properties, distribution, and functional significance, *Biol. Rev.* **47:**465.

Carter, N. D., 1972b, Carbonic anhydrase isozymes in *Cavia porcellus, Cavia aperea* and their hybrids, *Comp. Biochem. Physiol.* **43B:**743.

Dayhoff, M. O., 1972, *Atlas of Protein Sequence and Structure,* Vol. 5, National Biomedical Research Foundation, Silver Spring, Md.

DeSimone, J., Linde, M., and Tashian, R. E., 1973, Evidence for linkage of carbonic anhydrase isozyme genes in the pig-tailed macaque, *Macaca nemestrina, Nature (London) New Biol.* **242:**55.

Dickerson, R., 1971, The structure of cytochrome *c* and the rates of molecular evolution, *J. Mol. Evol.* **1:**26.

Ewens, W. J., 1972, The sampling theory of selectively neutral alleles, *Theor. Pop. Biol.* **33:**87.

Gentry, A. W., 1970, in: *Fossil Vertebrates of Africa,* Vol. 2, (L. S. B. Leaky and R. J. G. Savage, eds.), Academic Press, London.

Goodman, M., Moore, G. W., and Barnabas, J., 1974, The phylogeny of human globin genes investigated by the maximum parsimony method, *J. Mol. Evol.* **3:**1.

Henderson, L. E., Henriksson, D., and Nyman, P. O., 1973, Amino acid sequence of human erythrocyte carbonic anhydrase C. *Biochem. Biophys. Res. Commun.* **52:**1388.

Kannan, K. K., Liljas, A., Waara, I., Bergstén, P.-C., Lövgren, S., Strandberg, B., Bengtsson, U., Carlbom, U., Fridborg, K., Järup, L., and Petef, M., 1971, Crystal structure of human erythrocyte carbonic anhydrase C: Relation to other mammalian carbonic anhydrases, *Cold Spring Harbor Symp. Quant. Biol.* **36:**221.

Kannan, K. K., Notstrand, B., Fridborg, K., Lövgren, S., Ohlsson, A., and Petef, M., 1975, Crystal structure of human erythrocyte carbonic anhydrase B: Three-dimensional structure at a nominal 2.2 Å resolution, *Proc. Natl. Acad. Sci. USA* **72:**51.

Liljas, A., Kannan, K. K., Bergstén, P.-C., Waara, I., Fridborg, K., Strandberg, B., Carlbom, U., Järup, L., Lövgren, S., and Petef, M., 1972, Crystal structure of human carbonic anhydrase C, *Nature (London) New Biol.* **235:**131.

Lin, K.-T. D., and Deutsch, H. F., 1973, Human carbonic anhydrases. XI. The complete primary structure of carbonic anhydrase B, *J. Biol. Chem.* **248:**1885.

Lin, K.-T. D., and Deutsch, H. F., 1974, Human carbonic anhydrases. XII. The complete primary structure of the C isozyme, *J. Biol. Chem.* **249:**2329.

Lindskog, S., Henderson, L. E., Kannan, K. K., Liljas, A., Nyman, P. O., and Strandberg, B., 1971,

Carbonic anhydrase, in: *The Enzymes,* Vol. V (P. D. Boyer, ed.), pp. 587–665, Academic Press, New York.

Maren, T. H., 1967, Carbonic anhydrase: Chemistry, physiology, and inhibition, *Physiol. Rev.* **47:**595.

McKenna, M. C., 1969, The origin and early differentiation of therian mammals, *Ann. N.Y. Acad. Sci.* **167:**217.

Moore, G. W., Barnabas, J., and Goodman, M., 1973, A method for constructing maximum parsimony ancestral amino acid sequences on a given network, *J. Theor. Biol.* **38:**459.

Notstrand, B., Waara, I., and Kannan, K. K., 1974, Structural relationship of human erythrocyte carbonic anhydrase isozymes B and C, in: *Isozymes,* Vol. 1 (C. L. Markert, ed.), pp. 575–599, Academic Press, New York.

Ohta, T., and Kimura, M., 1971, On the constancy of the evolutionary rate of cistrons, *J. Mol. Evol.* **1:**18.

Osborne, W. R. A., and Tashian, R. E., 1974, Thermal inactivation studies of normal and variant human erythrocyte carbonic anhydrases using a sulphonamide binding assay, *Biochem. J.* **141:**219.

Osborne, W. R. A., and Tashian, R. E., 1975, An improved method for the purification of carbonic anhydrase isozymes by affinity chromatography, *Anal. Biochem.* **64:**297.

Pilbeam, P., 1968, The earliest hominids, *Nature (London)* **219:**1335.

Sciaky, M., Limozin, N., Fillippi-Foveau, D., Gulian, J. M., Dalmasso, C., and Laurent, G., 1974, Structure primaire de l'anhydrase carbonique erythrocytaire bovine CI, *C. R. Acad. Sci.* **279:**1217.

Simons, E., 1967, The earliest apes, *Sci. Am.* **217:**28.

Tanis, R. J., Ferrell, R. E., and Tashian, R. E., 1974, Amino acid sequence of sheep carbonic anhydrase C, *Biochim. Biophys. Acta* **371:**534.

Tashian, R. E., and Carter, N. D., 1976, Biochemical genetics of carbonic anhydrase, in: *Advances in Human Genetics* (H. Harris and K. Hirschhorn, eds.), in press, Plenum Press, New York.

Tashian, R. E., and Stroup, S. R., 1970, Variation in the primary structure of carbonic anhydrase B in man, great apes, and Old World monkeys, *Biochem. Biophys. Res. Commun.* **41:**1457.

Tashian, R. E., Shreffler, D. C., and Shows, T. B., 1968, Genetic and phylogenetic variation in the different molecular forms of mammalian erythrocyte carbonic anhydrases, *Ann. N.Y. Acad. Sci.* **151:**64.

Tashian, R. E., Tanis, R. J., and Ferrell, R. E., 1972, Comparative aspects of the primary structures and activities of mammalian carbonic anhydrases, *Alfred Benzon Symp.* **4:**353.

Tashian, R. E., Goodman, M., Tanis, R. J., Ferrell, R. E., and Osborne, W. R. A., 1975, in: *Isozymes,* Vol. 4 (C. L. Markert, ed.), pp. 207–223, Academic Press, New York.

Uzzell, T., and Pilbeam, P., 1971, Phyletic divergence dates of homonoid primates: A comparison of fossil and molecular data, *Evolution* **25:**615.

Toward a Genealogical Description of the Primates

17

MORRIS GOODMAN

1. Introduction

There is now a substantial body of molecular data on the genetic relationships of man and various primates to one another and to other animals. One of the ways to use these data is to deduce from them a phylogenetic classification or cladogram which describes the genealogical branchings of the order Primates. Such a framework for viewing the process of cladogenesis in primate phylogeny can provide an objective basis for a more detailed analysis of other processes underlying primate and human evolution. Like Emile Zuckerkandl (this volume), I consider anagenesis to be the principal process characterizing the emergence of man. Anagenesis is that form of progressive evolution which increases the level of molecular organization within organisms in such a way that the organisms gain greater independence from, and control over, the environment. Whereas Zuckerkandl emphasizes that aspect of anagenesis which accelerated evolutionary rates, I shall emphasize that aspect which decelerated evolutionary rates in proteins after higher levels of integration of molecular specificities had been achieved.

My current thinking on the processes of cladogenesis and anagenesis in the Primates has been influenced by the results of an investigation in which I use amino acid sequence data to construct genealogical trees by the maximum parsimony method. The molecular genealogies depict the phylogenetic relationships of man and other species, thereby testing the genealogical accuracy of the classification for Primates proposed from immunodiffusion data (Dene *et*

MORRIS GOODMAN • Department of Anatomy, Wayne State University School of Medicine, Detroit, Michigan 48201.

al., this volume). The molecular genealogies have the special value of providing information on rates of genetic change during earlier and later phylogenetic stages in the lineages represented by the sequences. Moreover, the specific mutational changes observed between the reconstructed sequence ancestors and descendants provide evidence for the role of natural selection on the rate of molecular evolution. The results indicate that rapid rates of protein evolution were produced by positive selection for increased efficiency in function and then were followed by slow rates because further major mutations were selected against. It seems that as anagenesis succeeded in improving the internal organization of organisms, stabilizing selection became more intense and more likely to restrain the rate of molecular evolution.

The findings highlight one of the central observations at issue in molecular anthropology, namely that in view of the seemingly large amount of morphological evolution in the Hominoidea, the amount of protein evolution distinguishing humans from other hominoids is surprisingly small. This is not a new observation. It goes back at least to the Burg Wartenstein Symposium on "Classification and Human Evolution" in 1962. I suggested then that "although morphological evolution appeared to accelerate with the emergence of higher organisms, molecular evolution decelerated." The very high antigenic similarity observed between man and the two African apes, chimpanzee and gorilla (Goodman, 1962, 1963*a,b*), indicated to me that a deceleration in molecular evolution had become particularly pronounced in hominoids. Because there are presently molecular genealogies for such proteins as cytochrome *c* and globins which encompass the whole span of vertebrate evolution, it is now possible to test more adequately the notion that a deceleration occurred in hominoids and not as an isolated phenomenon but as the culmination of a long-term evolutionary trend.

The idea that for a given amount of morphological or organismal evolution there is less molecular evolution in higher vertebrates than in lower vertebrates is supported by the recent work of Wilson *et al.* (1974). They found a striking difference between frogs and mammals in the potential for sibling species to hybridize in relation to the amount of serum albumin evolution separating the species. Frogs can show 10 times as much albumin evolution as mammals and still hybridize. Thus, to generalize, it appears that for roughly the same amount of organismal evolution frogs show 10 times the amount of protein sequence evolution as mammals. If mutations in regulator genes are more important in bringing about morphological evolution than mutations in the DNA coding for specific proteins, we can explain the seemingly greater morphological variability of mammals as compared to frogs for the same amount of variability in protein sequences as simply due to greater evolution in regulatory DNA in the mammals. Similarly, one does not have to believe that the trivial difference between *Homo* and *Pan* in protein sequences as compared to their seemingly large difference in morphology means that there has been a deceleration in protein evolution in hominoids. Instead, one can simply attribute the trivial sequence difference to a recent separation between *Homo* and

Pan (e.g., Sarich and Cronin, this volume) and the supposed large morphological difference to an accelerated rate of evolution in regulator DNA in the human lineage (e.g., King and Wilson, 1975).

While not denying the importance of regulatory evolution and the likely possibility that its rate indeed increased in mammals and in some mammals more than in others, I want to defend my view that a slowdown in molecular evolution in hominoids is a real phenomenon, and part of a long-term trend shaped by natural selection and connected specifically with the evolutionary process of anagenesis. In order to defend my view and present the evidence for it, I first describe below the approach used in reconstructing phylogeny from amino acid sequence data. Then I review relevant results on primate and vertebrate phylogeny from these reconstructions.

2. Phylogeny Reconstruction Approach

2.1. Maximum Parsimony Procedure

Moore (this volume) gives a clear account of the evolutionary concepts and mathematical foundation underlying the maximum parsimony procedure. Briefly, the object of the maximum parsimony or maximum homology procedure is to reconstruct both those hypothetical ancestral mRNA sequences and that genealogical network which require the fewest nucleotide changes (or mutations) to account for the descent of the contemporary amino acid sequences. Such a reconstructed genealogy maximizes the number of genetic similarities associated with common ancestry and minimizes convergent (parallel or back) changes. It obeys the phylogenetic reconstruction principles of Hennig (1966) in that patristic similarities (primitive features inherited unchanged from ancient common ancestors) and convergent similarities are distinguished from derived similarities and the latter (the forward mutations fixed in more recent common ancestors) determine the cladistic or genealogical relationships of the species in the parsimony tree. Moreover, evolutionary rates can vary markedly from one lineage to another and still not adversely affect the accuracy of the reconstructions.

If sequences in two lineages have already diverged from each other at an aligned position, the chances are that it would take millions of years of evolution before the same codon could again occupy this sequence position in some of the descendant species. By the time such a convergent change occurred in these species, new mutations fixed at other aligned positions would in all likelihood have increased the divergence of their sequences. Thus codon identities resulting from common inheritance are likely to be more frequent than those resulting from parallel or back mutation. If this is true, then a parsimony reconstruction based on adequate sequence data should come close to depicting correct genealogical relationships, even though the number of nucleotide replacements observed in the reconstructed genealogy may be less than the number which actually occurred in evolution.

A mathematically rigorous procedure for assigning the most parsimonious reconstructed ancestors to a given genealogy is used (Moore *et al.,* 1973). As there is no known method, short of an exhaustive search, for finding the most parsimonious genealogy, the approach followed is a heuristic one (Goodman *et al.,* 1974). An initial unweighted pair-group method (UWPGM) tree is improved repeatedly by a branch-swapping procedure. The search is stopped when it appears that a further swapping of branches will not reduce the length (i.e., the total number of nucleotide replacements) of the genealogy.

As stated before, the parsimony reconstruction process is not likely to count all superimposed nucleotide replacements which actually occurred during the descent of the contemporary sequences from their common ancestors. Undercounting will be more likely to occur in regions of the genealogy which are underrepresented in ancestors either because data have not been gathered on the proper range of extant species or because evolutionary side branches never existed or became extinct without passing on descendants to the present. Therefore, in the reconstruction process, whenever there are alternative maximum parsimony solutions, the solution chosen distributes mutations more often than the other alternatives to the underrepresented regions in the genealogy. This reduces but does not eliminate the bias toward more marked underestimation of evolutionary change in the sparser regions than in the denser regions of the genealogy. It cannot eliminate the bias because the basic condition causing it remains. There can never be more than one observed nucleotide replacement at any particular nucleotide position when two sequences are directly compared, but there can be and often is more than one observed replacement when two sequences are compared through intervening sequences. Thus, for the problem of measuring the full mutational change over a genealogy, an augmentation algorithm (Moore *et al.,* 1976; Goodman *et al.,* 1974) is used. The algorithm replaces the estimated lost nucleotide replacements in sparse regions of a genealogy on the basis of the distribution of nucleotide replacements in the dense regions. It yields results of the same order of magnitude as those obtained by the stochastic method of Holmquist (this volume).

2.2. Coping with Gene Duplications and Convergent Mutations

2.2.1. Distinction between Species and Gene Phylogenies

Protein chains in different animal species are homologous in the strict sense of genetic homology if they are coded for by genes descended from a single ancestral gene which existed in the most recent common ancestor of the species being compared. Protein chains are also homologous in the broader meaning of genetic homology as long as they are coded for by genes descended from a single ancestral gene. In this broader definition, separate gene lines would not always arise from species divergence but could sometimes arise from the ancestral gene's duplicating to produce two or more nonallelic genes. The terms "orthologous" and "paralogous," as used by Fitch (1970), distinguish between the strict and looser form of genetic homology. A gene phylogeny can contain both orthologous and paralogous sequences, but in determining an animal species phylogeny from protein sequences ideally only strictly orthologous sequences should be compared. In practice, to be effective and not throw away too many sequence data in constructing the species phylogeny, one must use "operational orthologous" sequences without being sure that all are strictly orthologous. Also, some data sets can include with the more rigorously determined sequences those inferred from amino acid compositions of peptide fragments by comparison with known homologous sequences. This makes available for analysis potentially useful information about proteins on which peptide patterns and amino acid compositions have been determined, but not the actual complete sequences.

2.2.2. Heuristic Strategy

An initial gene parsimony tree is constructed for each collection of homologous sequences and used to deduce the genealogical relationships of the species represented by those sequences which behave as though they were orthologously related. For example, in a genealogy of β-like hemoglobin chains, δ sequences from hominoids would group together first before they grouped with the branch of hominoid β sequences. Thus it would be obvious that δ and β chains are paralogously related even if each individual hominoid did not express both genes. Hominoid relationships could be determined by the branching pattern within the hominoid δ region and by the branching pattern within the hominoid β region but not by cross-comparison of δ's and β's.

In a collection of homologous sequences, the number of convergent similarities among a few sequences might happen to be large enough so as not to be recognized as such in the gene parsimony tree. Instead these sequences might be falsely grouped together and their similarities thereby depicted as derived from a common ancestor. This would lead to incorrect conclusions on phylogenetic relationships. To cope with this problem, more than one set of operational orthologous sequences are combined so as to expand the number of aligned sequence positions. Then the statistics of the sampling procedure will help in the search for the correct phylogenetic topology. The more aligned positions examined, the more certain it will be that the sequence identities from shared common ancestry will outweigh those from convergence. By expanding the sets of operational orthologous sequences, some degree of quasi-orthologous relationship in the expanded alignment should be endurable without producing an incorrect species genealogy.

Having used the initial gene parsimony trees to choose the operational orthologous sequences for constructing the species phylogeny, one can next use the species phylogeny to construct more accurate gene phylogenies. Again, the problem is that convergent mutations among some of the sequences in a homologous collection could reach a level whereby maximum parsimony depicted incorrect cladistic relationships in some regions of the putative genealogical tree. The likely way for the cladistic topology of genes to deviate from that of species in which the genes occur is by gene duplications producing gene splittings before the species splittings. Thus wherever a proposed gene phylogeny deviates from those branching features of the species phylogeny which are accepted as firmly established, there is a minimum number of probable gene duplications needed to account for the deviation. This number can be precisely calculated from the number of differences in branching topology between the respective species and putative gene phylogenies. The corresponding minimum number of deletions or regulator mutations (gene inactivations) which would be needed to account for the species differences in the expression of the postulated nonallelic loci can also be determined (e.g., see Hewett-Emmitt *et al.*, this volume). These numbers of probable gene duplications and inactivations (GDs plus GIs) can now be added to the number of nucleotide replacements (NRs) to get the total mutational score of the proposed gene phylogeny. At this stage, the most parsimonious of alternative topologies with respect to this total mutational score can be accepted as the closest approximation to the correct gene phylogeny.

3. Phylogenetic Relationships of Primates Deduced from Amino Acid Sequence Data

3.1. Species Phylogeny from Combined Sequences

On the basis of initially constructed gene parsimony trees, operational orthologous sequences of myoglobin and α- and β-hemoglobin chains were identified and combined into an expanded alignment for a data set representing six primates, five other placental mammals, two marsupials, and chicken. These aligned expanded sequences were then used to construct the species genealogy shown in Fig. 1. In this genealogy, the combined sequences for the platyrrhine Atelinae came from spider monkey (α and β chains) and woolly monkey (myoglobin). Otherwise, the three globins in each case came from members of the same genus if not the same species. Some of the sequences used in the expanded alignment, such as the pig β-hemoglobin chain, were inferred from amino acid compositions of peptide fragments. The genealogy shown in Fig. 1 is one of three found by the branch-swapping procedure. The two alternatives consist only of switching the position of gorilla and chimpanzee or of gorilla and human. I prefer keeping chimpanzee joined to human (the arrangement shown in Fig. 1) because parsimony analysis of α-hemoglobin sequences favors this arrangement. In particular, at position $\alpha23$ lorisiforms,

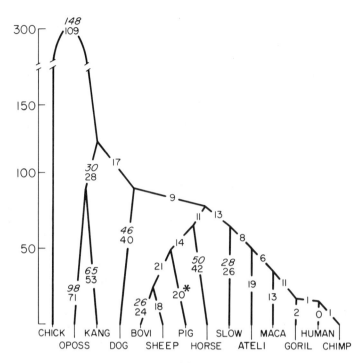

Fig. 1. Parsimony species genealogy, requiring 577 nucleotide replacements (NRs), for 20 vertebrates on using a combined myoglobin and α- and β-hemoglobin alignment. Link lengths are the
numbers of NRs between adjacent ancestral and descendant sequences; italicized numbers are the
link lengths augmented for missing superimposed replacements by the augmentation algorithm.
The ordinate scale, in million years, is inferred from fossil evidence on ancestral splitting times
of these vertebrate lineages. The myoglobin sequences are referenced in Romero-Herrera *et al.*
(this volume) (except the pig sequence, Floch *et al., Biochimie* 55:95, 1973). *Because of an incomplete myoglobin sequence, the pig is represented by 401 residues rather than 440 as are the other
vertebrates; thus the link to pig is probably slightly underrepresented in NRs. The α- and β-
hemoglobin sequences are referenced in Matsuda (this volume), Beard and Goodman (this volume),
and Goodman *et al.* (1974, 1975).

tarsier, ceboids, gibbon, and gorilla have aspartic acid whereas cercopithecoids and chimpanzee and human have glutamic acid. Thus, to have the fewest mutations in the descent of these α sequences, the identity between the
human and chimpanzee sequences and their one-nucleotide difference
from the gorilla sequence would have to result from a forward mutation of
aspartic acid (GAU/C) to glutamic acid (GAG/A), i.e., from a derived similarity. While this would also require a convergent mutation to glutamic acid in
the cercopithecoid ancestor, two such convergent mutations would be required if either human or chimpanzee joined gorilla rather than each other.

   The genealogy depicted in Fig. 1 generally parallels paleontological evidence on mammalian and primate phylogeny. The marsupials (kangaroo and
opossum) form one monophyletic group and the 11 placental mammals form

another. Within the latter, ungulates as one large branch and primates as another join together and are then joined by dog. This agrees with certain views of paleontologists on mammalian phylogeny (e.g., McKenna, 1975) but not with an older view that had condylarths (the ungulate ancestors) and creodonts (the carnivore ancestors) as closely related sister groups. In agreement with immunological data, a monophyletic Anthropoidea branches into Platyrrhini (Atelinae) and Catarrhini and the latter into Cercopithecoidea *(Macaca)* and Hominoidea (gorilla, human, and chimpanzee).

Earlier in this investigation, a parsimony reconstruction of a species genealogy had been carried out using an extended alignment of fibrinopeptides A and B, cytochrome *c*, myoglobin, and α- and β-hemoglobin sequences. Except for the exclusion of chicken, opossum, slow loris, Atelinae, and gorilla because of missing sequence data on these animals, exactly the same species genealogy as shown in Fig. 1 was found. In particular, primates and ungulates were closer to each other than to dog. However, in this earlier reconstruction it was just as parsimonious for pig and horse to join first before uniting with the bovid branch. Additional evidence on the pig's relationship has been provided by parsimony analysis of ribonuclease sequences (work of Beintema and Fitch reported in Gaastra, 1975). The results clearly place pig closer to bovids than to horse. Thus the weight of molecular evidence agrees with the parsimony tree in Fig. 1 as well as with traditional evidence concerning the pig's cladistic position.

More details about cladistic relationships among primates are revealed by the species genealogical tree constructed from combined α- and β-hemoglobin sequences (Fig. 3 in Beard and Goodman, this volume) and especially by the individual protein or gene phylogenies considered in the next section. In the work done with combined α- and β-hemoglobin sequences, less ambiguous reconstructions were obtained as the sequences used came from a denser collection of species. In earlier work (e.g., Fig. 4 in Goodman, 1976) when a smaller number of species were represented by the combined α and β sequences (data for platypus, opossum, tarsier, and mangabey were missing), there were alternative maximum parsimony topologies for the primate region of the tree. It was just as parsimonious either for the cercopithecines to first join *Homo* or for langur (*Presbytis*) to first join *Homo* as it was for cercopithecines and colobines (langur) to first join each other before joining *Homo*. Also it was just as parsimonious for the cebid *Ateles* to first join the catarrhines as for it to first join *Cebus*. After tarsier had been added to this data set, the only parsimonious arrangement for langur was to first join it to cercopithecines. However, the ambiguity with respect to the positioning of *Ateles* still remained. It could be joined first either to catarrhines or to *Cebus* (Figs. 1 and 2 in Beard and Goodman, this volume). Finally, with the addition of mangabey, opossum, and platypus (Fig. 3 in Beard and Goodman, this volume) this ambiguity also disappeared. The only parsimonious arrangement for *Ateles* was to join it first to *Cebus*. In fact, the only ambiguity in this denser data set concerns mangabey and *Cercopithecus*. Either macaques and mangabey must be joined first before being joined by *Cercopithecus* or, alternatively, macaques and *Cercopithecus* must be joined first before being joined by mangabey. The individual gene phylogenies (next section) show that the former arrangement is the most parsimonious one for β-hemoglobin sequences and the latter is the most parsimonious for α. The evidence from comparative anatomy, chromosomes, and immunological examination of albumins and transferrins (Sarich and Cronin, this volume) indicates that the β sequences provide the truer picture of cladistic relationships among these cercopithecines.

3.2. Individual Protein Genealogies

Individual protein genealogies have been constructed by the parsimony procedure for sequences of α-hemoglobin chains, β-type hemoglobin chains, myoglobins, fibrinopeptides, carbonic anhydrases I and II, and cytochromes c. The most recent reconstructions of the globin genealogies are now even denser in primate sequences than previously. Thus, as I hope to indicate, they contain both confirmatory and new information on primate phylogeny. The genealogy of carbonic anhydrases and its bearing on primate phylogeny are discussed by Tashian *et al.* (this volume). While the cytochrome c genealogy contains only a small piece of information on cladistic relationships among primates, namely that human and chimpanzee cytochromes c are identical and diverge from rhesus monkey cytochrome c by one mutation, it contains considerable suggestive information (reviewed later in this chapter) on the role of natural selection on the rate of molecular evolution in higher vertebrates.

As no new primate species other than the 12 discussed by Wooding and Doolittle (1972) have had their fibrinopeptides sequenced, the evidence on cladistic relationships among primates from my previously constructed parsimony tree of fibrinopeptides (Goodman, 1973) need be only briefly recapitulated here. In this tree, all 12 primate species grouped together and then subdivided into a monophyletic Anthropoidea (11 species) and slow loris. In turn, Anthropoidea divided into Platyrrhini (*Ateles* and *Cebus*) and Catarrhini (Old World monkeys and hominoids). All six genera of extant Hominoidea were represented. As in the immunodiffusion divergence trees shown by Dene *et al.* (this volume), a hylobatine branch to *Hylobates* (gibbon) and *Symphalangus* (siamang gibbon) separated from a hominid branch to *Pongo* (orangutan), *Pan* (chimpanzee), *Homo*, and *Gorilla*. In turn, the orangutan branch separated from a hominine branch containing identical chimpanzee, human, and gorilla sequences. In the Old World monkey branch, *Mandrillus* and *Cercopithecus* first joined and then were joined by *Macaca* (rhesus monkey). This finding contrasts with the carbonic anhydrase parsimony tree (Fig. 1 in Tashian *et al.*, this volume) in which *Papio* (on all accounts a very close relative to *Mandrillus*) groups with *Macaca* rather than *Cercopithecus*. The fibrinopeptide and carbonic anhydrase parsimony trees agree, however, in placing *Pan* cladistically closer to *Homo* than to *Pongo*. Thus these results support the primate classification proposed from immunodiffusion data (Table 10 in Dene *et al.*, this volume) in which the African apes are removed from their traditional grouping with orangutan in the subfamily Ponginae and placed instead with man in Homininae.

3.2.1. Myoglobin Sequences

The cladistic relationships among primates which can be deduced from myoglobin sequences are well summarized by Romero-Herrera *et al.* in Fig. 2 of their chapter in this volume. The relationships depicted by this cladogram are fully confirmed by the independent parsimony analysis carried out in my laboratory (Dene and Goodman, unpublished data). Our analysis does indicate, however, that it would be equally parsimonious for the strepsirhine branch (lorisoids and *Lepilemur*) to join the condylarthran branch (ungulates and cetaceans) as to stay in the Primates. In fact, it would cost two less nucleotide replacements but in turn would require an event of gene duplication to account for the paralogous relationship between Anthropoidea myoglobins on the one hand and strepsirhine and condylarthran on the other, as well as an event of gene inactivation to account for why different myoglobin loci are expressed. This interpretation follows from accepting the molecular evidence from other sources (e.g., immunological data on serum proteins) that lorisiforms and lemuriforms are primates. Also, as already noted, the parsimony species geneal-ogies from combined sequence data (e.g., Fig. 1) indicate that Lorisiformes must be a branch of Primates.

An observation worth emphasizing is that the myoglobin trees of lowest nucleotide replacement length always depict a monophyletic Anthropoidea essentially as shown in Fig. 2 of Romero-Herrera *et al.* (this volume). The only consistent ambiguity in this region of these trees is that it is equally parsimon-ious in nucleotide replacements for baboon and macaque to emerge out of a common branch already separated from the hominoid branch (the phylogenet-ically preferred solution) as it is for these two Old World monkeys to emerge as independent branches from the catarrhine stem.

3.2.2. α-Hemoglobin Sequences

The genealogy of the α chain of hemoglobin shown in Fig. 2 was con-structed for a collection of 29 contemporary sequences in which a majority of the residues in each chain had been placed by actual sequencing procedures. The collection represents ten primates (six catarrhines, two platyrrhines, tar-sier, slow loris), seven other eutherians, marsupials (opossum and kangaroo), monotremes (platypus and echidna), birds (chicken and duck), a snake, a newt, and two teleosts (carp and *Catostomus*). The tree shown in Fig. 2 requires 630 nucleotide replacements, four more than required for the tree of lowest nucleotide replacement length found after an extensive search by the branch-swapping procedure. The only difference in topology of this latter tree with 626 NRs is that the marsupial branch must be moved into the ungulate region

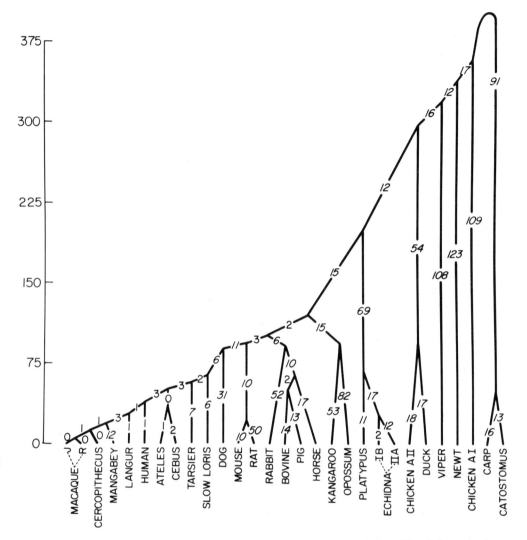

Fig. 2. Parsimony genealogy, requiring 630 NRs, for 29 α-hemoglobin chains with well-determined amino acid sequences. The ordinate scale is in million years. Link lengths are the NR numbers between adjacent ancestor and descendant sequences and are italicized when augmented for missing superimposed replacements by the augmentation algorithm. Original and corresponding augmented link lengths are as follows: 0-0, 1-1, 2-2, 3-3, 4-6, 5-7, 6-10, 7-11, 8-12, 9-13, 10-14, 11-15, 12-16, 13-17, 14-18, 15-31, 19-50, 20-52, 21-53, 22-54, 24-69, 37-82, 45-91, 56-108, 57-109, 64-123. Most sequences are referenced in Matsuda (this volume), Beard and Goodman (this volume), Hewett-Emmett *et al.* (this volume), and Goodman *et al.* (1975). In addition, there are sequences of rat (Chua *et al.*, *Biochem. J. 149*:259, 1975; Garrick *et al.*, *Biochem. J. 149*:245, 1975); opossum (Stenzel, *Nature 252*:62, 1974); platypus (Whittaker *et al.*, *Aust. J. Biol. Sci. 27*:591, 1974); echidna IIA (Thompson *et al.*, *Aust. J. Biol. Sci. 26*:1327, 1973); duck (Debouverie, *Biochimie 57*:569, 1975); viper (Duguet *et al.*, *FEBS Let. 47*:333, 1974); chicken AI (Takei *et al.*, *J. Biochem. 77*:1345, 1975).

and attached to horse. Inspection of Fig. 2 shows that in such an arrangement three branches of eutherian mammals would have to emerge at separate times from the therian stem before it reached the common ancestor of horse and Marsupialia. In other words, there would have to have been at least three events of gene duplication (GD) and three further events of gene inactivation (GI) because in the real animal species phylogeny (as shown in the overall molecular evidence, e.g., Fig. 1) the eutherians emerge from the therian stem as a monophyletic branch after diverging from the marsupial branch. Thus with respect to total mutational score (NR plus GD plus GI) the topology in Fig. 2 is the more parsimonious one. Also, no other topology with a lower total score was found. In the primate region of the α genealogy, Lorisiformes (slow loris) separates from Haplorhini and Haplorhini divides into the tarsier branch and Anthropoidea. Anthropoidea branches into Platyrrhini and Catarrhini, and the catarrhines separate into a human branch (Hominoidea) and Cercopithecoidea. In turn, the cercopithecoids split into colobines (langur) and cercopithecines (macaques, *Cercopithecus,* and mangabey). It is just as parsimonious in NRs for *Cebus* and *Ateles* to emerge as independent branches from the Anthropoidea stem as to emerge as they do in Fig. 2 from a monophyletic platyrrhine branch. No other alternative topologies of lowest NR length were found for the primate region of the tree.

Parsimony reconstructions were also carried out for an enlarged collection of 45 contemporary α-hemoglobin chain sequences, including not only the 29 used for Fig. 2 but also the sequences inferred at a majority of their residue positions by comparison of the amino acid compositions of peptide fragments to known homologous sequences. The best genealogy found for these 45 α sequences is shown in Fig. 3. Its NR length is 726. No other topology was found to have a lower total mutational score. Although 3 NRs could be saved by attaching the marsupial branch to the horse-donkey branch, this adds at least six mutations in the form of extra GDs and GIs to the total score, and thus produces a less parsimonious topology.

It is equally low in NRs for the tree shrew branch to emerge from the eutherian stem as shown in Fig. 3, or to be moved one step closer to the Primates by having it emerge from the stem of the primate-dog-rodent branch. Either way the results do not support inclusion of tree shrews in the Primates. No further NRs are added to the length of the genealogy either by switching of the dog and lemuriform (lemur and sifaka) branches or by joining them. Thus the α genealogy does not provide independent evidence for the primate placement of Lemuriformes. This α genealogy also raises questions about the phylogeny of Old World monkeys. For example, it adds 1 NR to join langur and *Colobus.* This does not agree, therefore, with the immunological data on serum proteins, which support the traditional taxonomic placement of *Colobus* with *Presbytis* (langur) and other leaf-eating monkeys in Colobinae rather than with cercopithecines. As Hewett-Emmett *et al.* (this volume) suggest, there might have been an α gene polymorphism common to the two stem populations of Old World monkeys, and the same allele might then have spread in the

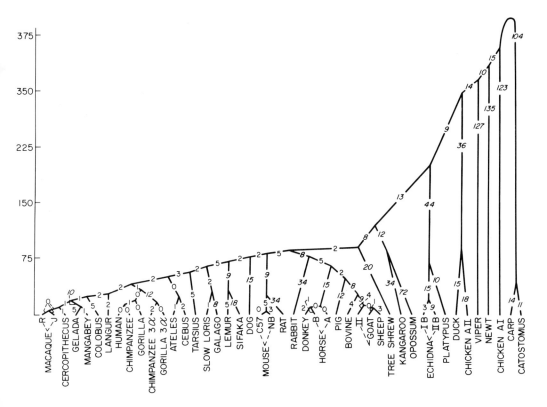

Fig. 3. Parsimony genealogy, requiring 726 NRs, for the enlarged series of 45 α-hemoglobin sequences. The additional sequences over those in Fig. 2 are referenced in Hewett-Emmett *et al.* (this volume) and Goodman *et al.* (1974). The ordinate scale is in million years. Original and corresponding augmented link lengths are as follows: 0-0, 1-1, 2-2, 3-3, 4-4, 5-5, 6-8, 7-9, 8-10, 9-11, 10-12, 11-13, 12-14, 13-15, 14-18, 16-20, 19-34, 20-36, 25-44, 38-72, 48-104, 54-123, 57-127, 65-135.

ancestral cercopithecine and *Colobus* lineages, whereas a different allele spread in the ancestral langur population.

The α genealogy does provide evidence, however, of the close cladistic relationship of mangabey *(Cercocebus)* to gelada baboon *(Theropithecus)*. The more extensive analysis of Hewett-Emmett *et al.* (this volume), which also includes sequences from *Mandrillus* and *Papio,* indicates that these four Papionini genera constitute a monophyletic branch within Cercopithecinae. Although the α genealogy does not place this baboon-like group with macaques to complete the tribe Papionini, only 1 extra NR is required to do so, i.e., to swap the positions of *Cercopithecus* and the mangabey-gelada branch.

Within the Hominoidea group, many alternative topologies were found to be just as parsimonious in NR length as that shown in Fig. 3. In particular, chimpanzee or human α could be joined to gorilla α rather than to each other, or alternatively the three hominines could form a trichotomy emerging from the same common ancestor. However, as pointed out earlier, if the gibbon α sequence were included in the analysis, the most parsimonious topology for the three hominine genera could only be the topology shown. Gibbon α was not included because only the first 31 residue positions at its N-terminal end have been sequenced (Boyer *et al.*, 1972).

The great power of the maximum parsimony procedure in revealing evolutionary relationships is worth commenting on at this juncture as this ability is well illustrated by the genealogy of α sequences. For example, the mangabey-gelada branch diverges much more from other cercopithecines than do the hominine α sequences; yet the trees of lowest NR length required that all the cercopithecines be grouped together as a monophyletic branch. A procedure based on a simple divergence model of evolution would not have captured this relationship. Similarly, the mouse α sequence diverges far less from primates than from rat, yet the parsimony procedure grouped mouse with rat. Also, the kangaroo diverges less from eutherian mammals than from opossum, but to achieve the lowest NR length the parsimony procedure required that kangaroo and opossum be cladistically closest to each other. Thus the empirical findings demonstrate that the maximum parsimony method can capture correct cladistic relationships even though the rates of evolutionary change in the diverging branches are greatly different. It is important to emphasize that these parsimony results sharply conflict with the clock model of protein evolution, which would have had such divergent branches as the opossum branch originating as paralogous gene lines from much more ancient ancestors in the genealogy.

3.2.3. β-Type Hemoglobin Sequences

As was done for α sequences, the genealogy of the β-type chain was reconstructed for a collection of only the more rigorously determined sequences and also for an enlarged collection including sequences inferred from amino acid compositions by homology with known homologous sequences. Most parsimonious genealogies for these two collections are shown, respectively, in Fig. 4 and 5. With respect to the sequences (28 of them) found in both collections, the only difference in topology between the two genealogies has to do with the positioning of the rabbit and elephant branches. In Fig. 4, these two branches emerge separately from the stem to the common ancestor of ungulates, whereas in Fig. 5 the rabbit branch emerges from the stem to the primates and the elephant branch from the eutherian stem to the common ancestor of primates, rabbit, and ungulates. These differences in topology have no effect on the conclusions which can be drawn about cladistic relationships within the order Primates.

The genealogy for the more rigorously determined β-type sequences, in contrast to that for α sequences, does not provide independent evidence against or for the primate positions of tarsier and slow loris. The genealogy shown in Fig. 4 supports the grouping of tarsier with Anthropoidea into Haplorhini and

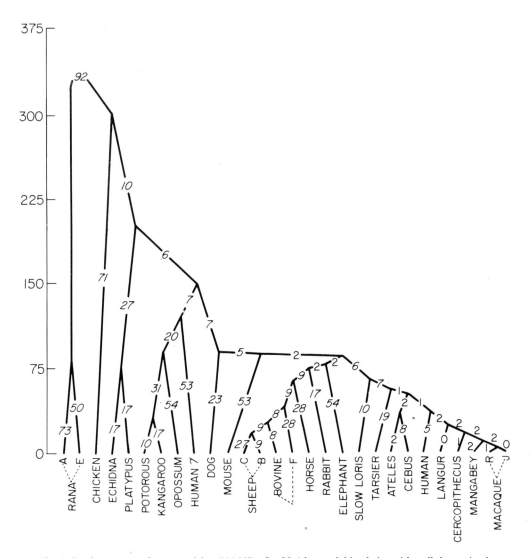

Fig. 4. Parsimony genealogy, requiring 514 NRs, for 28 β-hemoglobin chains with well-determined amino acid sequences. The ordinate scale is in million years. Original and corresponding link lengths are as follows: 1-1, 2-2, 3-5, 4-6, 5-7, 6-8, 7-9, 8-10, 9-17, 11-19, 12-20, 13-23, 15-27, 16-28, 17-31, 23-49, 26-53, 27-54, 33-71, 35-73, 43-93. Most sequences are referenced in Matsuda (this volume), Beard and Goodman (this volume), Hewett-Emmett et al. (this volume), and Goodman et al. (1975). In addition, there are sequences of elephant (Vedvick, Ph.D. thesis, University of Oregon Medical School, 1972); opossum (R. T. Jones, personal communication); platypus (Whittaker and Thompson, Aust. J. Biol. Sci. 28:353, 1975); bullfrog (Baldwin and Riggs, J. Biol. Chem. 249:6110, 1974).

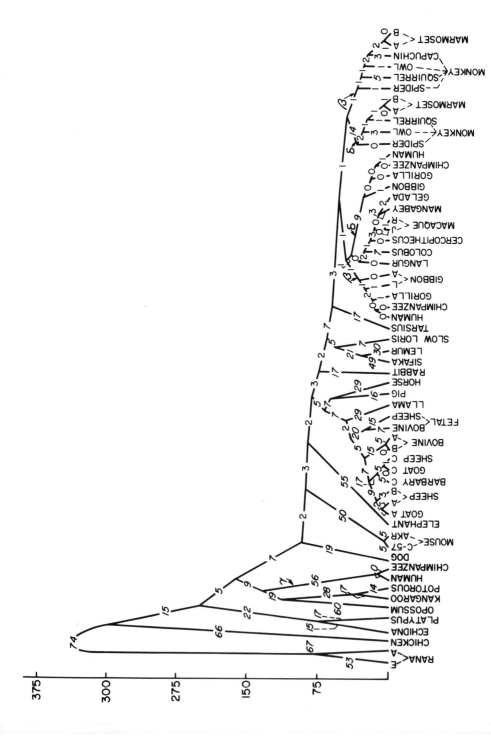

Fig. 5. Parsimony genealogy, requiring 666 NRs, for the enlarged series of 58 β-hemoglobin sequences. The additional sequences over those in Fig. 4 are referenced in Hewett-Emmett et al. (this volume) and Goodman et al. (1974). The ordinate scale is in million years. Original and corresponding augmented link lengths are as follows: 0-0, 1-1, 2-2, 3-3, 4-5, 5-7, 6-9, 7-14, 8-15, 9-16, 10-17, 12-19, 13-20, 14-21, 15-22, 17-29, 18-30, 20-49, 21-50, 24-53, 26-55, 27-56, 28-60, 34-66, 35-67, 42-74.

also the inclusion of Lorisiformes in the Primates. However, the NR length of this tree, 514, while the lowest discovered in the search for the most parsimonious topology, was also found for a number of other topologies which failed to group tarsier and slow loris specifically with other primates. For example, slow loris could be joined to rabbit and then this branch joined to Haplorhini, or alternatively the topology of Fig. 4 could be maintained except for moving the dog next to the Anthropoidea by having it join the tarsier. Of course, other molecular evidence from α chains, from combined sequences, and from immunological analyses shows that such alternative topologies would have larger total mutational scores than the genealogy in Fig. 4 in that gene duplications and inactivations would be required to position tarsier or slow loris with nonprimates. Furthermore, the genealogy of the enlarged collection of β-type sequences does support the primate status of tarsier; no topology was discovered with an NR length lower than 666, the length of the genealogy shown in Fig. 5, and no alternative topologies at this NR length placed tarsier with nonprimates, although one alternative was found which placed tarsier with the strepsirhines (lemur, sifaka, and slow loris). Only two other equally parsimonious topologies were discovered with respect to the most ancient branchings in the Primates. One placed the lemuriform branch cladistically closer to the Haplorhini (Anthropoidea and *Tarsius*) than to slow loris. The other joined the strepsirhine branch (slow loris and lemuriforms) to rabbit rather than the the haplorhine branch. Thus the parsimony reconstruction for the enlarged collection of β-type sequences, while consistent with the primate status of Lorisiformes and Lemuriformes and their grouping in Strepsirhini, does not provide additional evidence for these relationships in that the alternative topologies just mentioned were equally parsimonious.

The parsimony results on the β-type sequences always depict Anthropoidea *in toto* as a monophyletic group and Cercopithecidae as a monophyletic branch within the catarrhine division of Anthropoidea. Moreover, these β results, in agreement with the analysis of Hewett-Emmett *et al.* (this volume), always place members of the tribe Papionini together, with this tribe defined so as to include macaques as well as the more baboon-like cercopithecines. It adds 1 NR to the topology shown in Fig. 4 to switch the positions of the mangabey and *Cercopithecus* branches, i.e., place *Cercopithecus* next to *Macaca*, and it adds 3 NRs to the topology shown in Fig. 5 to move *Cercopithecus* next to *Macaca*. It also adds 3 NRs to join the mangabey-gelada branch to *Cercopithecus* rather than to *Macaca*. It may be recalled that with the enlarged collection of α sequences, although it was more parsimonious to join *Macaca* to *Cercopithecus* than to the mangabey-gelada branch, only 1 NR was added to the genealogy when the latter was done. Thus when the β genealogy is contrasted to the α genealogy the net weight of parsimony evidence depicts the Papionini as a monophyletic tribe. Although β and α genealogies differ concerning the Papionini, they both agree in placing *Colobus* closer to cercopithecines than to langur. In fact, it adds 2 NRs to the β genealogy to join *Colobus* to langur. Thus hemoglobin sequence data conflict with immunological data on serum proteins which place *Colobus* closer to langur and other colobines than to cercopithecines. Clearly, amino acid sequence data on other proteins from these animals are needed to resolve the issue.

The genealogy shown in Fig. 5 depicts two independent duplications of β-type genes in the early Anthropoidea, one in the stem catarrhines and the other in the stem platyrrhines. The hominoid δ branch is depicted as cladistically closer to catarrhine β than to the ceboid δ branch and conversely ceboid δ is depicted as cladistically closer to ceboid β than to hominoid δ; i.e., these two Anthropoidea δ branches might be paralogously rather than orthologously related. I must note, however, that it only costs 1 NR more to join hominoid and ceboid δ branches first and that this is balanced by the topology requiring 1 GD less. Thus parsimony does not tell us whether the two δ gene lines originated from separate duplications of β genes in early ancestors of platyrrhines and catarrhines or from a single duplication in early ancestors of Anthropoidea.

In both ceboid β and δ branches, the most ancient splitting separates spider monkey from the other ceboids. This supports a classification which would place New World monkeys and marmosets together in a single family, Cebidae. By excluding marmosets, the conventional Cebidae appears to be polyphyletic inasmuch as certain cebid genera are closer to marmosets than to other cebid genera.

Human δ differs from identical chimpanzee and gorilla δ chains by only 1 NR and from gibbon δ chain by 2. While the hominoid δ ancestor is separated by 6 NRs from its parent ancestor (thus demonstrating the gibbon's hominoid affinities), it is identical to the hominine δ ancester. That is, gibbon, human, gorilla, and chimpanzee δ sequences can all be originated simultaneously from a common ancestor. With respect to hominoid β sequences, there is a closer kinship between African apes and man than between these hominoids and gibbon in that 2–3 NRs separates the human-African ape ancestor from earlier hominoid ancestors from which the lines to gibbon β sequences descend. Gorilla β differs from that of the human-African ape ancestor by only one mutation, and human and chimpanzee sequences are identical to this ancestor. Thus gorilla, human, and chimpanzee lines can be originated simultaneously from the same common ancestor without costing mutations. One further piece of evidence from β-type sequences on the close relationship of *Homo* to *Pan* is that human and chimpanzee γ sequences are identical. In fact, a duplication of the γ gene in an earlier hominoid produced two paralogously related γ chains which are now found to diverge by one amino acid difference. Yet human beings and chimpanzees are identical with respect to each of these γ sequences (De Jong, 1971).

3.3. Genealogical Conclusions on Primates from Molecular Evidence

With regard to immunological data on serum proteins, flying lemur and tree shrew show strong affinities with Primates. No other molecular data exist for flying lemur, but tree shrews are better represented in that myoglobin and α-hemoglobin chain amino acid sequences have been inferred for *Tupaia*. Parsimony analysis of these sequence data favors excluding tree shrews from the order Primates. Primitive sequence homologies retained unchanged from ancient common ancestors in the Eutheria—i.e., patristic homologies—rather than derived homologies account primarily for the sequence similarities between *Tupaia* and Primates.

Within the Primates, parsimony analysis of the available amino acid sequence data fully supports the separation of Anthropoidea (Platyrrhini and Catarrhini) as a monophyletic branch distinct from Tarsioidea, Lorisiformes, and Lemuriformes. Moreover, this analysis (α- and β-hemoglobin sequences) agrees with immunodiffusion data that Tarsioidea and Anthropoidea are sister groups cladistically closer to each other than to any other primates and thus would also place them together as in Table 10 of Dene *et al.* (this volume) into the taxon Haplorhini. Inclusion of Lorisiformes and Lemuriformes in the taxon Strepsirhini as the other major cladistic division of Primates is supported by parsimony evidence from myoglobin sequences but opposed by that from α sequences. There are fragmentary DNA data which agree with the haplorhine position assigned to tarsier. From the standpoint of the divergent evolution model, the data on repetitive DNA sequences (Hoyer and Roberts, 1967) show Anthropoidea to be closer to tarsier than to slow loris, galago, lemur, tree shrew, mouse, hedgehog, and chicken (the other non-Anthropoidea species compared). They further show Anthropoidea to be closer to tree shrew than to mouse, hedgehog, and chicken, but as indicated above this could be due to patristic similarities, i.e., to the retention of primitive DNA sequences.

The divisions within the Anthropoidea proposed from immunodiffusion data are well supported in most instances by parsimony analysis of the available amino acid sequence data and also by divergence analysis of the data on repetitive DNA sequences (Hoyer *et al.*, 1964, 1965; Hoyer and Roberts, 1967; Martin and Hoyer, 1967) including satellite DNA (Jones, this volume) and on single-copy DNA (Kohne *et al.*, 1972; Hoyer *et al.*, 1972). To recapitulate, Anthropoidea divides into infraorders Platyrrhini and Catarrhini. Platyrrhini embraces just one superfamily, Ceboidea. Catarrhini subdivides into superfamilies Cercopithecoidea and Hominoidea. Cercopithecoidea contains one extant family, Cercopithecidae, with the immunological evidence dividing this family into the subfamilies Cercopithecinae and Colobinae and with the sequence evidence supporting the monophyly of Cercopithecinae but questioning that of Colobinae. Hominoidea splits into Hylobatidae and Hominidae. The subfamily Hylobatinae (*Hylobates* and *Symphalangus*) is the only extant taxon in Hylobatidae. The family Hominidae is redefined to include members of Pongidae as well as man. The traditional Pongidae as the family of great apes is eliminated. Instead, Hominidae divides into subfamilies Ponginae for *Pongo* and Homininae for *Pan*, *Homo*, and *Gorilla*. In general, the molecular data do not clearly show which two of these three hominines share the most recent common ancestor. However, the data on single-copy DNA and α-hemoglobin sequences, along with chromosome data (see Vogel *et al.*, this volume), suggest that *Homo* and *Pan* share the most recent common ancestor.

To have a truly effective genealogical description and classification of the Primates, the molecular findings need to be integrated with fossil evidence. The latter can provide a time scale, which in turn can serve as a yardstick for designating the hierarchical level of the groups to be described in the classification. It would help immensely if primate paleontologists joined with molecular evolutionists in developing such a description of the Primates. Certainly it is important to know whether *Ramapithecus* is cladistically closer to man than to chimpanzee and gorilla and if so to represent this in the classification. I submit that it is also important to find out if certain fossils already separated from the ancestry of orangutan could serve as common ancestors for the extant African apes and man, and, if such fossils are found, to so describe them in the classification.

In advocating a classification which is based as far as possible on cladistic relationships and, to the extent known, the times of ancestral separations, and which give no priority to the level of anagenetic advance, I am not downgrading the importance of anagenenesis in the evolution of man; rather, I am seeking an objective framework for viewing anagenesis. When the classification which is to serve as this framework offers no more than a genealogical description, there is less room for misinterpretations of what the classifications says, and no room for the subjective bias of giving taxonomic weight, as traditional classifications do, to considerations concerning an organism's ecological niche and type of adaptive specializations. The relationships determined by genealogy are genetic relationships. They reflect the most stable and fundamental qualities of organic beings and thus provide the real basis for a natural or truly evolutionary system of classification.

Although generally primate paleontologists have not been attracted by the idea that a formal zoological classification should be no more than a description of genealogy, they do agree on the necessity of having charts or cladograms which depict genealogy. It is encouraging that the cladistic branchings of Primates deduced from molecular data agree by and large with the picture of primate phylogeny derived by paleontologists from the comparative morphologies of fossil and living forms. This makes it possible to give dates in millions of years before the present (m.y. BP) to branching points on the molecular genealogies and to then calculate the rates of genetic change in earlier and later phylogenetic stages. In turn, the rate estimates, especially on the stages in the descent to man, provide pieces of information on the effects of anagenesis at the molecular level. I would now like us to consider this information.

4. Some Molecular Aspects of Anagenesis in Primate Phylogeny

4.1. Accelerations and Decelerations

Many examples of differences in evolutionary rates are found in the molecular genealogies. The simplest examples, a few of which were noted in the discussion of α-hemoglobin sequences, tell us mainly if certain lineages underwent more change than others since the time they shared a common ancestor. The more complex examples which can measure accelerations and decelerations depend on specified divergence times, i.e., on the actual dates in millions of years before the present assigned to ancestral nodes. These examples are potentially the most informative because they relate to a progression of different phylogenetic periods during descent and can thus point to the existence of evolutionary trends. The fact that the fossil evidence used to date particular branch points may be inadequate requires, of course, that findings suggestive of changes in evolutionary rates be cautiously interpreted. My feeling, however, is that taken as a whole the paleontological record provides a reasonable basis for imparting a time dimension to the phylogenies depicted in the parsimony genealogical reconstructions. The only other feasible approach is to calculate the divergence times between lineages according to the clock model of molecular evolution, but this model predetermines the nature of the findings for the process being investigated. Obviously, the divergence times calculated by a clock model can only conform to the expectation that divergences accumulate on the average at a constant rate. In the absence of an independent check from the fossil record, the results obscure any real trends that might exist toward accelerations or decelerations in rates.

What are the trends that come to light when we do use paleontological evidence to date divergence times? Let me refer first to the analysis of Fitch and Langley (this volume), who use the paleontological evidence to date divergence times between sequences from different species. They find, after combining sequences of a number of different proteins, an indication that a crude molecular clock of sorts does exist for the period of eutherian phylogeny, with one glaring exception, the slowdown in higher primates. The species genealogy which I constructed from combined myoglobin and α- and β-hemoglobin sequences (Fig. 1) leads to a similar conclusion in that clock calculations applied to certain regions of this genealogy such as the ungulate region would produce dates for divergence times which agree with paleontological evidence. This is also true for the primate region, except for the hominine area where much too recent dates for splitting times would be calculated. For example, if 65 m.y. BP were taken as the splitting time for the lorisiform and anthropoid branches, clock calculations would place the human-chimpanzee split at only about 1 m.y. BP. If a different topology for the hominines were used in the parsimony genealogical reconstruction so that a gorilla-chimpanzee branch separated from the human branch or so that the three lineages originated simultaneously from the hominine ancestor, there would be 3, 1, and 0 NRs, respectively, in the gorilla, chimpanzee, and human lines. In this case, the clock calculations would still place the human-African ape split no earlier than at about 3 m.y. BP. With fossil specimens of *Homo* now going back to close to 3 m.y. BP and *Australopithecus* to 5.5 m.y. BP, the human-chimpanzee or human-African ape split could hardly have occurred more recently than about 6 m.y. BP. If Simons (this volume) and Walker (this volume) are right about *Ramapithecus,* this split could not have occured until about 15 m.y. BP at the latest. The conclusion, of course, is that in all likelihood a marked deceleration in globin sequence evolution occurred in the final stages in the descent of *Homo*. It is worth noting that there was much less globin evolution in primates in the genealogy of Fig. 1 than in ungulates or in marsupials. For example, from the common bovine-human ancestor 39 NRs can be counted on the line to man compared to 71 NRs on the line to bovine. Similarly, from the opossum-human ancestor 65 NRs occurred on the line to man compared to 128 NRs on the line to opossum.

As indicated above for the Primates, the state of homoplasy in a region of a genealogical tree (i.e., of equal divergencies of lineages from their common ancestor) offers no assurance that clock calculations will produce correct splitting times. A striking example of how fallacious the times calculated by linear extropolation can be, even under conditions of homoplasy, is provided by the genealogical tree of carbonic anhydrases (Fig. 1 in Tashian *et al.*, this volume). The five carbonic anhydrase II lineages (two bovids, three catarrhine primates, and rabbit) all diverge to about the same degree from the common eutherian carbonic anhydrase II ancestor. Yet clock calculations place the human–Old World monkey split at only 2–3 m.y. BP, a ridiculously recent date especially as paleontological evidence places this split in the range of 30–40 m.y. BP. Clearly, there was a great acceleration of carbonic anhydrase II evolution on the stem line from the early eutherian ancestor to the catarrhine (cercopithecoid-hominoid) ancestor followed by a profound deceleration. This is the pattern one expects to see in protein evolution when there is first intense positive natural selection for new functions or improvements of function and then intense stabilizing selection to preserve the functional advances. The latter deceleration balances the earlier acceleration so that the overall amount of evolutionary change is comparable to that seen in other lineages where extreme rate changes from shifts in types of natural selection might not have occurred or, if they had, might not be evident to the observer.

In general, I would argue that just as morphological evolution speeds up during the earlier stages of an adaptive radiation and then slows down in each lineage after it acquires its unique specializations so too does molecular evolution accelerate and decelerate, respectively, in the earlier and later stages of the cladogenetic process. I would have the provision, however, that in the lineages undergoing anagenesis, and thereby giving rise to organisms with more complex internal organizations, molecular evolution would tend to decelerate at an increasingly steeper rate and in a more pervasive manner after each major advance. To see if such a pattern of molecular evolution actually occurred, we need to look at vertebrate phylogeny as a whole, not just at Primates and mammals. An opportunity to do this is provided by the genealogies for cytochrome *c* sequences and for the globin group of sequences. We can start with the cytochrome *c* phylogeny, which in fact encompasses the past billion or so years of metazoan evolution.

4.1.1. Cytochrome c Evolutionary Rates

The observations I want to call attention to come from an analysis of rates of cytochrome *c* evolution carried out in collaboration with Holmquist in which we estimated mutational change by both stochastic and augmented maximum parsimony methods (Moore *et al.,* 1976). Both methods demonstrated that cytochrome *c* evolved in a nonuniform manner over geological time and far more rapidly than previously estimated. The manner in which the nonuniform evolution occurred is clearly revealed in the parsimony reconstruction of cytochrome *c* phylogeny (Fig. 6). Rates were 3–4 times faster in protozoans, fungi, and plants and in most cold-blooded vertebrates than in birds and mammals (Table 1). In the descent of the stem line to the warm-blooded vertebrates from the ancient eukaryote ancestor at about 1000 m.y. BP to the amniote (bird-mammal) ancestor at about 300 m.y. BP cytochrome *c* appears to

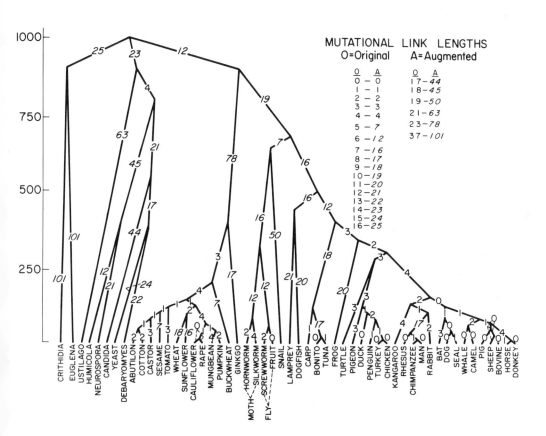

Fig. 6. Parsimony genealogy, requiring 521 NRs, for 53 cytochrome *c* amino acid sequences. The sequences are referenced in Dickerson and Timkovich (Table IX, p. 201, 1974). The ordinate scale is in million years. As in the other figures, it is based on paleontological views concerning the ancestral separations of the organisms from which the sequences came. Since I am unaware of fossil evidence for branch points among insects, among angiosperms, among fungi, and between *Crithidia* and *Euglena,* the times of these branch points were guessed at from the magnitudes of the link lengths in these regions of the genealogical tree and by interpolation between the points which were placed on the basis of paleontological views.

Table 1. Evolutionary Rates of Descending Lineages of Cytochromes c^a

Evolutionary period	Age (m.y. BP)	Nucleotide replacements per 100 codons per 10^8 yr
Eukaryote uni-multicell ancestor to protozoans	1000–0	12.2 (12.2–12.2)
Eukaryote uni-multicell ancestor to fungi	1000–0	8.6 (8.2–9.0)
Eukaryote uni-multicell ancestor to plants	1000–0	10.3 (9.7–11.5)
Eukaryote uni-multicell ancestor to metazoans	1000–0	7.4 (6.4–9.2)
Eukaryote uni-multicell to invertebrate-vertebrate ancestor	1000–680	9.4
Invertebrate-vertebrate ancestor to invertebrates	680–0	5.9 (5.0–8.1)
Invertebrate-vertebrate ancestor to vertebrates	680–0	6.6 (5.4–9.1)
Invertebrate-vertebrate to vertebrate ancestor	680–500	8.6
Vertebrate ancestor to anamniotes	500–0	7.6 (6.0–9.3)
Vertebrate ancestor to amniotes	500–0	5.3 (4.3–8.2)
Vertebrate to amniote ancestor	500–300	8.3
Amniote ancestor to turtle	300–0	1.9
Amniote ancestor to birds	300–0	3.0 (2.3–3.2)
Amniote ancestor to mammals	300–0	3.6 (1.6–8.1)
Amniote to therian ancestor	300–120	3.2
Therian to catarrhine primate ancestor	120–35	20.5
Catarrhine primate ancestor to catarrhines	35–0	1.8 (0.0–2.8)

[a] Evolutionary rates as nucleotide replacements per 100 codons per 10^8 years were calculated from the AD values on the links in the cytochrome c genealogical tree in Fig. 3 for the appropriate evolutionary periods, using a time scale based on paleontological views. When there is more than one lineage in an evolutionary period, the evolutionary rate shown is the average of the individual rates of the different lineages and the numbers in parentheses show the range in rates from slowest-evolving to fastest-evolving lineage. The approximate date of 1 billion years ago for the ancestral splitting of metazoan, plant, fungi, and protozoan branches follows from the views of Schopf *et al.* (1973) and Cloud (1974) that the first evidence of unicellular eukaryotes (interpreted as asexual with poorly developed mitotic apparatus) occurs in the fossil record at about 1.3 billion years ago. Moreover, Schopf (personal communication) indicated that the type of eukaryote which could have been the common ancestor of protozoans and multicellular eukaryotes probably did not exist before a billion years ago. The date of 1000 m.y. BP for these eukaryote splittings may be too ancient. Knoll and Barghoon (1975) in reassessing the evidence for Precambrian eukaryotic organisms conclude that "In short, there is no good evidence for the presence of eukaryotes in Bitter Springs cherts. Similarly, all reports of older eukaryotes do not withstand critical examination. It would be hazardous for us to state that eukaryotes did not exist 900 million years ago, but if they did, their remains have yet to be found. Alternatively, multicellularity as evidenced by the several known Ediacaran faunas may have evolved quite rapidly following the origin of the nucleated cell. That is, eukaryotic cells may not have existed until very near the end of the Precambrian." The approximate date of 680 m.y. BP as an upper limit for the ancestral arthropod-mollusc-chordate branch point follows from the views of Schopf (personal communication) and Cloud (1974). The dates of 500, 300, and 120 m.y. BP for the basal ancestors of vertebrates, amniotes (reptiles-birds-mammals), and therian mammals (marsupials and placentals), respectively, are based on views of Romer (1966) and Young (1962). In view of the topology of the mammalian region of the cytochrome c genealogy (Fig. 6), the date of 120 m.y. BP for the marsupial-placental split is taken from the ancestral node from which kangaroo and the rabbit-primate branch descend. The sequence at this node must therefore be considered either paralogously related to the other mammalian branch of bat, carnivores, cetaceans, and ungulates or related by way of an allelic polymorphism which predates the marsupial-placental split.

have evolved at the fairly steady rate of about 8–9 nucleotide replacements per 100 codons per 10^8 years (8–9 NR%), but then there was an abrupt slowing of rates to an average of about 3 NR% in the further descent to the present-day birds and mammals. In the stem line to the higher primates, however, a tremendous acceleration of rates occurred. From the amniote ancestor at about 300 m.y. BT to the therian (kangaroo-primate) ancestor at about 120 m.y. BP, the rate held at about 3 NR%. Then from this ancestor to the catarrhine (rhesus monkey–hominine) ancestor at about 35 m.y. BP cytochrome c evolved at the exceptionally fast rate of about 20 NR%. After this acceleration, a profound deceleration set in, with the higher primate lineages evolving at a rate of only 1.5–2 NR%.

For years, the marked sequence divergence between higher primate and other mammalian cytochromes c has been a mystery. It was usually considered an example of unchecked random drift of selectively neutral mutations in that no functional significance could be attributed to the sequence differences. This is no longer true. Ferguson-Miller *et al.* (1976) have recently found a striking species difference between cytochromes c of higher primates and other mammals in ability to interact with cytochrome oxidase. If the heart mitochondria used as the source of the oxidase are from a nonprimate mammal, higher primate cytochromes c have a very low electron transfer activity and inhibit the activity of nonprimate mammalian cytochromes c. However, if the heart mitochondria are from primates, primate cytochromes c are highly reactive and nonprimate cytochromes c are somewhat less reactive. In view of these findings, the most likely explanation for the accelerated rate of cytochrome c evolution in the early primate lineage is that it resulted from positive natural selection. Stabilizing selection then acted in the later primates to preserve the functional adaptations which were achieved during the earlier burst of change.

4.1.2. Globin Evolutionary Rates

A comparable pattern of fast rates in early cold-blooded vertebrates and of slow rates in later warm-blooded vertebrates is observed in globin phylogeny. Again the principal rate variations observed on using the maximum parsimony approach (Goodman *et al.*, 1975) are supported by the stochastic method (Holmquist *et al.*, 1976). Accelerated α and β evolution occurred in lineages of cold-blooded vertebrates, marsupials, and early placental mammals, while decelerated evolution occurred in the two lineages leading to birds and mammals and was especially pronounced in the higher primates. The stochastic method, in particular, shows that rates were much higher in the reptilian, amphibian, and teleostan branches than in the lineages to warm-blooded vertebrates.

With regard to the rapid rate of amphibian globin evolution, the parsimony genealogy of β sequences (Figs. 4 and 5) shows the divergence between two frogs in the same genus, *Rana esculenta* and *Rana catesbeiana,* to be much larger than between any two orders of placental mammals. Inasmuch as the basal members of the order Anura first appear in the fossil record in the mid-Jurrasic, i.e., about 150 m.y. BP, it is unlikely that one of its later branches, the genus *Rana,* is older than the Eutheria, the subclass for orders of placental mammals. Indeed, as far as I can tell from reading Romer (1966) *Rana* fossils are no older than the Eocene, whereas eutherian fossils go back into the Cretaceous. Thus we can calculate a rate of globin evolution in *Rana* to compare with the mammalian rate by assuming the split between the European form *R. esculenta* and the American form *R. catesbeiana* occurred in the range of 60 m.y. BP or so. This yields an average rate of evolution for the two *Rana* β sequences of about 72 NR% compared to the average placental mammalian rate of about 29 NR% in the genealogies depicted in Figs. 4 and 5. These findings agree with the view of Wilson *et al.* (1974) that for much more morphological or organismal evolution in mammals there is much less protein evolution than in frogs. The findings, however, do not support Wilson's further view that protein sequences evolve at the same rate in mammals and in frogs. Rather, it appears that sequence evolution is much slower in mammals.

The high rate of globin evolution in *Rana* is in the range of the most accelerated rates, the highest of which goes over 100 NR%, observed in early vertebrates on the stem lineage leading toward the ancestral stock of amniotes. As Matsuda (this volume) points out in reviewing our studies of globin evolution, these accelerated rates occurred during the evolutionary transition of hemoglobin from a monomeric protein to a sophisticated allosteric tetramer. Our most important finding was that during this transition the globin sequence sites which evolved at the most accelerated rate were precisely the sites responsible for the advance in functions, whereas after the transition these sites hardly changed at all in the hemoglobins of warm-blooded vertebrates. There seems little question, therefore, that the initial fast rates were due to positive natural selection and the subsequent slow rates to negative or stabilizing selection.

The present α- and β-hemoglobin genealogies (Figs. 2–5) are denser in sequences and more broadly represented in amniote vertebrates than the previous globin genealogy (Goodman *et al.*, 1975). Thus we can be more certain that the rates of globin evolution estimated in these amniote lineages approach the actual rates which occurred. What stands out are the following: (1) an abrupt slowing of the rate of β and α evolution so that between the amniote and eutherian ancestor (300–90 m.y. BP) it was only 8–10 NR%; (2) a four- to fivefold acceleration of rates between the eutherian and primate ancestors (90–65 m.y. BP) and then a deceleration which became particularly pronounced between the catarrhine ancestor and man (35–0 m.y. BP), especially in the α lineage where the rate fell to only about 2 NR%; (3) a much faster rate of β and α evolution in metatherians or marsupials than in eutherians, on following the lineages from their common therian ancestor to the present; (4) from the earlier basal mammalian (prototherian-therian) ancestor at about 200 m.y. BP to the present, a faster rate of α evolution in the descent of monotremes than the descent of eutherians, but a slightly slower rate of β evolution; and (5) between about 300 m.y. BP (the times of the basal amniote stock ancestral to birds and mammals) and the present, a rate of about 15 NR% for both α and β evolution in the lineage leading to birds and also in the lineage leading to eutherians. This contrasts with the preceding early vertebrate rate (calculated from Table 2 in Goodman *et al.*, 1975) which was in the range of 70 NR% between 500 and 300 m.y. BP. This almost 5 times faster rate occurred as the mutations selected produced hemoglobins which were more efficient at delivering oxygen to the respiring tissues of more mobile animals with larger bodies.

We have seen that fast evolutionary rates in the early vertebrates for cytochrome *c* and for α- and β-globins were followed by slow rates in the amniote lineages leading to birds and mammals. Myoglobin apparently also shows this pattern (e.g., compare the large number of NRs in the vertebrate lineage to therian myoglobins in Fig. 1 of Goodman *et al.*, 1975, to the much smaller number between birds and mammals in Fig. 2 of Romero-Herrera *et al.,* this volume). Dehydrogenases also appear to show deceleration of evolutionary rates in higher vertebrates (Fig. 6 in Rossmann *et al.,* 1974). These findings support the idea that the complex level of molecular coadaptations reached in higher vertebrates imposes greater selective restraints on the mutations which can be tolerated (Goodman, 1961, 1963*b*, 1965; Goodman *et al.,* 1971, 1975). For the idea to have merit, however, it should be able to predict the circumstances under which the postulated tendency toward slowing rates would be interrupted by periods of accelerated evolution. We have found that there were such periods during the radiations of placental mammals, and in two cases the cause of the faster rates could be traced directly to positive natural selection. One is the case already noted of exceptionally rapid cytochrome *c* evolution in the lineage from the basal eutherian ancestor to the higher primate ancestor. The other case dealing also with an exceptionally rapid rate of evolution, that of α-hemoglobin in the nonmacaque Papionini lineage, is described by Hewett-Emmett *et al.* (this volume). No doubt, more such cases will be found. It may be that positive natural selection acting on organisms is the main determinant of most periods of accelerated change. It would follow that a consequence of such selection is that organisms in certain lineages acquire increasingly more complex internal organizations and that eventually this shifts the balance of selection forces back to the stabilizing kind. Then the restraints on accepting mutations are more stringent (and rates of molecular evolution slower) than in any previous time.

I would now like in the concluding discussion to explore the ramifications of this idea.

4.2. Interdependence in Anagenesis of Organismal and Molecular Levels of Evolution

The emphasis Zuckerkandl (this volume) places on the accelerated evolutionary rates produced by anagenesis makes a great deal of sense because the organisms at the forefront of anagenetic advance are the most capable of invading new environments and serving as the source of new adaptive radiations. Thus these organisms tend to become subjected to intense positive natural selection, and, as we have seen, such selection produces accelerated evolution at both the organismal and the molecular level. Is there any reason, therefore, to expect a long-run trend toward deceleration of evolutionary rates? I believe there is. It is simply that the very success of natural selection in improving the fitness of competing organisms must, sooner or later, decrease the proportion of mutations which are advantageous and increase the proportion which are detrimental in each surviving and competing lineage. Let us consider the fate of mutations once the way was opened by the evolution of the genetic code for the origin of classes of proteins. In these early times, whenever mutations first produced a potentially useful protein, numerous possibilities for improving the protein's functions must have existed. Thus mutations in the gene coding for the protein would have often been advantageous and selected for, but after the protein had been perfected a much larger proportion of mutations would have been detrimental and selected against. Selection, of course, would not have acted on this or that protein in isolation from the others, but rather would have acted to coadapt the structures of many proteins for a cooperative execution of functions. As cooperative functioning depends on interactions between sites on the folded polypeptide chains constituting the globular proteins, natural selection over time would have tended to increase the density of stereospecific recognition sites possessed by each protein.* This is but another way of saying that organisms evolved to have more highly organized and complex internal environments.

It is notable that the same coding units for amino acids occur universally in the genes of all present-day forms of life. This means that once the genetic code had evolved in a common ancestor of bacteria and human beings about 3 billion years ago, natural selection preserved it against evolutionary change. It can be argued, as Tomkins (1975) does in his provocative article on metabolic codes, that biological networks which interconnect a number of crucial cellular processes tend to attain evolutionary stability, because mutations damaging one component of a system have pleiotropic (i.e., multiple) effects which might imperil the entire organism. Anagenesis intensified the threat from mutations, because the means for making organisms more independent from the external environment was through increases in the number of integrated internal molecular circuits for regulating physiological processes. It would follow that as the hierarchy of circuits became more complex a growing proportion of components within them would tend to become impervious to evolutionary change. Applying this reasoning to the multiple interactions among proteins on which physiological processes depend, the eventual trend toward a slowdown in rates of protein evolution is predicted.

A paradoxical situation emerges. An anagenetically higher species can invade a broader range of environmental zones then a lower species specialized for a particular niche and yet the members of the higher species, because of their greater mobility and greater degree of homeostasis, will, to use the terminology of Levins (1968) and Selander and Kaufman (1973), experience their surroundings as a succession of many interconnected conditions, i.e., as a "fine-grained" environment, whereas the members of the lower species will experience their more limited surroundings as sets of alternative, disconnected conditions, i.e., as a "coarse-grained" environment. The two kinds of species will pursue different adaptive strategies, with selection more strongly favoring

---

*Zuckerkandl (1975) has coined the apt term "functional density" to denote the proportion of sites in a protein involved in specific functions; the greater the functional density, the higher the proportion of such sites.

polymorphic genotypes in the coarse-grained, uncertain environment. A higher species will tend to be a K-selected one, whereas a lower species will tend to be an R-selected one. The distinction is that K-selection is for superiority in stable, predictable environments, whereas R-selection is for survivability by rapid rates of population increase in short-lived environments (Wilson, 1975). It is easy to envision that the R-selected species will tend to undergo more rapid protein evolution than the K-selected one. Take the effect of temperature on the rate of activity of enzymes. Clearly, cooling or warming trends in the environment will have much more drastic effects on the rate of evolution of enzymes in R-selected poikilotherms (the "cold-blooded" lower vertebrates) than in K-selected homeotherms (the "warm-blooded" higher vertebrates). Salthe (1975) describes what might be expected for the cold-blood species.

> Suppose we have a locus producing an enzyme in a poikilotherm: the environment is warm; the rate of activity of the enzyme is at an optimum. Now the environment becomes colder and the rate of activity of the enzyme is reduced to a suboptimum level. Any mutation that is capable of restoring the rate of activity to an optimal level is favored provided it has no untoward pleiotropic effects. Since more than one allotype is capable of showing the same kinetic rate, there are several to many possible, selectively equivalent alleles that can solve the adaptive problem. Furthermore, since altered rates of activity of other enzymes can compensate for a reduced rate in our enzyme, the adaptive solution is not restricted to mutations at a single locus. Later the environment becomes warmer again: now the enzyme in question has a superoptimal rate that disrupts homeostasis. Will a back mutation be most likely to restore the old allele? Or is it not more reasonable to invoke a more probable forward mutation to still a third allele whose product will be phenotypically like the original warm-adapted allotype? (It need not be exactly alike since other aspects of the intracellular milieu will have changed in the meantime.) We may observe this locus through a long period of time during which the environmental temperatures fluctuate and never once see a back mutation fixed so that, while the environment oscillates, the gene steadily becomes more and more different from its original, ancestral form—that is, it evolves.

I would add the comment that much slower rates of enzyme evolution would be expected in homeotherms as they have well-developed homeostatic mechanisms for adjusting physiologically to oscillations of the external environment. Perceiving their environment in a fine-grained manner, they would be subjected to less selection pressure for continual changes in enzyme structures. Positive natural selection clearly acted in a more fundamental way on the organisms in the mainstream of anagenesis. It acted to expand and perfect the physiological homeostatic mechanisms which give a measure of freedom to the individual from capriciousness in the environment.

A lineage of shrewlike primitive placental mammals which existed in the middle to late Cretaceous proved to be the source of the last great (and perhaps most far-reaching) adaptive radiation in vertebrate history and also the source of a new and still rising wave of anagenesis. Undoubtedly, as diverse environmental zones were invaded in this radiation, positive natural selection acted to bring about new adaptations in the emerging clades. Thus considerable evolution in regulatory DNA and in structural genes coding for proteins must have occurred, but according to my model less than in previous vertebrate radiations because a larger core of coadapted molecular specificities than existed in earlier vertebrates had already evolved in the basal eutherians before the radiation commenced.

The phyletic line in the Eutheria which primarily concerns us is the one which eventually gave rise to human beings. We have reviewed evidence that in this line, at the stages of late Cretaceous Eutheria and early Tertiary Primates, molecular evolution accelerated, at least in the genes for cytochrome c, hemoglobin, and carbonic anhydrase II. The cytochrome c results (Ferguson-Miller *et al.*, 1976) provide the clue that a major anagenetic advance at the biochemical level involving oxidative metabolism was taking place in this stem line to the higher primates. Another clue is that a shift from synthesis of A or skeletal muscle-type lactate dehydrogenase (LDH) to B or heart-type LDH was taking place in the tissues of the more advanced haplorhine primates and was especially pronounced in the neocortex (Goodman *et al.*, 1969; Koen and Goodman, 1969). The shift to synthesis of heart-type LDH favored more efficient and sustained energy metabolism in brain tissue. At the morphological or organismal level, profound changes were also occurring such as marked increases in the size and complexity of the central nervous system and advances in the mechanisms of placentation and gestation toward more effective nourishment of the fetus during an extended period of prenatal life. Such organismal changes might well have depended on an increase in the functional density of the proteins of the evolving anthropoids, i.e., on selection for further subtle coadaptive changes in a variety of proteins. The direction of anagenesis in the hominoids led to a marked prologation of generation times and a commensurate slowing of the rate of division in the germ cells. This, in turn, according to the analysis of Vogel *et al.* (this volume), slowed the per year rate of newly occurring mutations. Thus the present decelerated rates of protein evolution in hominoids can be attributed both to the slower mutation rate and to an intensification of stabilizing selection (possibly abetted by maternal immunological selection against some deviant fetal allotypes). It is notable that in the present wave of anagenesis the human species has radiated to all corners of the planet but is not in the process at the genetic level of splitting into separate groups.

5. Editorial Note

Two new pieces of molecular evidence for the genealogical grouping of *Pan* and *Gorilla* with *Homo* in the ape subfamily Homininae have now been reported. A paper in *Nature* by Romero-Herrera and his coworkers (1976) describes the amino acid sequence of orangutan myoglobin and its relationships, as revealed by the method of parsimony, to human, chimpanzee, gorilla, and gibbon myoglobin sequences. The most parsimonious reconstructions place the African ape myoglobin sequences closer to human myoglobin than to either gibbon or orangutan myoglobin. The orangutan myoglobin sequence, in fact, separate from the chimpanzee, gorilla, and human sequences even before the gibbon sequence do, a rather enigmatic finding in view of most other molecular evidence which places the orangutan closer to the African apes and man.

The other new piece of molecular evidence deals with the nonrepetitive cellular DNAs of the various hominoids and is reported by Benveniste and Todaro (1976) in the same issue of *Nature* as the myoglobin evidence. Benveniste and Todaro (1976), confirming and extending earlier work of Kohne *et al.* (1972) and Hoyer *et al.* (1972), find that the evolutionary tree based on degrees of nucleic acid sequence homology places man, chimpanzee, and gorilla much closer to one another than to orangutan and closer to orangutan than to gibbon or siamang. The extent of nucleic acid sequence divergence among chimpanzee, man, and gorilla is about one half that between any of these three Homininae and orangutan, and is slightly less than that found between the tribes Papionini and Cercopithicini within the old world monkey subfamily Cercopithecinae.

ACKNOWLEDGMENTS

I wish to thank Dr. G. William Moore for stimulating discussions of relevant issues, Mr. Walter Farris and Miss Elaine Krobock for expert assistance, and Drs. E. O. Thompson and R. T. Jones for providing me with information on several globin amino acid sequences before publication. The research from my laboratory is supported by NSF grant GB 36157.

6. References

Benveniste, R. E., and Todaro, G. J., 1976, Evolution of type c viral genes: evidence for an Asian origin of man. *Nature* **261**:101.

Boyer, S. H., Noyes, A. N., Timmons, C. F., and Young, R. A., 1972, Primate hemoglobins: Polymorphisms and evolutionary patterns. *J. Hum. Evol.* **1**:515–543.

Cloud, P., 1974, Evolution of ecosystems, *Am. Sci.* **62**:54–66.

De Jong, W. W. W., 1971, Chimpanzee foetal haemoglobin: Structure and heterogeneity of the γ-chain, *Biochim. Biophys. Acta* **251**:217–226.

Dickerson, R. E., and Timkovich, R., 1974, in: *The Enzymes* (Oxidation-Reduction Volume: Cytochrome c), Part 11 (P. D. Boyer, ed.), Academic Press, New York.

Ferguson-Miller, S., Brautigan, D. L., Chaviano, A. R., and Margoliash, E., 1976, Kinetic control of electron transfer in primate and nonprimate cytochromes c, *Fed. Proc.* **35**:1605.

Fitch, W. M., 1970, Distinguishing homologous and analogous proteins, *Syst. Zool.* **19**:99–113.

Gaastra, W., 1975, Giraffe pancreatic ribonuclease, thesis, University of Groningen, Netherlands.

Goodman, M., 1961, The role of immunochemical differences in the phyletic development of human behavior, *Hum. Biol.* **33**:131–162.

Goodman, M., 1962, Immunochemistry of the Primates and primate evolution, *Ann. N.Y. Acad. Sci.* **102**:219–234.

Goodman, M., 1963a, Serological analysis of the systematics of recent hominoids, *Hum. Biol.* **35**:377–436.

Goodman, M., 1963b. Man's place in the phylogeny of the primates as reflected in serum proteins, *Classification and Human Evolution* (S. L. Washburn, ed.), pp. 204–234, Aldine, Chicago.

Goodman, M., 1965, The specificity of proteins and the process of primate evolution, *Protides of the Biological Fluids—1964* (H. Peters, ed.), pp. 70–86, Elsevier, Amsterdam.

Goodman, M., 1973, The chronicle of primate phylogeny contained in proteins, *Symp. Zool. Soc. London* **33**:339–375.

Goodman, M., 1976, Protein sequences in phylogeny, *Molecular Evolution* (F. J. Ayala, ed.), pp. 141–159, Sinaur Associates, Sunderland, Mass.

Goodman, M., Barnabas, J., Matsuda, G., and Moore, G. W., 1971, Molecular evolution in the descent of man. *Nature (London)* **233**:604–613.

Goodman, M., and Moore, F. W., 1971, Immunodiffusion systematics of the primates. I. The Catarrhini, *Syst. Zool.* **20**:19–62.

Goodman, M., Syner, F. N., Stimson, C. W., and Rankin, J. J., 1969, Phylogenetic changes in the proportions of two kinds of lactate dehydrogenase in primate brain regions, *Brain Res.* **14**:447–459.

Goodman, M., Moore, G. W., Barnabas, J., and Matsuda, G., 1974, The phylogeny of human globin genes investigated by the maximum method, *J. Mol. Evol.* **3**:1–48.

Goodman, M., Moore, G. W., and Matsuda, G., 1975, Darwinian evolution in the genealogy of haemoglobin, *Nature (London)* **253**:603–608.

Hennig, W., 1966, *Phylogenetic Systematics,* pp. 263, University of Illinois Press, Urbana, Ill.

Holmquist, R., Jukes, T. H., Moise, H., Goodman, M., and Moore, G. W., 1976, The evolution of the globin family genes: Convergence of stochastic and augmented maximum parsimony genetic distances for alpha hemoglobin, beta hemoglobin, and myoglobin phylogenies, *J. Mol. Biol.* (in press).

Hoyer, B. H., and Roberts, R. B., 1967, Studies of nucleic acid interactions using DNA-agar, *Molecular Genetics: Part II* (H. Taylor, ed.), pp. 425–479, Academic Press, New York.

Hoyer, B. H., McCarthy, B. J., and Bolton, E. T., 1964, A molecular approach in the systematics of higher organisms, *Science* **144**:959–967.

Hoyer, B. H., Bolton, E. T., McCarthy, B. J., and Roberts, R. B., 1965, The evolution of polynucleotides, in: *Evolving Genes and Proteins* (V. Bryson and H. J. Vogel, eds.), pp. 581–590, Academic Press, New York.

Hoyer, B. H., van de Velde, N. W., Goodman, M., and Roberts, R. B., 1972, Examination of hominid evolution by DNA sequence homology, *J. Hum. Evol.* **1**:645–649.

King, M.-C., and Wilson, A. C., 1975, Evolution at two levels in humans and chimpanzees, *Science* **188**:107–116.

Knoll, A. H., and Barghoon, E. S., 1975, Precambrian eukaryotic organisms: A reassessment of the evidence, *Science* **190**:52–54.

Koen, A. L., and Goodman, M., 1969, Lactate dehydrogenase isozymes: qualitative and quantitative changes during primate evolution, *Biochem. Genet.* **3**:457–474.

Kohne, D. E., Chiscon, J. A., and Hoyer, B. H., 1972, Evolution of primate DNA sequences, *J. Hum. Evol.* **1**:627–644.

Kohne, D. E., Chiscon, J. A., and Hoyer, B. H., 1972, Evolution of primate DNA sequences, *J. Hum. Evol.* **1**:627.

Levins, R., 1968, *Evolution in Changing Environments,* Princeton University Press, Princeton, N.J.

Martin, M. A., and Hoyer, B. H., 1967, Adenine plus thymine and guanine plus cytosine enriched fractions of animal DNA as indicators of polynucleotide homologies, *J. Mol. Biol.* **27**:113–129.

McKenna, M. C., 1975, Towards a phylogenetic classification of the Mammalia, in: *Phylogeny of the Primates: A Multidisciplinary Approach* (W. P. Luckett and F. S. Szalay, eds.), pp. 21–46, Plenum Press, New York.

Moore, G. W., Barnabas, Jr., and Goodman, M., 1973, A method for constructing maximum parsimony ancestral amino acid sequences on a given network, *J. Theor. Biol.* **38:**459–485.

Moore, G. W., Goodman, M., Callahan, C., Holmquist, R., and Herbert, M., 1976, Estimation of superimposed mutations in the divergent evolution of protein sequences: Stochastic vs. augmented maximum parsimony method—cytochrome *c* phylogeny, *J. Mol. Biol.* (in press).

Romer, A. S., 1966, *Vertebrate Paleontology,* University of Chicago Press, Chicago.

Romero-Herrera, A. E., Lehmann, H., Castillo, O., Joysey, K. A., and Friday, A. E., 1976, Myoglobin of the orangutan as a phylogenetic enigma. *Nature* **261:**162.

Rossmann, M. G., Moras, D., and Olsen, K. W., 1974, Chemical and biological evolution of a nucleotide-binding protein, *Nature (London)* **250:**194–199.

Salthe, S. N., 1975, Problems of macroevolution (molecular evolution, phenotype definition, and canalization) as seen from a hierarchial viewpoint, *Am. Zool.* **15:**295–314.

Schopf, J. W., Haugh, B. N., Molnar, R. E., and Satterthwait, D. F., 1973, On the development of metaphytes and metazoans, *J. Paleontol.* **47:**1–9.

Selander, R. K., and Kaufman, D. W., 1973, Genetic variability and strategies of adaptation in animals, *Proc. Natl. Acad. Sci. USA* **70:**1875–1877.

Tomkins, G. M., 1975, The metabolic code, *Science* **189:**760–763.

Wilson, A. C., Maxson, L. R., and Sarich, V. M., 1974, Two types of molecular evolution: Evidence from studies of interspecific hybridization, *Proc. Natl. Acad. Sci. USA* **71:**2843–2847.

Wilson, E. O., 1975, *Sociobiology,* Belknap Press, Cambridge, Mass.

Wooding, G. L., and Doolittle, R. F., 1972, Primate fibrinopeptides: Evolutionary significance, *J. Hum. Evol.* **1:**553–563.

Young, J. Z., 1962, *The Life of the Vertebrates,* 2nd ed., Oxford University Press, Oxford.

Zuckerkandl, E., 1975, The appearance of new structures and functions in proteins during evolution, *J. Mol. Evol.* **7:**1–57.

Multigene Families and Genetic Regulation in the Evolution of Man

V

Comparative Aspects of DNA in Higher Primates

18

K. W. JONES

1. Introduction

The DNA of a species documents its evolutionary history. For this reason, the analysis of DNA is of fundamental interest to the science of anthropology. Moreover, an understanding of how DNA evolves contains valuable clues to its functions, a point which is becoming increasingly apparent to molecular biologists. Thus the two sciences converge to their mutual benefit.

DNA is now known to be comprised of a number of different classes of nucleotide sequences differing in the extent to which each is represented in the genome (Britten and Kohne, 1968). Those present as single copies, unique or nonrepetitive DNA, contain the information necessary for protein synthesis. Those present in more than a few copies, repetitive DNA, contain the information necessary to construct the machinery for the translation of proteins: the genes for ribosomal RNAs (see Birnstiel *et al.,* 1971) and transfer RNAs (see Clarkson *et al.,* 1973) as well as the genes coding for histone synthesis (Kedes and Birnstiel, 1971). In addition to containing genes of known function, repetitive DNA contains nucleotide sequences of unknown function of which the most distinctive are the so-called satellite or simple-sequence DNAs. These are composed of very short, highly repetitive, nucleotide sequences, making them somewhat exceptional in their nucleotide composition as well as in their kinetic properties. They are therefore easily isolated and purified since they may be induced to separate from the rest of the genomal DNA under condi-

K. W. JONES • Epigenetics Research Group, Department of Animal Genetics, Edinburgh University, Edinburgh, EH9 3JN, Scotland.

tions of isopycnic gradient centifugation in solutions of dense salts such as cesium chloride.

Unique DNA evolves comparatively slowly in terms of its rates of nucleotide substitution. Man and chimpanzee, for example, differ by only 1.0%* of their single-copy DNA (Kohne, 1970). Given such a sample of human and chimpanzee DNA, it would be difficult to tell them apart. Satellite DNA on the other hand is changing very rapidly—so rapidly that each species tends to possess its own complement of satellite DNAs. The rapidity of evolution of satellite DNA is presumably based on a noncoding function. However, evolution of a class of DNA in step with the evolution of species poses the question of a possible evolutionary function for this type of DNA. This question provided the basis for our interest in satellite DNA and stimulated our studies on such DNA in the higher primate group. Unfortunately, it is difficult to obtain DNA samples from higher primates, with the exception of man, so progress is slow, and much of what I have to discuss will be in the nature of a review of recent work with the addition of a few unpublished details and speculations.

2. Human Satellite DNA

Human DNA contains a number of satellites, of which three have been studied cytologically in the context of higher primate evolution (Jones et al., 1972, 1974). These are called satellite DNAs I, II, and III, in order of their discovery by Corneo and his colleagues (see review by Jones, 1973). By in situ hybridization (John et al., 1969; Gall and Pardue, 1969), these DNA satellites have been directly mapped on human chromosomes (Jones and Corneo, 1971; Jones et al., 1973, 1974). All appear to be concentrated within centromeric heterochromatin, and chromosomes with prominent amounts of this chromatin contain the highest concentrations. However, each satellite DNA is not present in equivalent concentration; certain human chromosomes possess, characteristically, more of one sequence than of another (Jones et al., 1974). By in situ hybridization, most of the human chromosome complement contains an amount of satellite DNA commensurate with the amount of centromeric heterochromatin which is visible in stained preparations. Thus this type of DNA constitutes a quantitatively significant component of centromeric heterochromatin. An exceptional chromosome, in which the satellite DNA is concentrated in a distal piece of heterochromatin as well as at the centromere, is the human Y, which contains a high concentration of satellite DNA I (Jones et al., 1974) as well as some satellite DNA III (Moar et al., 1975). All of the chromosomal regions containing high concentrations of satellite DNA show size variation within the population. Polymorphisms of this sort are heritable (Craig-Holmes

---

*Data of Kohne adjusted by factor 1.5°C T_m = 1% divergence in base sequence (McCarthy and Farquhar, 1971).

et al., 1973) and constitute the most obvious morphological difference between the chromosomes of individual humans. Changes in these regions are
also among the most prominent which have occurred in the evolution of
the karyotypes of man and the chimpanzee. What I now have to say concerns,
principally, the molecular basis of such evolutionary changes at the DNA level.

3. The Satellite DNA of Chimpanzee

Chimpanzee DNA contains two prominent satellite DNAs (Prosser *et al.,*
1973), which have been called satellites A and B. Satellite A is virtually identical
in its physical properties with human satellite DNA III and furthermore, shows
sufficient base sequence homology to form heteroduplexes (cross-hybridize)
with it (Jones *et al.,* 1972). Cytologically, the sites of cross-hybridization *in situ*
are identical with the sites of concentration of each satellite DNA sequence
within the chromosomes of its species of origin (see Fig. 1). There is therefore
little doubt that the heterochromatin containing these nucleotide sequences in
man and chimpanzee has originated from a common ancestral form and has
been conserved within the derived species. Chimpanzees have more of the
centromeric heterochromatin conserved than do humans. Little is yet known
about the distribution or molecular homologies of the B satellite in chimpanzee.
Physically, it resembles human satellite I, but it does not appear to show base
sequence homology by hybridization.

4. Satellite DNA in Other Higher Primates

Less is directly known about the satellite DNAs of the gorilla, orangutan,
and gibbon. Human satellite DNA III and chimpanzee satellite A, however,
both hybridize interchangeably *in situ* with the chromosomes of the gorilla and
orangutan (Jones *et al.,* 1972; Jones, 1974). A satellite DNA has been isolated
from the orangutan which is related in base sequence to that of human III
(Jones, unpublished). It may be concluded that man shares satellite DNA
with chimpanzee, orangutan, and gorilla through a common ancestral form.
The gibbon, however, appears to have had a different immediate ancestry in
that it does not seem to show heterochromatic blocks rich in the satellite
sequence which is common to the other higher primates (Jones, unpublished);
its chromosomes, moreover, differ considerably from those of the other higher
primates (Dutrillaux *et al.,* 1975; Tantravahi *et al.,* 1975).

Human satellite DNA III does not hybridize with the DNAs of any of the
lower primates so far tested. These include the baboon, macaque, marmoset,
and vervet monkey. The heterochromatin rich in this sequence therefore

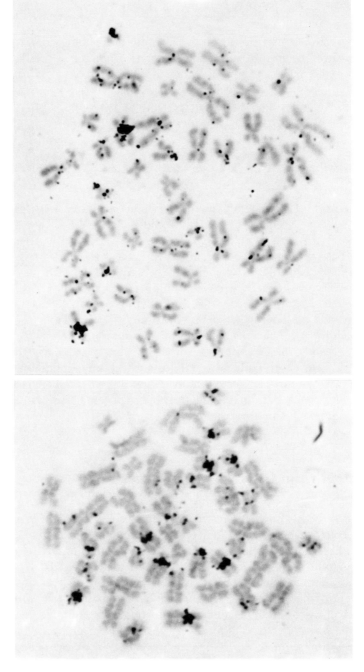

Fig. 1. Autoradiographs of chromosomes of human male (upper) and chimpanzee (lower) showing the locations of *in situ* hybridized human satellite III cRNA. In man, the concentrations of silver grains lie over the centromeric heterochromatin of, principally, the No. 9 and the Y chromosomes, with minor locations elsewhere. In the chimpanzee, about five pairs of chromosomes show high concentrations of a satellite DNA (chimpanzee satellite A) which is sufficiently related in base sequence to form heteroduplexes with the human probe.

probably arose in evolution at a time subsequent to the gibbon divergence. Giemsa at pH 11 (see Pearson, 1972) stains this particular heterochromatin bright magenta, differentiating it from other heterochromatin which stains red in man, chimpanzee, orangutan, and gorilla. The gibbon, however, does not exhibit this magenta staining reaction (Bobrow, personal communication; Dutrillaux *et al.*, 1975), thus confirming its different heterochromatin structure. Cytologically, the heterochromatin containing the related sequences in the three higher primates is more prominent than in man and contains more of the related DNA as determined by hybridization *in situ*. Satellite A is also present in detectably greater amount in isopycnic gradients of chimpanzee DNA than is satellite III in similar preparations of human DNA. Evidently there has been appreciable modification and possibly rearrangement of this DNA in the course of evolution. The karyotypic distribution of the sequence in the gorilla is strikingly similar to that in the orangutan, the main emphasis in both being on the acrocentric group of chromosomes, with only one or two minor localizations elsewhere. The distribution in the chimpanzee is quite different, with many metacentric chromosomes involved (Jones, 1974). In man, the Y chromosome contains satellite III by *in situ* hybridization (Moar *et al.*, 1975), and it stains positively by the Giemsa II method (Bobrow *et al.*, 1972), as does the Y of the chimpanzee (Bobrow and Madan, 1973). It therefore seems likely that the chimpanzee Y chromosome contains related satellite DNA, although this has not yet been directly demonstrated.

5. Restriction Enzymes

Evolutionary changes in DNA can also be studied by means of restriction endonucleases (see Nathans and Smith, 1975). These enzymes include many which cut DNA at defined short nucleotide sequences. The arrangement of such restriction sites is characteristic in simple genomes such as those of bacteriophages and viruses and in simple-sequence DNAs (DNA satellites). When digested with a suitable enzyme, these DNAs are cut into a series of fragments of particular length which may be separated into discrete bands by electrophoresis on agarose gels. In satellite DNAs, the distribution of the fragments reflects the basic repeating unit of the DNA (Southern and Roizes, 1974; Botchan *et al.*, 1974). More complex genomes such as those of primates are also digested to yield a series of fragment lengths, but these, when separated on gels, yield a continuous distribution reflecting the overlapping distribution of the fragments. Nevertheless, within such a distribution particular sequences will be found in a more or less constant location and may be demonstrated by hybridization, using a suitable probe. By this means, the interspersion pattern of satellite DNA among the other DNA sequences in the genome can be determined. In species such as man and chimpanzee which share related satellite DNA, the patterns can be compared to determine, in a general way, the changes in arrangement which may have occurred in evolution. Using this method, we have initiated a study of the relative distribution of human satellite III related DNA in man and chimpanzee (Fig. 2). Under the conditions of the experiment shown, hybridization occurred to a higher value in the chimpanzee, reflecting its greater amount of the related satellite. The interspersion pattern shows some marked differences in that DNA fragments of high, intermediate, as well as low molecular weight contain related sequences in the chimpanzee,

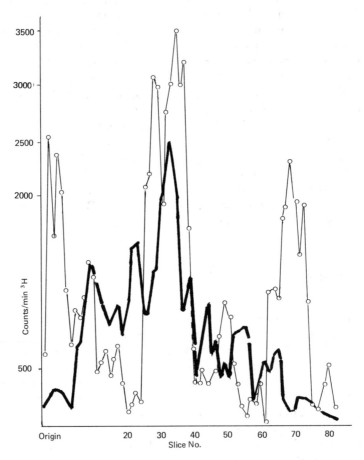

Fig. 2. Patterns of hybridization of human satellite DNA III cRNA with restriction fragments of human (bold line) and chimpanzee (light line) DNA. The DNAs were digested to completion under identical conditions with Hae III, electrophoresed at identical concentrations on 1% agarose gels, transferred to Millipore filter strips, and hybridized in the same solution of cRNA. The strips were cut into 2-mm slices and counted in toluene scintillator. Hybridization has occurred to fragments in the intermediate molecular weight range in both cases. In the chimpanzee DNA, however, hybridization has occurred with DNA fragments in the highest as well as the lowest molecular weight range, and similar hybrids are absent from human DNA. This difference possibly reflects rearrangements in the pattern of related sequences in evolution such as may be seen in *in situ* hybridization (Fig. 1).

whereas in man mainly the fragments of intermediate molecular weight contain such sequences. This shift in the interspersion pattern must reflect, among other things, the karyotypic rearrangements which have appeared in the related satellite DNA as revealed by *in situ* hybridization. This approach to studying DNA structure is applicable to other DNA sequences, including unique genes, and it offers the possibility of comparing the DNA of closely related species in great detail.

6. Human Satellite DNAs I and II in Evolution

As outlined in the previous sections, human satellite III DNA and the cytological features associated with it have their homologues in the chromosomes and DNA of the other great apes, excluding the gibbon. The other human satellites, I and II, however, are apparently less widely distributed, as are the cytological features with which they are associated. Human satellite I is prominently visible on the fluorescent distal region of the Y chromosome (Jones, 1974; Jones *et al.*, 1974), a feature absent from the Y of chimpanzee. Satellite DNA I hybridized with chimpanzee chromosomes shows only a random pattern of autoradiographic grains, indicating that there are no concentrations of a related sequence. Hybridization of satellite I with chimpanzee DNA likewise is negative. Although the chimpanzee Y chromosome lacks the distal fluorescent region which is associated with satellite DNA I in man, the gorilla Y chromosome possesses a prominent fluorescent distal segment (Pearson *et al.*, 1971). It seems unlikely that this feature would have arisen independently on two occasions in evolution. Unfortunately, however, satellite DNA has not yet been studied in this animal. The difference in the Y chromosome in man and chimpanzee probably reflects a structural rearrangement (a pericentric inversion?) and the addition of a relatively large amount of heterochromatin rich in satellite DNA I.

The chimpanzee No. I chromosome, like the Y chromosome, lacks a prominent heterochromatic feature which is present on its human counterpart (human chromosome No. I). This particular block of centromeric heterochromatin contains a high concentration of human satellite DNA II. Here again the addition of a structural feature to the chromosome apparently has involved the appearance of a new satellite DNA in evolution. Satellite DNA II will hybridize to a limited extent with satellite III (Melli *et al.*, 1975), although this cross-reaction is not apparent *in situ*, which suggests that this sequence may have arisen from part of the preexisting nucleotide sequence of satellite III. The DNA of the chimpanzee does not contain a satellite DNA at the buoyant density of human satellite DNA II. On the assumption that satellite DNAs have not simply been lost in the course of chimpanzee evolution, one interpretation of these observations is that satellites I and II have evolved subsequent to the chimpanzee-human divergence.

In accordance with this suggestion, it has been shown that the nucleotide divergence (ΔT_m) of these satellites is decreased relative to that of satellite DNA III (and chimpanzee satellite A). In terms of the decrease in divergence, the satellites may be ranked in order III, I, II, suggesting that they arose in similar order in evolution. Human III and chimpanzee A are similar in their degree of divergence, as could be expected from their related nucleotide sequences (Prosser *et al.*, 1973; Jones *et al.*, 1972). Assuming that these satellite DNAs have evolved by similar mechanisms and have diverged at similar rates and that satellite III (chimp A) arose subsequent to the gibbon divergence, a tentative age may be assigned to each of the human satellite DNAs according to the time scale of the fossil record (Simons, 1964). Thus satellite III would be approximately 30 m.y., satellite I approximately 18 m.y., and satellite II approximately 6 m.y., based on ΔT_m's of approximately 10, 6 and 2°C, respectively. The existence of a relatively recent human satellite DNA (II) offers the possibility of determining even more recent aspects of human evolution. For example, as suggested by Sarich (personal communication), it may be possible to study the divergence of this satellite DNA among existing racial groups in order to resolve the question of a recent as opposed to a more ancient separation between them.

7. Significance of the Satellite DNAs

The significance of the origin of the human satellite DNAs at different periods in evolution is presently unknown, but it is possible to offer a plausible explanation based on the distribution of satellite DNA in animals generally. Within a given species, satellite DNA seems to be characteristic and is usually different from that of other species in much the same way that karyotypes of different species tend to differ. Alterations in satellite DNA therefore appear to correlate with speciation. The most likely explanation of this seems to me to be that selection pressure operates against alterations in satellite DNA during the lifetime of a species, in much the same way that it operates against karyotypic changes. In fact, since satellite DNA is concentrated within cytologically prominent chromosomal structures, changes in it may be tantamount to structural alteration in the karyotype. Most karyotypic alterations, except those which accompany speciation, are lethal or extremely disadvantageous to the individual concerned. Speciation therefore probably occurs under conditions in which there is altered selection pressure that favors certain chromosomal changes. These changes will very likely include the loss or gain of heterochromatic regions, involving a concomitant loss or gain of satellite DNA.

If this interpretation of the general features of satellite DNA is correct, it follows that when satellite DNA is generated (or lost) in evolution it documents changes in the line of descent of the organism of the type which normally accompany speciation. This would mean that in man there have been at least two speciation events involving karyotypic changes, with tentative dates around 18 and 6 m.y., since the divergence of the chimpanzee line. Fossil evidence indeed supports the view that such events have occurred. Thus *Ramapithecus*, considered by some anthropologists to be an early hominid (see Simons, 1964), has been dated by potassium/argon measurements at 14 m.y. It must therefore have originated before this time (see Walker, this volume). Later dates are suggested for *Australopithecus*, around 3–5 m.y. (Walker, this volume; Simons, 1972). The concordance between these dates and those derived from the molecular measurements using satellite DNA, allowing the necessary assumptions, are provocative. They prompt the idea that satellite DNAs may offer a way to gain insight into more recent evolutionary events in a manner which would complement studies done using fossils.

It is necessary to mention that satellite DNA may evolve by other mechanisms in addition to base substitution. These may involve the saltation (tandem reduplication) of part of an existing satellite DNA to produce a new sequence family (Southern, 1970). This new DNA would have a very small ΔT_m. If it originated from a much mutated part of the original sequence, the effect might be to give rise to a new satellite DNA because of the possibility of its overall base composition being different from that of the original. New satellite DNA may well originate by such a mechanism since the same bit of heterochromatin sometimes possesses many different satellite DNAs. However, if the newly saltated derivative did not differ appreciably in base composition it would copurify with the original satellite DNA and its presence would interfere with estimates of the age of the whole. Caution is therefore indicated in basing age purely on ΔT_m measurements.

8. Evolution of Ribosomal Genes in Primates

There is some evidence that the genes coding for ribosomal RNA have diverged at a fairly steady rate over a wide range of species from bacteria to mice (Bendich and McCarthy, 1970). This offers the possibility of estimating the relative divergence times within the primate group from a study of their ribosomal genes. Comparison of the T_m's of heteroduplexes formed between the DNAs of various primates and human 28 S rRNA (Table 1) shows a decrease correlated with increasing evolutionary distance. Rates of divergence of these sequences calculated from the fossil time scale show a reasonably constant value. Data on a wider range of species will, however, be necessary in order to evaluate this promising approach to primate phylogeny.

Progress in this area, although encouraging, is frustratingly slow because of the scarcity of tissue for research. This is largely a question of organization, since births, affording placental material, are now not infrequent in captive

Table 1. Evolution of 28 S Ribosomal Genes in Primates[a]

Species compared	$T_m(°C)$	ΔT_m	Percentage diverged[b]	Years $\times 10^{-6}$ diverged[c]	Percentage divergence per 10^6 years
Human/human	85.5	0	0	0	0
Human/chimp	83.5	2.0	1.330	30	0.044
Human/baboon	82.5	3.5	2.330	60	0.046
Human/marmoset	80.0	5.5	3.660	90	0.040

[a] From Jones and Purdom (1976). Human 28 S ribosomal RNA was hybridized at T_{opt} with total DNA from each of the species shown. The T_m's of the hybrids were then determined in 0.1 × SSC. The T_m's show a progressive decrease with increasing evolutionary distance (see text).
[b] 1% altered base pairs = 1.5°C T_m.
[c] From Simons (1964).

colonies of higher primates and deaths must also occur from time to time. Perhaps I may finish by taking this opportunity to publicize our needs as a step toward encouraging those with such material not to waste it.

9. Summary

It appears that two classes of repetitive DNA, one not transcribed or translated, namely satellite DNA, and the other transcribed but not translated, namely 28 S ribosomal DNA, offer means to investigate primate evolution. The satellite DNAs in particular pose questions concerning the role of heterochromatin in evolution and offer a means for investigating more recent evolutionary events since they appear to be evolving faster than other molecular systems. The ribosomal genes offer the possibility to study longer-range phyletic relationships since they appear to be evolving at a relatively steady rate.

ACKNOWLEDGMENTS

Work described in this chapter was carried out with the financial assistance of the Medical Research Council, the Cancer Research Campaign, and the Muscular Dystrophy Group of Great Britain.

10. References

Bendich, A. J., and McCarthy, B. J., 1970, Ribosomal RNA homologies among distantly related organisms, *Proc. Natl. Acad. Sci. USA* **65**:349.
Birnstiel, M. L., Chipchase, M., and Spiers, J., 1971, The ribosomal RNA cistrons, *Prog. Nucleic Acid Res. Mol. Biol.* **11**:351.
Bobrow, M., and Madan, K., 1973, A comparison of chimpanzee and human chromosomes using the giemsa-II and other chromosome banding techniques, *Cytogenet. Cell Genet.* **12**:107.

Bobrow, M., Madan, K., and Pearson, P. L., 1972, Staining of some specific regions of human chromosomes particularly the secondary constriction of No. 9, *Nature (London) New Biol.* **238:**122.

Bonner, W. M., and Laskey, R. A., 1974, A film detection method for tritium-labelled proteins and nucleic acids in polyacrylamide gels, *Eur. J. Biochem.* **46:**83.

Botchan, M., McKenna, G., and Sharp, P. A., 1974, Cleavage of mouse DNA by a restriction enzyme as a clue to the arrangement of genes, *Cold Spring Harbor Symp. Quant. Biol.* **38:**383.

Britten, R. J., and Kohne, D. E., 1968, Repeated sequences in DNA, *Science* **161:**529.

Clarkson, S. G., Birnstiel, M. L., and Purdom, I. F., 1973, Clustering of transfer RNA genes in *Xenopus laevis, J. Mol. Biol.* **79:**411.

Craig-Holmes, A. P., Moore, F. B., and Shaw, M. W., 1973, Polymorphism of human C-band heterochromatin. I. Frequency of variants, *Am. J. Hum. Genet.* **25:**181.

Dutrillaux, B., Rethore, M. O., Aurias, A., and Goustard, M., 1975, Analse du caryotype de deux especes de Gibbons (*Hylobates lar* et *H. concolor*) par differentes techniques de marquage, *Cytogenet. Cell Genet.* **15:**81.

Gall, J. G., and Pardue, M. L., 1969, Formation and detection of RNA-DNA hybrid molecules in cytological preparations, *Proc. Natl. Acad. Sci. USA* **63:**378.

John, H. A., Birnstiel, M. L., and Jones, K. W., 1969, RNA-DNA hybrids at the cytological level, *Nature (London)* **223:**582.

Jones, K. W., 1973, Annotation: Satellite DNA, *J. Med. Genet.* **10:**273.

Jones, K. W., 1974, Repetitive DNA sequences in animals, particularly primates, in: *Chromosomes Today,* Vol. 5 (P. L. Pearson and K. R. Lewis, eds.), Wiley, New York, Israel Universities Press, Jerusalem.

Jones, K. W., and Corneo, G., 1971, Location of satellite and homogeneous DNA sequences on human chromosomes, *Nature (London) New Biol.* **223:**268.

Jones, K. W., and Purdom, I. F., 1976, The evolution of defined classes of human and primate DNA, in: *Proceedings of the Society for the Study of Human Biology: Symposium on Chromosome Variation* (A. J. Boyce, ed.), Taylor and Frances Press, London.

Jones, K. W., Prosser, J., Corneo, G., Ginelli, E., and Bobrow, M., 1972, Satellite DNA constitutive heterochromatin and human evolution, in: *Modern Aspects of Cytogenetics: Constitutive Heterochromatin in Man,* Vol. 6 of *Symposia Medica Hoechst* (R. A. Pfeiffer, ed.), pp. 45–74, F. K. Schattauer Verlag, Stuttgart, New York.

Jones, K. W., Corneo, G., Ginelli, E., and Prosser, J., 1973, The chromosomal location of human satellite DNA III, *Chromosoma* **42:**445.

Jones, K. W., Purdom, I. F., Prosser, J., and Corneo, G., 1974, The chromosomal localisation of human satellite DNA I, *Chromosoma* **48:**161.

Kedes, L. H., and Birnstiel, M. L., 1971, Reiteration and clustering of DNA sequences complementary to histone messenger RNA, *Nature (London) New Biol.* **230:**165.

Kohne, D. E., 1970, Evolution of higher organism DNA, *Quart. Rev. Biophys.* **3:**327.

McCarthy, B. J., and Farquhar, M. N., 1971, The rate of change of DNA in evolution, *Brookhaven Sym. Quant. Biol.* **23:**1–43.

Melli, M., Ginelli, E., Corneo, G., and Di Lernia, R., 1975, Clustering of the DNA complementary to repetitive nuclear RNA of HeLa cells, *J. Mol. Biol.* **93:**23.

Moar, M. H., Purdom, I. H., and Jones, K. W., 1975, Optimum hybridization conditions for the chromosomal mapping of AT-rich satellite DNA, *Chromosoma* (Berl.) **53:**345–359.

Nathans, D., and Smith, H. O., 1975, Restriction endonucleases in the analysis and restructuring of DNA molecules, *Annu. Rev. Biochem.* **44:**273.

Pearson, P. L., 1972, The use of new staining techniques for human chromosome identification, *J. Med. Genet.* **9:**264.

Pearson, P. L., Bobrow, M., Vosa, C. G., and Barlow, P. W., 1971, Quinacrine fluorescence in mammalian chromosomes, *Nature (London)* **231:**326.

Prosser, J., Moar, M., Bobrow, M., and Jones, K. W., 1973, Satellite sequence in chimpanzee (*Pan troglodytes*), *Biochim. Biophys. Acta* **319:**122.

Simons, E., 1964, The early ancestors of man, *Sci. Am.* **211:**50.

Simons, E., 1972, *Primate Evolution: An Introduction to Man's Place in Nature,* Macmillan, New York.

Southern, E. M., 1970, Base sequence and evolution of guinea pig satellite DNA, *Nature (London)* **227:**794.

Southern, E. M., 1975, Detection of specific sequences among DNA fragments separated by gel electrophoresis, *J. Mol. Biol.* **98:**503–518.

Southern, E. M., and Roizes, G., 1974, The action of restriction enzymes on higher organism DNA, *Cold Spring Harbor Symp. Quant. Biol.* **38:**429.

Tantravahi, R., Dev, V. G., Firschein, I. L., Miller, D. A., and Miller, O. J., 1975, Karyotype of the gibbons *Hylobates lar* and *H. moloch, Cytogenet. Cell Genet.* **15:**92.

Evolutionary Origin of Antibody Specificity

19

N. HILSCHMANN, H. KRATZIN, P. ALTEVOGT,
E. RUBAN, A. KORTT, C. STAROSCIK, R. SCHOLZ,
W. PALM, H. U. BARNIKOL,
S. BARNIKOL-WATANABE, J. BERTRAM,
J. HORN, M. ENGELHARD, M. SCHNEIDER,
and L. DREKER

1. Introduction

Expressions like "immunological learning" and "immunological memory" indicate that the mechanism of antibody formation has always been considered to be a very peculiar problem of protein synthesis. The capacity of the immune system to respond to an enormous number of antigens, even to antigens which normally do not exist in nature, has made it difficult to believe that the specificities of antibodies are genetically determined in the same way that the specificities of the proteins are.

The "genetic" concept has only recently made progress, mainly through determination of the primary structure of myeloma proteins, which are immunoglobulins produced by plasma cell tumors. Unlike antibody preparations, myeloma proteins are usually monoclonal, i.e., chemically homogeneous. These advantageous circumstances are due to the capacity of the individual plasma cell to synthesize only one type of immunoglobulin with one unique amino acid sequence (Burnet, 1959).

N. HILSCHMANN *ET AL.* • Max Planck-Institut für experimentelle Medizin, Göttingen, West Germany.

This has also made the antibody problem an excellent model system for the study of cell differentiation. Since the immunological stem cell has the potential to synthesize about 10^6 different antibodies with different specificities and different amino acid sequences, there must be a mechanism acting during differentiation which reduces the multiple possibilities to *one* in the antibody-producing plasma cell. It has been shown in the past that the study of antibody structure can provide us with deep insights into the genetic mechanisms underlying these differentiation processes (Hilschmann, 1969; Edelman and Gall, 1969; Milstein and Pink, 1970).

2. The Heterogeneity Problem

No other protein class exhibits such a high degree of polymorphism as the immunoglobulins. Several classes of immunoglobulins can be distinguished: IgG, IgM, IgA, IgD, and IgE. The IgM immunoglobulin molecule, the basic structure of which has recently been determined (Watanabe *et al.*, 1973; Putnam *et al.*, 1973*a*), is depicted in Fig. 1. It is the nature of the H chain which determines the class to which an immunoglobulin belongs. The H chain is of γ, μ, α, δ, or ϵ type. The L chain can be of either κ or λ type, independent of the H chain class. Heterogeneity of that type has no influence on antibody specificity. This type of polymorphism is comparable to that of the hemoglobins.

3. The Specificity Region

It is a peculiarity of the immunoglobulins that the sections which bind the antigens are limited to the *N*-terminal variable parts (V parts) of the molecule. The V parts, which contain about 107–120 amino acid residues, differ mark-

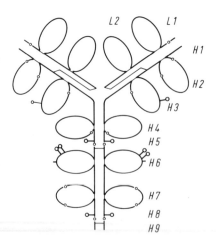

Fig. 1. Subunit of IgM GAL ($\mu_2\kappa_2)_5$ (Watanabe *et al.*, 1973). Six interpeptidal S-S bridges connect L to H chains and connect the subunits in homology region C3 and the μ chains to each other. Intrapeptidal S-S bridges form loops in V parts and C parts. L, Light chain; H, heavy chain; V, variable part; C constant part; O, carbohydrate; –o–, methionine.

edly from the rest of the molecule. They are characterized by great variability in the amino acid sequence, contrary to the amino acid sequence of the C-terminal constant part (C part), which has a unique and specific sequence for every chain type or subtype. In L chains, the V and the C parts are of equal length.

Figure 2 demonstrates the variability of the V parts by a comparison of 18 κ chains whose sequence had been determined. This comparison shows that the variability of the V parts is caused by multiple amino acid exchanges and single deletions. The basic sequence is still the same for the V part of each immunological chain.

Figure 2 also indicates that the amino acid substitutions and deletions causing variability are not randomly distributed in the V part, but rather linked. These linked amino acid exchanges are indicated by light boxes. On the basis of these linked amino acid exchanges, we can divide the κ chains into four subgroups. We have recently described the fourth subgroup, which appears in the form of the protein LEN. The deletions after position 31 agree with this division into subgroups. Subgroup I has with the exception of one protein a deletion of six amino acids after this position and subgroup III a deletion of five amino acids after this position. Amino acid exchanges which so far do not fit into these subgroups are indicated by heavy boxes. As soon as more data are available, further subdivisions may be recognized. This is already the case within subgroup I, where we observe an additional linkage group in positions 24, 28, 50, 73, 83, and 94 (Watanabe and Hilschmann, 1970; Schiechl and Hilschmann, 1971; Schneider and Hilschmann, 1975).

This demonstrates that linked amino acid exchanges are a general principle which can likely be extended to all exchanges. The principle becomes clearer as more proteins are investigated. The C part of the κ chains is depicted in Fig. 7, which will be discussed later.

This principle of regularity is valid not only for κ chains but also for λ (Fig. 3) and H (Fig. 4) chains. The nature of the exchange is exactly the same, only in the H chains, contrary to κ and λ chains, one finds a peculiarity insofar as the V parts are not chain-type specific. The V parts of the H chains are derived from γ, μ, or α chains, but in spite of this the variable parts can be arranged in three subgroups independent of chain type. This means that all H chains are using the same set of V parts (Köhler et al., 1970). This peculiarity and the structure of the C parts (depicted in Fig. 7 for the γ_1, μ, α_1, and ϵ chains) will be discussed later.

4. Variability and Specificity

Antigens are bound to the combining site of an antibody molecule. This combining site is constituted by the variable parts of the L and H chains. The question arises as to which residues of the variable part participate in the combining site.

It is obvious from Figs. 2, 3, and 4 that the amino acid exchanges which

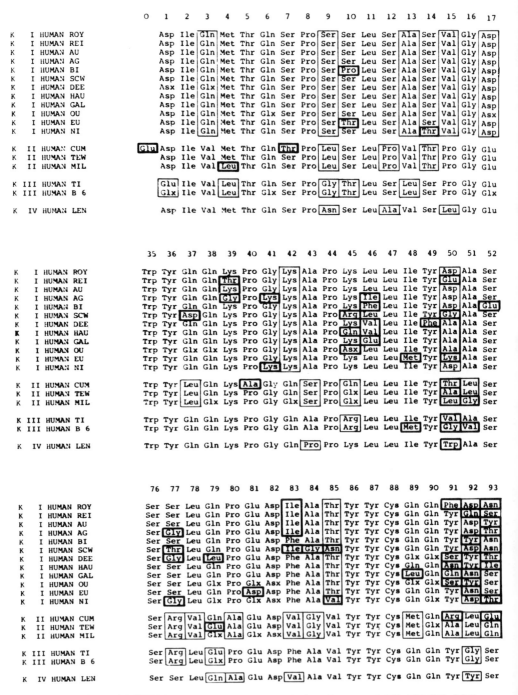

Fig. 2. Comparison of the variable parts of human L chains of κ type: ROY, AU, AG, SCW, BI, DEE, HAU, OU, EU, CUM, MIL, TI, B6 (Hilschmann *et al.*, 1972), REI (Palm and Hilschmann, 1973), GAL (Laure *et al.*, 1973), NI (Shinoda, 1973). TEW (Putnam *et al.*, 1973*b*),

18	19	20	21	22	23	24	25	26	27	28	29	30	31	a	b	c	d	e	f	32	33	34
Arg	Val	Thr	Ile	Thr	Cys	Gln	Ala	Ser	Gln	Asp	Ile	Ser	Ile	---	---	---	---	---	---	Phe	Leu	Asn
Arg	Val	Thr	Ile	Thr	Cys	Gln	Ala	Ser	Gln	Asp	Ile	Ile	Lys	---	---	---	---	---	---	Tyr	Leu	Asn
Arg	Val	Thr	Ile	Thr	Cys	Gln	Ala	Ser	Gln	Asp	Ile	Ser	Asp	---	---	---	---	---	---	Tyr	Leu	Asn
Arg	Val	Thr	Ile	Thr	Cys	Gln	Ala	Ser	Gln	Asp	Ile	Asn	His	---	---	---	---	---	---	Tyr	Leu	Asn
Ser	Val	Thr	Ile	Thr	Cys	Gln	Ala	Ser	Gln	Asp	Ile	Arg	Asn	---	---	---	---	---	---	Ser	Leu	Ile
Arg	Val	Thr	Ile	Thr	Cys	Gln	Ala	Ser	Gln	Asp	Ile	Arg	Lys	---	---	---	---	---	---	His	Leu	Asn
Arg	Val	Thr	Ile	Thr	Cys	Arg	Ala	Gly	Gln	Ser	Val	Asn	Lys	---	---	---	---	---	---	Tyr	Leu	Asn
Arg	Val	Thr	Ile	Thr	Cys	Arg	Ala	Ser	Gln	Ser	Ile	Ser	Ser	---	---	---	---	---	---	Tyr	Leu	Ser
Arg	Val	Thr	Ile	Ile	Cys	Arg	Ala	Ser	Gln	Gly	Ile	Arg	Asn	---	---	---	---	---	---	Asp	Leu	Thr
Arg	Val	Thr	Ile	Thr	Cys	Arg	Ala	Ser	Glx	Thr	Ile	Ser	Ser	---	---	---	---	---	---	Tyr	Leu	Asx
Arg	Val	Thr	Ile	Thr	Cys	Arg	Ala	Ser	Gln	Ser	Ile	Asn	Thr	---	---	---	---	---	---	Trp	Leu	Ala
Arg	Val	Thr	Leu	Leu	Cys	Glu	Ala	Ser	Gln	Ser	Val	Leu	Glu	Ser	Gly	Asn	---	---	Thr	Phe	Leu	Ala
Pro	Ala	Ser	Ile	Ser	Cys	Arg	Ser	Ser	Gln	Ser	Leu	Leu	Asp	Ser	Gly	Asp	Gly	Asn	Thr	Tyr	Leu	Asn
Pro	Ala	Ser	Ile	Ser	Cys	Arg	Ser	Ser	Gln	Ser	Leu	Leu	His	Ser	---	Asp	Gly	Phe	Asp	Tyr	Leu	Asn
Pro	Ala	Ser	Ile	Ser	Cys	Arg	Ser	Ser	Gln	Asn	Leu	Leu	Glx	Ser	Asx	Gly	Asx			Tyr	Leu	Asp
Arg	Ala	Thr	Leu	Ser	Cys	Arg	Ala	Ser	Gln	Ser	Val	Ser	Asn	Ser	---	---	---	---	---	Phe	Leu	Ala
Arg	Ala	Ala	Leu	Ser	Cys	Arg	Ala	Ser	Gln	Ser	Leu	Ser	Gly	Asn	---	---	---	---	---	Tyr	Leu	Ala
Arg	Ala	Thr	Ile	Asn	Cys	Lys	Ser	Ser	Gln	Ser	Val	Leu	Tyr	Ser	Ser	Asn	Ser	Lys	Asn	Tyr	Leu	Ala

53	54	55	56	57	58	59	60	61	62	63	64	65	66	67	68	69	70	71	72	73	74	75
Lys	Leu	Glu	Ala	Gly	Val	Pro	Ser	Arg	Phe	Ser	Gly	Thr	Gly	Ser	Gly	Thr	Asp	Phe	Thr	Phe	Thr	Ile
Asn	Leu	Gln	Ala	Gly	Val	Pro	Ser	Arg	Phe	Ser	Gly	Ser	Gly	Ser	Gly	Thr	Asp	Tyr	Thr	Phe	Thr	Ile
Asn	Leu	Glu	Ser	Gly	Val	Pro	Ser	Arg	Phe	Ser	Gly	Gly	Gly	Ser	Gly	Ala	His	Phe	Thr	Phe	Thr	Ile
Asn	Leu	Glu	Thr	Gly	Val	Pro	Ser	Arg	Phe	Ser	Gly	Ser	Gly	Phe	Gly	Thr	Asp	Phe	Thr	Phe	Thr	Ile
Asn	Leu	Glu	Ile	Gly	Val	Pro	Ser	Arg	Phe	Arg	Gly	Ser	Gly	Ser	Gly	Thr	Asp	Phe	Ala	Leu	Ser	Ile
Thr	Leu	Glu	Thr	Gly	Val	Pro	Ser	Arg	Phe	Ser	Gly	Ser	Gly	Ser	Gly	Thr	Asp	Phe	Thr	Leu	Thr	Ile
Ser	Leu	Lys	Ser	Gly	Val	Pro	Ser	Arg	Phe	Ser	Gly	Ser	Gly	Ser	Gly	Thr	Asp	Phe	Thr	Leu	Thr	Ile
Ser	Leu	Pro	Ser	Gly	Val	Pro	Ser	Arg	Phe	Ser	Gly	Ser	Gly	Ser	Gly	Thr	Asp	Phe	Thr	Leu	Thr	Ile
Asn	Leu	Gln	Ser	Gly	Val	Pro	Ser	Arg	Phe	Ser	Gly	Ser	Gly	Ala	Gly	Thr	Glu	Phe	Thr	Leu	Thr	Ile
Asx	Leu	His	Ser	Gly	Val	Pro	Ser	Arg	Phe	Ser	Gly	Ser	Gly	Ser	Gly	Ser	Asx	Phe	Thr	Phe	Thr	Ile
Ser	Leu	Glu	Thr	Gly	Val	Pro	Ser	Arg	Phe	Ile	Gly	Ser	Gly	Ser	Gly	Thr	Glu	Phe	Thr	Leu	Thr	Ile
Asn	Leu	Glu	Thr	Gly	Val	Pro	Ser	Arg	Phe	Ser	Gly	Ser	Gly	Ser	Gly	Thr	Asp	Phe	Thr	Phe	Thr	Ile
Tyr	Arg	Ala	Ser	Gly	Val	Pro	Asp	Arg	Phe	Ser	Gly	Ser	Gly	Ser	Gly	Thr	Asp	Phe	Thr	Leu	Lys	Ile
Asn	Arg	Ala	Ser	Gly	Val	Pro	Asp	Arg	Phe	Ser	Gly	Ser	Gly	Ser	Gly	Thr	Asp	Phe	Thr	Leu	Lys	Ile
Asn	Arg	Ala	Ser	Gly	Val	Pro	Asn	Arg	Phe	Ser	Gly	Ser	Gly	Ser	Gly	Thr	Asx	Phe	Thr	Leu	Lys	Ile
Ser	Arg	Ala	Thr	Gly	Ile	Pro	Asp	Arg	Phe	Ser	Gly	Ser	Gly	Ser	Gly	Thr	Asp	Phe	Thr	Leu	Thr	Ile
Ser	Arg	Ala	Thr	Gly	Ile	Pro	Asp	Arg	Phe	Ser	Gly	Ser	Gly	Ser	Gly	Ala	Asp	Phe	Thr	Leu	Thr	Ile
Thr	Arg	Glu	Ser	Gly	Val	Pro	Asp	Arg	Phe	Ser	Gly	Ser	Gly	Ser	Gly	Thr	Asp	Phe	Thr	Leu	Thr	Ile

94	95	96	97	98	99	100	101	102	103	104	105	106	107	108	191	Inv	214
Leu	Pro	Leu	Thr	Phe	Gly	Gly	Gly	Thr	Lys	Val	Asp	Phe	Lys	Arg	Leu	Inv a	Cys
Leu	Pro	Tyr	Thr	Phe	Gly	Gln	Gly	Thr	Lys	Leu	Gln	Ile	Thr	Arg	Leu	Inv a	Cys
Leu	Pro	Trp	Thr	Phe	Gly	Gln	Gly	Thr	Lys	Val	Glu	Ile	Lys	Arg	Val		Cys
Leu	Pro	Arg	Thr	Phe	Gly	Gln	Gly	Thr	Lys	Leu	Glu	Ile	Lys	Arg	Val	Inv b	Cys
Leu	Pro	Tyr	Thr	Phe	Gly	Gln	Gly	Thr	Lys	Leu	Glu	Ile	Lys	Arg	Val		Cys
Val	Pro	Ile	Thr	Phe	Gly	Gln	Gly	Thr	Arg	Val	Glu	Asn	Lys	Gly	Leu		Cys
Thr	Pro	Tyr	Thr	Phe	Gly	Pro	Gly	Thr	Lys	Val	Glu	Met	Thr	Arg	Val		Cys
Thr	Pro	Thr	Ser	Phe	Gly	Gln	Gly	Thr	Arg	Val	Glu	Ile	Lys	Arg	Val		Cys
Tyr	Pro	Arg	Ser	Phe	Gly	Gln	Gly	Thr	Lys	Val	Glu	Ile	Lys	Arg	Val		Cys
Ser	Pro	Thr	Thr	Phe	Gly	Glx	Gly	Thr	Arg	Leu	Glx	Ile	Lys	Arg	Val		Cys
Asp	Ser	Lys	Met	Phe	Gly	Gln	Gly	Thr	Lys	Val	Glu	Val	Lys	Gly	Val	Inv b	Cys
Leu	Pro	Ser	Thr	Phe	Gly	Val	Ala	Ser	Lys	Val	Glu	Ser	Lys	Arg	Val		Cys
Ile	Pro	Tyr	Thr	Phe	Gly	Gln	Gly	Thr	Lys	Leu	Glu	Ile	Arg	Arg	Val	Inv b	Cys
Ala	Pro	Ile	Thr	Phe	Gly	Gln	Gly	Thr	Arg	Leu	Glu	Ile	Lys	Arg	Val		
Thr	Pro	Leu	Thr	Phe	Gly	Gly	Gly	Thr	Asn	Val	Glu	Ile	Lys	Arg	Val		Cys
Ser	Pro	Ser	Thr	Phe	Gly	Gln	Gly	Thr	Lys	Val	Glu	Leu	Lys	Arg	Val		Cys
Ser	Pro	Phe	Thr	Phe	Gly	Gln	Gly	Ser	Lys	Leu	Glu	Ile	Lys		Val		
Thr	Pro	Tyr	Ser	Phe	Gly	Gln	Gly	Thr	Lys	Leu	Glu	Ile	Lys	Arg	Val		Cys

and LEN (Schneider and Hilschmann, 1974). Subgroup-specific residues are marked by light boxes, single point mutations by heavy boxes, deletions by dashed lines.

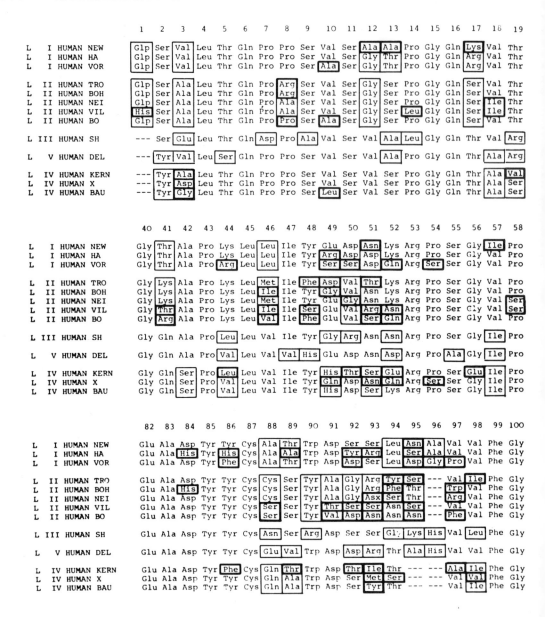

Fig. 3. Comparison of the variable parts of human L chains of λ type: NEW, HA, BO, VIL, NEI, SH, KERN, X, BAU, (Hilschmann *et al.*, 1972), VOR (Englehard *et al.*, 1974), TRO (Scholz

```
      20  21  22  23  24  25  26  27   a   b   c   28  29  30  31  32  33  34  35  36  37  38  39

      Ile Ser Cys Ser Gly Gly Ser Thr Asn Ile --- Gly Asn Asn Tyr Val Ser Trp His Gln His Leu Pro
      Ile Ser Cys Ser Gly Gly Ser Ser Asn Gly Thr Gly Asn Asn Tyr Val Tyr Trp Tyr Gln Gln Leu Pro
      Ile Ser Cys Ser Gly Gly Asn Phe Asp Ile --- Gly Arg Asn Ser Val Asn Trp Tyr Gln Gln Val His Pro

      Ile Ser Cys Thr Gly Thr Ser Ser Asp Val Gly Ala Tyr Asn Ser Val Ser Trp Tyr Gln Gln His Pro
      Ile Ser Cys Ala Gly Thr Ser Ser Asp Val Gly Gly Asn His Phe Val Ser Trp Tyr Gln Gln His Pro
      Ile Ser Cys Thr Gly Thr Thr Ser Asp Val Gly Ser Tyr Asn Phe Val Ser Trp Tyr Gln Gln Asn Pro
      Ile Ser Cys Thr Gly Thr Ser Ser Asp Val Gly Gly Tyr Asn Tyr Val Ser Trp Phe Gln Gln His Pro
      Ile Ser Cys Thr Gly Thr Ser Ser Asp Val Gly Asp Asn Lys Tyr Val Ser Trp Tyr Gln Gln His Pro

      Ile Thr Cys Gln Gly Asp Ser Leu --- --- --- Arg Gly Tyr Asp Ala Ala Trp Tyr Gln Gln Lys Pro

      Ile Thr Cys Gly Gly Asp Gly Ile --- --- --- Gly Gly Lys Ser Val His Trp Tyr Gln Gln Lys Pro

      Ile Thr Cys Ser Gly Asp Asn Leu --- --- --- Glu Lys Thr Phe Val Ser Trp Phe Gln Gln Arg Pro
      Ile Thr Cys Ser Gly Asp Lys Leu --- --- --- Gly Asp Lys Asp Val Cys Trp Tyr Gln Gln Arg Pro
      Ile Thr Cys Ser Gly Asp Lys Leu --- --- --- Gly Glu Gln Tyr Val Cys Trp Tyr Gln Gln Lys Pro
```

```
      59  60  61  62  63  64  65  66  67  68  69  70  71  72  73  74  75  76  77  78  79  80  81

      Asp Arg Ile Ser Ala Ser Lys Ser Gly Thr Ser Ala Thr Leu Gly Ile Thr Gly Leu Arg Thr Gly Asp
      Asp Arg Phe Ser Gly Ser Lys Ser Gly Thr Ser Ala Ser Leu Ala Ile Ser Gly Leu Gln Ser Glu Asp
      Asp Arg Phe Ser Gly Ser Lys Ser Gly Thr Ser Ala Ser Leu Ala Ile Ser Gly Leu Gln Ser Glu Asn

      Asp Arg Leu Ser Gly Ser Lys Ser Gly Asp Thr Ala Ser Leu Thr Ile Ser Gly Leu Arg Ala Asp Asp
      Tyr Arg Phe Ser Gly Ser Lys Ser Gly Asn Thr Ala Ser Leu Thr Ile Ser Gly Leu Gln Ala Glu Asp
      Asp Arg Phe Ser Gly Ser Lys Ser Gly Lys Thr Ala Ser Leu Thr Ile Ser Gly Leu Gln Ala Glu Asp
      Asp Arg Phe Ser Gly Ser Lys Ser Ala Asn Thr Ala Ser Leu Thr Ile Ser Gly Leu Ala Ala Glu Asp
      Asp Arg Phe Ser Gly Ser Lys Ser Asp Asn Thr Ala Ser Leu Thr Val Ser Gly Leu Arg Ala Glu Asp

      Asp Arg Phe Ser Gly Ser Ser Ser Gly His Thr Ala Ser Leu Thr Ile Thr Gly Ala Gln Ala Glu Asp

      Glu Arg Phe Ser Gly Ser Asn Ser Gly Asn Thr Ala Ala Leu Thr Ile Ser Arg Val Glu Ala Gly Asp

      Glu Arg Phe Ser Gly Ser Ser Ser Gly Ala Thr Ala Thr Leu Thr Ile Ser Gly Ala Gln Ser Val Asp
      Glu Arg Phe Ser Gly Ser Asn Ser Gly Asn Thr Ala Thr Leu Thr Ile Ser Gly Thr Gln Ala Met Asp
      Glu Arg Phe Ser Gly Ser Asn Ser Gly Thr Thr Ala Thr Leu Thr Ile Ser Gly Thr Gln Ala Met Asp
```

```
      101 102 103 104 105 106 107 108 109            145        154        173        191        214

      Gly Gly Thr Lys Val Thr Val Leu Gly            Ala        Ser        Lys        Arg        Ser
      Gly Gly Thr Gln Leu Thr Val Leu Arg            Ala        Ser        Lys        Arg        Ser
      Gly Gly Thr Lys Val Thr Val Leu Gly            Ala        Ser        Lys        Lys        Ser

      Gly Gly Thr Lys Leu Thr Val Leu Gly            Ala        Ser        Lys        Arg        Ser
      Gly Gly Thr Asn Leu Thr Val Leu Gly
      Gly Gly Thr Arg Val Thr Val Leu Ser            Ala        Ser        Lys        Arg        Ser
      Gly Gly Thr Lys Leu Thr Val Leu Gly            Ala        Ser        Lys        Arg        Ser
      Gly Gly Thr Lys Leu Thr Val Leu Arg            Ala        Ser        Lys        Arg        Ser

      Gly Gly Thr Lys Leu Thr Val Leu Gly            Ala        Ser        Lys        Arg        Ser

      Gly Gly Thr Lys Leu Thr Val Leu Gly

      Gly Gly Thr Lys Leu Thr Val Leu Ser            Ala        Gly        Lys        Arg        Ser
      Gly Gly Thr Arg Leu Thr Val Leu Ser            Ala        Ser        Lys        Arg        Ser
      Gly Gly Thr Lys Leu Thr Val Leu Gly            Ala        Ser        Lys        Arg        Ser

                                                     Val(Mz)    Gly        Asn(Mz)    Lys
                                                                Kern+                 Oz+
```

and Hilschmann, 1975, BOH (Köhler *et al.,* 1975), and DEL (Eulitz, 1974). For symbols, see Fig. 2.

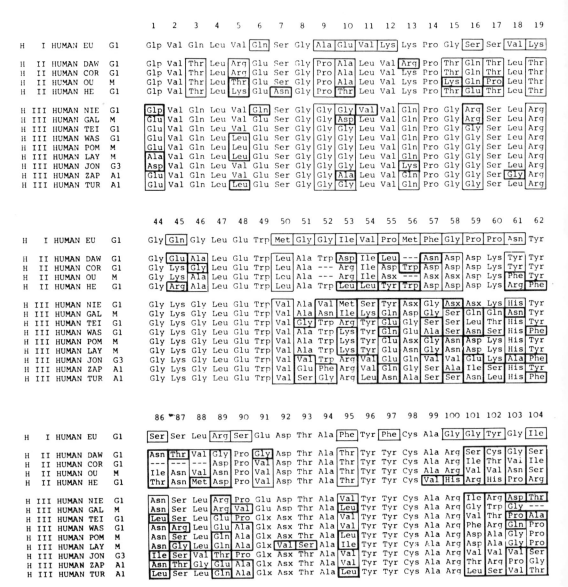

Fig. 4. Comparison of the variable parts of human H chains of different chain types: EU, DAW, COR, OU, HE, NIE (Hilschmann *et al.*, 1972), GAL (Watanabe *et al.*, 1973), TEI, WAS, POM, LAY,

cause variability are not equally distributed over the chains but are accumulated in three sections of the V part. These hypervariable sections which appear in positions 28–34, 50–56, and 91–96 in κ chains were for a long time thought to be constituents of the combining site. Affinity-labeling experiments pointed in the same direction: the labeled residues belonged to those hypervariable regions (Givol *et al.*, 1972).

```
   20  21  22  23  24  25  26  27  28  29  30  31  32  33  34  35  36  37  38  39  40  41  42  43

  Val Ser Cys Lys Ala Ser Gly Gly Thr Phe Ser Arg Ser Ala Ile --- --- Ile Trp Val Arg Gln Ala Pro

  Leu Thr Cys Thr Phe Ser Gly Phe Ser Leu Ser Gly Glu Thr Met Cys Val Ala Trp Ile Arg Gln Pro Pro
  Leu Thr Cys Thr Phe Ser Gly Phe Ser Leu Ser Ser Thr Gly Met Cys Val Gly Trp Ile Arg Gln Pro Pro
  Leu Thr Cys Thr Phe Ser Gly Phe Ser Leu Ser Thr Ser Arg Met Arg Val Ser Trp Ile Arg Arg Pro Pro
  Leu Thr Cys Thr Leu Ser Gly Leu Ser Leu Thr Thr Asp Gly Val Ala Val Gly Trp Ile Arg Gln Gly Pro

  Leu Ser Cys Ala Ala Ser Gly Phe Thr Phe Ser Arg Tyr Thr Ile --- --- His Trp Val Arg Gln Ala Pro
  Leu Ser Cys Ala Ala Ser Gly Phe Asx Leu Asx Phe Asx Gly Met --- --- Thr Trp Val Arg Gln Ala Pro
  Leu Ser Cys Ala Ala Ser Gly Phe Thr Phe Ser Thr Ser Ala Val --- --- Tyr Trp Val Arg Gln Ala Pro
  Leu Ser Cys Ala Ala Ser Gly Phe Ser Phe Ser Thr Asp Ala Met --- --- Tyr Trp Val Arg Gln Ala Pro
  Leu Ser Cys Ala Ala Ser Gly Phe Thr Phe Ser Ser Ser Ala Met --- --- Ser Trp Val Arg Gln Ala Pro
  Leu Ser Cys Ala Ala Ser Gly Phe Thr Phe Ser Ala Ser Ala Met --- --- Ser Trp Val Arg Gln Ala Pro
  Leu Ser Cys Ala Ala Ser Gly Phe Thr Phe Ser Thr Ala Trp Met --- --- Lys Trp Val Arg Gln Ala Pro
  Leu Ser Cys Ala Ala Ser Gly Phe Thr Phe Ser Thr Thr Ser Arg --- --- Phe Trp Val Arg Gln Ala Pro
  Leu Ser Cys Ala Ala Ser Gly Phe Thr Phe Ser Arg Val Leu Ser --- --- Ser Trp Val Arg Gln Ala Pro
```

```
   a  63  64  65  66  67  68  69  70  71  72  73  74  75  76  77  78  79  80  81  82  83  84  85

  --- Ala Gln Lys Phe Gln Gly Arg Val Thr Ile Thr Ala Asp Glu Ser Thr Asn Thr Ala Tyr Met Glu Leu

  --- Gly Ala Ser Leu Glu Thr Arg Leu Ala Val Ser Lys Asp Thr Ser Lys Asn Gln Val Val Leu Ser Met
  --- Asx Thr Ser Leu Glu Thr Arg Leu Thr Ile Ser Lys Asp Thr Ser Arg Asn Gln Val Val Leu Ile Met
  Trp Ser Thr Ser Leu Arg Thr Arg Leu Ser Ile Ser Lys Asn Asp Ser Lys Asn Gln Val Val Leu Thr Met
  --- Ser Pro Ser Leu Lys Ser Arg Leu Thr Val Thr Arg Asp Thr Ser Lys Asn Gln Val Val Leu Thr Met

  --- Ala Asp Ser Val Asn Gly Arg Phe Thr Ile Ser Arg Asn Asp Ser Lys Asn Thr Leu Tyr Leu Asn Met
  --- Val Asp Ser Val Lys Gly Arg Phe Thr Ile Ser Arg Asp Asp Ala Lys Asn Ser Leu Tyr Leu Gln Met
  --- Ala Val Ser Val Gln Gly Arg Phe Thr Ile Ser Arg Asn Asp Ser Lys Asn Thr Leu Tyr Leu Gln Met
  --- Ala Asp Thr Val Asn Gly Arg Phe Thr Ile Ser Arg Asn Asp Ser Lys Asn Thr Leu Tyr Leu Leu Met
  --- Ala Asp Ser Val Asn Gly Arg Phe Thr Ile Ser Arg Asn Asp Ser Lys Asn Thr Leu Tyr Leu Gln Met
  --- Ala Asn Ser Val Asn Gly Arg Phe Thr Ile Ser Arg Asn Asp Ser Lys Asn Thr Leu Tyr Leu Gln Met
  --- Ala Asp Ser Val Gln Ala Arg Phe Thr Ile Ser Arg Asn Asp Ser Lys Asn Thr Leu Tyr Leu Gln Met
  --- Ala Val Ser Ala Gln Gly Arg Phe Thr Ile Ser Arg Asn Asp Ser Lys Asn Thr Leu Tyr Leu Gln Met
```

```
   a   b   c   d   e   f   g   h  105 106 107 108 109 110 111 112 113 114 115 116 117 118 119

  --- --- --- --- --- --- --- --- Tyr Ser Pro Glu Glu Tyr Asn Gly Gly Leu Val Thr Val Ser Ser

  Gln --- --- --- --- --- --- --- Tyr Phe Asp Tyr Trp Gly Gln Gly Ile Leu Val Thr
  Pro Ala Pro Ala Gly --- --- --- Tyr Met Asp Val Trp Gly Arg Gly Thr Pro Val Thr
  Val Met Ala Gly Tyr Tyr Tyr Tyr Tyr Met Asp Val Trp Gly Lys Gly Thr Thr Val Thr Val Ser Ser
  Thr Leu --- --- --- --- --- --- Ala Phe Asp Val Trp Gly Gln Gly Thr Lys Val Ala Val Ser Ser

  Ala Met --- --- --- --- --- --- Phe Phe Ala His Trp Gly Gln Gly Thr Leu Val Thr Val Ser Ser
  --- --- --- --- --- --- --- --- Gly Gly Asp Tyr Trp Gly Gln Gly Thr Leu Val Thr Val Ser Thr
  Ala Ala Ser Leu Thr --- --- --- Phe Ser Ala Val Trp Gly Gln Gly Thr Leu Val Thr
  Phe Val Gln --- --- --- --- --- Phe Phe Asp Val Phe Gly Gln Gly Thr Leu Val Thr
  Tyr Val Ser Pro Thr --- --- --- Phe Phe Ala His Tyr Gly Gln Gly Thr Leu Val Thr
  Tyr Val Ser Pro --- --- --- --- Phe Phe Ala His Trp Gly Gln Gly Thr Leu Val Thr
  Thr --- --- --- --- --- --- --- Ser Met Asp Val Trp Gly Gln Gly Thr Pro Val Thr
  Gly Tyr --- --- --- --- --- --- Phe Ser Asp Val Trp Gly Gln Gly Thr Leu Val Ser
  Ala Val --- --- --- --- --- --- Ala Phe Asp Val Trp Gly Gln Gly Thr Lys Val Ser
```

JON, ZAP, and TUR (Capra and Kehoe, 1974). For symbols, see Fig. 2.

Clear-cut answers, however, can only be expected after X-ray crystallography of these molecules. Although a model of an entire immunoglobulin has not yet been obtained, we do have X-ray data of high resolution for biologically important parts and fragments of the molecule.

Figure 5a shows the schematic model of the V part of the κ chain REI, which crystallizes in the form of a V-part dimer (Palm, 1970). The model could be

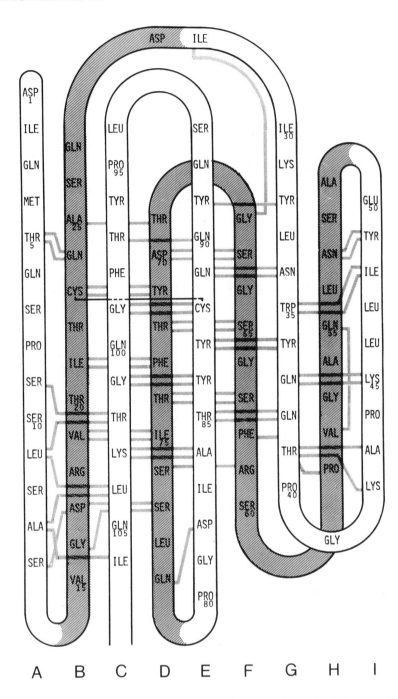

Fig. 5. Tertiary structure of the V part of the κ chain REI (Palm and Hilschmann, 1975). (a) The chain is arranged in two levels (upper level white, lower level hatched). Fifty percent of the structure is that of a β-pleated sheet. Segments are connected by hydrogen bonds. (b) Hypervari-

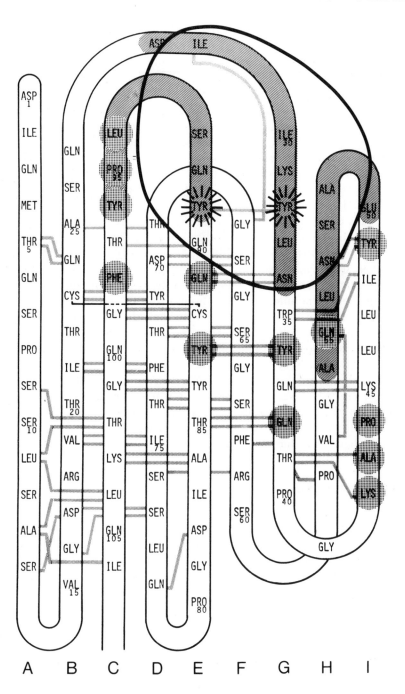

able parts (hatched) are located in one end of the molecule. Marked amino acids give contacts in the dimer, forming a pocket of 15 Å diameter (encircled). This is probably the combining site.

constructed because both the X-ray data up to a resolution of 2.8 Å (Epp *et al.*, 1974) and the primary structure of this protein are known (Palm and Hilschmann, 1973). Valuable conclusions concerning the combining site could be drawn from these studies. The molecule, which has dimensions of 40 by 25 by 28 Å, exhibits a high degree of regularity. At least 50% of the structure is that of a β-pleated sheet, the polypeptide chain being composed of singular segments. These segments are arranged in two levels, five segments in the upper level and four in the lower level. Neighboring segments in the same level run antiparallel and are connected by hydrogen bonds.

This type of folding is also found in λ chain dimers (Schiffer *et al.*, 1973) and in an F (ab) fragment (Poljak *et al.*, 1973). As analyzed so far, both the V part of the H chain and the C parts of the L and H chains show a similar folding. This is not surprising, since all sequences are homologous.

In Fig. 5b, the hatched areas indicate the hypervariable regions of this V part, which accumulate at one end of the molecule. In addition, those residues are marked which make contacts between the two monomers in the dimer. Thus a pocket of 15 Å in diameter results, the bottom of which is formed by tyrosine-36 and glutamine-89 and the walls of which are formed by the hypervariable regions (Epp *et al.*, 1974). It is very likely that this pocket is that part of the combining site to which the L chain contributes. Similar results have also been obtained with other proteins. Only in protein REI, however, is the location of the residues exactly determined.

It is not presently known which antigen is specific for this combining site. The specificity is, however, certainly determined by the side chains which protrude into the combining site, e.g., the tyrosines in positions 32 and 91. It is conceivable that if these two tyrosines are replaced by other residues then the specificity changes, as has been shown for other proteins (Fig. 2).

5. The Constant Part

As already mentioned, the V and C parts in the L chains are of equal length, as are the V parts of the H chains (Fig. 6). The greater length of the H chains is caused solely by their constant parts. In human H chains, the constant part of the γ_1 and α_1 chains is 3 times as long as the L chain C part and that of

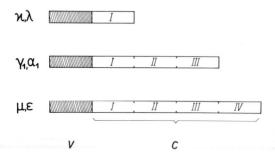

Fig. 6. Chain length of human L chains (κ and λ) and H chains (γ_1, α_1, μ, and ε). Variable parts are hatched.

Fig. 7. Structural comparison of the constant parts of κ, λ, γ_1 (Ponstingl and Hilschmann, 1972), μ (Watanabe *et al.*, 1973), α_1 (Kratzin *et al.*, 1975), and ϵ (Bennich and von Bahr-Lindenström 1974) chains. All chains are homologous, indicating their common evolutionary origin. The constant part of the γ_1 and α_1 chains is 3 times longer than that of the I. chains and the constant part of the μ and ϵ chains is 4 times longer.

the μ and ε chains is 4 times as long (see also Fig. 1). The integer ratio is caused by the fact that the C parts are composed of several homology regions, which correspond to an L chain C part in length (Fig. 6). All chains or homology regions are homologous to each other, but to a different extent (Fig. 7).

This type of homology indicates that all immunoglobulin chains are derived through evolution from a common precursor containing about 110 residues. First the L and H chain genes separated from the common ancestor. Such chain elongation is caused by gene duplication and subsequent fusion, which also occurs with other proteins such as the haptoglobins. There are two exceptions to this rule: (1) the hinge region which follows the end of the C1 homology region and (2) the C-terminal end of the α_1 and μ chains. The hinge region originated from a partial gene duplication of the eight C-terminal amino acid residues of the C1 region. This can be nicely demonstrated in the α_1 chain, where the last gene duplication must have occurred very recently, since the amino acid sequence is not only homologous but also still identical (Fig. 7). Genetically important is the fact that the Inv and Gm factors are located on the constant sections and express themselves in the form of single amino acid exchanges. Since they are inherited as alleles, they indicate that the constant sections of each immunoglobulin chain are controlled by one single gene.

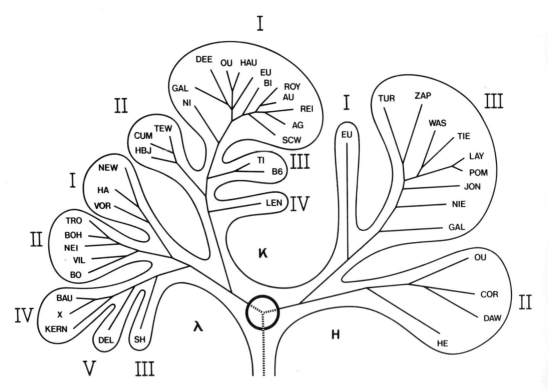

Fig. 8. Phylogenetic tree of V parts of κ, λ, and H chains. The main branches correspond to sub-groups, the finer ramifications to the individual proteins. Tree construction according to Dayhoff (1972) and Engelhard and Hilschmann (1976).

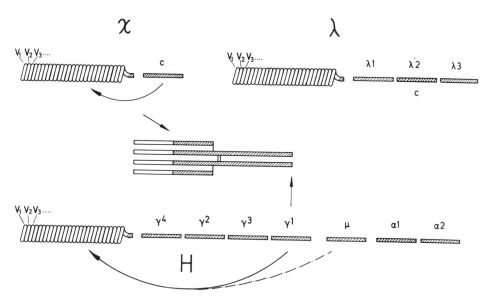

Fig. 9. Genetic control of immunoglobulin structure. There are three sets of V genes—one for κ, one for λ, and one for H chains—containing hundreds or thousands of V genes. There is, however, only one gene for the constant part of each chain type or subtype. The undifferentiated cells become differentiated as soon as the C gene fuses with one of the V genes.

6. Separate Genetic Control of the V and C Parts

Although there has been general agreement on how the C part is genetically controlled, contradictory hypotheses have been proposed for the V part. During the last few years, it became clear, however, that the V part is controlled by many genes, which have arisen by gene duplications and subsequent mutations, beginning hundreds of million years ago (Hilschmann, 1969; Hilschmann *et al.*, 1972). Only in this way can the pattern of variability in the variable parts be explained. Subgroup-specific exchanges or deletions (Fig. 2, 3, and 4) can be interpreted as old mutations which have been acquired in early ancestral genes and which have been carried on in evolution through numerous serially duplicated copies of those genes. Singular amino acid exchanges are more recent mutations which have not yet had the chance to be serially copied. All V genes have *one* common ancestral gene for all chain types. The different degrees of homology and the evolutionary pattern can be demonstrated in a phylogenetic tree (Fig. 8). The number of V genes is most likely in the thousands. There are three V gene sets: one for κ, one for λ, and one for all H chains (Hilschmann *et al.*, 1972; Köhler *et al.*, 1970) (Fig. 9).

The concept of "many genes for the variable part and one gene for the constant part" has a necessary consequence. There must be somatic fusion of the C part with one of the V parts during differentiation. This fusion probably occurs at the DNA level. It is assumed that this fusion occurs twice during immunization in the H chain genes. First a particular V gene is connected to a C gene of μ type, then the same V gene to a C gene of γ or α type. This way specificity remains unchanged, although antibody production switches from IgM to IgG or IgA (Wang *et al.*, 1970).

7. Immunoglobulins and Cell Differentiation

As stated in the introduction, plasma cells synthesize only one type of antibody with one specificity and one amino acid sequence. A cell can switch from IgM to IgG production, while the sequence of the V parts remains the same.

Comparative sequence analyses of V parts (Hilschmann, 1969; Hilschmann *et al.*, 1972) and also DNA-RNA hybridization experiments (Storb, 1972) show more and more convincingly that the information for the specificity of antibodies is genetically fixed. This information is stored in sets of hundreds or thousands of V genes. Since, however, only one V gene is activated for each chain type or subtype, and since neighboring V genes are most likely very similar on the basis of their evolutionary origin, there must be a mechanism acting during differentiation which is able to discriminate between very similar V genes. The Jacob-Monod mechanism proposed for the regulation of enzymes is probably not discriminating enough.

Antibody V parts and enzymes have a number of properties in common. In both systems, changes in specificity are acquired by point mutation. The forces acting between enzyme and substrate and antibody and antigen are of the same nature and magnitude. The number of genes, however, is greater in the immunoglobulin system, which is understandable when one compares the function of antibodies and enzymes. Enzymes have a fixed position within a metabolic chain and one specificity. Antibodies, on the contrary, have to be specific for every possible antigen. With enzymes as well as with the C parts of the immunoglobulins, selection operates in the direction of the structure best suited for one function. With antibodies, it is the opposite. Selection favors not one particular structure but a large number of different V parts, which guarantees that there is more than one specific immunoglobulin for every single antigen. The differences are quantitative in nature, not qualitative. A qualitative difference, however, from every other known protein class exists in the separate genetic control of the V and C parts. While usually the *one gene–one protein* dogma is valid, the antibodies obey the rule *two genes–one polypeptide chain*. We are convinced that this peculiarity is connected with the differentiation of the antibody-forming cells. Through fusion of the C gene with one of the V genes, it will be decided which V gene will be switched on. This fusion turns a multipotent but undifferentiated stem cell into a unipotent but differentiated antibody-synthesizing cell. This happens without the action of the antigen. It assures that every cell produces only one species of antibodies and, furthermore, that a great variety of different specific antibodies originate which are built into the cell wall, where they function as receptors. The antigen selects that cell which synthesizes the best-fitting antibody. The immune system is probably the best model for investigation of differentiation processes on a molecular level.

8. References

Bennich, H., and von Bahr-Lindström, H., 1974, *Progress in Immunology II,* Vol. 1, (L. Birent and L. Holborow, eds.), North-Holland, Amsterdam.

Burnet, F. M., 1959, *The Clonal Selection Theory of Acquired Immunity,* Vanderbilt University Press, Nashville, Tenn.

Capra, D. J., and Kehoe, M. J., 1974, *Proc. Natl. Acad. Sci. USA* **71:**4032.

Dayhoff, M. O., 1972, *Atlas of Protein Sequence and Structure,* National Biomedical Research Foundation, Silver Spring, Md.

Edelman, G. M., and Gall, W. E., 1969, *Annu. Rev. Biochem.* **38:**415.

Engelhard, M., and Hilschmann, N., 1975, *Hoppe-Seyler's Z. Physiol. Chem.* **356:**1413.

Engelhard, M., Hess, M., and Hilschmann, N., 1974, *Hoppe-Seyler's Z. Physiol. Chem.* **355:**85.

Epp, O., Colman, P., Fehlhammer, H., Bode, W., Schiffer, M., Huber, R., and Palm, W., 1974, *Eur. J. Biochem.* **45:**513.

Eulitz, M., 1974, *Eur. J. Biochem.* **50:**49.

Givol, D., Wilchek, M., Eisen, H. N., and Haimovich, J., 1972, *FEBS Symp.* **26:**77.

Hilschmann, N., 1969, *Naturwissenschaften* **56:**195.

Hilschmann, N., Ponstingl, H., Watanabe, S., Barnikol, H. U., Baczko, K., Braun, M., and Leibold, W., 1972, *FEBS Symp.* **26:**31.

Köhler, H., Shimizu, A., Paul, C., Moore, V., and Putnam, F. W., 1970, *Nature (London)* **227:**1318.

Köhler, H., Rudofsky, S., and Kluskens, L., 1975, *J. Immunol.* **114:**415.

Kratzin, H., Altevogt, P., Ruban, E., Kortt, A., Staroscik, K., and Hilschmann, N., 1975, *Hoppe-Seyler's Z. Physiol. Chem.* **356:**1337.

Milstein, C., and Pink, J. R. L., 1970, *Prog. Biophys. Mol. Biol.* **21:**209.

Palm, W. H., 1970, *FEBS Lett.* **10:**46.

Palm, W., and Hilschmann, N., 1973, *Hoppe-Seyler's Z. Physiol. Chem.* **354:**1651.

Palm, W., and Hilschmann, N., 1975, *Hoppe-Seyler's Z. Physiol. Chem.* **356:**167.

Poljak, R. J., Amzel, L. M., Avey, H. P., Chen, B. L., Phizackerley, R. P., and Saul, F., 1973, *Proc. Natl. Acad. Sci. USA* **70:**3305.

Ponstingl, H., and Hilschmann, N., 1972, *Hoppe-Seyler's Z. Physiol. Chem* **353:**1369.

Putnam, F. W., Florent, G., Paul, C., Shinoda, T., and Shimizu, A., 1973a, *Science* **182:**287.

Putnam, F. W., Whitley, E. J., Paul, C., and Davidson, J. N., 1973b, *Biochemistry* **12:**3763.

Schiechl, H., and Hilschmann, N., 1971, *Hoppe-Seyler's Z. Physiol. Chem.* **352:**111.

Schiffer, M., Girling, R. L., Ely, K. R., and Edmundson, A. B., 1973, *Biochemistry* **12:**4620.

Schneider, M., and Hilschmann, N., 1974, *Hoppe-Seyler's Z. Physiol. Chem.* **355:**1164.

Schneider, M., and Hilschmann, N., 1975, *Hoppe-Seyler's Z. Physiol. Chem.* **356:**507.

Scholz, R., and Hilschmann, N., 1975, *Hoppe-Seyler's Z. Physiol. Chem.* **356:**1333.

Shinoda, T., 1973, *J. Biochem. (Tokyo)* **73:**433.

Storb, U., 1972, *J. Immunol.* **108:**755.

Wang, A. C., Wilson, S. K., Hopper, J. E., Fudenberg, H. H., and Nisonoff, A., 1970, *Proc. Natl. Acad. Sci. USA* **66:**337.

Watanabe, S., and Hilschmann, N., 1970, *Hoppe-Seyler's Z. Physiol. Chem.* **351:**1291.

Watanabe, S., Barnikol, H. U., Horn, J., Bertram, J., and Hilschmann, N., 1973, *Hoppe-Seyler's Z. Physiol. Chem.* **354:**1504.

Programs of Gene Action and Progressive Evolution

20

EMILE ZUCKERKANDL

This work is dedicated to Professor Linus Pauling on the occasion of his seventy-fifth birthday.

1. Introduction

Structural genes and their products have captured much of the attention of a number of contributors to this volume on molecular anthropology. It seems to me that, in addition, we must not fail to discuss the role of gene *regulation* and of genetic regulatory networks in evolution, in particular with respect to conditions that paved the way to man.

This way, so far, has been on an upward slope: the animal to which conferences at Burg Wartenstein are devoted *is,* in some sense, the most highly evolved, in spite of the philosophical hypochondriacs who deny that such a statement can be objective. If we hope, tomorrow, to analyze how molecules behave in climbing this slope, we may as well ask ourselves today what the evolution from lower to higher organisms might mean in general molecular terms. To do this, we must at first discuss at some length gene regulation in general. After placing some elements of gene regulation in eukaryotes in their probable mutual relationship, we shall examine the various ways in which programs of gene activity are modified during evolution. An analysis of a possible molecular foundation of parallel evolution should allow us to further define the workings of regulatory relationships. Last, the question of progressive evolution will be discussed.

EMILE ZUCKERKANDL • Marine Biological Laboratory, Woods Hole, Massachusetts, 02543, Directeur de Recherche, CNRS, Paris, and nonresident fellow, Linus Pauling Institute for Science and Medicine, Menlo Park, California.

387

2. Units of Gene Regulation in Eukaryotes

2.1. Genes and Functional Units of Gene Action

We cannot get into the topic of gene regulation without first defining our vocabulary. This implies a short analysis of the elements we are dealing with.

A structural gene is a section of DNA that governs the amino acid sequence of a polypeptide chain, and therefore a section of DNA that is both transcribed and translated.

Alterations in structural genes have regulatory effects. Some of them seem to be autoregulatory (Calhoun and Hatfield, 1973; Goldberger, 1974). One effect of this type—namely the change in rate of synthesis in a polypeptide chain following a mutation affecting that chain—was described a long time ago by Itano (1957) in a pioneering article. We shall not discuss such structure rate effects here. A direct regulatory role of the translation product of the gene, with a specific DNA sequence as target, may be expected to be very rare. A polypeptide chain endowed with a "cellular" function, metabolic or structural—the typical product of a structural gene—can hardly be expected to be at the same time a regulator protein that combines with a receptor gene. Specificity requirements are very high in proteins for both "cellular" and regulator functions. If united in the same polypeptide chain, these functions are likely to be seriously conflicting.

A regulator polypeptide chain and a "cellular" polypeptide chain might combine with one another, however. A "cellular" polypeptide chain might also combine with a small molecule that intervenes in the allosteric regulation of a regulator polypeptide chain. Further, a functional or quantitative deficiency in enzymes participating in hormone synthesis will lead to physiological, morphological, and psychological consequences that implicate altered rates of activity of a whole array of structural genes. In these and other fashions, the regulation of the transcriptional activity of structural genes may depend on other structural genes. It may also depend on sections of DNA that are only transcribed, not translated, and it does depend on others that do not act through their transcription product and may mostly not be transcribed.

The regulatory regions of DNA that may not be transcribed, and whose transcription products, if they exist, do not act on messages sent to other parts of the genome, may be of two kinds. They may be rather short sequences, functioning as targets of macromolecules that interact with them specifically. A good name for them is *receptor genes.* Or they may be represented by long sequences endowed with the property of reversible folding and unfolding. I have given the name *transconformational DNA* to such sections whose hypothetical role is to control the activity of functional units of gene action *(fugas)* that are either included in such a region or apposed to it (Zuckerkandl, 1974).

Some transcribed yet nontranslated stretches of DNA may be viewed as structural genes, and thus provide exceptions to the statement that structural genes are segments of DNA that are both transcribed and translated. This applies to genes for RNA molecules that are involved with cellular functions in the same sense as enzymes or structural proteins are. Examples are transfer RNA and ribosomal RNA. In a symetrically corresponding fashion, stretches of DNA that are transcribed *and* translated will be considered as regulator genes, when the polypeptide product apparently has no function other than to interact specifically with DNA for regulatory purposes, like the lac-operon repressor in *Escherichia coli.*

The last category of regulatory DNA to be considered *a priori* is represented by sections that are transcribed but not translated, giving rise to RNA molecules that interact directly with receptor DNA in the capacity of regulator RNA. The existence of such regulator RNA has not been demonstrated unequivocally. It cannot, however, be ruled out as yet.

The classification of regulatory sections of DNA implies that practically all DNA—i.e., the whole genome—is regulatory in some fashion. However, different sections of DNA exert this general activity in different ways and to different extents. The distinction between structural genes and regulatory DNA still makes sense, provided that it is not viewed in a simplistic way.

Whereas the extent to which structural genes also act as regulator genes is not known and probably is variable, at least one aspect of the situation seems unambiguous at first sight. Namely, there are regions of DNA—and very extensive ones—that do not function as structural genes under any definition of the term. To add to the complexity, not to say confusion, of the picture, we must recognize the possibility that in heterogeneous nuclear RNA (hnRNA) parts of the DNA transcripts may function neither as messengers for the synthesis of protein—this is certain—nor as RNA endowed with regulatory functions—this is uncertain. The idea cannot be excluded that parts of hnRNA are destroyed after their synthesis without exerting any functional role other than one that might be inherent in this synthesis itself.

If there is an ambiguity in the function of different categories of genes with respect to gene regulation, there should at least be no ambiguity in the definitions used (Fig. 1). *Regulator genes* are a particular category of structural genes and the RNA or proteins synthesized under the control of these genes are *regulator molecules.* The regulator molecules exert either activitating or inhibitory action on other structural genes and these are simply called *structural genes,* even though regulator genes in a sense are structural genes, also. The regulator molecules will act by combining specifically with target sections of DNA near the structural genes, and these sections, as mentioned, are *receptor genes.* Regulator and receptor genes are of equal rank with respect to their participation in the control of the activity of structural genes, and when we want to refer to regulator and receptor genes at once, we shall call them *controller genes.* Finally one might unite all regions of DNA giving rise to translation and/ or transcription products under the designation of *emittor genes.* These include *inter alia* classical structural as well as regulator genes.

It has become apparent that a significant part of the eukaryote genome is

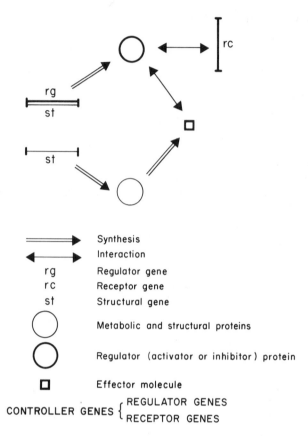

Fig. 1. Control of gene activity: definitions.

organized, as mentioned, into functional units of gene action (e.g., Judd *et al.*, 1972), abbreviated fugas (Zuckerkandl, 1974), whereof the chrommeres in polytene chromosomes of Diptera may be a somewhat ambiguous model, the more so as pecularities in the organization of DNA in *Drosophila* have lately been uncovered (Manning *et al.*, 1975). Polytene chromosomes have nonetheless become essential for our nascent understanding of regulation in higher eukaryotes (cf. Beerman, 1972*a*). We may suppose that a chromomere corresponds to a fuga.

A fuga may be thought to contain (1) structural genes and associated sequences transcribed in mRNA, apparently usually one pcr fuga and per chromomere (one that might, in fact, sometimes be a regulator gene) (Hochman, 1973; Lefevre, 1971; Judd *et al.*, 1972; Beerman, 1972*b*; Lewis *et al.*, 1975,* (2) the sequences transcribed in hnRNA, and (3) any connected further sequences of DNA that are not transcribed but are involved in the control of transcription within the fuga. Some such sequences seem to occur in compara-

---

*However, in at least some cases there seem to be several structural genes per fuga (Falk, 1970).

tively large-sized sections of DNA outside of fugas, in the form of heterochromatin (see Zuckerkandl, 1974), others may occur with a periodicity equal to that of DNA sections of types (1) and (2) and represent effectively parts of fugas. The relationship between a fuga and a complementation group will not be discussed here.

The fuga is thus a continuous section of DNA containing functionally distinguishable parts that form together a unit of transcription and transcriptional control. It is the "operon" of the eukaryote and probably greatly differs in structural and functional anatomy from the classical operon.

2.2. A Unit Regulatory Cell?

In 1964, a model for the regulation of the activity of human hemoglobin genes was proposed (Zuckerkandl, 1964a). It was based on the elements of gene regulation that had been worked out by Jacob and Monod (1961) for the prokaryotes. It used only repressor and not activator molecules as regulators. It was able to tentatively explain a switch from γ chain to β chain production, but when the embryonic ϵ chain came into the picture at about that time the relation of its synthesis to those of the β and γ chains could not be satisfactorily integrated into the scheme. Non-α hemoglobin genes display a cascadelike succession of activities. All genes of the non-α series are repressed except one or a group of a few that are active at any one time. Such a time pattern of transcriptional activities is well accounted for by the simultaneous use of positive *and* negative regulator effects of the kind shown in Fig. 2.

Strides are being made in the understanding of gene regulation in eukaryotes, particularly thanks to the work of Ashburner *et al.* (1973; Ashburner, 1974). In the light of their analysis of the action of the molting hormone ecdyson on puffing in *Drosophila* polytene chromosomes, some basic features of this regulation, although still in part hypothetical, are now better substantiated than before.

From the point of view of the time sequence in puffing, Ashburner *et al.* (1973) distinguish early and late puffs. Their results suggest that it might be of some interest to consider three groups of puffs, the early (23 E, 74 EF, 75 B), late (62 E, 78 D), and late late (63 E, 22 C, 82 F). These are represented by A^1, B^1, and C^1, respectively, in Fig. 2.

From their data, Ashburner *et al.* (1973) derive the view that the early puffs, through a process of transcription and translation, autoinhibit themselves. Such a hypothesis seems the more legitimate as autoregulation is now a widely acknowledged phenomenon (Calhoun and Hatfield, 1973; Goldberger, 1974). It might nevertheless be proposed, as an alternate hypothesis, that the inhibition of the early puffs is caused by the activity of puffs that, timewise, are next in line.

Complete autoinhibition is possible if it is mediated by a protein and if the messenger is stable. On the other hand, if the inhibiting species were ribonucleic acid or a protein depending on an unstable messenger, a dynamic equilibrium

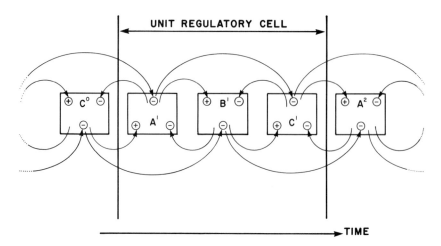

Fig. 2. A proposal for a unit regulatory cell in eukaryotes. The rectangles designate groups of fugas (see text) that are activated nearly simultaneously or, at the limit, individual fugas. No statement is made about the interactions between fugas within one group. Arrows designate the origin and target of regulator proteins, namely activators (+) and repressors (−). Three fugas or groups of fugas, initiated in succession, are necessary and sufficient, according to this scheme, for sequential activations and inactivations of transcription during development. The possibility that such a unit regulatory cell is repeated a certain number of times is suggested by the diagram.

between the puff-promoting and the puff-inhibiting species should be reached and the puff should not disappear. In fact, it seems that if the early puffs autoinhibited themselves, the onset of this process should be earlier than is the case. The puffs start regressing only after a few hours. This might, admittedly, be explained by delayed translation. It may nevertheless be pointed out that early puff 74 EF, for instance, starts regressing after 4 hr, whereas after 1 hr it already had reached about 70 % of its maximum size. There should normally be enough messenger and enough corresponding protein made by then for regression to be well under way. One may consider that the lag time in the regression of the early puffs points to a later puff as the likely causal factor. This view is supported by a further observation of Ashburner *et al.* (1973). Since a process of autoinhibition seems to imply that more and more autoinhibitory protein is produced from stable messenger, autoinhibition would be expected to be accelerated with time. In reality, the rate of decrease of early puffs is not greater terminally than initially. Finally, autoinhibition should start the earlier, the greater the initial rate of increase in puff size, if this rate reflects the increase in rate of transcription.* It is true that at higher rates of transcription more repressor protein molecules are probably needed for autoinhibition.

---

*This may be the case, even though puff formation and transcription seem to be two independently triggered processes. Puff formation appears to be only a condition, not the initiation of transcription. Essential features of a puff may appear even when RNA synthesis is completely inhibited (Pelling, 1964; Berendes, 1968, 1972). Nevertheless, RNA synthesis certainly increases puff size (Berendes, 1968).

But this should be more than compensated for by the fact that the production of mRNA leads to a multiplicative increase of the amount of protein product. Thus, up to a saturation point, the addition of more ecdyson should lead not only to a more rapid increase in the size of early puffs but also to a more rapid decrease. The opposite is found: the larger the puff that is formed, the longer it takes the puff to regress completely.

We shall then assume here that an early puff is inhibited, not by itself, but by a later puff that its activity induces (cf. Fig. 2).

Within one group of fugas that are transcriptionally active at about the same time, mutual interactions may exist. "Early" puff 23 E of Ashburner *et al.* (1973) furnishes an example of a puff that is activated secondarily, even though it is part of a group of puffs that appear almost simultaneously. However, at this time, it does not seem warranted to consider the activation and inactivation of individual chromomeres. It is more advisable to consider small groups of chromomeres together, and this is done in Fig. 2. Each such group is comprised of physically unconnected bands that are activated within a given restricted period of developmental time. It is indeed not possible, at the moment, to estimate how many different fugas are involved in the emission and reception of controller molecules relevant to the interaction between the time-*groups* of puffs. Nor can we tell which of the postulated regulator proteins may actually be identical molecular species that change their affinities and action through some allosteric transition.

The central part of Fig. 2 shows three interacting groups of fugas or, at the limit, three interacting fugas, A^1, B^1, and C^1. The central fuga or group of fugas, B^1, is pictured as being at once the target and the emittor of regulator molecules. As a target, it is subjected to three distinct actions. At first, it is inhibited by a repressor emanating from a set of fugas two steps back in the time sequence of activities, namely from link C^0 in the preceding unit regulatory cell. It is then activated by an activator protein emanating from the set of fugas one step back in time (A^1), which overcomes the inhibition by direct or indirect competition and leads to the release from its DNA receptor of the repressor molecule from C^0. Later B^1 is inhibited again by a regulator molecule synthesized under the control of the set of fugas one step ahead in time, C^1. In this scheme, the absence of transcriptive activity of a fuga is always due to active inhibition, not just to the absence of activation. Likewise, activation is effectively active, and not just due to the absence of inhibition. Each gene is at once the target of positive and negative control.

Each set of fugas is thought to contain at least three key receptor sequences. If two of the emittor effects are attributable to identical molecules, then two adjacent sets of fugas always have one receptor sequence in common, although combination with the regulator molecule leads to inhibition in one case and to induction in the other.

Briefly, this tentative scheme may be justified as follows. Retroinhibition is indicated by two observations: (1) the addition of ecdyson leads to the regression of the puffs of the preceding intermolt stage, i.e., of the puffs corresponding to group C^0 on the diagram, and backward-directed inhibition is therefore operating; (2) the regression of the early puffs (group A^1) is blocked by cycloheximide. This is interpreted as meaning (see discussion above) that the group B^1 of fugas is prevented from determining the synthesis of the repressor protein of the A^1 fugas.

Forward activation is supported by the fact that all late puffs (group B^1) are inhibited by puromycin or cycloheximide. This is interpreted as meaning that the activator protein that is normally emitted under the control of the early group of fugas (A^1) cannot be synthesized. Further, early puff formation (A^1) is dependent on the combination of a preexisting protein with ecdyson. This protein must have been synthesized under the control of a preceding group of puffs, presumably C^0, and represents therefore an activator protein with forward-directed action.

Forward inhibition is suggested by the following observation: if ecdyson, after having been allowed to act and to induce early puff formation, is washed out during the first 4 hr, premature induction of late late puffs (group C^1) results. According to an already adopted feature of the system, late late puffs are considered to be induced by an activator protein synthesized under the control of late puffs (group B^1). The washing out of the ecdyson brings about a regression of the early puffs. However, the late puffs, which are already induced through the initial action of the early puffs, induce in turn the late late puffs. The fact that they can do it at this stage suggests that the early puffs normally control the synthesis of an inhibitor of the late late puffs. In fact, upon reexposure of the washed salivary glands to hormone, the prematurely induced late late puffs regress again. This is precisely what would be expected if the early puffs synthesized again the repressor of the late late puffs. As the early puffs A^1 regress for the second time, under the action of the late puffs B^1, the late late puffs C^1 are again free to be active under the action of the late puffs.

The regulatory relationships as suggested by the data of Ashburner *et al.* may be extrapolated to further stages of development and schematized as in Fig. 2. Even if this diagram correctly interpreted gene regulation in *Drosophila* as it occurs over a certain developmental phase, it is obvious that many other regulatory relationships may obtain. There is no basis for speculating how many times a "unit regulatory cell," such as depicted, might in fact be repeated.

The results of Ashburner *et al.* (1973; Ashburner, 1974) suggest that simultaneous positive and negative control is the basic mode of gene control in eukaryotes. Interestingly, mixed positive and negative control systems have been found in prokaryotes also (e.g., Englesberg *et al.*, 1969; Englesberg and Wilcox, 1974). In this respect, eukaryotes may have essentially taken over a heritage from prokaryotes.

The integrative features of gene control as provided by Britten and Davidson's model of gene control in eukaryotes, namely the type of relationship that they picture between regulator, receptor, and structural genes (in our terminology), has considerable potential for explaining the relationships between partially overlapping gene programs, such as we shall consider in a later section of this chapter. However, only positive specific control is considered by the authors, negative control being abandoned to nonspecifically combining histones (Davidson and Britten, 1973). According to the present scheme, negative specific control is also inherited from the prokaryotes. At the same time it is contemplated that an *additional,* transconformational mechanism of gene control, involving nonspecifically bound histones, is operative in eukaryotes (Zuckerkandl, 1974).

A further remark about the Britten-Davidson model seems to be in order. It also applies to the Georgiev (1969) model and to the Elder (1973) revision of the Britten-Davidson model. In the regulatory system, as worked out by these last authors and pictured for instance in Fig. 6,3 of the article by Davidson and Britten (1973), structural genes are flanked by contiguous receptor genes. Linked regulator genes, which would be of the "emittor" type, active in *trans,* are missing. They should frequently be found within fugas, and specifically within the transcribed area of fugas. Because of the absence of these emittor sequences, a necessary element of circularity for equilibria of gene activities and shifts in such equilibria seems to be lacking.* The system seems to go only one way, from the binding of an allosterically activated protein by a "sensor" gene (i.e., in the terminology used here, a receptor gene linked to and controlling the transcriptive activity of a regulator gene) to the synthesis of messenger RNA under the control of a structural gene. The transcriptive activity of the structural gene fails to be directly associated with another transcriptive activity directed at interaction with receptor sequences, notably the sensor genes of Britten and Davidson. If such a provision is not made, it is difficult to understand, for example, why at a particular moment of development a certain protein and not another is present to combine with a sensor gene. Many fugas must be able to send out controller messages besides receiving such messages.† The postulated emittor sequence would, in a number of cases, exert a negative control.

Although some nuclear types of RNA molecules of unknown destination are begging for a function (Weinberg and Penman, 1969; Moriyama *et al.,* 1969; Holmes *et al.,* 1972), the bulk of experimental evidence has so far been in favor of protein being the regulator species (e.g., Paul *et al.,* 1974; Ashburner, 1974).

It was pointed out above that regulatory interactions such as those pictured in Fig. 2 probably represent only one aspect of gene regulation in eukaryotes. Stenogenic control and eurygenic control of gene activity have been defined, the first as relative to a narrowly determined pattern of transcriptive activities,

---

*Regulator genes might have been considered a special case of the Britten-Davidson "producer genes" (a synonym for structural genes), and thus the Britten-Davidson model would accommodate formally the presence of regulatory emittor sequences within fugas, if the definition of the producer gene, as given by the authors (1969), did not explicitly exclude species engaged directly in genome regulation.

†In Fig. 6,3 of the article by Davidson and Britten (1973), an "integrator" (regulator) sequence has been placed in the vicinity of a structural gene. It is, however, supposed to act there as a receptor element, and thus does not satisfy the requirement of an emittor sequence put forth here.

relevant, for instance, to the question of whether an erythroid cell synthesizes fetal or adult hemoglobin, and the second, to which the first is subordinate, as encompassing several alternate stenogenic states and determining, for instance, whether a cell is of the erythroid type or not (Sallei and Zuckerkandl, 1975; Zuckerkandl, 1975). Eurygenic control thus concerns tissue differentiation. At the moment, these notions cannot help but shift between two alternative states. The question is whether the control of transcriptive activity in the genomes of eukaryotes is essentially monistic or dualistic.

It would be monistic if throughout all phases of differentiation only one type of mechanism were effective, or, perhaps a combination of mechanisms that were, however, intertwined in similar fashion over all developmental stages in a way that did not allow them to be considered as alternate occurrences. In this case, eurygenicity and stenogenicity would be only relative concepts. Each more narrowly determined state of differentiation would be stenogenic in relation to a preceding less narrowly determined, more eurygenic state, which would include the stenogenic state among its potentialities.

Gene control in eukaryotes would be dualistic, on the other hand, if a control involving a shift between certain higher and lower orders of DNA structure, relevant to transcriptive activity, came into play only at certain times of tissue determination; whereas another type of control, taken over from the prokaryote heritage and giving rise to regulatory patterns more or less similar to the one pictured in Fig. 2, determined the precise constellations of gene activities. For these, the more general process would provide the openings but not the triggers for actual occurrence. In this case, the distinction between eurygenicity and stenogenicity would be absolute. The notions would apply to two mechanisms that act in succession and would perhaps be congruent with the distinction between determination and differentiation.

Transconformational events are expected to accompany the combination between regulator and receptor genes, the progression of polymerase molecules along the DNA, etc. Transconformation in DNA thus would undoubtedly play an important role in stenogenic processes as well as in the eurygenic processes, even if the two were basically distinct. But in eurygenic processes the transconformational events in DNA would concern shifts between higher orders of structural states, and in stenogenic processes between lower orders.

A justification for the dualistic view, which I personally favor, is not to be presented here. The following nevertheless needs to be stated: even if integrative aspects of the Britten-Davidson model and certain elements of the present model were verified by future research, it may be that only a partial answer to the problem of the mechanisms of cell differentiation in eukaryotes would thereby be provided.

2.3. Effects of Mutations in Controller Genes

Point mutations—the class of mutations most frequently accepted by evolution—do not have the same effects on controller genes as they do on structural genes. Two distinctions may be made. One relates to the range of the quantitative effects of missense mutations and the other to the qualitative character of the effects.

The distinctions are made clear in Fig. 3. First, individual point mutations in emittor genes (structural genes and regulator genes) may have a very wide range of effects on the functionality of the polypeptides controlled by these genes, from undetectable to totally inactivating. On the other hand, the substitution of a single base pair in receptor DNA may be expected to have mostly only limited and incremental effects. Indeed, in the receptor DNA sequence, a single base substitution should not as a rule lead to very notable effects in secondary and tertiary DNA structure. Although a single base substitution that occurs at random may be expected to decrease the affinity between regulator protein and receptor DNA, this decrease should only occasionally be radical. On the other hand, it could be expected, as in the case of human hemoglobins, that some point mutations totally inactivate the molecules functionally and others destabilize them and thus decrease their life span.

Second, single amino acid substitutions in polypeptide chains, when they have any detectable functional effect at all, easily lead to qualitative functional changes. These may amount to a different combination of quantitative physicochemical characteristics, to the abolition of some functional characteristics while

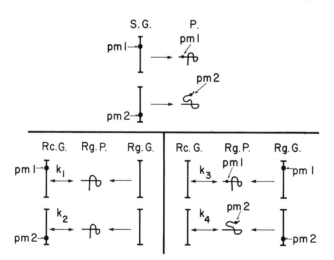

Fig. 3. The effects of mutations in controller genes are quantitative, not qualitative. S.G., structural gene; P., protein, Rg.G., regulator gene; Rg.P., regulator protein; Rc.G., receptor gene; pm, point mutation; k_1-k_4, affinity constants for interaction between regulator molecule and receptor DNA. The upper part of the diagram shows that different point mutations in a structural gene can lead to qualitative differences in the protein product. The lower part of the diagram emphasizes that, with respect to interactions between regulator and receptor molecules, the effects of point mutations will be only quantitative (change in affinity constant). This is so whether the mutation occurs in the receptor gene (lower left) or in the regulator gene (lower right), even though the regulator gene also has the character of a structural gene.

others are maintained, or, more exceptionally, to the appearance of new functional characteristics. To this we may oppose the absence of any possible qualitative functional change in the interaction between a given set of controller molecules (as long as the set remains the same, and no translocation has taken place). At the level of the receptor gene, all that can change is the value of the affinity for the regulator molecule. This is a quantity. It may of course have important qualitative repercussions at higher levels of biological integration. At the molecular level, the nature of the function is not altered.

The distinction between two classes of average quantitative characteristics of the effects of point mutations probably parallels the distinction between receptor and emittor genes. The latter, as mentioned, are comprised of both "classical" structural and of regulator genes. Thus the distinction here is not exactly between the regulating and the regulated molecules. On the other hand, with respect to the presence or absence of a qualitative functional effect, the dividing line does pass between the controller genes (comprising receptor *and* emittor controller genes) on the one hand and classical structural genes on the other.

The expectation that the quantitative effects of single point mutations on the functionality of receptor genes will in general be weak and incremental is borne out by available evidence on the lactose operator in *Escherichia coli* (Smith and Sadler, 1971). Likewise, the expectation that point mutations in regulator genes will have stronger, more often totally inactivating, effects is supported by experimental evidence relating to the repressor protein of the lactose operon in *E. coli* (Shineberg, 1974). However, in over one half of the sample, the binding of the repressor to the operator was only impaired, not eliminated. Suppressed nonsense mutations, on the other hand, lead to functional impairment most of the time (Miller *et al.*, 1975) (but see Weber *et al.*, 1972).

As to a structural gene that is part of the same lactose operon, the gene for β-galactosidase, the proportion of functionally impairing missense mutations is much lower than one would anticipate (Langridge, 1974). This finding, however, must be judged in the light of the following circumstances: the capacity of the system to lead to increased enzyme synthesis in the case of functional impairment of the enzyme molecule, and perhaps the size of the polypeptide chain (one amino acid substitution in a chain of 100 may have on the average a larger functional effect than one in a chain of 1000). Moreover, it remains to be seen whether the weighted functional density (Zuckerkandl, 1976*a*) of β-galactosidase is not below that of the average protein. If it were, this would account in part for the fact that a large proportion of the amino acid substitutions are innocuous. The example of β-galactosidase or of other enzymes capable of compensating for molecular dysfunction by increased rates of synthesis cannot as yet be considered as representative.

For the purpose of illustrating what effects mutations in the controller system will have, let us consider just two of the elements of the unit regulatory cell pictured in Fig. 2, namely the forward activator and the backward repressor. Let us assume that these are regulator proteins which, in acting on a fuga, compete with each other, as the results of Ashburner *et al.* (1973) suggest. Such competition need not imply that the receptor sequences are the same. It would be sufficient for receptor sites to be linked so that one will undergo an allosteric transition as a consequence of a regulator-receptor interaction at the next receptor site. Let us further assume that the mutations considered change the conditions of this competition slightly enough so that a newly appearing mutated regulator molecule will, as before mutation, displace completely the regulator molecule with which it competes, only at higher concentrations than it did before, or at lower concentrations; further that, if higher concentrations are required, these are indeed reached in the system, but at a somewhat later time than are the concentrations that are normally effective. Finally, let us assume that the maximal rates of synthesis of the regulator molecules remain unaffected. Different assumptions could of course be made, and different conditions will apply in many cases, but there is no point in covering a wider range of *a priori* possibilities to show the effects that will be obtained by single mutations in such a system. In view of certain changes in gene programming to be discussed below, the interest here is in mutations with moderate, incremental effects on affinity constants between regulator proteins and receptor DNA.

Consider, then, a fuga, say B^1 in Fig. 2, that is the target, first of an activator protein controlled by a fuga A^1 whose activity peak came at an earlier time, and later of a repressor controlled by a fuga C^1 whose activity period is next. The following cases can be listed:

1. *Increase in affinity between activator and target DNA.* Gene products from target fuga B^1 will appear earlier than normal, since a smaller concentration of activator from A^1 will be effective. The B^1 gene products will be synthesized over a longer time because, as the repressor from C^1 begins to appear, it has to be synthesized over a longer time to reach a concentration sufficient for displacing the activator from A^1. Under our assumptions, the time of synthesis and the total amount of gene product from B^1 will be increased on these two counts, therefore rather notably.

2. *Decrease in affinity between repressor and target.* The competition of the repressor from C^1 with the activator from A^1 for target B^1 will be less efficient. The total time of synthesis and amount of gene product from B^1 will again be increased, but on one count only.

3. *Decrease in affinity between activator and target.* Gene products from B^1 will appear later, since an accumulation of a larger amount of activator molecules from A^1 is required for the full activation of B^1. Furthermore, the gene products will be synthesized for a shorter time, since a smaller concentration of competing repressor from C^1 is necessary for achieving repression. The total time of synthesis and the total amount of gene product from B^1 will therefore be decreased on two counts.

4. *Increase in affinity between repressor and target.* Synthesis of gene product from B^1 will discontinue earlier, since smaller concentrations of repressor from C^1 will be effective. Time of activity of B^1 and total amount of gene product from it will be decreased again.

These changes, a fraction of those that might occur, are represented in Table 1. More generally, mutational changes in such a system may be expected to affect
   —the time in relation to a developmental stage at which a gene product is made,
   —the duration of its period of synthesis,
   —the rate of its synthesis,
   —the total output of gene product.

Similar results can be brought about, basically, in two ways. One has just been considered, namely mutations in the gene controlling the regulator molecule (activator or repressor) or in the receptor sequence, with an ensuing change in the affinity between the interacting species. The other process leaves the affinity constants unaltered, but alters the quantity of either activator, repressor, or effector. Since regulator genes are also structural genes of a kind and as such are subject to regulation, the rate and quantity of synthesis of regulator molecules can be altered by a mutation in the gene for a regulator of the regulator. (In fact, it is conceivable that fugas exist that contain no "structural gene" other than a regulator gene.) Alterations in the structural genes coding for enzymes necessary for the synthesis of an effector or in controller genes relating to these enzymes are obvious alternatives. Long chains of reactions, implicating any one among a great many genes, and among them any one among a number of structural genes, may thus be implicated in explaining similar end effects. This will allow us, later, to interpret parallel evolution and other phenomena.

The mutational effects envisaged have a bearing on the timing of gene activity, but basically only in the sense of a delay or an acceleration in the succession of the developmental phases. It would, however, be rash to say that the developmental *pattern* of an organism cannot be affected by a simple delay and a reduced output of gene product. In ontogeny, induction must indeed depend at least indirectly on the synthesis of informational macromolecules. If,

Table 1. Anticipated Effects of Mutations in Controller Genes, under Conditions Defined in the Text

	Change in affinity with DNA target	Consequences		
		Onset of synthesis	Arrest of synthesis	Total time of synthesis and total amount of gene product
Repressor	↑	Delayed	Normal	↓
	↓	Advanced	Normal	↑
Activator	↑	Advanced	Delayed	↑ ↑
	↓	Delayed	Advanced	↓ ↓

on account of a developmental delay in one tissue, the right inducer is not present at the right time, the target tissue may lose its state of competence and induction, rather than being delayed, may be prevented.

Such an event would amount to a change in the *topology* of the regulatory circuits of an organism that are actually effective. Generally, however, accelerations and retardations in regulatory effects will occur under constant topology of regulatory circuitry. This is an important distinction for considerations of changes in gene programming. Changes in topology of regulatory circuitry may be brought about either by the indirect effect of changes in affinity constants in controller molecules or by the direct effect of topology-modifying mutations, such as a modification in the relative position of controller elements* and structural genes, through translocation events. The basic types of changes in gene programming can thus be grouped under the two headings:

1. Pattern of regulatory circuitry maintained constant.
   Changes expected in
   —time of onset and duration of transcriptional activities.
   —rate of transcriptional activities.
2. Modification of regulatory circuitry.
   Changes expected in
   —developmental (temporal) context of transcriptional activity.
   —tissular (spatial) context of transcriptional activity.

At the limit, mutational changes of type (1) are turned into mutational changes of type (2). This should happen particularly when the change in an affinity constant is radical. When such a radical change is maintained by natural selection, we have an evolutionary event consisting of a change of regulatory topology. The affinity of a regulatory molecule for its receptor may have become so great that the regulator cannot be dislodged by a competing regulator. Or else, more commonly (at least more commonly at the mutational level), a regulator-receptor interaction will cease, the interaction constant dropping below the threshold of functionality. As a consequence, in both cases, certain events of activation or repression will not occur at all in the organism. Ensuing effects on the spatial insertion of gene programming will, however, probably be confined to changes in location of certain developmental (e.g., morphogenetic) processes within one type of tissue (e.g., homoeosis). Such a restriction surely does not apply to changes in spatial context of coordinated gene activities when the changes in circuitry are due to translocations.

Processes of type (1), namely relatively moderate changes in affinity constants within an established multipolar controller system, *should account for the control of continuously variable characters, such as morphological characters.* Processes of type (2), changes in regulatory topology, on the other hand, lead to discontinuous variation.

Processes of type (1) can also lead to discontinuous variation by the

*Including transconformational DNA, such as constitutive and nonconstitutive heterochromatin (Zuckerkandl, 1974).

"flipping over" of certain equilibria between parts of the controller system, a topic that will be briefly discussed below.

One of the most frequently asked biological questions is how an organism can determine that the right components be made at the right time and in the right place. A likely general answer is: by adjusting the mutual relationships between affinity constants characterizing regulator-receptor relationships. A great many different mutations may, in this respect, lead to similar results. Thus the mutational resources for modulating regulator-receptor relationships and relationships between regulator-receptor couples and groups of such couples must be considerable. There is no *a priori* difficulty in understanding, tentatively and in most general terms, how such relationships can vary in any desired direction, including reversals to anterior states, or how similar phenotypic results can be achieved through different genetic means. Such reversals, or such parallel evolution of phenotypic characters do not necessitate the appearance or reappearance of identical situations with respect to the mutations involved, even when the same regulatory systems (regulator-receptor couples) are involved. This assertion is founded on the purely quantitative aspects, pointed out above, of mutational changes affecting regulator-receptor functional relations.

3. Impact of Gene Regulation on Evolution

Invoking changes in gene regulation as a dominant factor in evolutionary transformations is not novel today, especially in view of the recent and current work of Wilson and his associates (Wilson *et al.*, 1974*a,b*). In doing so, however, I am quoting not only others but also myself. I remember first addressing myself to the subject at a Burg Wartenstein conference in 1962 (Zuckerkandl, 1963). At that time, one of the members of the conference, the anatomist Otto Schultz, stated that many anatomical differences between man and the anthropoid apes can be reduced to a matter of differential growth rates. "Not in a single species," said Schultz, "occur only retardations . . . nor exclusively accelerations, since every single feature can independently shift its place in the sequence of ontogenetic processes in either direction."

This suggested to me that mutations in a class of rate-controlling genes might be responsible for the observed effects, that the organization of living systems may be changed by a variation not so much in structure as in dosage of gene products, without the introduction of any new types of genes, and that such a variation might be ascribable to DNA that presides over the temporal and quantitative regulation of the activity of structural genes rather than to structural genes. In 1962, this possibility was evoked only as one among others. Yet an experimental test of this hypothesis was proposed in general terms, and this test was of the nature of one that has now been applied by Wilson *et al.* (1974*b*) in their work on the comparison of rates of evolutionary change in chromosome numbers on the one hand and in structural genes on the other. In

1965 (Zuckerkandl and Pauling, 1965), we stated again that many phenotypic differences may be the result of changes in the patterns of timing and rate of the activity of structural genes rather than of changes in functional properties of the polypeptides due to changes in amino acid sequence. Consequently, we said, two organisms may be phenotypically more different than they are on the basis of the amino acid sequences of their polypeptide chains. In 1968, this hypothesis was elaborated on further. It was considered that reproducible morphogenesis may depend on the constancy of genic regulation more than on the constancy of genic structure (Zuckerkandl, 1968). The same phenotypes, it was said, may thus be controlled by different structural genes, provided that the regulatory mechanisms remain constant, whereas different phenotypes may be produced by the same structural genes when the ratio of their synthetic activities and/or the timing of their activities is altered. During ontogenesis, which is characterized by multiple transformation of an organism's morphology, the same structural genes may contribute to the different forms that succeed each other. The difference in the successive results of the interaction between genes, it was proposed, may be attributed to changes in relative rates of synthesis of polypeptide chains under the control of these structural genes, as well as to the intervention of previously inactive genes and to the repression of previously active ones.

It may now be useful to proceed to a general survey of types of changes in the programming of the transcription of structural genes. Mutational changes offer potential models for evolutionary change. We may thus examine types of mutational and of evolutionary transformations in the programming of individual as well as of coordinated sets of structural genes. Limiting illustrative examples to the animal kingdom will be a regrettable bias, but one in keeping with the theme of this volume.

3.1. Changes in Control of Individual Structural Genes

Consider first changes in gene control that implicate individual structural genes. One example of such a change in control is a redistribution of variant duplicates of a gene over different phases of ontogenesis. This is an example of mutation in temporal regulation. Embryonic, fetal, and adult hemoglobin chains illustrate such a case. The first published molecular phylogenetic tree (Zuckerkandl and Pauling, 1965)—drawn simply on the basis of numbers of amino acid differences between chains, because we considered in 1964 that the genetic code was not yet sufficiently well established to use it—showed that in the ancestry of cattle the adult non-α gene presumably underwent duplication, and that one duplicate was apparently transferred to the control of a fetal gene complex. This duplicate has functioned since as the fetal hemoglobin gene along the artiodactyl line of descent. Such "temporal translocations," involving different developmental stages, seem to imply a change in the topology of regulatory circuitry, even though they do not necessarily imply spatial translocations.

Under the same heading come the numerous cases where the activity of gene duplicates is confined to different tissues, even though the evolutionary history of duplicates necessarily suggests that they were not always so distributed. Hemoglobin and myoglobin genes offer one example. Homologous genes that are active in different tissues seem never to be found to be, and may never actually be, closely linked. Such homologous genes have arisen either by gene duplication, followed by the translocation of a gene duplicate, or by chromosome duplication (Ohno, 1973). There is no doubt that the transcriptive activity of closely linked structural genes often is *temporally* and quantitatively varied, as illustrated by the human non-α hemoglobin genes (for the linkage and order of these genes, see Huisman *et al.,* 1972; Kendall *et al.,* 1973). Yet different duplicates found in a linkage group may be unable to be involved individually in the determination of different *tissues.*

If further data confirm these extrapolations, *tissue differentiation would be seen to involve blocks of contiguous genes,* i.e., segments of chromosomes, rather than an organization of activities of individual genes scattered all over the genome. At the same time, tissue differentiation would implicate a certain number of such blocks, and these would indeed be scattered over the genome. Eurygenic differentiation would be able to act only on whole linkage groups, and not on individual structural genes that are members of such groups, whereas stenogenic differentiation can act on individual genes or small groups of genes *within* a linkage group. The suggestion would then be that the distinction between eurygenic and stenogenic differentiation is more likely to be an absolute than merely a relative one; in other words that we are dealing with two distinct underlying mechanisms.

3.2. Alterations in Activity Ratios of Genes

A second kind of change in genetic programs consists of alterations of activity ratios within a given activity complex of structural genes. This kind of change occurs under constant topology of regulatory circuitry. Phenotypic alterations and adaptations induced by the environment are undoubtedly based on the intervention of external factors at various points of the regulatory network, through their presence or absence. It may be possible to distinguish among such phenotypic alterations those that must be considered as changes in state of differentiation from others that may only represent modified rates of activities within one given state of differentiation. When ratios of rates of synthesis of polypeptide chains, probably reflecting transcriptive activity,* are altered secondarily by environmental factors, the slope of the regression line of the ratio of two such rates on one of the rates seems not to be altered environmentally and may characterize a given state of gene regulation (Fig. 4).

---

*Some experimental evidence, involving the use of actinomycin D (Sallei and Zuckerkandl, in preparation), has been confirmed by the work of Lanyon *et al.* (1975), who showed that translational control is not a major control mechanism in the expression of human globin genes.

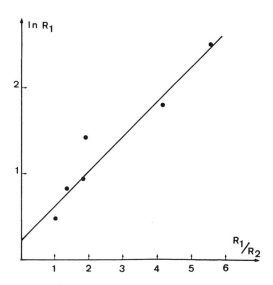

Fig. 4. The regression of the ratio between two rates of gene transcription on one of the rates may characterize a state of differentiation. Synthesis of human β- and γ-hemoglobin chains: the factor of stimulation of the most actively synthesized chain in relation to the ratios of the factor of stimulation of the two chains. From Sallei and Zuckerkandl (1975).

A change in the slope of such a regression line would characterize a change in state of differentiation. (On the other hand, different eurygenically related stenogenic states might yield regression lines with similar slopes.)

Environmentally induced (e.g., in the brine shrimp *Artemia salina*) as well as evolutionarily fixed changes in morphology can very probably be attributed to changes in the activity ratios of genes, structural and regulatory. There is considerable evidence to the effect that *extensive changes in morphology can take place in the absence of supplementary or modified structural genes* (see below the cases of bithorax mutants in *Drosophila* and *E* gene mutants in *Bombyx*). Functional innovation at the morphological level does not appear to require any functional innovation at the molecular level. One may wonder whether the formation, in the course of evolution, of a limb of a land vertebrate from the fin of a fish requires the appearance of new proteins endowed with novel functions. I am inclined to believe that it does not. To be sure, there may be types of proteins in a limb that were not present in a fin. The point is that such proteins, if and when they appear in an appendage, were not "invented" for that occasion, but preexisted in other parts of the organism. The required structural genes preexisted, but gene programming had to be modified. Evolution is largely, and morphological evolution is mainly, innovation in gene programmation.

Turning a limb, for instance, into an antenna, which is a sense organ, is one instance when the activity of new genes, presumably already present in the animal, is added to parts of the "limb gene complex." In terms of the vocabulary to be proposed below, a certain "appended program," which defines the sensory

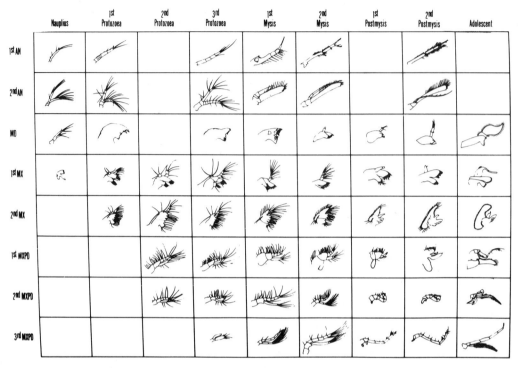

Fig. 5. The variations over the spatial and temporal dimensions of the shape of appendages in a cephalocarid crustacean. The columns relate to different larval stages. The rows designate the most anterior appendages in their anatomical order: first and second antenna, mandible, first and second maxilla, first to third maxilliped. From Sanders (1963), by kind permission of the author.

organ, is being connected through an integrating master switch to elements of the appended program represented by the "limb gene complex."

The evolutionary time required for the transformation of fin to limb is perhaps considerable.* The amino acid sequence of the proteins involved will therefore undoubtedly have varied, and the functional characteristics of these proteins will in turn have varied to some extent. Yet the functions and tertiary structures of the proteins involved will probably have been maintained basically constant in the face of conservative amino acid substitutions, and no structural genes will have been added to those that the organism already contained, with the exception of some duplicates of preexisting genes. Such duplicates will be either identical or very similar and will not represent functional innovations at the molecular level. Differences between two appendages or other parts of the anatomy that morphologists consider to be homologous probably amount largely to differences in activity ratios between kinds of structural genes that are common to both morphological formations.

The variety of types of appendages that can be produced in a single

*It may have taken less than the duration of the Devonian, 65 million years (e.g., Frazzetta, 1975).

arthropod, such as in higher crustaceans (e.g., see Korschelt, 1944; Sanders, 1963), has a spatial as well as a temporal dimension: it is manifest over the body segments at any one time of development, and for any one body segment throughout the developmental stages (Fig. 5). Does each of these different appendages then depend upon a different set of structural genes, with perhaps some overlaps? Among all possible hypotheses, this may well be the most unlikely, because it is the most uneconomical. We may expect that the proteins present in the differently shaped appendages, at different times and in different locations, will be found to be mostly the same. Changes in the form of appendages can be analyzed in terms of allometric relationships. Evolutionary, as well as developmental, changes in allometric relationships may be traceable to alterations in activity ratios of identical structural genes.

In this respect, the famous findings of Thompson (1917), reported in his book *On Growth and Form,* are particularly suggestive. He considered a number of related organisms and inscribed their outlines between cartesian coordinates. As he changed the coordinates, the outline of one organism appeared as a simple mathematical expression of the outline of another. One particular mode of variation of the coordinates was sometimes prominent throughout the entire organism. In comparing, for instance, the porcupine fish (*Diodon*) with the sunfish (*Orthagoriscus*) (Fig. 6), the treatment accounts not only for the change in overall shape of the fish, but also for all the apparently separate and distinct external differences between them. The skeleton follows the same relation. Such generalized coordinated effects can be due only to controller genes or to structural and controller genes that determine the synthesis of effector molecules. And one is led to infer that *a few interacting master regulator genes extend their morphogenetic effects to the whole organism.*

Morphological transformations of this kind are most likely obtained through progressive, directional evolutionary series. The fossil record offers many examples of such series. The purely quantitative character of changes in the interaction between regulator molecules and receptor genes has been pointed out above. This circumstance allows one to tentatively interpret directional evolutionary series in simple terms. *Any progressive, directional evolutionary trend may be based on the progressive decrease in affinity constants between regulator molecules and receptor genes.* The decrease will be incremental especially in the case of mutations in receptor genes. Such a decrease may obviously be expected to be a universal mutational trend. This trend could be made use of in oriented (orthogenetic) evolution. Over long evolutionary periods, it must necessarily be balanced, along most lines of descent, by corresponding increases in affinity constants, which occur perhaps elsewhere in the genome. Taken together, the decreases and increases probably account for most of the evolutionary transformations of organisms. Yet the supposition can be made that the decreases correlate more generally with directional evolution, since they correspond to a spontaneous, thermodynamically oriented, mutational trend. Repeated mutational "proposals," if they are not notably deleterious, probably eventually become fixed in populations.

As more regulator-receptor gene pairs come to interact at the level of the

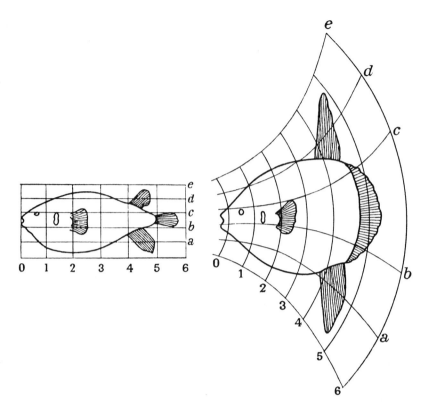

Fig. 6. Coordinated evolutionary changes in shape are considered to be attributable to mutations in controller genes, or to mutations changing the availability of effector molecules. Left: *Diodon,* the porcupine fish. Right: *Orthagoriscus mola,* the sunfish. From Thompson (1917).

regulation of general morphology, the mathematical relationship between shapes of related organisms, of the kind uncovered by Thompson, should become more complex. Vermeij (1973) states that the potential versatility of a given higher taxon or body plan of organization is determined by the number and range of independent parameters controlling form. The increase in the number of controlling parameters may be tentatively interpreted in terms of an increase in regulator-receptor gene pairs that interact in general morphogenetic determination. The relative simplicity of Thompson's mathematical transforms suggests that this number is modest.

We converge here with a theoretical finding made independently by King (personal communication) namely that mutations in controller sections of DNA are likely to account for the progressively variable quantitative effects attributed by geneticists to so-called polygenes—obscure entities which now seem to find a plausible and concrete substratum. This substratum is represented by the cumulative effects of mutations in controller sections of the genome. Mutational alterations in the affinity between regulator molecules and receptor DNA should be typical "polygenic" events, but mutations in other sections of DNA associated with a structural gene within the confines of a fuga, such as in the hypothetical transconformational DNA, may also influence the transcription of a structural gene in an incremental, oriented fashion.

Decays in affinity constants that are accepted by evolution are not expected to usually proceed to the point of a breakdown of integration patterns. They will mostly only modulate the equilibria between interacting groups of controller genes. Increases in affinity between regulator molecules and receptor genes may be a prominent feature during phases of buildup of genome organization, of extensions and recombinations of gene programs, and of shifts in regulatory topology. Decreases in affinity may be prominent when shifts in regulation occur within a constant topology of the regulatory network.

The refined mutual adjustment of parts of an organism, through all evolutionary physiological and morphological transformations, need not be considered a miracle, as some feel compelled to do. A notable part of internal coadaptations no doubt occur *automatically*. Established interactions between controller genes are likely to see to it that all spaces be properly filled and all topographical relationships between cells and tissues be of the proper kind. Mutual adaptation between cell groups, tissues, organs, etc., surely does not involve any special measure on the part of the biological system; it is a necessary projection of the processes by which this system was built up in the first place. Mutations that affect morphogenesis probably produce coordinated modifications of inductive effects such that the new state of the system respects the rules of tissular relationships that obtained in the preceding state. Such rules must have been established during very remote ancestral evolutionary phases and are probably expressed by spontaneous coordination within a system of regulator–effector–receptor relationships when any part of such a system is modified. Thus new mutations are just an occasion for the manifestation of old relationships. Vascularization of a tumor is an example. When a mutation brings about, say, a change in the shape of a limb, which contains bones, muscles, nerves, blood vessels, and so on, it is not appropriate, in order to account for the coherence, the coadaptation, the functionality of the modified structure, to postulate adaptive mutations in "bone genes," "nerve genes," "muscle genes," etc., or in controller genes corresponding to each of these kinds. In fact, there undoubtedly *is* no such thing as a specific bone gene or muscle gene or nerve gene, etc., beyond the structural genes that determine proteins filling physiological functions found exclusively in bone, muscle, or nerve. The *morphology* of bone and muscle presumably results from an interplay of kinds of genes many of which function also in other organs and tissues, but within different setups of regulatory coordination. In the homozygote of the E^N mutation in the E gene pseudoallelic complex of the silk moth *Bombyx mori,* all abdominal segments of the larva (the caterpillar) are transformed into thoracic segments. Internally, the tracheation, the nerve cord distribution, and the other anatomical arrangements are typical for thoracic segments (see Counce, 1973). One single mutation brings about all these changes simultaneously.

3.3. Shifts of Gene Programs

This leads us to the topic of shifts of the integrated action of whole gene complexes. Gene programs can be shifted in time or in space.

Consider first *shifts in time*. Such shifts have probably occurred very frequently during evolution. The importance and generality of the introduction of adult gene programs into preadult stages have been emphasized by Jägersten (1972) under the name of "adultation." A certain number of cases of temporal shifts of programs are quoted by Rensch (1960). Shifts of this kind occur in fact in any direction and involve various developmental stages. Gene programs may be transferred from adult to larva, without being eliminated from the adult. Thus it would seem that in several phyla, such as Cnidaria and Bryozoa, crowns of tentacles were first evolved in the adult and then appeared secondarily in the larva (Jägersten, 1972). When adultation of an earlier stage is followed by the evolutionary loss of later stages, it is called "neoteny." This can happen when the gene program transferred to an earlier stage is that of sexual maturity. If the animal reproduces as a larva, the adult stage may become "unnecessary."

The best-known case of this kind is that of a urodele amphibian, the axolotl (see Abeloos, 1956). The axolotl becomes sexually mature before metamorphosis. It does, however, occasionally metamorphose into the adult form, called *Ambystoma mexicanum*. Thus its gene program for metamorphosis is still intact. But it has been established that the threshold of sensitivity to the effector molecule that sets the process going, namely thyroxin, is heightened. Like ecdyson, thyroxin presumably combines with an activator protein that thus undergoes an allosteric transition and now is able to combine with a receptor gene. Even though a mutational decay has apparently occurred in some affinity constant within the controller system, this constant has changed only moderately. In individual axolotls, whose natural level of thyroxin, for genetic or environmental reasons, is particularly high, this level is sufficient for bringing about metamorphosis.

Neoteny has been established independently, by parallel evolution, in five different families of urodeles. In some, a beginning of metamorphosis takes place, in others not. In some, injection of thyroxin will bring about a measure of metamorphosis, in others even high doses of thyroxin are not able to do so. These species continue to synthesize thyroxin endogenously: when their thyroid glands are transplanted into a urodele that normally metamorphoses, metamorphosis is brought about in the host.

A regulator protein is likely to be involved in the neotenic species that respond to thyroxin, rather than receptor DNA. If a receptor gene had mutated, increasing the dose of thyroxin should not make any difference. The species in which partial metamorphosis occurs after administration of thyroxin offer evidence to the effect that decay in controller gene interactions occurred not only at the level of the master switch that sets off the whole process of metamorphosis but further down the line as well. After sexual adultation, gene programs leading to the adult amphibians may decay progressively in those parts that are not also involved in earlier stages of development.

Spatial shifts of gene programs—programs carried out at unusual locations in the body—have been called "homeosis." The term refers only to shifts in morphological features. An example is furnished by the well-known *bithorax* mutants in *Drosophila,* studied by Lewis (1951, 1963). Through a single mutation in any one of five closely linked genes ("pseudoalleles"), or through combined mutations in this series of genes, various extensive morphological changes are brought about that represent either an "anteriorization" of posterior segments or a "posteriorization" of anterior segments. In some cases, schematically stated, the halters—organs used for keeping balance during flight—are replaced by wings or wings by halters on segments that normally are reserved for carrying one or the other.

Another case of spatial transposition of gene programs is presented by the so-called E locus of *Bombyx mori,* which, as mentioned, is very likely also to represent a small genic region, rather than a single locus, and to be comprised of several pseudoallelic genes. The normal caterpillar of the silk moth carries a pair of legs on each of the three thoracic segments, and four pairs of a different type of appendages also used for locomotion, although not homologous with the thoracic legs, on four of the 11 abdominal segments. Heterozygotes for the Ca and the N mutants (E^{Ca}/e^{Ca} and E^N/e^N, where e stands for a wild-type locus of the ill-defined pseudoallelic series and E for a mutant locus) produce normal caterpillars. The morphological abnormalities appear in the homozygotes and are thus due to the absence of certain functional gene products. In the absence of the wild-type alelle e^{Ca}, no abdominal "legs" are made. They are not replaced by other appendages. In the absence of e^N, even though e^{Ca} is normal and therefore should activate the abdominal leg program, the thoracic leg program takes over and all abdominal segments bear thoracic legs. Thus, when the thoracic leg program is not inhibited in the abdomen, it has precedence over the abdominal leg program activated by e^{Ca}. In fact, the thoracic program that is inhibited in the abdomen by the normal product of the e^N locus is more than an appendage program, it is a thoracic segment program. In the absence of a functional e^N allele, it transforms the abdominal segments into thoracic segments, also with respect to stigmata, nerve commissures, tracheae, etc. Many of the data were obtained by Itakawa (see Tazima, 1964). Their simplest interpretation seems to be furnished by the diagram of Fig. 7, which shows the abdominal leg program to be at once under positive and negative control, in agreement with the scheme based on the *Drosophila* data from Ashburner's laboratory.

Pseudoallelic genes may have arisen by duplication of an original gene. Lewis (1951) believes this to be the case in the bithorax region of *Drosophila,* and it may apply as well to the E region in *Bombyx.* In turn, we assume that the products of these genes are regulator molecules, presumably regulator proteins. On this basis, it would appear that not only normal structural genes but also regulator genes arise by duplication and diverge subsequently, as predicted by Britten and Davidson (1969). Since at least some of the products controlled by these genes must be thought to interact with different receptor sequences, it would follow that receptor sequences also originate by duplication from a

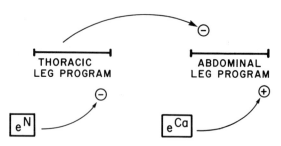

Fig. 7. Proposed regulatory relationships at the *E* locus of *Bombyx mori*. Regulation in the abdominal segments of the caterpillar. (+) and (−), activator and repressor molecules. See text.

common ancestral sequence and subsequently diverge in correlation with the divergence of the regulator genes. Finally, it would follow from this model that, in the case of the *E* region, the positive (activator) and negative (repressor) controller actions are in turn due to homologous if not sometimes identical molecules—proteins and receptor DNA sections. The situation might thus not be without some analogy with that found within the both positively and negative controlled arabinose operon in *Escherichia coli* (Englesberg *et al.,* 1969).

Itakawa (see Tazima, 1964) classified the different *E* mutants and showed that some produced shifts of posterior programs toward the front and others produced shifts of anterior programs toward the rear. Such kinds of effects are by no means limited to insects. Russel (1966) irradiated mice with X-rays and looked for resulting morphological abnormalities. Among other results, she found shifts in the borders between regions of the vertebral column. The number of vertebrae in one region was sometimes increased at the expense of the number of vertebrae in the adjacent region, and such shifts occurred in either the anterior or the posterior direction. Again, the result should depend on how the equilibrium of the interaction of activators, repressors, and receptors is disturbed. Related work by Jacobsen (1968) suggests, in preliminary fashion, that another basic deduction made from observations on insects is also of more general validity, namely the contention that the same morphological character is determined at once by master switches that induce and others that repress. Jacobsen found that in the thoracic and postpelvic regions of mice irradiation produced increased numbers of vertebrae in some individuals and reduced numbers in others. This is easily interpretable in terms of the model adopted in this chapter: sometimes a mutation somewhat lowers the affinity between an inducer molecule and a receptor gene, at other times between a repressor molecule and a receptor gene. In body regions where, on account of the presence of a *gradient,* the equilibrium between inducer and repressor is close to a point of reversion, the observed effect is obtained. A single point mutation in controller genes will, on the average, have a much more drastic outcome, morphologically and otherwise, in the body region of lability or near lability than elsewhere. It cannot, of course, be claimed that observations such as those quoted here constitute evidence in favor of the model, but only that the evidence does not contradict it.

During development, a whole gene complex geared to manufacturing a certain organ can be turned off and then on again, but may, the second time, be instated in a different tissue, even derived from a different germ layer. A gene complex may thus be functioning within different larger constellations of gene controls. This happens, for instance, in a group of Bryozoa, the Membranipora, in which the gut temporarily disappears during development and is reestablished later (cf. Siewing, 1963). In evolution, such developments are no doubt secondarily acquired. They bring out the fact that *discrete units of integrated gene action do have an existence as definable entities,* since they can be handled globally by mutation and by evolution. It is true that, in the case of the Membranipora, we do not know whether the original "gut program" has been transposed as such to another tissue, or whether a new "gut program" has, in this other tissue, been set up so to speak from scratch. It would be surprising if a whole set of different structural genes were used for making the second gut. If many of the same structural genes are used, many of the same controller genes are undoubtedly also used, and the first alternative would apply. We would then have a case of "reawakening," during ontogenesis, of a given eurygenic gene complex. With respect to an ancestral state, the opening in a second tissue of an access to the same eurygenic master switch would represent an innovation in the topology of the regulatory network. It would obviously be of interest to establish whether the proteins composing the first gut and the second gut are identical or not. Many might be slightly different, if, within the same eurygenic state, the stenogenic controls are different.

Although the manipulation of whole gene programs under the command of a master switch is demonstrated in mutation, by cases such as the bithorax mutants in *Drosophila,* it is less directly demonstrated in evolution, especially when major changes in circuit topology occur. The alternative in evolution would be, in such cases, the tendency for individual structural genes to hook the control of their activity onto other genes active during different developmental phases. Such a flow of individual structural genes toward and their crowding around certain controller elements seem quite unlikely, however. It is doubtful that in neoteny, for example, the different genic components of sexual functionality have been shifted one by one to an earlier developmental stage. Natural selection can hardly be expected to sponsor such a process. Gene activities associated with sexual maturity must either be shifted all at once or, at best, by packages, each of which displaces in time a whole subcomplex of gene activities.

In general, the evidence referred to in this section points strongly to the fact that there are flexibility and variability in the shifting around of discrete gene programs during ontogenesis, and that, in the process, these programs behave as coherent entities, and may therefore be identified conceptually as such, even though the regulatory network must of course be a continuum. The discreteness and individuality of gene programs are understandable if they are under the command of master switches. The following generalization is thus suggested: programs, subprograms, subsubprograms, etc., are under the control of master switches of different rank that can be turned either on or off and that are in relations of interdependence with respect to master switches belonging to other hierarchies of partial programs. Through such connections, a master switch in one gene complex can become indirectly the activator of a master switch in another gene complex, and thus carry the organism over to a new developmental phase.

In this fashion, one may conceive of dendrograms of master switches where each commands a smaller or larger number of structural genes according to its rank (Fig. 8). The distinction may be made between a *backbone genetic program,* leading from the fertilized egg to the adult, and *appended genetic programs* which are collateral and may show variants and special developments without affecting the connections of the backbone program. The criterion whereby appended programs are thought to be distinguishable from backbone programs is that appended programs can be dispensed with, shifted, expanded, reduced, or otherwise changed, while subsequent phases of development are left unaffected. The distinction may in practice turn out to be somewhat artificial, but it may help organize the picture.

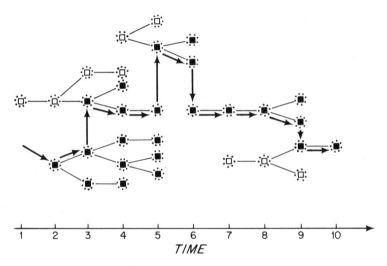

Fig. 8. Gene programs during development. Each horizontal dendrogram represents a gene program composed of subprograms and programs of still lower rank. Squares: master switch genes commanding a certain number of structural genes. Dots: structural genes. Between a master switch gene and the structural genes that depend on it there are further controller genes, not shown. Each solid square represents a master switch gene to be activated during the developmental sequence considered in the diagram. Empty squares are switches not so activated. Sequence of activation is according to time vector. Each master switch acts positively on the switches next in time. Negative (inhibitory) actions are thought to form a superimposed network, but are not represented. Which part of each dendogram of master switches will be active and when depends on the connections with other dendrograms. Heavy arrows: sequence of activation of master switches that represents the backbone program (see text). Solid squares connected by light lines only: appended programs (see text). Empty squares connected by light lines: programs not in use during phase of development shown.

The backbone program may be viewed as the line that best links successive partial dendrograms to others through master switches of different rank (Fig. 8). In each partial dendrogram, the section of the gene program that is hierarchically lower than the master switch through which the backbone line is traced will be an appended program. The part that is hierarchically higher (here drawn to the left of the master switch) may have been eliminated by an evolutionary adjustment that directed the backbone line to a master switch of lower rank. In this case, it is destined to decay in further evolution, except if it is integrated into a "live" program through other circuits. It is hardly necessary to recall that various kinds of circuits of interacting genes have been drawn starting with Monod and Jacob (1961), to some extent by Waddington (1962), and most especially by Britten and Davidson (1969).

The figure purports to represent only some aspects of the succession of gene activities during development. The confinement to certain tissues of some of the activities and the generality, throughout the organism, of others are not indicated. The dendrogram would be different if we considered programs in individual tissues during ontogeny. Nor are repressor actions shown. Only activation is represented. The figure thus does not suggest what proportion of the dendrogram is shut off at each stage of development. It has been shown in mice (Church and Brown, 1972) that, early stages of development excepted, the diversity of RNA transcripts of DNA increases as development proceeds. Therefore, in the organism as a whole, the number of different structural genes that are "turned on" exceeds in this case the number of those that are "turned off." The generality of this finding is not established.

Finally, in Fig. 8, the succession in time of gene programs is represented as purely linear, when we have reasons to believe that loops do exist and are important. "Earlier" master switches may be turned on again through actions mediated via "later" master switches, thus causing a tissue to reiterate a certain succession of gene activities. Also, the same or a different tissue may go through variants of earlier activities that include the activation of genes not previously transcribed (empty squares of Fig. 8).

3.4. Addition of New Programs

Another type of change in gene programming is, of course, the addition of new programs to preexisting ones. Given the number of successive regulatory steps in the backbone programs in contemporary organisms, this number, to reach its present magnitude, must have increased some time in the past. Such growth in developmental phases between ovum and adult might take place largely without the addition of new kinds of structural genes, but it does require the addition of complementary gene programs. This addition will be either intercalary or terminal or both.

Processes of terminal program additions imply that contemporary larval stages should be closely related to ancestral adult stages (such as the trochophore larva of annelids to the adult ancestors of ctenophors, cf. Sharov, 1966) and present adult stages, strictly speaking, should have no equivalent in the distant past. Many evolutionists are opposed to such a view. It seems, however, difficult to escape it altogether. Terminal addition of gene programs should be particularly "easy," in that they can hardly interfere significantly with the programs for preceding developmental stages. The specific development of the human cerebral cortex is an example.

However, when intermediate stages are free larval forms, they are as much as the adult forms submitted to solicitations of adaptational change. The degrees of freedom for transformations, losses, or additions of programs for intermediate phases should be much more limited than the corresponding degrees of freedom that apply to a terminal stage. Given the interconnections between gene activities, many changes that might be of interest to larval forms will tend to be eliminated because of their interference with regulatory processes relevant to later phases. The earlier the ontogenetic stage concerned, the greater are the chances of such interference.

In fact, many organisms have solved this problem. Among them, the arthropods and particularly the holometabolic (fully metamorphosing) insects are outstanding. It seems that the main evolutionary scenario is as follows. Terminal additions of programs occur. Their acquisition is later followed by a retardation of their onset during development (see, for instance, Hüsing, 1963, p. 13). A whole set of late gene programs thus becomes a "package" under the control of a high-rank master switch, which is activated by a hormone (ecdyson in insects, thyroxin in amphibians). By this temporal isolation, all specific adult circuits of gene activity are effectively shut off, prior to their activation, from the repercussions of other gene actions. The adult circuits are thus protected from any interference by newly introduced ontogenetically earlier gene programs. A special provision is necessary for allowing this device, namely the early shutdown of programs leading to the adult, to fill its purpose: at a given point in development, the secondarily acquired intercalary (larval) patterns of gene activity must be turned off at the same time that the package of adult programs is turned on. This is what metamorphosis provides.

In holometabolic insects, such as the flies, the adult gene programs are held in reserve in groups of cells, the so-called imaginal disks, whose future activity is determined in early development. Within the body of the insect, the adult backbone program as well as the correlated appended programs are thus kept so to speak in a frozen state, throughout multiple, adventurous developments of intermediary larval forms. These larval developments indeed correspond to more than an intermediate phase of a single backbone program, although they do preserve one intermediate-phase characteristic in that they eventually command the synthesis of the hormone that triggers metamorphosis. In all other respects, the larval program, given its independence, may be considered a second backbone program on its own. Any discontinuation of a backbone program, accompanied by the unfolding of another, is a metamorphosis (Fig. 9).

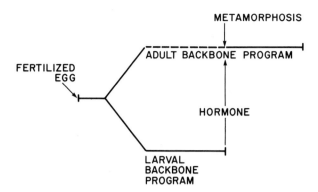

Fig. 9. Metamorphosis involves the coexistence of two backbone programs within one organism. The horizontal dimension, from left to right, is time. The adult backbone program is stored away while the larval backbone program is active. The last step in the larval backbone program, the synthesis of a hormone, brings about its own discontinuation and sets the adult backbone program going.

How independent larval programs may be is strikingly illustrated by certain species of Lepidoptera (butterflies). Species of Lepidoptera are known which in the adult forms can hardly be distinguished from one another, but which have very different larvae (*Acronyta psi* and *A. tridens,* quoted by Wigglesworth, 1954).

We can see why Haeckel's rule of recapitulation is at once true and not true. It has been criticized on the grounds that the gill slits of a mammalian embryo recapitulate the gill slits of an *embryonic* fish rather than of an adult fish. Yet there is no doubt that the gill slits of an embryonic fish are much closer, in developmental relationship—i.e., in programming of gene activities—to an adult fish than to an adult mammal. Even if the mammal recapitulates only an embryonic fish, not an adult fish, it is a fish that is recapitulated. Whereas the "adultness" in gill slit character may be controlled and defined by an appended program rather than by a backbone program, what the mammal has preserved is likely to be a part of the fish's backbone program; otherwise, it should long since have been lost. The example of direct development in higher crustaceans (crayfish) given below will show that, within the egg, only stumps of the appendages corresponding to the appendages of ancestral free-swimming larvae are preserved. Yet no one will deny that these stumps "recapitulate" the homologous appendages of the free-swimming larvae. The fully elaborated morphology of the appendages would be determined by an "appended" program that could indeed be dispensed with by the embryo of the crayfish. But the distinct *ébauche* of the appendages could not be dropped: it probably corresponds to a part of the backbone program. The case of the gill slits in the mammalian embryo is surely similar.

Yet, as we began to say, the rule of recapitulation is at once true and not true. It is true because terminal additions to gene programs are an important and very general evolutionary process, as the observed cases of recapitulation themselves strongly suggest. It is not true because the secondary development of extensive intercalary gene programs is also a frequent evolutionary process. Such ontogenetically younger stages are then also phylogenetically younger than some ontogenetically older stages—which is the contrary of recapitulation.

A number of very visible innovations in adult development, especially in segmented animals, may not require new kinds of structural genes, and may not even require new programs of integrated gene action. This should apply to the multiplication of metamers (segments) in annelids and arthropods. A single controller gene may suffice for determining their number, on the basis of the same program whose expression may be modulated by certain gradients in the controller system. Such an extreme case—when even in the realm of regulation only a minimal amount of novelty is required for adding a conspicuous developmental element to the adult stage—may help us realize that program extensions will use essentially the same genes that came into play before.

When this applies to intercalary gene programs, it cannot have been brought about without the establishment of loops and anastomoses within the pattern of interconnected gene programs. Development of additional gene programs, at whatever stage they are inserted, may often consist in the formation of new loops in the pattern of regulatory interactions, such that subprograms of various ranks already used during some other ontogenetic phase are called upon again in a new combination and under conditions of certain shifts in the equilibria between the elements of the controller system.

Thus a large part of the organism's structural genes may be activated over again during development in different constellations of preestablished subprograms of different rank. Preestablished subprograms may be among the chessmen evolution plays its game with.

3.5. Elimination of Programs

We may continue this condensed review of types of evolutionary changes attributable to changes in gene regulation by listing the elimination of gene programs. Cases of this are known to all, and were implied in some of the examples already given. An illustration is, for instance, the loss of the limbs in snakes or lizards. The structural genes necessary for making limbs probably continue to be used in these organisms for other purposes. For not only are there no bone structural genes or nerve structural genes specially affected to the morphogenesis of a limb, as I proposed; but also it is likely that there are no structural limb genes at all, specifically limited to action in limbs. Therefore, the "limb structural genes" in snakes are presumably being maintained functionally intact by natural selection. But the network of regulator genes that engage structural genes in a limb program should decay if the abandoned limb program is not "reawakened" in some descendant organism before much evolutionary time has gone by.

The elimination of a program may be terminal, namely concern the adult state, or affect some intermediary developmental state. When it is terminal, it may or may not be accompanied by program substitution. We already saw that neoteny, the shifting of sexual maturity to a formerly immature stage of development, tends to be accompanied by the elimination of the formerly adult stage. Cases of program elimination in the adult, accompanied by program substitution, occur in transformations from free-moving to sessile habits of life and in cases of parasitism. The rhizopod crustacean *Sacculina* offers a renowned, striking example of what changes parasitism can promote in an organism. The larva of *Sacculina* is free swimming like that of its nonparasitic relatives. The adult *Sacculina,* on the other hand, is described as essentially a sac filled with genital products and rootlike extensions that grow inside the crab. It seems quite unnecessary to postulate that *Sacculina* had to develop a number of distinctly original structural genes in order to be able to become the parasite that it is. Rather, the working hypothesis can be put forth that in such cases what happens is essentially the elimination of numerous genetic programs and the reorganization of programs on the basis of structural genes that preexist.

Program elimination may affect intermediary developmental stages. This happens when development as it occurs in the free-living condition, through a succession of progressively modified larval stages, becomes partly or totally "direct," whereby is meant that it now occurs within the egg, up to a more or less advanced larval stage or up to the adult stage. In decapod crustaceans, for instance, the more ancestorlike *Penaeus* goes through a series of larval stages, whereas the related crayfish hatches as a full-fledged small adult. During the development of the crayfish within the egg, certain remains of ancestral larval stages can be recognized. But much—such as the formation of functional larval appendages—has been lost (Fig. 10).

As appended programs are totally or partly abandoned in "direct" development—sometimes perhaps simply by not being allowed the *time* necessary for complete unfolding before being inhibited by a subsequent phase—the backbone program itself may or may not be somewhat condensed. That it can be condensed in timing is not to be doubted. Whether it can also be condensed in structure, namely whether certain phases of the backbone program can be eliminated, as one can cut a piece of wire and rejoin its extremities—or a piece of DNA, for that matter—cannot be decided at this stage. It would seem that the earlier the embryonic stage that corresponds to a formerly, but no longer, free-living form, the more reduced are the traces of the ancestral appended programs for that form.

As mentioned, adaptation to parasitic life is accompanied by program elimination, We saw, with *Sacculina,* a case of elimination and transformation of terminal programs of the adult stage. There is a corresponding example for the elimination of larval programs. Another primitive crustacean, a copepod, *Haemocera danae,* has a free-swimming nauplius larva, but then becomes the parasite of a serpulid worm during subsequent larval stages, and at this time its larval programs are largely abolished. The organism is reduced to a small nondiffer-

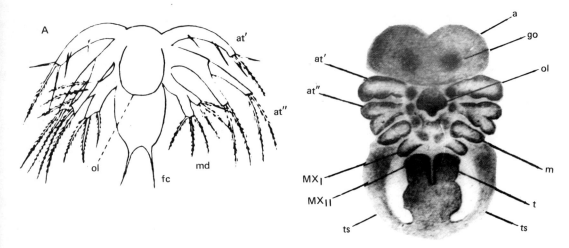

Fig. 10. The "direct" development within the egg of the crayfish goes through stages inherited from ancestral free-living larval forms. Such an ancestral form is approximated by the contemporary *nauplius* larva from *Penaeopsis stebbingi* (left) (according to Gurney). In the embryo of the crayfish (right) (according to H. Reichenbach), corresponding to the *metanauplius* stage, the gene programs leading to the formation of the larval appendages have reduced but still recognizable effects. at', antennula; at", antenna; md (left) or m (right), mandible; mx$_I$, maxillula; m$_{II}$, maxilla; ol, labrum; t, telson. From Korschelt (1944).

entiated mass of cells. From this mass, the future free-swimming adult is built up within the host (cf. Smith, 1909). Likewise, the nauplius and cypris larva of *Sacculina* contains a mass of undifferentiated—but probably already determined—cells. It is this small mass of cells that is injected into the host, the crab, and that develops into the adult parasite. Whether programs are added or subtracted, the storage of programs in predetermined nondifferentiated cells permits independent evolution at different stages. Nondifferentiated stored cells presumably are "noncompetent" with respect to most regulatory actions (their DNA presumably is nonaccessible to most regulator molecules). Thus regulatory processes can vary mostly without affecting these cells.

Such masses of nondifferentiated cells, which contain in a "frozen" state the program for a whole organism, are to be compared on the one hand to the ovum and on the other to stem cells, as known in vertebrates (e.g., those of the hematopoietic system), and all those precursor cells that give rise at various stages of ontogenetic development, no longer to complete organisms, but to differentiated cell types. All these precursor cells are "frozen" in a certain stage of determination, until a specific stimulus elicits their differentiation. *There probably is no fundamental difference in the mechanisms involved in these various situations.* The set of frozen programs is more or less extensive and is kept dormant over longer or shorter developmental times. Also, the cells involved are kept together in groups or dispersed. As to the definition of the underlying mechanisms, it is evidently linked to our understanding of the difference between determination and differentiation, between the eurygenic and steno-genic controls of gene activities.

A final category, *the recovery of lost functions* (such as enzymatic functions) *and of corresponding gene programs,* has been examined recently (Zuckerkandl, 1975) and will not be discussed again.

The cases of evolution of gene programs considered here were supposed to be based essentially, if not exclusively, on the reorganization of activity patterns of preexisting structural genes. The major part of evolutionary processes was deemed to be ascribable to the reutilization, under different conditions of regulation, of already acquired structural-genic material—already acquired, except for relatively minor, though numerous, variations and for gene duplication. Gene duplication is always probable during evolution; so is always probable, on the other hand, the loss of similar gene duplicates. (This question has been discussed in Petit and Zuckerkandl, 1976.)

If it were possible to take judiciously chosen structural genes and put them together in the right relationships with regulatory elements, it should be possible to make any primate, with some small variations, out of human genes, provided that the synthetic organisms be granted vitamins that the natural organisms do not require. Likewise, it should be possible to make any crustacean out of the genes of higher crustaceans. This fairy tale may be conservative. It is told only to emphasize that structural genes are building stones which can be used over again for achieving different styles of architecure, and that *evolution is mostly the reutilization of essentially constituted genomes.* We may not generally have been tempted to take such a view, impressed as we were by the constant structural divergence of genes throughout evolution. Yet, by and large, this divergence resembles Brownian motion (cf. Zuckerkandl, 1976*b*). With notable, all-important exceptions, the essential properties of the proteins are not changed.

When from time to time a novel protein function does arise, it will be controlled by a modified gene duplicate (Lewis, 1951; Pauling and Zuckerkandl, 1963). Initially this function probably will not be integrated fittingly into the concert of the already established functions, perhaps neither temporally with respect to the phases of ontogenesis, nor quantitatively with respect to the rate of synthesis of the new protein, nor spatially in being present in or confined to the tissue where the new function is useful. Whereas a temporal or quantitative lack of fitness may not always be prohibitive and progressive adjustments may be made, it seems that the appearance of the gene in a "wrong" tissue would lead to its immediate elimination. Think of silk appearing in the brain. . . . Even if the gene were not eliminated, its translocation to the "right" tissue would be an unlikely event (Zuckerkandl, 1975).

In fact, one might expect a functionally novel gene to appear in the "best" tissue much of the time. Indeed, as a duplicate, albeit a modified one, it is not improbable that it will still be part of the eurygenic gene program within which its ancestor arose and that it will perform a function which is a mere development, in line with those already carried out. At times the new gene will thus be accommodated by the tissue of origin itself. If it is not, a new tissue may be formed as a differentiating offspring of the old, which would be simultaneously maintained. The gene program of the new tissue would be identical with that of the parent tissue in essential parts, modified in a few others. In this fashion, *structural genes endowed with new functions may bring about the formation of new tissues*—of new gene programs. It is to be expected that one single modified gene duplicate can set this process going. Gene duplication followed by gene divergence, when the duplicate gene finally performed a new function, may thus have been accompanied by "tissue duplication" followed by tissue diversification. Tissue deletion may also occur (Zuckerkandl, 1975).

The evolutionary differentiation of tissues should be expected to be based quite generally on the functional differentiation of the polypeptides controlled by structural genes. The generation of a new tissue is a very rare, but obviously essential, event in evolution. The variation of structural genes is thus endowed with an evolutionary function of prime importance. Clearly, this must not be forgotten when we observe that the quantitatively predominant kinds of events among those that have an evolutionary impact, and many of evolution's most striking features, depend mainly on changes in gene regulation.

4. The "Intrinsic" Directionality of Evolution

4.1. Parallel Evolutionary Change in Molecules and Organisms

The data of Romero-Herrera *et al.* (this volume) on myoglobin evolution show through striking quantitation that parallel evolution is a very common event at the level of amino acid substitutions in proteins. So is parallel morphological evolution. One finds it wherever one turns. It is surprising that it is not commonly considered one of evolution's basic and pervasive features.

A connection between the two parallelisms, at the molecular and at the morphological level, should, however, be denied. Parallelism in amino acid substitutions is a side effect of what has been called "evolutionary noise" (Zuckerkandl, 1976a,b). In different but related proteins, coincidence of amino acid substitutions at highly variable sites is likely, because the set of frequently accepted substituents at such sites is limited.

There are no grounds for identifying coincidence of substitutions with structural and functional convergence. Convergence of amino acid sequence is rather unlikely even when there is convergence of two polypeptide chains toward identical functions (Zuckerkandl, in preparation). Indeed, once two homologous polypeptide chains have diverged by a certain amount, if natural selection tends to reestablish some identical functional property in the two it is likely to arrive at this functional convergence by way of two different sequence solutions. After divergence, each polypeptide chain sequence starts out from a different situation. The problem is thus posed in different sequence terms to the two chains. As long as accepted amino acid substitutions are limited to one or very few single steps, a polypeptide chain might retrace its evolutionary pathway and recover an original state. When this very small number of substitutions is exceeded, Dollo's rule of the irreversibility of evolution must apply at the molecular level as well. It is a rule relevant to structure, not to function. It applies to the primary structures of polypeptide chains, not to their functions.

However, if one considers *groups* of functions, the rule of Dollo must also apply. Any one polypeptide chain has multiple functional characteristics. After structural and functional divergence of homologous chains, any one or any very small set of functional characteristics of the same molecule may be recovered. A larger set of functional features may not. During the evolution of a given type of protein, while sequences diverge and do so at a relatively regular rate (see below), diverse functional characteristics will present kaleidoscopic combinations. In this process, the appearance of sequence identities is likely to be an accidental by-product of either convergence or divergence in any particular function.

Molecular coincidence can be understood in terms of a neutralist model of substitutions. It is equally well interpretable in terms of a selectionist model (Zuckerkandl, 1976b). Even on the basis of the selectionist model, parallel evolution at the protein site level must be understood to have a random component (Zuckerkandl, 1976b).

On the other hand, as we shall see, morphological and other systemic parallelisms in the evolution of organisms are probably not random. What determines these parallelisms? I believe this question should not be skipped here, in view of its importance to anthropology. Reluctance or readiness to assume a high frequency of parallel evolutionary developments leads to very different views on how and when higher primates evolved. This has been evident in the present volume.

Many cases of parallel evolution have been considered as somewhat of a mystery. It happens indeed that the same new, apparently anodine, morphological characters appear independently in a number of related groups under what are believed to be such strongly divergent conditions of environment and ethology that the existence of common selective pressures in favor of these characters is hardly a plausible assumption (Grassé, 1973). An outstanding example of such a situation is the recurrent trend in fossil reptiles to develop mammalian characters, as detailed by Grassé. Within the group of theriodonts, only one or two orders are considered to have eventually led to the mammals, but several of them developed mammalian characters to different degrees and in different combinations, even though representatives of these orders probably lived in very different environments.

Possible changes in structure depend on structure. By no pathway could mammalian characters be introduced into a worm. On the other hand, structural properties shared by the reptiles offer a common ground for innovations of the mammalian type. Yet it is considered that the odds are against having some structures change in the same direction independently a number of times, when myriads of different morphological changes are possible and especially when it is not obviously reasonable to assume common selection pressure in favor of the same character.

We take a step forward, I believe, if we presume that parallel evolution in morphology represents parallel evolution in gene regulation.

4.2. Reduced Diversity of Mutational Effects in Controller Gene Systems

The first point to be considered here is that the diversity of heritable phenotypic changes likely to be obtained by chance mutation in controller genes is probably more reduced than the range of effects of mutations in structural genes on the function of their products.

Whether controller DNA acts via transcriptional and translational products or whether it acts as a specific target for other molecules, point mutations in controller DNA, as we saw, can only lead to an increase or a decrease of molecular affinities. There is no possible *qualitative* modification of the controller gene here. It will function more or less efficiently; its functional "mission" will remain the same. Whereas, in products of structural genes, point mutations induce quantitative *and* qualitative effects, their effects on controller genes will vary only quantitatively. Thereby *many different mutations are liable to converge on similar functional or morphological effects.* Jack L. King, whose mind and mine must share controller genes, judging from continuous parallel evolution, made independently a similar point (personal communication, 1975).

The effects of point mutations in controller genes will not only often be qualitatively similar but presumably will also mostly be quantitatively moderate and incremental, at least when the mutations are in the receptor genes (see above).

There are then no doubt many mutations in controller genes that lead to only slight phenotypic expression. And many such mutations may be of nearly neutral selective value.

Moreover, because of the interlinking of regulatory processes, similar morphological effects will be obtained not only through different mutations in a given controller gene but also through mutations in different controller genes. This has been shown in the "bithorax" mutants in *Drosophila* (Lewis, 1963).

On the other hand, when a mutation results in a radical reduction of the affinity of a regulator protein for its target, as for instance deletions and insertions would cause, the effect may the more surely be highly deleterious or lethal the higher the rank of the affected switch gene and therefore the larger the number of structural genes under its control (Fig. 8). Pleiotropic effects are the rule here. Therefore, if and when mutations in controller genes do not have a moderate, incremental effect, they then should have a larger chance of being deleterious than mutations in structural genes. Mutations in the regulatory system thus should be, it seems, at once more often more innocuous and more often more deleterious than mutations in most structural genes.

How about mutations that affect the topological relationships between controller genes and structural genes? A reshuffling of regulatory circuitry, if all theoretically possible combinations acceptable to natural selection were considered, would again lead to a relatively limited number of, this time, consistently discontinuous phenotypic expressions. A certain limitation of the variety of possible phenotypic effects due to alterations in gene regulation would again be expected.

Considering the very large number of distinct mutations that must affect regulatory sections of DNA, it thus seems that the number of phenotypic expressions that are *manifest* (not eliminated by lethality or disadvantage) and that are *distinct* (not too weak, and differentiated from other effects) should be relatively reduced.

The striking directionality of evolution, as illustrated in particular by parallel evolution, has led to many passionate and extravagant views. It is probably due, in part, to the intrinsic limitation of viable and morphologically detectable mutations in controller DNA. The recurrence of the same morphological changes in independent lineages may be considered as far more probable than would be the case if the situation were not as described. Similar morphological changes may thus spread in several related lineages, even when not clearly advantageous and at the sole condition of not being clearly unfavorable. Accordingly, the vectors of evolution may be in part of intrinsic origin and only in part of selective origin. The intrinsic component should be linked to the characteristics of phenotypic effects of mutation in controller genes. More generally, it should be based on the fact that, in the succession of genomes by filiation, the state that follows has its possibilities restricted by the state that precedes.

The intrinsic vector of evolution is not, in fact, a line, but a vectorial zone. This oriented zone is defined by the limited possibilities of new "orchestrations" of gene activities, starting from the orchestration that exists. Within the vectorial zone, accidental environmental circumstances, as well as circumstances of internal environment (Whyte, 1965), will trace the course of evolution as it actually occurs. There is no reason to amend neo-Darwinian theory, but it may be completed.

When new phenotypic features are not significantly disadvantageous under a rather wide spectrum of environmental circumstances, and when their mutational recurrence is practically ensured, the chance of their representing a "characteristic" development in a large taxon is high. Such phenotypic features may be generalized in a population by Sewall Wright's random genetic drift, or as nearly neutral phenotypes according to the rules established by Kimura and by his associate Ohta (e.g., Kimura and Ohta, 1971). Although the number of possible different mutations is exceedingly large, we propose that, on account of some particular traits of gene regulation, the range of possible *phenotypic* characters is limited enough for evolutionary channeling to become apparent.

Whyte (1965) gives the history of this general idea, and insists upon its importance. To Whyte, possibilities of organic developments are restricted because many of them, as they arise, are incompatible with the internal mutual adaptation of functions. This is one side of natural selection, and I am not aware that anyone contests it today.

A different point is also made by Whyte (1965), but not in focus. It is present notably in a quotation from H. Spurway, dating back to 1949. Many structural and functional developments will not occur at all, because they are incompatible with the existing genetic constellation. There will thus be no opportunity for them to be selected against. The material offered to natural selection is limited in advance. This is the point made here. And we add that, on the same basis, other developments will recur repeatedly.

4.3. Lability and Stability of Phenotypic Characters

Some observations in the field of teratology, which deals with the appearance of morphological anomalies, help us make these statements more precise. When we hear about teratogenesis, either as observed in nature or as experimentally induced, we may be struck by the fact that some types of anomalies recur with a relatively high frequency, others are rare, whereas an untold number of what would appear to be *a priori* possibilities of abnormal formations are never reported. For instance, the unicorn has never been observed as a mutant, except by artists in the Middle Ages, whereas other morphological anomalies, such as a supernumerary digit, are seen again and again. Some normal characters are found to be *more labile* than others, and we may ask ourselves what this lability is likely to mean in molecular terms. To give an example, when mice are irradiated with X-rays, one of the characters that proves to be particularly prone to variation is the number of ribs (Russel, 1966). In terms of the model of gene regulation, a situation of lability should arise when the competition between activator and repressor, with respect to a given structural gene, is close to the point where the equilibrium can "flip over" and the activator win out where previously the repressor did, or the repressor win out where previously the activator did. As we saw, there are many different mutational ways to shift the equilibrium in the competition between an activator and a repressor. It is only to be expected that in a certain fraction of cases a small change in affinity constant or a small change in concentration of a regulator species will suffice for an equilibrium to flip over.

Now, since the elements of the complex regulatory network are eminently interdependent it is also natural that a change in such an equilibrium that affects one given functional unit of gene action will have consequences for other equilibria and will bring some other previously stable equilibria near the brink of reversion.

This allows us to conceive of the appearance of new types of instabilities as evolution proceeds. Evolution no doubt often consists in making a labile character stable by fixing a new concentration ratio or a new relation in affinity constants between activators and repressors. However, each such stabilization is likely to lead to *some* labilization of other activator-repressor equilibria. Most of these will still remain within the boundaries of stability, but some may be brought into the region of instability. A slight further change in the same direction will then produce the flipping over of the equilibrium to activation or inactivation of the corresponding functional unit of gene action.

However, inactivation of a fuga is not even necessary for a "flip-over" of a character to occur at another level. We emphasized that changes in affinity constants within the controller system, or in quantity of regulator or effector molecules, will lead to only quantitative effects at the level of the regulator-receptor pair of molecular species. These effects will also be purely quantitative at other levels of biological integration, e.g., the morphological level, as long as the changes in affinity constants take place within certain limits. When these limits are transgressed, the effects, at the higher levels of integration, must become qualitative. These qualitative changes are expressed there as either the presence or the absence of a character, or the presence of one of two possible characters. The "flip-over point" for such binary alternatives of phenotypic expression may correspond, in any one given case, to different values for regulator-receptor equilibria in different regulator-receptor pairs that all contribute to the same end effect. This end effect is cell differentiation or morphogenesis.

It may be proposed that in a closed system, comprised of a finite number of controller elements that are all engaged in equilibria as described and all interacting with each other, *there is a certain maximum proportion of highly stable equilibria,* in terms of the alternatives with respect to the end effects, *and a certain minimum proportion of unstable equilibria.* If so, and if, in the past, the system as a whole tended toward maximum stability, then any new stabilization of an equilibrium situation within the system (in terms of the end effects) must lead to a certain labilization of some other equilibria as a necessary consequence.

When the affinity of one regulator molecule for the corresponding receptor changes, the affinity constant pertaining to another regulator-receptor couple may not have to physically change in order to have a significance that is now different from the one it had before, in terms of the end result at a higher level of integration (say the morphological level); this constant may now be closer to the value that corresponds to the flip-over within a binary morphological alternative. To give an example, suppose the total quantity of gene product is decreased in a given fuga, fuga I, on account of a mutation within a regulatory system such as the one considered in this chapter (Fig. 2 and Table 1). Suppose that the gene product is an activator of another fuga, fuga II. Fuga II may still be activated, because the affinity of the repressor of fuga II for its receptor is not high enough for the repressor to "compete out" the activator, in spite of the lowered activator concentration. However, the repressor is now *close* to being able to have this effect on fuga II, whereas before the mutation in fuga I the affinity of this repressor for the corresponding receptor was *far* from being high enough. The situation in fuga II has been destabilized—has been brought into the immediate vicinity of a "flip-over point," even though no mutation occurred in this fuga.

Thus a change in a single affinity constant somewhere in the system should alter the significance, e.g., in terms of the stability of a morphological character, of a certain number of other affinity constants. Because of multiple interactions within the regulatory network, a mutation that has a phenotypically undetectable effect on the stability of one morphological character may simultaneously have such a cryptic effect on that of another morphological character as well.

In this fashion, a stabilization of one morphological character can and, on our assumption, will bring about a destabilization of some other morphological character, because in terms of this other character the equilibrium between some regulators and receptors will now be somewhat closer to the flip-over point of this character, at the brink of this flip-over point, or beyond this brink.

We may think of such labilizations as the principal source for the parallel appearance of similar morphological or other changes in related forms. Parallel evolution thus may well have its principal foundation in the sharing of the same lability of a character and the same subsequent stabilization of this character. Stabilization means that henceforth the equilibrium among activator, repressor, and their target is further removed from the zone of reversal of the phenotypic character under consideration. This stabilization effect can certainly be obtained by a large number of different mutations in the implicated controller genes, one such mutation being potentially sufficient.

Thus the evolutionary labilization of a character as a consequence of an evolutionary stabilization of some *other* character will be the source for the repeated appearance of a similar phenotypic effect in related forms, and this may provide the materials for, the "offer" made to, natural selection or genetic drift for parallel evolution. Indeed, the "other" character will be an ancestral character common to several independent lines of descent.

Parallel evolution will occur if natural selection or genetic drift retains and stabilizes, in different lineages, the same characters among those that are prone to appear and reappear because of character labilization, and because of the ensuing relatively restricted variety of phenotypic offers that are made by mutations in controller genes. *The directionality behind parallel evolution appears as a path through common labilities turned into similar stabilities.*

4.4. Reduced Randomness of Effects of Mutations at Higher Levels of Integration

How can evolution be based on random mutations? Anti-Darwinists of all degrees of mysticism, from zero to infinity, ask this question with the gleeful expectation that it can receive no satisfactory answer. The answer is that although the mutations are indeed largely random within certain limits (mutational trends, restrictions imposed by the genetic code and by the base and amino acid sequences that are initially present), *effects* at higher levels of integration of most mutations are not predominantly random.

The regulatory part of the genome no doubt greatly exceeds in numbers of base pairs the nonregulatory part—the part that represents structural genes. Furthermore, the average rate of acceptance of mutations in the regulatory compartment (we refer as before to the regulation of transcription) must be equal to or, more probably, distinctly higher than the rate of acceptance in structural genes. This can be deduced, for instance, from the finding that non-structural-genic unique-sequence DNA evolves on the average at a faster rate than structural genes do (Rosbash *et al.,* 1975). Repetitive DNA may be expected to accept mutations at an average rate at least equal to, or higher than, non-structural genic unique DNA. The faster rates may not apply to controller (regulator plus receptor) genes, which may in part be found among the medium-repetitive DNA. However, as stated earlier, it is to be expected that, beyond controller genes, further sections of DNA (e.g., "transconformational DNA") probably have important functions in the control of gene activity. These further sections may indeed exceed in quantity the (unknown) total fraction of DNA committed to controller genes. It follows that the majority of the accepted mutations are likely to occur in the regulatory part of the genome, as also pointed out by Jack King (personal communication). Yet, because of the characteristics of the system of gene regulation, the very large number of distinct mutations in the regulatory section of the genome will be channeled into a very much smaller number of phenotypic effects. One may therefore say that the randomness of the mutational effects, in the sense of the range of effects on which randomness can operate, has been limited by the regulatory system.

Under these conditions, extraordinary adaptations also become less improbable. Some cases of mimicry, for instance, may appear close to miraculous if they depended on a few particular base substitutions that must occur within a relatively short evolutionary time span. The miracle fades away if 1,000,000 or more different combinations of base substitutions can lead to the same effect.

Many mutations in the regulatory system no doubt have no phenotypic effect at all. But all such phenotypically cryptic mutations, as well as the phenotypically expressed ones, change the genetic situation of the organism, perhaps slightly, but on a large front, and establish intrinsic propensities for and against certain further phenotypic developments. What French physicians call the "terrain" proper to a human individual, the particular makeup of his physical organization against which the effect of any external factors, such as drugs, is to be viewed, may be in part due to a "personalized" combination of alleles of structural genes, but probably to a larger extent to cryptic differences between the states of the regulatory systems of different individuals. *Any mutation within the regulatory system establishes a set of trends*, a hierarchy in the likelihood of the very large but finite *number of phenotypic effects of mutations that the system allows to happen next.* Chance will play a role, but so will necessity. There thus is necessity, not just in natural selection, but even in the offer of mutation to natural selection. The extent of this necessity becomes clear only through considering systems of gene regulation. I believe it to be one of the main components of the necessity of the evolutionary process.

The classical coupled processes—random mutation, followed by deterministic fixation (selection)—are of course to be preserved in evolutionary theory as far as the processes themselves go: mutation followed, or not, by fixation. *But the equally classical opposition in the amount of randomness that was thought to be associated with the two processes no longer holds.* A large amount of randomness has diluted the deterministic character of fixation (cf. Kimura, 1968; King and Jukes, 1969). Likewise, I think we should now recognize that a large amount of nonrandomness probably limits the randomness of the mutational *effects* themselves.

A basis for an internal directionality in the effects of mutations was tentatively defined here. This directionality may lead to fixation in populations by either selection or drift. Since selection—to the extent that its pressures are, over certain periods, exerted in the same general direction—can act directionally also on undirected effects of mutations, the extent of the contribution of this internal directionality to the overall directionality in evolution may be difficult to assess. The extraordinarily high incidence of parallel evolution apparently in all types of organisms is compatible with the concept that this contribution is great. The predominant role of gene regulation in morphological changes argues in favor of the same view.

There is no reason why parallel evolution in the close fossil relatives of man should not be as probable as it generally is among related taxa. It would seem risky to fully trust conclusions on human ascent when they are based on the assumption that certain similarities in skulls indicate close phyletic relation rather than parallel evolution. Under these conditions, one may indeed wonder what proportion of the conclusions regarding evolutionary lineage and rate, based solely on the morphology of hominid and anthropoid fossil finds, may ever be considered as firm.

Help from the molecular level would be most welcome. Indeed, parallel substitutions in proteins no doubt do not obscure the picture of evolutionary divergence to the point of falsifying phyletic relationships in a significant proportion of cases. Unfortunately, even if fossil bones contained some informational macromolecules in a relatively unmodified state, the molecular species involved might not have diverged significantly enough over the evolutionary periods in question. Yet anthropology needs molecular anthropology to achieve in this respect all it can.

5. Anagenesis

5.1. Acceleration of Anagenesis

Anagenesis, progressive evolution, has a functional as well as an organizational basis. Its organizational features are astonishingly well understood in their broadest outlines, and astonishingly little understood in what these outlines actually contain and cover. It is commonplace now to refer to the hierarchical superposition in organisms of different levels of integration, each of which displays its characteristic elements and characteristic relationships between these elements. The hierarchy goes from the subatomic level or levels to societies of higher organisms. A hierarchy of sizes accompanies it from the electron to the elephant, but this is not very revealing. Neither do we grasp the content of organizational levels with the help of the notions of the individual and of societies, since there are individual molecules and societies of molecules, individual cells and societies of cells, individual higher organisms and societies of these. Yet we are dealing in each of these cases with very different levels of organization. The difficulty in measuring organization seems to have precluded an understanding of this hierarchy beyond, or much beyond, some initially illuminating but soon inadequate generalities.

The point to be recalled here is, however, that these various levels of organization of matter have not always existed in the solar system, and that those characteristic of living things have only progressively emerged on the earth. More highly integrated levels of organization eventually appeared, while preserving, each time, within themselves, all the lower levels. The actual processes in these are redirected so as to fit the processes at the higher levels. The combination of evidence from contemporary nature and from the fossil record leaves no doubt about these generalities. They suffice to establish the existence of what may be called progressive evolution. The progression is not defined in anthropomorphic terms, in terms of what is good or bad, desirable or undesirable. It is an observed fact, like a star.

Within the realm of living systems, what might be the driving force behind this progression? Huxley (1942) defines anagenesis as an increased independence from, and control over, the environment. All other aspects of anagenesis, such as those analyzed by Rensch (1960), may be viewed as subordinate to this essential "motivation." It is this feature that may be conceived as providing a handle for natural selection to direct the derivation of higher forms from lower forms.

Control over and independence from the environment must be linked to complexity of organization, since these basic features of anagenesis imply at once the two apparently contradictory requirements of a stabilization of the living system with respect to external perturbations (homeostasis) and an increased sensitivity and power of discrimination of the system with respect to external perturbations (sense organs). Such sensitivity and power of discrimination are required for an evolutionary increase in precision, variety, and adaptability of organized response to the perturbations. No simple organization could satisfy these demands.

Again, measurements of complexity and of hierarchical organization are difficult. At the molecular level, Gatlin (1966, 1974) and Reichert (1976) proceeded to measure features of DNA and protein organization. They found oriented processes in sequence variation seemingly correlated with the position of organisms on the phylogenetic scale. The features in question might be related to "functional density" in DNA as well as in proteins, i.e., to the proportion of sites involved in specific functions (Zuckerkandl, 1976a). A certain increase in average functional density of informational macromolecules might conceivably occur during anagenesis, a point that it should soon be possible to verify. Such an increase would have a bearing, at the organismal level, on the "increase in independence from and control over" the environment. The rationale for this correlation would no doubt lie in the frequency of feedback and other control mechanisms that affect the synthesis as well as the functioning of proteins. Functional density would in effect represent density in control.

It is of interest to observe the rate of emergence of ever higher levels of organization. At this point, statements about this question remain largely impressionistic, in the absence of quantitative descriptions of organizational levels. Such a description is difficult enough at the molecular and macromolecular levels, and far as I know, it does not exist for hierarchically higher states of organization. Nevertheless, some overall features of the rates are apparent.

Along individual lines of descent, an acceleration of the rate of evolution seems to accompany their establishment (Simpson, 1953). A deceleration follows later. This observation is compatible with the molecular evidence adduced by Goodman *et al.* (1975), relative to rates of evolution of hemoglobin molecules. The underlying generality, valid probably at all levels of biological integration, seems to be that the appearance of a new function leads to the temporary acceleration of evolutionary rates. This, incidentally, may be tantamount to saying that acceleration of evolutionary rates accompanies the incipient adaptation to new environmental factors, since the handling of new such factors, or a novel way of handling familiar factors, requires new functions.

If we take a bird's eye view of evolution at the organismal level, we see, at each point in time, some rates accelerated, others slowing down or constant, while the relative contributions of these regimes may vary. A different picture emerges if, instead of rates of evolution along individual lines of descent, we consider the successive times at which salient features of progressive evolution appeared here and there. To be sure, during this ascent of some organisms to "higher" levels, most forms continued to exist and to change without deviating significantly from their own level.

Prokaryotes may have been in existence for 2 billion years before any eukaryotelike cell appeared. The first multicellular organism clearly distinct from the protistans must have followed after a decidedly shorter span. The evolution of the most primitive metazoans and metaphytes into fully characterized representatives of most extant or extinct phyla is considered by many as "explosive," which means in this context that it apparently took less than one hundred million years. From this rather primitive state to the ultimates of anagenesis, only a little over a half-billion years elapsed.

During this last phase, anagenesis in animals, whenever it occurred and however far it went in a given line of descent, may be considered as characterized chiefly by the evolution of central nervous systems and by the appearance of higher forms of mental organization. Versatility and creativity of the reactions to perception reached unknown heights with the class of mammals. Within only 80 million years, very high forms of sensitivity and intelligence developed, independently and by parallel evolution, along lineages such as those that led to the dog, the porpoise, and the apes.

At the organismal level, and in an overall way, the following picture seems to emerge: Along individual lines of descent, anagenesis, when it occurs at all, proceeds for different amounts of evolutionary time, up to different levels, reaches different plateaus, and occasionally decreases again, as in evolution to parasitism (recall the example of *Sacculina,* given previously). If, however, we correlate the highest level of anagenesis reached by any living form with geological time, it seems that the rate of reaching ever higher levels of organization, in the 4 billion years or so that life has been in existence, was consistently accelerated.

A particularly important feature of anagenesis, viewed as an increase in control over the environment, is the increase in control over the part of the environment represented by other species. This increase was achieved by plants to only a small extent and characterizes the evolution of animals. In an ecological niche shared or partly shared by two species, if one preys on the other, the victimized species has at least three strategies at its disposal for responding to the challenge. It must either leave the niche, increase its reproductive rate, or become more intelligent. Leave the niche it cannot, except by becoming another species. But it may vacate a section of its niche, if there is overlap and not congruence of habitats. As to increased reproduction, most species, including humans, find it easier than increasing intelligence. Yet an increase in reproductive rate may be ineffective in some systems, when it leads to increased numbers of predators, or impractical, for reasons of organic structure, physiology, or resources. The latter factor may be limiting especially for larger species, and this could be one of the reasons why the highest forms of intelligence appeared exclusively in larger animals. Another probable reason of course is the requirement of a large number of nerve cells.

In the presence of predators, when no other riposte is practical, development of a central nervous system and of intelligence should be favored. When the structural and functional basis for such a process exists in one organism, it is likely to be present also in related but competing organisms. Hence, once an animal takes the lead in this field, it should induce a similar development in related animals, provided that predator-prey relationships exist between them. This implies that the evolution of central nervous systems could be autocatalytic within a certain group of species—a statement that might possibly apply also to the acceleration of anagenesis that we seemed to note for evolution at large. In a more restricted sector of evolution, when a superiority in sensory and motor organs as well as in the computing of sensory information and in the process of decision-making appears in one single species, this event should increase selection pressure in the same direction for other animals whose ecological niche is overlapping and whose evolutionary potential is adequate.

Thus intelligence should proliferate for the simple reason that it first appeared. It provides indeed the best chance of escaping predators as well as of catching prey, when the antagonistic species—predator or prey—is itself endowed with intelligence. The superior central nervous sytem of contemporary cephalopod molluscs might not have evolved if fish and/or other cephalopod molluscs had not already been around. Biological progress is comparable to the "progress" of armaments in technically advanced nations.

leads to the search for and the finding of a proper countermeasure, and in some circumstances only intelligence is an appropriate countermeasure to intelligence.

Plants, inasmuch as they are the prey of animals, also invent countermeasures, e.g., by developing thorns. Countermeasures as prey are compatible with the vegetal character. Countermeasures in the capacity of prey *and* predator characterize animals. To be sure, the animal trend may exist in plants (insectivorous plants) and the vegetal trend in animals (sea urchins covered with spines), as secondary acquisitions.

No IQ measurement valid for the whole animal kingdom has as yet been proposed, nor is such a proposal likely to be imminent. On the other hand, brain volume is a measurable quantity that seems to be roughly correlated with intellectual capacities in primates. Fig. 11 reproduces a graph by Stebbins (1971). It illustrates the acceleration of increase in brain volume during primate anagenetic evolution. An autocatalytic course of this evolution, at least during certain phases—an autocatalysis of intelligence—is not excluded by these data, although they are insufficient for demonstrating it. The first part of the curve, practically linear in appearance, of course cannot be so, since otherwise, by extrapolation, vertebrates would have started with zero brain capacity about 70 million years ago.

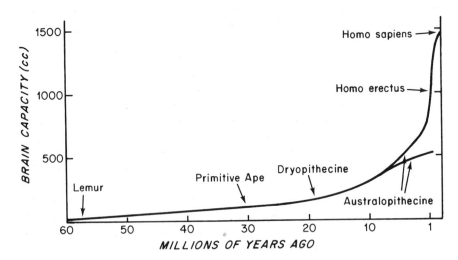

Fig. 11. An illustration of accelerated evolution during anagenesis. The increase in cranial capacity during primate evolution. From Stebbins (1971), reproduced by kind permission of the author and the publisher.

At the molecular level, what are the main mechanisms whereby such accelerated processes of evolution can be brought about?

5.2. Relation between Molecular and Organismal Rates of Evolution

If a correlation existed between anagenesis and the evolution of structural genes, should the evolution of structural genes not be accelerated in the lines of descent where anagenesis takes place?

If we pool all changes in proteins that were accepted by evolution—a naive approach to our problem—we seem to witness that they conform roughly to the prediction of the existence of an approximate evolutionary clock. The cautiousness of this statement is prompted by the past and present history of the problem. The variance in the ticking of the clock has been found to be approximately twice that of radioactive decay (Fitch, 1976), so that 2 times as many data points are necessary for determining the rate of ticking of the molecular evolutionary clock within the same approximation as are necessary for determining this rate in radioactive decay. This clock of course is the same for anagenetic as for approximately "isogenetic" lines of descent (i.e., for lines of descent along which no significant change in level of organization occurs). If it were not so, there would be no clock at all.

There have been so far three phases to the history of the idea of a molecular evolutionary clock since it was first proposed.

It met at once with utter skepticism on the part of all reasonable biologists—and only the reasonable ones were vocal. They could not see why molecular evolution should display regularity, when the picture of rates of biological evolution at large was that of irregularity.

During a second phase, the molecular clock hypothesis gained momentum, on the basis of accumulating molecular data (Sarich and Wilson, 1967; Dickerson, 1971; see also Derancourt *et al.*, 1967), through theoretical developments (Kimura, 1968; King and Jukes, 1969), and through the use of methods for building molecular phylogenetic trees that introduced a bias in favor of the clock (Helmut Vogel, personal communication; Fitch and Langley, this volume).

The third phase began recently. Various ways of analyzing the data seemed to lead to the conclusion that the clock did not work (Jukes and Holmquist, 1972; Romera-Herrera *et al.*, 1973; Goodman *et al.*, 1975).

It seems, however, that the only clock that has been knocked down definitively is a perfect clock. Yet it was expected from the start (Zuckerkandl and Pauling, 1962) that the molecular clock should be, at best, only approximate. The question is: how approximate? Or, to use Walter Fitch's phrase: how "sloppy" a clock is one prepared to call a clock?

There are good reasons why the clock should not be exact. Yet each time significantly more amino acid or nucleotide substitutions than expected are found in a section of a phylogenetic tree we must keep in mind the possibility that the homology relationships might be duplication dependent (Zuckerkandl and Pauling, 1955)—"paralogous," as they are called (Fitch, 1970)—a possibility that, I fully agree with Fitch, deserves emphasis. If two genes result from a duplication that took place long before the divergence of two species, yet we assume that the genes started diverging at the same time as the species, the fallacious conclusion will of course be reached that the two genes have diverged much more rapidly than they should have.

Related to this illusion would be one that might be created if, by virtue of the founder principle (cf. Mayr, 1963), one and the same species gave rise, as it were by budding, to two new species. In such cases, two different variants of a given polymorphic locus of the original species may have been communicated to the daughter species. Assume that, in the original species, these polymorphic alleles happened to differ by several substitutions. In that case the homologous polypeptide chains found in the two daughter species will spuriously appear to have diverged more than they should have since the time of the common ancestor of the two. By analogy with Fitch's terminology, one might perhaps call such an evolutionary relationship "paramorphic," or, by analogy with our own terminology, call it a polymorphism-dependent homology. Such a situation, although infrequent, appears like a further possible cause of spuriously accelerated molecular evolution. It would be a special case of orthologous (duplication-independent) homology tending to imitate the apparent effects on rate of paralogous (duplication-dependent) homology.

The ticking of the clock may also be really, and not just apparently, accelerated. An increase in fixation rate following functional changes in proteins has already been mentioned. It was predicted by Pauling and Zuckerkandl (1963). The new function, or version of a function, has indeed little chance of being immediately optimal. Intramolecular adaptations in the direction of functional optimization should occur as fast as the mutation rate will allow. In this process, the number and kind of Fitch and Markowitz's (1970) covarions (concomitantly variable codons) will probably be shifted. As was mentioned also, a slowdown will occur later and quasiconstant rates of molecular evolution will be maintained until the next important functional change. Goodman *et al.* (1975) have obtained important evidence in this respect through their analysis of the evolution of hemoglobin chains.

Another mechanism whereby true acceleration of the clock might be achieved is a marked increase, along one line of descent, in the frequency of recurrence of random genetic drift in small populations. Under such conditions, new polymorphic variants of a gene could be transferred repeatedly to founder populations and become dominant, particularly if the founder situations alternate with rapid population expansions. Because of their ecological and ethological traits, some types of organisms may be more prone than others to such cycles of drift. They might be identified by the simultaneous occurrence, along their lines of descent, of abnormal evolutionary rates in a number of proteins.

Large apparent deviations from expected evolutionary rates in individual informational macromolecules could obviously be due to genic transduction. For example, there is a question as to how a hemoglobin gene got into a leguminous plant. Among the several possible hypothesis to account for it, transduction might not be the least likely. According to the phylogenetic tree of Dayhoff *et al.* (1972), plant hemoglobin is much closer to insect hemoglobin than insect hemoglobin to the globins of higher vertebrates. On the basis of a limited variability of rates of amino acid fixation, this is not standard molecular divergence.

Apparent decelerations in rate will often be due to coincident molecular evolution. As has been pointed out several times over the years (e.g., Zuckerkandl, 1964*b*) and is demonstrated with particular relief by Romero-Herrera *et al.* (this volume), independent mutations may often lead to the fixation of the same amino acid at a site.*

Even a relatively poor molecular clock of course implies that the rate of sequence evolution cannot correlate with highly irregular or divergent rates of morphological evolution. The clock implies in particular that along lines of descent leading to lower contemporary animals the number of amino acid substitutions in proteins will not be smaller than along the lines of descent leading to higher contemporary animals.

That rates of evolution at higher levels of biological integration, such as the morphological level, must *somehow* be reflected in informational macromolecules seems, however, to be a necessary assumption. There are two ways for highly variable rates of organismal evolution to be reconciled with the existence of an approximate molecular clock.

One possibility is that rates of organismal evolution indeed depend on rates of substitution in structural genes. But critical substitutions would represent only a small fraction of the total number of accepted substitutions. Such rare critical substitutions could occur quite irregularly. They would not interfere significantly with a molecular evolutionary clock.

A second possibility is that rates of organismal evolution are correlated with accepted mutational events in controller sections of DNA rather than in structural genes and that, again, few of these events are individually of notable functional significance. In controller sections, a summation of oriented incremental effects, or individual critical occurrences such as translocations, would be systematically eliminated by natural selection during some evolutionary periods and along some lines of descent: if over a long time interval, then we end up with a living fossil; and if with respect to backbone programs of gene action, but not to appended programs, then we end up with an organism that has changed but remains true to a general type (remains lower if it was lower to start with). In other eras, namely when the organisms undergo important adaptive changes, many incremental mutations that respect the established regulatory topology as well as some critical mutations leading to changes in regulatory topology may be accepted in relatively short succession. This no doubt occurs at times of "explosive radiation" of higher organismal taxa.

Fixations in *controller* sections of DNA are likely to be implicated in most

---

*Helmut Vogel and I (unpublished) observed shuttle motions between proline and alanine residues along single lines of molecular descent.

morphological evolution, as pointed out above, and in physiological evolution, inasmuch as physiology is an expression of morphology. In addition, fixations in structural genes probably also intervene here, notably in genes that control cell surface recognition sites and the system that conveys messages relating to cell division from the cell surface to the nucleus.

On the other hand, fixations in *structural* genes must be intimately linked with physiological evolution to the extent to which physiology expresses *biochemical* characteristics. Simultaneously, mutational events in controller genes must be part of this process also, since the temporal, spatial, and quantitative aspects of new biochemical functions must be regulated.

The two genetic pathways for organismal change that are in principle compatible with the existence of a molecular evolutionary clock thus appear to be intimately linked. We may ask here which of these pathways is likely to be used predominantly in anagenesis.

5.3. A Search for the Molecular Basis of Anagenesis

I have discussed in another article (Zuckerkandl, 1975) why structural and functional innovation in proteins must necessarily be based on limited, progressive modifications of preexisting structurally homologous and functionally efficient protein molecules. Except perhaps in the case of some simple-sequence "periodic" proteins, a structural gene that did not fill the function of a structural gene before cannot have arisen *de novo* from DNA. Other exceptions, which are really only variants, should be furnished by novel recombinations of parts of preexisting genes. It can thus be contended that there is no protein in the human brain, or elsewhere in the human body for that matter, whose *very* close homologue does not also exist in the anthropoid apes (or at least did not in the apelike ancestors of man) and whose close homologue does not also exist in other organisms much farther removed from man on the phylogenetic tree.

The acceleration of evolution of the primate nervous systems is thus probably accompanied by only a very limited amount of innovation at the level of the structural genes. Some might presume that such limited amounts of structural change are compatible with significant functional evolution of corresponding proteins. This is unlikely to be so. I shall submit that the human brain, in comparison with that of the apes if not also of the Old World monkeys, may well not contain a single protein endowed with a truly novel function.

The reason for this assertion is as follows. The differences in amino acid sequence of corresponding proteins from anthropoid apes and man seem to range, according to the types of proteins, from zero to a very small number. Such a small number of substitutions is capable of changing the precise functional characteristics of proteins, but presumably not of replacing one *kind* of function by another. The chances of this statement's being correct are enhanced by the fact that most if not practically all substitutions that differentiate very closely related proteins occur at highly variable protein sites, on the protein surface, where the functional impact of the substitutions is mostly close to its minimum. This means that if, in a protein, a less variable and functionally more critical site is to be altered, it will be usually one out of many substituted sites or, at best, one out of several, so that, if only several substitutions have occurred, there will be no more than one such critical substitution among them. Barring very exceptional cases, we may say that one single substitution, even when functionally critical, will be able only to alter a protein function, not to transform it into another, novel function.

I am thus tempted to conclude that the human brain, for performing even its most specifically human functions, must use only those proteins that are structurally closely related to the proteins of even distant animals, and functionally very closely related to corresponding proteins *at least* in the anthropoid apes, if not in more remote relatives.

The uniqueness of the human brain must be due to structural and functional variants of preexisting proteins and to variations in the quantity, timing, place, and coordination of the action and interaction of these proteins.

There is certainly room, within a small evolutionary range such as the one that leads from the monkey stage through the ape stage to man, for a quantitative increase in *specificities* of molecular interaction. The way this can, in principle, be achieved is through duplication of structural genes followed by slight structural variations, and through the process used for multiplying antibody specificities, whatever this process be.

Increased resolution of stimuli and increased versatility of response to stimuli undoubtedly sometimes required the appearance of novel protein functions through mutations in structural genes. This may occur, for instance, when a new spectral domain is to be included, or when the organization of message transmission is to be supplemented by a new hormone. Generally, however, refinements in the sense organs, central nervous system, or effector organs may not require new types of protein functions, but rather an extension of the range of specificities of interaction within an already established system of protein types. A corresponding multiplication of and variation in structural genes is in order.

Of course, neither variation of distinct interaction specificities involving the products of structural genes nor even their numerical increase is likely to offer a sufficient basis for explaining the process of anagenesis. Alteration and diversification of functional characteristics of proteins can hardly lead by themselves to a novel hierarchical integration of cellular and intercellular processes. Rather, they will characterize variants of processes as they occur within the framework of preestablished relationships.

To determine what types of genetic change are most centrally implicated in anagenesis, we must ask what types of genetic change bring about increased independence from and control over the environment. The physiological processes that lead to such a result are improved homeostatic systems and increased powers of discernment—increased spectral resolution—of environmental stimuli, combined with increased versatility of response to these more finely analyzed stimuli.

With the exception of the improvements in homeostatic mechanisms, such as physiological buffering against salinity changes in aquatic animals or against temperature changes in terrestrial animals, anagenesis is thus seen to be associated exclusively with the development of the nervous system and of its receptor and effector organs.

Buffering capacity in relation to environmental change also exists in a bacterial spore, where, however, it is passive, not active buffering. Yet even at its most sophisticated, homeostasis is only the control of environmental actions whereof the organism is the target. It is not the control by the organism of the environment as a target. Anagenesis would be much more limited than it is if it were confined to the first type of achievement. According to definition, a euryhaline crab is a "higher" animal than a stenohaline crab, but this superiority remains rather confined. It seems clear that the major steps of anagenesis, the control of the environment as an entity that is to be acted upon and not just successfully suffered, although helped and perhaps rendered possible by homeostasis, are indeed due to the development of the nervous system and its dependencies. Plants are thus excluded from major anagenesis.

Whether we deal with a novel protein function or with gene duplication and moderate structural diversification of the duplicates, anagenetic invention depends on integration of the mutant structural genes in organismal time and space. Thus anagenesis should typically imply modifications of regulatory circuitry: additions of new "limbs" to the regulatory network, formation of new regulatory loops, or introduction of new regulatory anastomoses, as well as severance of others.

It is not implied that anagenesis rests solely on such events. Nor is it implied that topological alterations in patterns of gene regulation do not occur also in "isogenesis" (no change in organizational level) and "catagenesis" (change toward lower levels as in parasitism). In isogenetic adaptation, changes in the distribution of preexisting alleles within a population and within individuals—of structural as well as, no doubt, of regulator genes—are prominent; changes in affinity between controller molecules are likely to be much more frequent than changes in regulatory circuitry. Anagenesis as well as catagenesis, on the other hand, probably requires such changes in circuitry.

Changes in gene regulation essentially within the boundaries of preestablished regulatory patterns are implicated here in morphological evolution. Morphological changes, precisely, may accompany or precede anagenesis and create proper conditions for it, but they are not among its essential features. For instance, we should not, in my opinion, call anagenesis the substitution of a limb for a fin, even though such substitutions have marked important steps in the "ascent" of animals. In fact, such kinds of substitution are but conditions for true processes of anagenesis that follow suit. By acquiring a limb, the organism masters a new environment. However, as it adapts better to land, it adapts less well to water. The control over and the independence from the environment are not extended. They are only shifted. Such shifts should be often compatible with simple changes in the rates of transcriptional activity of preexisting genes.

It is another matter for the organism to provide the regulatory circuitry for specificity and variety of recognition and for specificity and variety of response, to set up the organization of ever higher resolution in the discrimination of stimuli and for an ever more discerning and flexible adjustment of the reactions. Obviously, anagenesis in the processing of input information requires circuits for increasingly refined analysis and synthesis. The evolution of computers and of their circuitry should provide an interesting parallel to biological anagenesis.

Proteins are unique in offering inexhaustible resources for varying interaction specificity, for setting up interaction switches through precise transconformational change, and for modulating interaction intensity (i.e., frequency of interaction occurrences and duration of any single occurrence based on the kinetic and thermodynamic characteristics of the interaction). Nucleic acids seem to have these properties to a somewhat lesser extent. The immediate molecular instruments for discrimination must therefore be proteins, perhaps in association, at some levels, with nucleic acids. A predominant reason that proteins consistently seem to be the regulator molecules in genic regulation, rather than nucleic acids, probably lies in their transconformational properties. These properties allow a change in nature of the signal that is conveyed and ensure the necessary precision, stability, and lack of ambiguity of this switch. Since anagenesis is based on developments in processes of specific recognition and specific translation of the recognized information, it must be based on the specificity of interaction and transconformation of macromolecules, primarily proteins.

In the best-worked-out case of multiple specific interactions as carried out by one type of protein, the case of the immunoglobulins (probably related phylogenetically to recognition proteins on cell surfaces), it does not seem to have been definitively established that the variety of variable regions of the polypeptide chain is wholly accounted for by heritable duplicates of structural genes. If part of the multiplicity of antibody specificities were obtained through some somatic process of variation, this part might escape the direct action of gene regulation. It is thus perhaps not to be excluded at this point, even for the nervous system, that a decisive trait of anagenesis, developments in specific recognition, implies a process that is partly beyond gene control proper, and is controlled by function itself. In such a case, any heritable anagenetic potential should nevertheless have to be acquired and developed on the basis of mutational alterations in a preexisting regulatory framework.

I personally would not be surprised—nor would Carl Woese (see Woese, 1974)—if it turned out that the main basis for anagenesis, besides adjustments in regulatory circuitry, were represented by the mechanism that so far we know only (inasmuch as we know it) from the immunochemical system. It will perhaps be discovered that a related system is involved in cell recognition, in particular in the recognition between nerve cells and in the stimulation within nerve cells of the synthesis of specific macromolecules.

It can be legitimately asked—though probably not legitimately answered at this point—whether the trend toward higher levels of organization, in biological systems, is not intrinsic. Such intrinsic trends have been recognized at all lower levels. Subatomic particles form atoms, atoms form molecules, molecules macromolecules, and, as Sidney Fox's work shows, macromolecules form some sort of protocells, *under certain circumstances*. By analogy, it is possible that, under certain circumstances, a given biological system inherently tends toward a more highly organized system. This hypothesis is purely mechanistic. It is as free of vitalism as the opinion that hydrogen and oxygen are intrinsically determined to form water. We note that natural selection can accept or refuse a biological event, never produce it. Natural selection must therefore be expected to merely retain anagenetic evolutionary steps as, under some circumstances, they spontaneously occur. It can be answered that no single evolutionary step is anagenetic by itself, it becomes anagenetic through entering into certain relationships, and natural selection may be totally responsible for the choice of these relationships and therefore for anagenesis itself. No doubt. But this does not exclude the existence of an intrinsic trend toward the kinds of relationships that are then accepted by selection. Do we need to postulate such an intrinsic trend? Today it cannot be said whether we do or not, but it is important not simply to assume that we do not. It is plain that the impact of natural selection on evolution, especially at the molecular level, has still not been measured unambiguously.

All that can be said is that the spontaneous organization of material systems has been shown to prevail up to a high level of the prebiotic field, namely, up to the level of Fox's microspheres, if not well into the biological field, viz., Tracey Sonneborn's self-reproducing structures in Ciliates. It would be rather strange that self-organization of matter be a pervasive principle—much emphasized by Manfred Eigen—up to a limiting level and that further increases in organization should be based on a totally different and independent principle. Strange, but admittedly possible.

Living matter is a system such that the controlled subsystems partake themselves in the controlling. Each level, from the highest to the lowest, is double-faced in that it is at once controlling and controlled. A new level of integration is the appearance of a new level of control. What a new control amounts to is a *selection* among energetically possible events according to a new bias. Each level indeed has its own structural and functional trends. It is these intrinsic trends that are sometimes improperly called controlling "forces." The term force, which should keep exclusively the meaning that physicists give it, should not be so used. A new level of integration does not introduce new forces but a new distribution of dominant directions of exercise of preexisting forces. It is this selection of new dominant directions, as manifest in living systems, that might be attributable to intrinsic organizational determination to a larger extent than neo-Darwinism has been ready to admit. The question remains open.

6. Concluding Remarks

The understanding of the role of gene regulation in evolution, as outlined here, is unsatisfactory, since not only is it tentative but even in this capacity it is too general. Yet, as many realize, this is a sector of the unknown from which, right now, we must apply ourselves to gnawing off pieces as best we can. Perspectives alone will not do, nor will the laboratory bench alone. The present analysis leads one to reemphasize the widely recognized, dominant importance of understanding the molecular basis of the functioning of nervous systems and brains and in particular their genic control. Putting the question this way is not disapproved by anyone. This is no longer so when we restate the phrase by speaking of the molecular basis of the mind and its genetic control.

We touch here upon a controversial area, where convictions precede knowledge. I submit, as a free-lance concluding remark, that it is the vocation of the scientist not to shun this area, but to establish the truth, whatever it may be, about the heritability of human mental characteristics of different kinds. To state the problem correctly, one should not say heritability of mental characteristics, but, of course, heritability of the potential for mental characteristics.

The beliefs that humans have about themselves in this respect are to a large extent the foundations for their philosophies of social organization and social action. It seems to me that the most important *practical* task of molecular anthropology (I need not emphasize its theoretical interest) is to provide at long last a scientifically sound foundation for social action, which has been wrongly claimed already to exist. Values are something else, but they need knowledge for their proper insertion into reality. In view of the power we have today over ourselves and over the environment we create for ourselves, we no longer can dispense with certain kinds of knowledge about ourselves. With knowledge, man may do tremendous harm to man. Without knowledge, he cannot fail to do so. Knowledge is dangerous, but it is also the only hope.

May I express the wish that 10 years from now or earlier another symposium on molecular anthropology be organized at Burg Wartenstein, namely in the area of the heritability of dispositions of mind and kinds of perceptiveness. I do hope that the participants in such a conference will not feel compelled to come to the round table dressed in the armors that are in the hall!

ACKNOWLEDGMENTS

I thank Mr. Peter R. C. Gascoyne for his kind assistance, Professor Stephen J. Gould for his comments, and the American Philosophical Society for financial support of part of this work.

7. References

Abeloos, M., 1956, *Les Métamorphoses,* Collection Armand Collin, Paris.

Ashburner, M., 1974, Sequential gene activation by ecdyson in polytene chromosomes of *Drosophila melanogaster.* II. The effects of inhibitors of protein synthesis, *Dev. Biol.* **39:**141.

Ashburner, M., Chihara, C., Meltzer, P., and Richards, G., 1973, Temporal control of puffing activity in polytene chromosomes, *Cold Spring Harbor Symp. Quant. Biol.* **38:**655.

Beerman, W., ed., 1972*a, Developmental Studies on Giant Chromosomes,* Vol. 4 of *Research Problems in Cell Differentiation,* Springer-Verlag, Heidelberg.

Beerman, W., 1972*b,* Chromomeres and genes, *Res. Probl. Cell Differ.* **4:**1.

Berendes, H. D., 1968, Factors involved in the expression of gene activity in polytene chromosomes, *Chromosoma* **24:**418.

Berendes, H. D., 1972, The control of puffing in *Drosophila hydei, Res. Probl. Cell Differ.* **4:**181.

Britten, R. J., and Davidson, E. H., 1969, Gene regulation for higher cells: A theory, *Science* **165:**349.

Calhoun, D. H., and Hatfield, G. W., 1973, Autoregulation: A role for a biosynthetic enzyme in the control of gene expression, *Proc. Natl. Acad. Sci. USA* **70:**2757.

Church, R. B., and Brown, I. R., 1972, Tissue specificity of genetic transcription, *Res. Probl. Cell Differ.* **3:**11.

Counce, S. J., 1973, in: *Developmental Systems: Insects,* Vol. II (S. J. Counce and C. H. Waddington, eds.), p. 133, Academic Press, New York.

Davidson, E. H., and Britten, R. J., 1973, Organization, transcription, and regulation in the animal genome, *Quart. Rev. Biol.* **48:**565.

Dayhoff, M. O., Hunt, L. T., McLaughlin, P. J., and Jones, D. D., 1972, Gene duplications in evolution: The globins, in: *Atlas of Protein Sequence and Structure 1972,* Vol. 5 (M. O. Dayhoff, ed.), p. 17, National Biomedical Research Foundation, Washington, D.C.

Derancourt, J., Lebor, A. S., and Zuckerkandl, E., 1967, Sequence des acides amines, sequence des nucléotides et évolution, *Bull. Soc. Chim. Biol.* **49:**577.

Dickerson, R. E., 1971, The structure of cytochrome *c* and the rates of molecular evolution, *J. Mol. Evol.* **1:**26.

Elder, D., 1973, A multiple promoter model for transcriptional control in differentiated organisms, *J. Theor. Biol.* **39:**673.

Englesberg, E., and Wilcox, G., 1974, Regulation: Positive control, *Annu. Rev. Genet.* **8:**219.

Englesberg, E., Squires, C., and Meronk, F., 1969, The 1-arabinose operon in *Escherichia coli* B/r: A genetic demonstration of two functional states of the product of a regulator gene, *Proc. Natl. Acad. Sci. USA* **62:**1100.

Falk, R., 1970, Evidence against the one-to-one correspondence between bands of the salivary gland chromosomes and genes, *Drosophila Inform. Serv.* **45:**112.

Fitch, W. M., 1970, Distinguishing homologous from analogous proteins, *Syst. Zool.* **19:**99.

Fitch, W. M., 1976, Molecular evolutionary clocks, in: *Molecular Evolution* (F. J. Ayala, ed.), p. 160, Sinauer Associates, Sunderland, Mass.

Fitch, W. M., and Markowitz, E., 1970, An improved method for determining codon variability in a gene and its application to the rate of fixation of mutations in evolution, *Biochem. Genet.* **4:**579.

Frazetta, T. H., 1975, *Complex Adaptations in Evolving Populations,* Sinauer Associates, Sunderland, Mass.

Gatlin, L. L., 1966, The information content of DNA, *J. Theor. Biol.* **10:**281.

Gatlin, L. L., 1974, Comments on papers by Reichert and Wong, *J. Mol. Evol.* **3:**233.

Georgiev, G. P., 1969, On the structural organization of operon and the regulation of RNA synthesis in animal cells, *J. Theor. Biol.* **25:**473.

Goldberger, R. F., 1974, Autogenous regulation of gene expression, *Science* **183:**810.

Goodman, M., Moore, G. W., and Matsuda, G., 1975, Darwinian evolution in the genealogy of haemoglobin, *Nature (London)* **253:**603.

Grassé, P. P., 1973, *L'Evolution du Vivant*, Albin Michel, Paris.

Hochman, B., 1973, Analysis of a whole chromosome in *Drosophila, Cold Spring Harbor Symp. Quant. Biol.* **38**:581.

Holmes, D. S., Mayfield, J. E., Sander, G., and Bonner, J., 1972, Chromosomal RNA: Its properties, *Science* **177**:72.

Huisman, T. H. J., Wrightstone, P. N., Wilson, J. B., Schroeder, W. A., and Kendall, A. G., 1972, Hemoglobin Kenya, the product of fusion of γ and β polypeptide chains, *Arch. Biochem. Biophys.* **153**:850.

Hüsing, J. O., 1963, *Die Metamorphose der Insekten,* A. Ziemsen Verlag, Wittenberg.

Huxley, J. S., 1942, *Evolution, the Modern Synthesis,* Allen and Unwin, London.

Itano, H. A., 1957, The human hemoglobins: Their properties and genetic control, *Adv. Protein Chem.* **12**:215.

Jacob, F., and Monod, J., 1961, On the regulation of gene activity, *Cold Spring Harbor Symp. Quart. Biol.* **26**:193.

Jacobsen, L., 1968, *Low Dose X-Irradiation and Teratogenesis,* Munksgaard, Copenhagen.

Jägersten, G., 1972, *Evolution of the Metazoan Life Cycle,* Academic Press, London.

Judd, B. H., Shen, M. W., and Kaufman, T. C., 1972, The anatomy and function of a segment of the X chromosome of *Drosophila melanogaster, Genetics* **71**:139.

Jukes, T. H., and Holmquist, R., 1972, Evolutionary clock: Nonconstancy of rate in different species, *Science* **177**:530.

Kendall, A. G., Ojwang, P. J., Schroeder, W. A., and Huisman, T. H. J., 1973, Hemoglobin Kenya, the product of a γ-β fusion gene: studies of the family, *Amer. J. Hum. Genet.* **25**:548.

Kimura, M., 1968, Evolutionary rate at the molecular level, *Nature (London)* **217**:624.

Kimura, M., and Ohta, T., 1971, *Theoretical Aspects of Population Genetics,* Princeton University Press, Princeton, N.J.

King, J. L., and Jukes, T. H., 1969, Non-Darwinian evolution, *Science* **164**:788.

Korschelt, E., 1944, Ontogenie der Dekapoden, in: *Bronn's Thierreich,* Vol. V (1,7), p. 671, Akademische Verlagsgesellschaft, Leipzig.

Langridge, J., 1974, Mutation spectra and the neutrality of mutations, *Aust. J. Biol. Sci.* **27**:309.

Lanyon, W. G., Ottolenghi, S., and Williamson, R., 1975, Human globin gene expression and linkage in bone marrow and fetal liver, *Proc. Natl. Acad. Sci. USA* **72**:258.

Lefevre, G., 1971, Salivary chromosome bands and the frequency of crossing over in *Drosophila melanogaster, Genetics* **67**:497.

Lewis, E. B., 1951, Pseudoallelism and gene evolution, *Cold Spring Harbor Symp. Quant. Biol.* **16**:159.

Lewis, E. B., 1963, Genes and developmental pathways, *Am. Zool.* **3**:33.

Lewis, M., Helmsing, P. J., and Ashburner, M., 1975, Parallel changes in puffing activity and patterns of protein synthesis in salivary glands of *Drosophila, Proc. Natl. Acad. Sci. USA* **72**:3604.

Lochhead, J. H., 1963, in: *Phylogeny and Evolution of the Crustacea* (H. B. Whittington and W. D. I. Rolfe, eds.), p. 109, Museum of Comparative Zoology, Cambridge, Mass.

Manning, J. E., Schmid, C. W., and Davidson, N., 1975, Interspersion of repetitive and nonrepetitive DNA sequences in the *Drosophila melanogaster* genome, *Cell* **4**:141.

Mayr, E., 1963, *Animal Species and Evolution,* Harvard University Press, Cambridge, Mass.

Miller, J. H., Coulondre, C., Schmeissner, U., Schmitz, A, and Lu, P., 1975, Altered lac repressor molecules generated by suppressed nonsense mutations, Proceedings of the Tenth FEBS Meeting, Federation of European Biochemical Societies, p. 223.

Monod, J., and Jacob, F., 1961, Teleonomic mechanisms in cellular metabolism, growth, and differentiation, *Cold Spring Harbor Symp. Quant. Biol.* **26**:389.

Moriyama, Y., Hodnett, J. L., Prestayko, A. W., and Busch, H., 1969, Studies on the nuclear 4 to 7s RNA of the Novikoff hepetoma, *J. Mol. Biol.* **39**:335.

Ohno, S., 1973, Ancient linkage groups and frozen accidents, *Nature (London)* **244**:259.

Paul, J., Gilmour, R. S., Affara, N., Birnie, G., Harrison, P., Hell, A., Humphries, S., Windass, J., and Young, B., 1974, The globin gene: Structure and expression, *Cold Spring Harbor Symp. Quant. Biol.* **38**:885.

Pauling, L., and Zuckerkandl, E., 1963, Chemical paleogenetics: Molecular restoration studies of extinct forms of life, *Acta Chem. Scand.* **17:**S9.

Pelling, C., 1964, Ribonucleinsäure—Synthese der Riesenchromosomen, *Chromosoma* **15:**71.

Petit, C., and Zuckerkandl, E., 1976, *Evolution, Génétique des Populations, Evolution Moléculaire*, Editions Hermann, Paris.

Reichert, T. A., Yu, J. M. C., and Christensen, R. A., 1976, *J. Mol. Evol.* **8:**41.

Rensch, B., 1960, *Evolution above the Species Level*, Columbia University Press, New York.

Romero-Herrera, A. E., Lehmann, H., Joysey, K. A., and Friday, A. E., 1973, Molecular evolution of myoglobin and the fossil record: A phylogenetic series, *Nature (London)* **246:**389.

Rosbash, M., Camp. M. S., and Gummerson, K. S., 1975, Conservation of cytoplasmic poly(A)-containing RNA in mouse and rat, *Nature (London)* **258:**682.

Russel, L. B., 1966, X-ray induced developmental abnormalities in the mouse and their use in the analysis of embryological patterns, *J. Exp. Zool.* **131:**329.

Sallei, J. P., and Zuckerkandl, E., 1975, The *in vitro* cellular synthesis of hemoglobin components in various media and its relevance to the concept of stenogenic and eurygenic states of differentiation. *Biochimie* **57:**343.

Sanders, H. L., 1963, The Cephalocarida, *Mem. Conn. Acad. Arts Sci.* **15:**1.

Sarich, V. M., and Wilson, A. C., 1967, Rates of albumin evolution in primates, *Proc. Natl. Acad. Sci. USA* **58:**142.

Sharov, A. G., 1966, *Basic Arthropodan Stock*, Pergamon Press, Oxford.

Shineberg, N., 1974, Mutations partially inactivating the lactose repressor of *Escherichia coli*, *J. Bacteriol.* **119:**500.

Siewing, R., 1963, in: *Phylogeny and Evolution of the Crustacea* (H. B. Whittington and W. D. I. Rolfe, eds.), p. 108, Museum of Comparative Zoology, Cambridge, Mass.

Simpson, G. G., 1953, *The Major Features of Evolution*, Columbia University Press, New York.

Smith, G., 1909, The Crustacea, in: *The Cambridge Natural History*, Macmillan, London.

Smith, T. F., and Sadler, J. R., 1971, The nature of lactose operator constitutive mutations, *J. Mol. Biol.* **59:**273.

Stebbins, G. L., 1971, *Processes of Organic Evolution*, 2nd ed., Prentice-Hall, Englewood Cliffs, N.J.

Tazima, Y., 1964, *The Genetics of the Silkworm*, Prentice-Hall, Englewood Cliffs, N.J.

Thompson, A. W., 1917, *On Growth and Form*, Cambridge University Press, Cambridge.

Vermeij, G. J., 1973, Biological versatility and earth history, *Proc. Natl. Acad. Sci. USA* **70:**1936.

Waddington, C. H., 1962, *New Patterns in Genetics and Development*, Columbia University Press, New York.

Weber, K., Platt, T., Ganem, D., and Miller, J. H., 1972, Altered sequences changing the operator-binding properties of the *Lac* repressor: Colinearity of the repressor protein with the i-gene map, *Proc. Natl. Acad. Sci. USA* **69:**3624.

Weinberg, R. A., and Penman, S., 1969, Metabolism of small molecular weight monodisperse nuclear RNA, *Biochim. Biophys. Acta.* **190:**10.

Whyte, L. L., 1965, *Internal Factors in Evolution*, Braziller, New York.

Wigglesworth, V. B., 1954, *The Physiology of Insect Metamorphosis*, Cambridge University Press, Cambridge.

Wilson, A. C., Maxson, L. R., and Sarich, V. M., 1974a, Two types of molecular evolution: Evidence from studies of interspecific hybridization, *Proc. Natl. Acad. Sci. USA* **71:**2843.

Wilson, A. C., Sarich, V. M., and Maxson, L. R., 1974b, The importance of gene rearrangement in evolution: Evidence from studies on rates of chromosomal, protein, and anatomical evolution, *Proc. Natl. Acad. Sci. USA* **71:**3028.

Woese, C. R., 1974, The custom fitting problem and the evolution of developmental systems, *J. Mol. Evol.* **3:**109.

Zuckerkandl, E., 1963, Perspectives in molecular anthropology, in: *Classification and Human Evolution* (S. L. Washburn, ed.), p. 243, Aldine, Chicago.

Zuckerkandl, E., 1964a, Controller gene diseases: The operon model as applied to β-thalassemia, familial fetal hemoglobinemia and the normal switch from the production of fetal hemoglobin to that of adult hemoglobin, *J. Mol. Biol.* **8:**128.

Zuckerkandl, E., 1964b, Further principles of chemical paleogenetics as applied to the evolution of hemoglobin, in: *Protides of the Biological Fluids, 1964* (H. Peeters, ed.), p. 102, Elsevier, Amsterdam.

Zuckerkandl, E., 1968, Hemoglobins, Haeckel's biogenetic law, and molecular aspects of development, in: *Structural Chemistry and Molecular Biology* (A. Rich and N. Davidson, eds.), p. 256, Freeman, San Francisco.

Zuckerkandl, E., 1974, A possible role of "inert" heterochromatin in cell differentiation: Action of and competition for "locking" molecules, *Biochimie* **56:**937.

Zuckerkandl, E., 1975, The appearance of new structures and functions in proteins during evolution, *J. Mol. Evol.* **7:**1.

Zuckerkandl, E., 1976a, Evolutionary processes and evolutionary noise at the molecular level. I. Functional density in proteins *J. Mol. Evol.* **7:**167.

Zuckerkandl, E., 1976b, Evolutionary processes and evolutionary noise at the molecular level. II. A selectionist model for random fixations in proteins, *J . Mol. Evol.* **7:**269.

Zuckerkandl, E., and Pauling, L., 1962, Molecular disease, evolution, and genic heterogeneity, in: *Horizons in Biochemistry* (M. Kasha and N. Pullman, eds.), p. 189, Academic Press, New York.

Zuckerkandl, E., and Pauling, L., 1965, Evolutionary divergence and convergence in proteins, in: *Evolving Genes and Proteins* (V. Bryson and H. J. Vogel, eds.), p. 97, Academic Press, New York.

Glossary

The following terms and definitions are not meant to be inclusive. They were selected mainly to aid the reader in understanding some of the specific points that are discussed in this volume and should be considered in terms of their context. Nevertheless, some basic terms have also been included where it was felt that they would be helpful to the nonspecialist. For a more comprehensive listing of genetic and evolutionary terms, one should consult such sources as the glossary in E. Mayr: *Animal Species and Evolution* (Belnap Press of Harvard University Press, Cambridge, Mass., 1963) or R. C. King: *A Dictionary of Genetics,* second edition (Oxford University Press, New York, 1972).

Additive hypothesis. See hypothesis, additive.

Adultation. The evolutionary transfer to an immature stage of a morphological character first developed in the adult.

Algorithm. A step-by step statistical strategy for solving a given class of problems.

Algorithm, branch swapping. A systematic and computer-aided removal and relocation of branches of a phylogenetic tree, with a view to generating alternative trees whose merits (e.g., path lengths) can be compared. One of the many available algorithms for stepwise improvement of an initial, maximum parsimony estimate.

Algorithm, Fitch. A procedure for determining the minimum number of times that a character has changed during its evolution, given the state of that character in several taxa and a phylogenetic relationship among those taxa. The character used is a physical entity such as a nucleotide, an amino acid, or other trait that is easily categorized.

Anagenesis. Progressive evolution, i.e., evolution toward increased independence from, and control over, the environment, accompanied by the appearance of higher levels of biological integration. Opposite of catagenesis.

Appended genetic program. A program of gene activities involving groups of structural genes and corresponding controller genes whose activities can be expanded, reduced, or changed with respect to their ratios, without affecting the backbone genetic program. *See also* Backbone genetic program.

Aromatic cluster. A localized region in the tertiary structure of a protein molecule made up of "aromatic" and cyclic amino acid residues (i.e., tyrosine, phenylalanine, tryptophan, histidine, and proline). These clusters are believed to be important in stabilizing the three-dimensional structures of proteins.

Augmented distance. An estimate, inferred by the maximum parsimony method and corrected for by the fact that the phylogenetic tree is not everywhere equally dense, of the total number, countable plus superimposed, of fixed nucleotide mutations separating two nucleic acid sequences. *Compare* Phylogenetic density.

Back mutation. A mutation that occurs if a nucleic acid or protein sequence, or part thereof, has a given sequence at time t, diverges away from that sequence to a different sequence and returns to the original sequence at time $t + \Delta t$, e.g., $A \rightarrow T \rightarrow G \rightarrow A \rightarrow C \rightarrow A$ is a series of events with two back mutations.

Backbone genetic program. A program of gene activities involving groups of structural genes and corresponding controller genes implicated in the basic pathway for the development (ontogenesis) of an organism through successive activations and repressions of transcriptive activities. *See also* Appended genetic program.

Bohr effect. The shift in oxygen affinities of hemoglobin when pH alone is altered.

Branch swapping algorithm. See Algorithm, branch swapping.

Camin–Sokal hypothesis. See Hypothesis, Camin–Sokal.

Catagenesis. A neologism designating regressive evolution, where an organism loses some of its independence from and control over the environment and abdicates the highest levels of biological integration that its ancestors have reached. Examples of catagenesis are found in parasitic descendants of nonparasitic forms. Opposite of anagenesis.

Clade. A monophyletic, phylogenetic branch whose members share a more recent common ancestry with one another than with the members of any other clade.

Cladistics. The ancestral branching arrangement of a set of species; the branching order aspects of a phylogeny.

Cladogenesis. The evolutionary process of phylogenetic branching which produces different clades. This process encompasses speciation (formation of new species), but is more extensive in its time span in that its results are revealed by the emergence of new orders, classes, or even phyla.

Cladogram. The unique treelike branching arrangement which depicts the actual order of ancestral branching (i.e., the order of recency of common ancestry) as it occurs in nature. While the actual cladogram is rarely available to us (except in breeding experiments), we can make inferences about the cladogram using observed data and evolutionary hypotheses. *Compare* Dendrogram.

Coarse-grained environment. An ecological concept based on the way organisms with poor internal homeostatic mechanisms and low behavioral plasticity and mobility experience their environment. Their limited surroundings are experienced as sets of alternative, disconnected conditions. *Compare* Fine-grained environment.

Contradiction, method of. A technique in mathematics in which a statement, S, is proved true by demonstrating that its negation, "not-S," leads to a contradiction.

Contradiction and induction, method of. A technique in mathematics for proving the truth of a collection of statements, ordered as the first, the second ... the nth, etc. One assumes that a statement is false, states the negation of the statement and shows that this leads to a contradiction. *Compare* Method of induction.

Covarions. Concomitantly variable codons. Those codons that at any one time could change the amino acid encoded in a organism without a significantly deleterious effect and hence are not evolutionarily invariable. *Compare* Varion.

Dendrogram. Any branching arrangement which depicts the order of ancestral tertiary branching, real or hypothetical, for a collection of organisms, species, or higher taxonomic units. *Compare* Cladogram.

Dendrogram, exterior point of. A point on a dendrogram which has no descendants.

Dendrogram, interior point of. A point on a dendrogram which has both an ancestor and descendants.

Dendrogram, root of. The single most ancestral point in a dendrogram.

Dollo's rule. The rule that states that evolutionary change is irreversible.

Dutrillaux's hypothesis. See Hypothesis, Dutrillaux's.

Eurygenic gene control. The broad control of transcriptive gene activity patterns acting on whole linkage groups and responsible for cellular differentiation at the tissue level. *Compare* Stenogenic gene control.

Fine-grained environment. An ecological concept based on the way organisms with good internal homeostatic mechanisms and high behavioral plasticity and mobility experience their environment. Their extensive surroundings are experienced as a succession of many interconnected conditions. *Compare* Coarse-grained environment.

Fitch algorithm. See Algorithm, Fitch.

Founder effect. The possibility that biological species (or populations within a species) can result from geographical isolation of a small sample of the ancestral or founder species, which, because of its size, does not adequately represent the gene frequencies present in that ancestral species.

Fuga. Functional unit of gene action. A unit is supposed to contain one to several structural genes (namely genes controlling the structure of polypeptide chains), some of which may function as regulator genes, when the polypeptide chains are proteins or parts of proteins that exert a specific regulatory function on the transcriptive activity of other structural genes.

Functional density (of a protein). The proportion of amino acid residues in the folded polypeptide chain or chains of a protein which are involved in specific functions, especially contact functions. Such functions depend on the interactions of stereospecific recognition sites formed by amino acid side chains in the folded polypeptide chains.

Gradient centrigufation. See Isopycnic gradient centrifugation.

Genetic drift. See Random genetic drift.

Heteroduplex. A double-stranded nucleic acid molecule in which the paired strands have derived from different sources.

Heterologous species. In immunology, a term used for those species whose degrees of similarity to an homologous species are assayed by the reaction of the antiserum to the homologous species. The term is also used in comparative studies of DNA sequences for a species whose DNA is compared to an homologous species. *Compare* Homologous species.

Homologous proteins. Proteins which have a common evolutionary origin and structure or function similar to one another.

Homologous species. In immunology, a term used for the species which was the source of the immunizing material against which antiserum was produced. The term is also used in studies concerned with how closely the DNA sequences of different species match each other. *Compare* Heterologous species.

Hydrophobic core. The interior of a protein molecule where the nonpolar side chains of amino acid residues (e.g., alanine, valine, leucine, isoleucine) are held together by hydrophobic bonds. The force of these hydrophobic interactions acting on proteins in water is one of the chief determinants of the three-dimensional structure of proteins.

Hypothesis, additive. An evolutionary hypothesis which states that there is a value corresponding to each link on the true evolutionary tree, and that the observed distance between each pair of contemporary taxonomic units equals the sum of the values for the links connecting the pair of taxonomic units. Ultimately, it says that any parallelisms and convergences in protein and nucleic evolution are randomly distributed among the various lineages in the study.

Hypothesis, Camin–Sokal. A special instance in the maximum homology hypothesis in which back mutations to prior character states are not permitted.

Hypothesis, Dutrillaux's. An hypothesis which assumes the following steps of primate evolution: (1) The ancestors of the present-day gorilla separated from the ancestors of humans and chimpanzees. (2) Substantially later, human and chimpanzee ancestors separated. (3) Afterward, repeated hybridizations occurred between gorilla and chimpanzee ancestors. (4) Finally, reproductive barriers were established prohibiting these hybridizations.

Hypothesis, evolutionary. A statement of mathematical properties which relates real data on organisms to the most likely pattern of ancestral branching which actually occurred in nature.

Hypothesis, maximum homology. An evolutionary hypothesis which states that the best reconstruction of ancestors and the evolutionary tree is that which maximizes identity due to common ancestry, as indicated by homologous genetic coding sites. Also called the Red King hypothesis.

Hypothesis, maximum parsimony. An evolutionary hypothesis which states that the best reconstruction of ancestral character states is that which requires the fewest mutations in the evolutionary tree to account for contemporary character states. As such a tree would maximize identity due to common ancestry, the maximum parsimony hypothesis is equivalent to the maximum homology hypothesis.

Hypothesis, uniform convergence. An evolutionary hypothesis which states that convergence in all regions of a tree take place at a uniform rate, even though evolutionary change may not do so.

Hypothesis, uniform rate. An evolutionary hypothesis which states that any two lines of descent in an evolutionary tree diverge at a constant rate with respect to one another. *See also* Unweighted pair group method.

Immunological distance. Generally, a difference between two antigens measured immunologically; specifically, for microcomplement fixation results, it is 100 times the log of the ratio of amounts of antibody added to fix equal amounts of complement in the heterologous and homologous reactions.

Induction, method of. A technique in mathematics for proving the truth of a collection of statements, ordered as the first, the second, the nth, etc. The first statement is proved, and then it is proved that for any k, the $(k+1)^{th}$ statement is true whenever the kth statement is true.

Information theory. An attempt to answer the intuitive question: How much information is transmitted by or stored in a given information carrier? Classical theory is concerned primarily with the technical capacity to transmit any message and studiously avoids any evaluation of the meaning of a given message. Modern theory boldly attempts evaluation of the usefulness or value or complexity of a particular message.

Isogenesis. A neologism designating evolution accompanied by neither anagenetic nor catagenetic change, i.e., evolution without any essential change in levels of biological integration and in independence from, and control over, the environment.

Isopycnic gradient centrifugation. Literally, same density. A gradient of heavy salts (e.g., cesium chloride) in which macromolecules are separated at their characteristic neutral buoyant densities.

Isozyme. As a working definition, isozymes are different molecular forms of an enzyme in an organism which catalyze the same reactions. More accurately, however, "true" isozymes are the products of two or more nonallelic genetic loci which were formed by gene duplication and exist either as single polypeptide chains or monomers (e.g., the carbonic anhydrase isozymes, CA I and CA II) or as polymers, the subunits of which are the products of more than one locus (e.g., lactate dehydrogenase isozymes: LDH 1–5).

Jaynesian formalism. The mathematical formulation developed by Edwin Thompson Jaynes for implementing the method of maximum entropy inference. *See also* Maximum entropy inference.

Macroevolution (of proteins). A process of rapid evolutionary change in which new major functional adaptations in proteins are brought about by intense positive selective pressures. *Compare* Microevolution (of proteins).

Markov process. A stochastic process where future development is completely determined by the present state and is independent of the way in which the present state was developed. *See also* Stochastic process.

Maximum entropy inference. A quantitative generalization of Occam's razor; an objective criterion allowing one to assign numerical probabilities to each possible outcome, if an event can have any one of N different outcomes and it is not known which outcome will actually occur. Operationally this criterion determines those probabilities which deviate minimally from equiprobable values and which also satisfy known observations about the system.

Maximum homology hypothesis. See Hypothesis, maximum homology.

Maximum parsimony hypothesis. See Hypothesis, maximum parsimony.

Maximum parsimony problem. A mathematical problem whose major part deals with ancestor reconstruction consistent with the maximum parsimony hypothesis without knowledge of the branching arrangement. It is an important, unsolved mathematical problem. A minor part of the problem deals with ancestor reconstruction consistent with the maximum parsimony hypothesis, given a knowledge of the branching arrangement. This part of the problem has been solved. *See also* Hypothesis, maximum parsimony.

Method of contradiction. *See* Contradiction, method of.

Method of contradiction and induction. *See* Contradiction and induction, method of.

Microevolution (of proteins). The minor evolutionary changes which occur in proteins after their functional adaptations have been perfected and stabilized. For example, the substitution of one amino acid for another which is similar in physical and chemical properties. *Compare* Macroevolution (of proteins).

Minimum mutation distance. In amino acid sequences, the difference between two kinds of homologous proteins shown by the minimum number of mutations of the genetic codon.

Molecular anthropology. The study of the evolution of the molecular constituents of human beings and other primates, especially informational macromolecules, i.e., proteins and polynucleotides.

Molecular clock model of evolution. An hypothesis proposing that proteins and genes or DNA sequences accumulate point-mutational changes in a sufficiently regular fashion over evolutionary time to permit the dating of branching points in phylogeny. "Clock" calculations assume that a direct proportionality exists between the degree of molecular divergence separating two phylogenetic branches and the time of ancestral separation of the two branches.

Monophyletic assemblage. An assemblage in which all phylogenetic branches have a closer kinship (i.e., share a more recent common ancestor) with one another than with any phylogenetic branch outside of the assemblage. A less restrictive definition would simply have all branches in the assemblage descend from the same phylogenetic source. *Compare* Polyphyletic assemblage.

Natural theory of evolution. This theory has acquired some variability in use, but its originator clearly meant that it is any genetic change whose cumulated expression at all levels (replication, transcription, translation, and phenotype) would produce a change in fitness sufficiently small

that chance rather than selection would be more likely to determine its evolutionary fate.

Network (dendrogram). A dendrogram whose root has been removed and whose remaining interior points, therefore, have no temporal direction.

Node. A branch point in an evolutionary tree. *See* Dendrogram, interior point of.

Nonrandom event. An event whose *a priori* probability of occurrence is 0 (zero) or 1 (unity). *Compare* Random event.

Nucleotide divergence. Degree of alteration of DNA sequences resulting from nucleotide substitutions.

Orthologous homology. Homologous structures in different taxa whose common ancestral gene(s) existed in the most recent common ancestor of the taxa from which they derive (from ortho, meaning exact, from the precise correspondence between the phyletic history of the structures and the taxa). Opposed to paralogous homology.

Paleogeography. The study of the geographic conditions of earlier evolutionary times in which attempts are made to reconstruct the habitats of organisms that were living at that time and later uncovered as fossils.

Paralogous homology. Homologous structures commonly present in the same taxon that arose through gene duplication(s). Their common gene ancestor is, typically, much more remote than the most recent common ancestor of the taxa from which they derive (from para, meaning in parallel from their concurrent evolution in a single line of descent). Opposed to orthologous homology.

Paramorphic homology (also called polymorphism-dependent homology). A neologism designating the presence, in two distinct species of organisms, of polypeptide variants whose ancestral forms were polymorphic alleles in the ancestral species.

Patristic feature. A feature primitive in the sense that it is inherited by all species collectively from their ancient common ancestor. The collective inheritance should represent a fairly large phylogenetic branch containing anciently separated lineages. Distinguished from a derived feature shared by more recently separated lineages.

Phenogram. A branching arrangement obtained by applying an objective (usually computerized) algorithm to a given set of phenotypic data which may or may not reflect the actual state of nature.

Phyletic gradualism model of evolution. An evolutionary model in which the species change through time in a long and imperceptibly graded series. Inasmuch as the fossil record, according to this model, should show a long sequence of intermediate forms, morphological discontinuities are attributed to an imperfect record. *Compare* Punctuated equilibrium model of evolution.

Phylogenetic density. A term referring to the topological quality of a phylogenetic tree or subregion thereof. If the region in question contains many closely related sequences throughout it is said to be dense; if it does not, then it is said to be sparse. A region is maximally dense when it accounts for all (countable plus superimposed) fixed mutations. The denser a tree, the more adequate a reflection of the actual evolutionary history it represents.

Polymorphism dependent homology. See Paramorphic homology.

Polyphyletic assemblage. An assemblage made up of branches from different phylogenetic sources, such that the kinship of at least one of these branches is closer to a phylogenetic branch outside of the assemblage. *Compare* Monophyletic assemblage.

Punctuated equilibrium model of evolution. In this model, stable species are expected to be relatively long-lived; when new species appear, they do so abruptly in the geological record and are recognizable as new, stable forms. Thus, many of the sharp morphological discontinuities in the fossil record between related forms are thought to be real, i.e., due to rapid evolution of new forms, rather than to imperfections of the record. *See also* Phyletic gradualism model of evolution.

Random event. An event whose *a priori* probability of occurrence is other than 0 (zero) or 1 (unity). *Compare* Nonrandom event.

Random evolutionary hits (REH's). An estimate, inferred by the interactive model of the total number, countable plus superimposed, of fixed nucleotide mutations separating two nucleic acid sequences.

Random genetic drift. Changes in allele frequency in a population caused by chance variations from generation to generation. The

effect is greater the smaller and more isolated the population.

Red King hypothesis. See Hypothesis, maximum homology.

Redundancy index. A dimensionless fraction between zero and one, which measures the total departure from the maximum entropy state. It is thus a measure of total nonrandomness in terms of the entropy concept. For example, the redundancy of an alternating binary sequence, 010101 . . . is one. Hence such a sequence has maximal nonrandomness under entropy measures. The *organization of redundancy index*, RD2, which is also a dimensionless fraction, expresses the fraction of the redundancy contributed by D_2. This is a measure of the structure or organization of the redundancy. Vertebrates always have an RD2 index of $\approx 0.6-0.8$.

Repetitive DNA. DNA in which the same or related nucleotide sequences are repeated in tandem a great number of times. Often called repeated DNA. *See also* Satellite DNA.

Saltation. A sudden evolutionary change. A term applied here to a possible process whereby a nucleotide sequence becomes tandemly reduplicated in evolution.

Salt-bridge. An interactive attraction between two amino acid side-chains of opposite charge. Salt-bridges can be important in determining the folding of a protein as witnessed by the unfolding that takes place when salt-bridges are disrupted by high salt concentration. Also termed ionic bond.

Satellite DNA. Minor buoyant density fractions of DNA which appear under appropriate conditions as "satellites" to the main fraction of, usually, genomal DNA. Characteristically, very highly repeated, simple sequence DNA. *See also* Repetitive DNA.

Sequence affinity. The clustering in a phylogenetic tree of certain biological species because of the similarities of their amino acid sequences.

Stenogenic gene control. The control of certain limited patterns of transcriptive gene activity, such as the changes in the expression of globin genes in erythroid cells. *Compare* Eurygenic gene control.

Stochastic process. A random phenomenon which follows a time course of development described by probabilistic laws.

Structural free degree. The structural limits of a

protein within which changes can occur without affecting its function.

Tandem reduplication. Repetition of nucleotide sequences in an adjacent manner along the DNA molecule.

Transconformational DNA. DNA whose main function, hypothetically, is to bring about condensation and decondensation in certain regions of chromatin, thus providing one of the mechanisms for the control of transcription and cell differentiation in eukaryotes.

Triangle inequality. A relationship which measures of distance must satisfy if phylogenetic trees are to be constructed. The triangle inequality states that $d_{AB} \leq d_{AC} + d_{BC}$, where A and B are two protein or nucleic acid sequences descendant from a common ancestral sequence C. The relationship between any two of these sequences is demonstrated by $d,$ a well-defined metric measure of distance.

Uniform rate hypothesis. See Hypothesis, uniform rate.

Uniform convergence hypothesis. See Hypothesis, uniform convergence.

Unweighted Pair Group Method. A tree-building, clustering procedure which joins the smallest branches of a dendrogram first and the two largest branches last. Each clustering cycle groups together that pair of singleton species or previously joined species whose two members diverge less from each other than the two members of any other pair in the collection.

Varion. In a pair of homologous protein sequences which have diverged from a common ancestor over a given time period (Δt), the varions are those specific amino acid residues, in either sequence, each of which has been free to fix substitutions during some part of that time period. *Compare* Covarions.

Index